D1426319

ENGINEERS

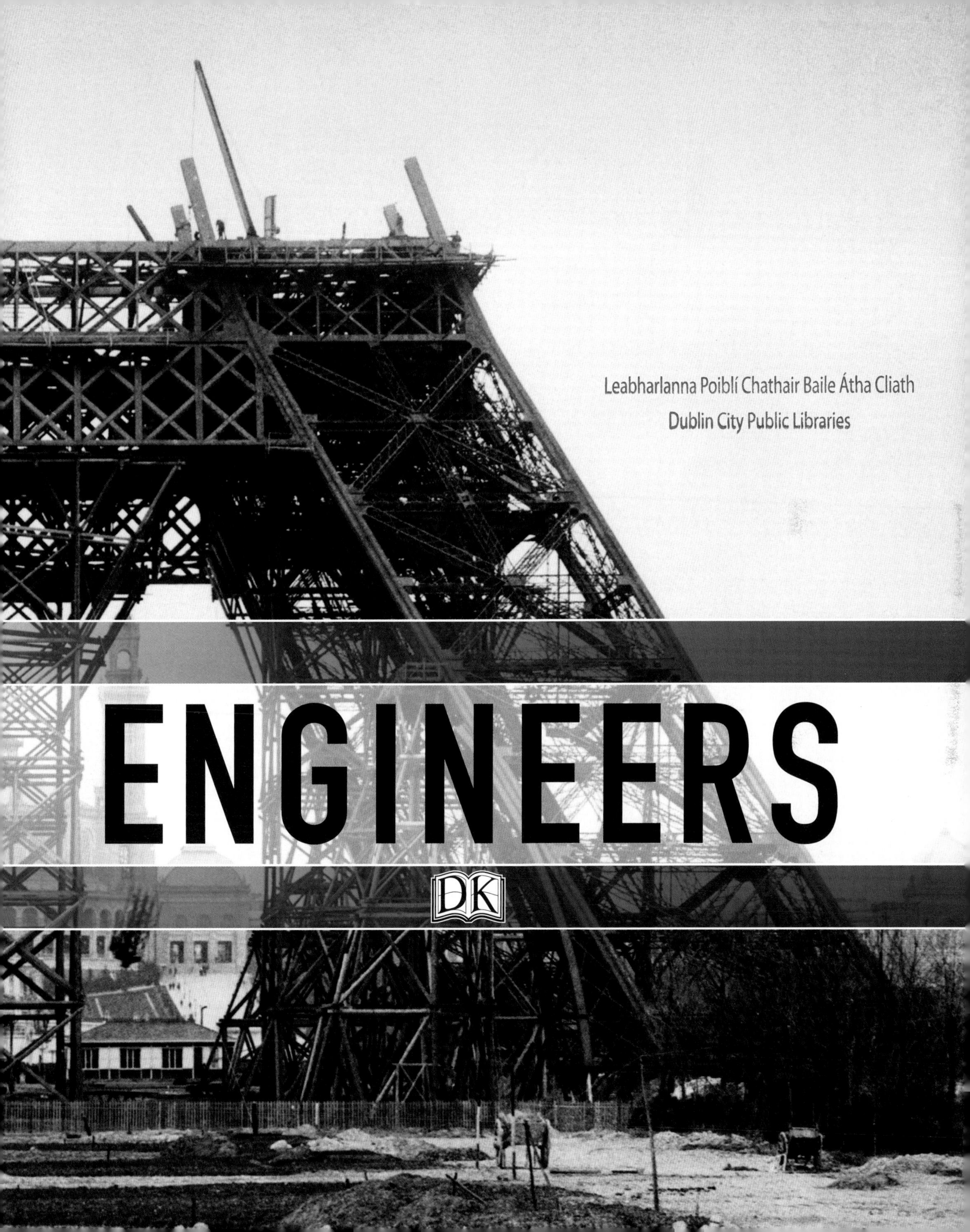
Leabharlanna Poiblí Chathair Baile Átha Cliath
Dublin City Public Libraries
ENGINEERS
DK

Editor-in-Chief Adam Hart-Davis

Senior Editor Bob Bridle
Senior Art Editor Sharon Spencer
Jacket Designer Chris Gould
Production Editors Tony Phipps, Rebekah Parsons-King
Production Controller Louise Minihane
Managing Editor Stephanie Farrow
Managing Art Editor Lee Griffiths

DK INDIA
Project Editor Neha Gupta
Project Art Editor Govind Mittal
Senior Editor Priyanka Nath
Editors Suneha Dutta, Tina Jindal
Art Editors Mahipal Singh, Isha Nagar
DTP Designers Anita Yadav, Mohammad Usman, Arvind Kumar
Managing Editor Saloni Talwar
Managing Art Editor Romi Chakraborty
CTS Manager Balwant Singh
Production Manager Pankaj Sharma

TALL TREE LTD
Managing Editor David John
Senior Designer Ben Ruocco
Picture Researcher Louise Thomas

Written by John Farndon, Hugh Ferguson, Joe Fullman, Carrie Gibson, Ian Graham, Sally MacGill, Philip Parker, Sally Regan, and Marcus Weeks

This edition published in 2017
First published in Great Britain in 2012 by
Dorling Kindersley Limited
80 Strand, London, WC2R 0RL

10 9 8 7 6 5 4 3 2
002–305348–Apr/2017

A CIP catalogue record for this book is available from the British Library.
ISBN: 978-0-2412-9882-4

Printed and bound in China

A WORLD OF IDEAS:
SEE ALL THERE IS TO KNOW

www.dk.com

CONTENTS

THE INDUSTRIAL REVOLUTION 112

THE MACHINE AGE 182

MODERN TIMES 272

INTRODUCTION

Engineers have always been heroes of mine. From Archimedes and his hydraulics to Brunel and his railways, I have enjoyed both learning about them and telling their stories in print and on radio and television. So when I was asked whether I would like to edit this book, I was delighted, although slightly daunted by the task of choosing the world's best 200 or so.

Most people don't realize the importance of engineers. Without them we would have none of the technology of the modern-day world: no computers, no phones, no televisions, no railways, no aircraft, no cars, nor even proper roads. In this book we have not tried to be comprehensive, for that would take thousands of pages. Nor have we extended the range much beyond 1940; so most of our heroes are dead. We have tried instead to pick out the pioneers and the leaders from the days when it was still common for one man, or occasionally, one woman, to design and build some great scheme. Further, we have focused on great infrastructure – bridges, ships, and skyscrapers – at the expense of more mundane things. We have also, reluctantly, left out Christopher Wren and other great architects who we decided were designers rather than engineers.

Putting all these engineers together has produced an intriguing collage of the history of technology over the last few millennia. The ancients constructed impressive bridges, aqueducts, tunnels, and buildings. Then, along came millwrights, clockmakers, and builders of canals. Steam engines created a new playing field, and, as new materials became available, engineers were able to

think bigger and better: Portland cement, cast iron, and steel allowed them to reach further under the water and into the sky. Meanwhile, their constructions gradually evolved. More than 2,000 years ago, Roman engineer Vitruvius wrote that buildings should be strong, useful, and aesthetically pleasing. Those objectives are still desirable today, and are apparent in many of the structures in this book, from the Pantheon in Rome to the Eddystone Lighthouse, and the Manhattan Bridge.

War has always been a driver of engineering. Any country going to war suddenly needs better weapons and robust defences, and many of the engineers in this book spent at least a part of their lives in military service: Vitruvius served under Julius Caesar, while agricultural engineer William Tritton built the first tank. Most of the people in this book were primarily professionals, but in the 18th and 19th centuries, there appeared a new breed – the gentleman-tinkerer, who found time from his routine life to take on some extraordinary project in his garden shed; this is how Sir George Cayley invented the aircraft and Ferdinand von Zeppelin the airship.

Needless to say, it is hard to overstate the importance of engineers. We need engineers desperately today to solve the modern world's problems, including energy production, global warming, and water supply. I hope this book will inspire some young people to study first physics and mathematics, and then engineering, for this profession will provide some of the most interesting and challenging jobs that can be found in the world today, just as they have in the past.

ADAM HART-DAVIS

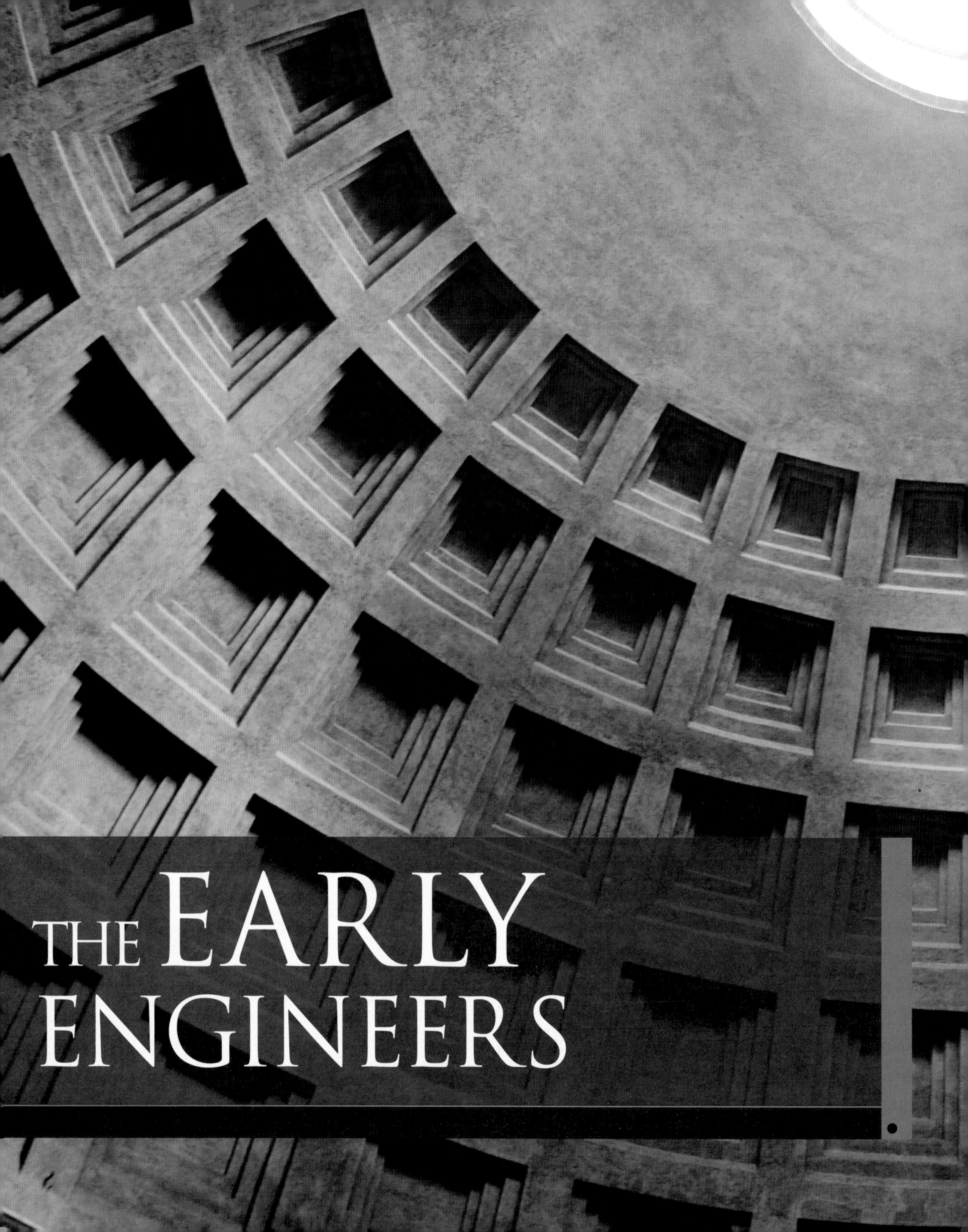

THE EARLY ENGINEERS

1

THE EARLY ENGINEERS

3500BCE | 2500BCE | 700BCE

▲ *c.*3500BCE **The potter's wheel**, which uses the rotational power generated by a treadle to shape clay pots, is invented in the Near East.

*c.*2800BCE The earliest known **dam is built at Helwan**, south of modern-day Cairo, Egypt – it contains 23,000 cu m (810,000 cu ft) of stone.

*c.*2625BCE **Imhotep** designs and builds the Step Pyramid at Saqqara in Egypt as a tomb for the Egyptian pharaoh Djoser (see pp.14–15).

▼ *c.*2530BCE **The Great Pyramid** of Khufu is built by Hemiunu at Giza in Egypt (see p.15). It remains the tallest building in the world until the 19th century CE.

▼ *c.*2500BCE A circle of bluestones is erected **at Stonehenge** in southwest England on an earlier earthwork. Two hundred years later, 26 giant sarsen stones are added to the monument.

*c.*2500BCE The **city of Mohenjo-Daro** is founded on a site in present-day Pakistan. It is the first to have a system of public drains.

2100BCE The building of the **Great Ziggurat of Ur** begins under King Ur-Nammu. It is a tiered structure, with a base of 65m (213ft) × 45m (148ft), dedicated to the Moon goddess Nanna.

*c.*1700BCE Construction of the first stages of the great **Minoan palace at Knossos** on the island of Crete begins. It is so complex that it gives rise to the legend of a "labyrinth".

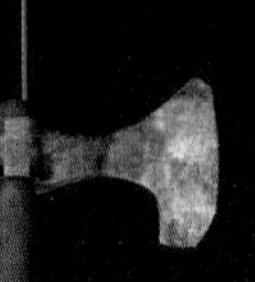

◀ *c.*1200BCE **Ironworking** begins in central Anatolia and soon spreads through the Middle East, creating a more durable metal that can be used in building and in portable artefacts.

▶ *c.*700BCE **The Assyrian army uses siege** equipment, including ladders and battering rams, in its attacks on Near Eastern cities.

*c.*530BCE **Eupalinos** of Samos builds a water tunnel through Mount Kastro in Samos, Greece (see pp.16–17). Starting at opposite ends of the mountain, the tunnel meets almost perfectly in the middle.

◀ 479BCE **Hippodamus of Miletus** is the first known urban planner, arranging new Greek cities including Miletus (left) on a grid system.

▼ *c.*470BCE This Greek temple at **Agrigento** shows evidence of the use of iron beams for supporting the structure of the building.

*c.*400BCE The screw is said to have been invented by **Archytas of Tarentum** (428–347BCE).

◀ **PP.08–09** The oculus in the dome of the Pantheon, Rome, admits the building's only light.

HE EARLIEST MONUMENTAL BUILDINGS that appeared the 3rd millennium BCE demonstrate that there must ve been individuals whose role was to supervise their nstruction, even though the modern distinction between e architect, who draws up the design plans for a building, d the engineer, who oversees its actual construction, was ot as well-defined as it is today. The key challenge for ancient engineers was to multiply the force that individuals could direct, enabling them to lift heavier loads or even to automate some processes entirely. Transporting construction materials for such huge monuments was a big challenge. Their search for solutions to this problem led them to discover the power of levers and pulleys and the ways in which water and steam could be harnessed to power machinery.

300BCE

c.330BCE
exander the Great es arrow-shooting tapults during s campaign of nquest against e Persian Empire.

312BCE
The **Via Appia**, first of the great Roman roads, is built between Rome and Capua (see pp.12–13).

c.280BCE
Sostratus of Cnidus completes the **Pharos** of Alexandria in Egypt, a great lighthouse that is numbered among the Seven Wonders of the World (see pp.18–19).

c.250BCE
e mathematician **rchimedes** invents e Archimedes rew – an innovative ethod for pumping ater (see pp.38–41).

120BCE

▼ c.120BCE
Hipparchus of Nicaea (*c*.190–120BCE), the founder of trigonometry, makes the first accurate calculations of the motion of the Moon.

c.100BCE
The Romans begin using *opus caementicum* (**"concrete"**) for the construction of buildings on a large scale.

c.33BCE
Roman architect **Vitruvius** writes a manual for architects. It includes research on the merits of materials used in construction and also describes a type of water wheel (see pp.20–21).

c.80CE
Hero of Alexandria describes a number of innovative machines, including a wind operated organ, a fire engine, and the aeolipile – a rotating sphere driven by steam power (see pp.32–33).

◀ 102CE
Emperor Trajan's army uses the **cheiroballistra** – a mobile catapult – operated by a crew of two men (see pp.32–33).

105CE

▲ 105CE
The **bridge at Alcantara** in Spain is built of stone arches. Emperor Trajan ordered a bridge that would "last forever".

105–106CE
Roman engineer **Apollodorus of Damascus** (see pp.24–25) builds a 1,100m- (3,600ft-) long bridge across the River Danube. Soon after, he designs Trajan's Forum.

118–128CE
The Pantheon in Rome is rebuilt on the orders of Emperor Hadrian – its 43m- (140ft-) wide dome is not exceeded until modern times (see p.23).

c.130CE
Chinese engineer **Zhang Heng** invents an armillary sphere – a model of the heavens – in which the planets actually move through the use of hydraulic power (see pp.34–35).

13–138CE
Zhang Heng builds the world's first seismograph – it can determine the direction of an earthquake several hundred kilometres away (see pp.34–35).

250CE

c.250CE
Chinese engineer **Ma Jun** invents the south-pointing chariot – a device for determining the direction of south using differential gears (see pp.44–45).

c.360CE
The Greek medical writer **Oreibasius** describes machines for use in the resetting of broken and dislocated limbs.

c.370CE
The Roman author of the *De rebus bellicis* describes a **paddle-ship driven by oxen**. He also designs a four-wheeled ballista pulled by armoured horses.

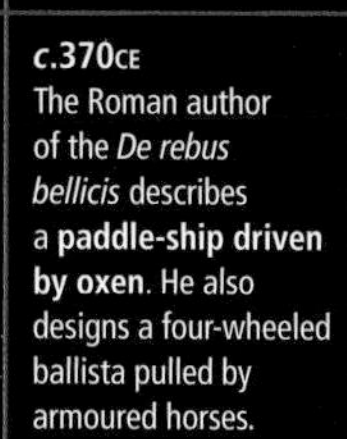

▲ 537
Isidore of Miletus (*c*.474–558) and **Anthemius of Tralles** (532–37) build the church of Hagia Sophia in Constantinople. Its unsupported 32m- (105ft-) wide dome is the first of its kind.

BUILDING THE ANCIENT WORLD

DESPITE THE MANY CHALLENGES THEY FACED IN SOURCING AND MOVING MATERIALS, AND IN DESIGNING STRUCTURES OF SUFFICIENT STRENGTH AND DURABILITY, ANCIENT ENGINEERS CREATED SUBSTANTIAL BUILDINGS, SOME OF WHICH HAVE EVEN SURVIVED TO THE PRESENT DAY.

ANCIENT ZIGGURAT
The four levels of the Chogha Zanbil ziggurat, built in Persia in about 1250BCE by Untash-Napirisha, the king of Elam, were constructed around an original square stone tower.

The earliest engineers whose names we know came from ancient Egypt, beginning with Imhotep (see pp.14–15) who constructed the Step Pyramid of Djoser in about 2625BCE. During the New Kingdom, Egypt's period of greatest wealth and political power, inscriptions tell of engineers such as Senenmut, who was responsible for the Temple of Queen Hatshepsut at Luxor in about 1460BCE. He also set up twin obelisks at the entrance to the Temple at Karnak.

The quarrying, transport, and lifting of the stones required for such enormous monuments was a principal challenge facing ancient engineers. Moving these blocks from quarries east of the Nile required the use of rollers, sledges, and rafts, and an organizational effort on a grand scale. Also, to be able to lift them, engineers would have needed knowledge of mechanical devices such as the windlass – a machine consisting of a rope wound around a drum that, when rotated by means of long spokes pushed by men, can be used to lift a weight.

GREEK TECHNOLOGY

From the 9th century BCE onwards, iron and other useful new materials became increasingly available for engineers. By Classical Greek times, wrought iron beams were used to support the heaviest statues set inside the Parthenon, and in the Greek temples at Agrigento in Sicily, built about 470BCE, beams of wrought iron were used to support the architraves. Once more, the lifting of massive stones to great heights must have required the use of substantial cranes; the architraves of the Parthenon were made up of sections weighing 9 tonnes (10 tons), which each had to be lifted to a height of 10.5m (34ft). As well as the ability to construct large-scale buildings, the Greek engineering achievement extended to planning their setting. The first evidence of town planning is by Hippodamus of Miletus, who, in 433BCE, laid out the new colony of Thurii in Italy on a geometric grid.

LIFTING BLOCKS
This Roman crane, of a type described by Vitruvius, is operated by a pulley and has a grabbing arm to fix to heavy weights such as building blocks.

ADVANCEMENTS IN CHINA

Such precise urban planning had long been practised in China, where cities were laid out on square grids in the early Zhou Dynasty (1046–256BCE) and imperial capitals, such as Chang'an and Luoyang, were built according to a geometrical grid pattern. In China, engineers were typically part of the imperial bureaucracy. Among these was Zhang Heng (see pp.34–35), who invented the first seismograph, but the Chinese also put military engineering skills to good use, such as those of Meng Thien (*c.*221BCE), an army general who created the forerunner of the Great Wall.

> **The earliest buildings built in the 3rd century BCE clearly show that there must have been people whose role was to supervise their erection, even though the distinction between an architect and an engineer was not as clear as it is today.**

ROMAN PIONEERS

Of all the peoples of the ancient world, the Romans have become identified with their skill as practical engineers. It is from them that we derive our word for "engineer", although the Latin *ingenuum* originally referred to a mechanical device. One of the earliest known Roman engineers was Appius Claudius (*c.*340–273BCE), who constructed the great Via Appia, a road from Rome to Capua, in 312BCE.

The Romans also made use of advanced building materials, such as concrete mixed from *pozzolana* – a volcanic earth found in the Alban Hills near Naples – to build the great bath-houses and amphitheatres. Although many of the engineers are known to us, such as Apollodorus of Damascus (see pp.24–25), who supervised the construction of the Forum of Trajan in the early 1st century CE, the name of the engineer responsible for the spectacular Colosseum in Rome has been lost. The building, consisting of 100,000 cu m (1,000 sq ft) of travertine stone, with the weight supported on a complex series of arches, had the capacity for about 45,000 spectators who thronged there to watch gladiatorial and other games. It is a fitting memorial to all the engineers of antiquity, both known and unknown.

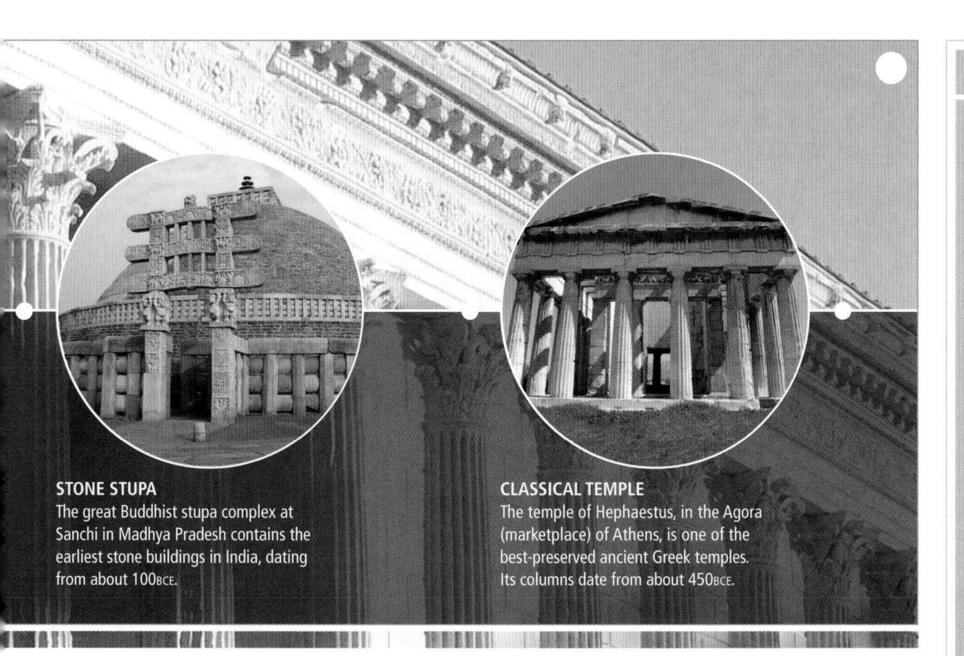

STONE STUPA
The great Buddhist stupa complex at Sanchi in Madhya Pradesh contains the earliest stone buildings in India, dating from about 100BCE.

CLASSICAL TEMPLE
The temple of Hephaestus, in the Agora (marketplace) of Athens, is one of the best-preserved ancient Greek temples. Its columns date from about 450BCE.

SETTING THE SCENE

- Before 3500BCE loads to be transported had to be carried or dragged by people or animals. **The invention of wheels and carts** revolutionizes transport by greatly increasing its ease, efficiency, and speed. A horse can carry 500kg (1,100lb) on its back, but can pull 10,000kg (22,000lb) in a cart.
- Evidence of **engineering skills and planning exist from the earliest times**. The Treasury of Atreus in Mycenae, Greece, is carefully planned (*c.*1250BCE) with a subterranean tunnel leading to a *tholos* – a beehive-shaped interior grave chamber supported by corbelling.
- By the 3rd century BCE in China, **suspension bridges** are developed.
- Between 703 and 690BCE, Sennacherib, king of Assyria, orders the construction of a large-scale series of **water-supply works** to carry water to Nineveh, his capital, from the River Khosr. This includes **channelling irrigation** water from 18 streams in the mountains to the northeast, and the building of a huge dam on the River Atrush.
- The Etruscans, the predecessors of the Romans, **pave roads** as early as the 6th century BCE, with roads as wide as 15m (49ft) in the town of Marzabotto (near modern-day Bologna).

BUILT FOR ETERNITY
The exterior of the Great Pyramid of Khufu at Giza was originally faced with smooth white limestone blocks that gave it a dazzlingly bright appearance in sunlight.

IMHOTEP

ANCIENT ENGINEER WHO BECAME A GOD

EGYPT **2655–2600BCE**

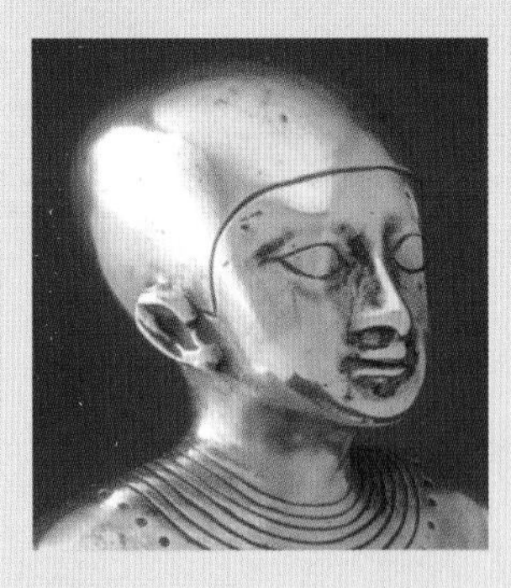

THE PYRAMIDS OF EGYPT are perhaps the most visible engineering feat surviving from the ancient world. Their construction, requiring thousands of cubic metres of stone, was a stunning achievement. The man behind the construction of the earliest of these, Imhotep, was history's first engineer; and the scale of his accomplishments led to him being worshipped as a god.

A LIFE'S WORK

- Imhotep **oversees the construction of the Step Pyramid**, Egypt's first monumental stone tomb, laid out over a labyrinth that served as the king's palace in the afterlife
- Built under his supervision, the **Step Pyramid** takes about **18 years** to build, almost the **entire length of Djoser's reign**
- He is **recognized as the author of a medical treatise** that includes details on how to treat head wounds and how to prevent wounds from becoming infected
- As well as being regarded as a god by the Egyptians, **Imhotep is adopted as a deity by the Greeks**, who revere him as Asklepios, their god of medicine
- His **tomb** is believed to be located in Saqqara, but has **never been found**

By about 2700BCE, ancient Egyptian civilization bore many of the hallmarks that would characterize it for the next 2,000 years. Pharaohs ruled over a united kingdom that depended on the annual Nile floods for its massive prosperity, and their achievements were memorialized by inscriptions in hieroglyphic writing. Yet although these early Egyptian rulers built themselves grand palaces and elaborate tombs, they did so using wood and mud-brick, both of which are perishable materials.

It took the remarkable partnership of Djoser (2630–2611BCE), the second king of the Third Dynasty, and his Chancellor and Great Seer, Imhotep, to begin the transformation of Egypt into a land of monumental stone buildings. Djoser ordered an extensive building programme throughout the land, including temples at Heliopolis in Lower Egypt and at Gebelein in Upper Egypt.

However, it was Imhotep who achieved more lasting fame than his master through his role in overseeing the construction of the Step Pyramid at Saqqara on a plateau some 30km (18 miles) south of modern-day Cairo. Intended as the centrepiece of Djoser's mortuary complex, its ground-breaking form and its place as the earliest substantial stone building in Egypt mean that Imhotep can be considered the first engineer and architect in history. Inscriptions bearing Imhotep's name carry a striking range of titles, including "chief sculptor" and "chief carpenter" – an acknowledgment of his role in Djoser's building projects – and, more importantly *htmw-bity* ("royal seal-bearer"), which was considered the most important role in the royal administration.

IMHOTEP'S CLIENT
This seated limestone statue of the pharaoh Djoser was found sealed in the northeast corner of the Step Pyramid complex that was supposed to house the *ka* (soul) of the dead ruler.

THE FIRST PYRAMIDS

Construction of the Step Pyramid began in about 2625BCE. At first, Djoser's tomb was erected in a traditional style as a *mastaba* – a low, flat brick-built structure, about 28m (92ft) in length, which covered the pharaoh's tomb shaft. The outer wall, which was faced with mud-brick, enclosed a variety of rooms such as chapels, and included niches for cult objects. The one unusual feature was the way in which Imhotep laid out this original structure on a square base, although *mastabas* were usually rectangular.

Some years later, a major adaptation of the *mastaba* was made by building a core of rough stones and casing this in limestone – a ready supply of which was available from nearby quarries. The workmen began to build a series of four "steps" leaning inwards, made up of larger limestone blocks. Finally, a third modification was made, with extensions built to the north and west (making the pyramid's base rectangular) and the addition of two further "steps" to make a total of six. In its final form, the Step Pyramid was about 60-m (200-ft) high on a base 121m × 109m (400ft × 360ft). Its construction required the use of about 300,000 cu m (11 million cu ft) of stone.

The tools available to Imhotep's workmen to achieve such a feat of engineering and construction were not sophisticated. The main devices to ensure that the building lines of the walls and foundation platforms ran straight were simple set squares, to lay out right angles, and plumb

lines, to ensure true verticals. Observation of the stars may also have been used to achieve precise alignments of the pyramid walls. The walls are precisely north-south and east-west.

These alignments can also be found by another method – by bisecting the angle between sunrise and sunset to give a true south, and then using set squares. To cut out blocks of stone, quarrymen used pounders made of dolerite, a hard quartz-like rock, and the blocks were probably transported to the site on wooden rollers, to be finished and smoothed off with copper drills and saws.

HIDDEN MASTERPIECE

Although the Step Pyramid is the most visible of Imhotep's works at Saqqara, beneath it lies the largest subterranean complex ever built in Egypt. Consisting of almost 6km (3.5 miles) of tunnels, galleries, and about 400 rooms – creating a veritable underground palace to surround Djoser's burial vault – the complex is intended to mirror what he enjoyed in life. Together with the "South Tomb", a less elaborate structure, possibly the resting place for Djoser's *ka* ("soul"), these constituted an engineering and architectural feat of astonishing proportions.

Imhotep may have lived on until the reign of Huni (2599–2575BCE), though whether he was involved in the pharaoh's construction projects or those of Djoser's successor Sekhemkhet (2611–2603BCE) is unknown. His posthumous reputation as a great engineer and the founder of Egyptian medicine – with an early treatise on the subject attributed to him – grew over time. He came to be seen as a personification of wisdom and was deified, as the son of Ptah, the principal Egyptian god of wisdom. A cult grew up at Memphis dedicated to him, a temple to him was built at Philae, and there is even evidence for his worship in the kingdom of Meroë to the south of Egypt. As well as being the first recorded engineer, Imhotep is the only one to have become a god.

HEMIUNU

EGYPT C.2570BCE

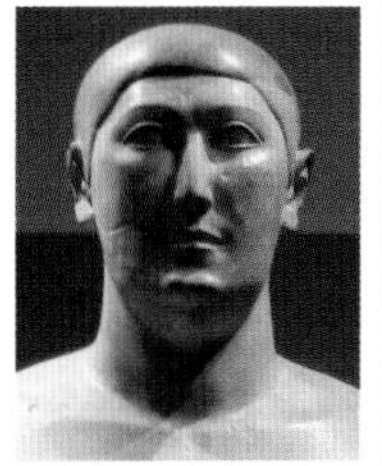

The architect responsible for the construction of the Great Pyramid at Giza, Hemiunu held the title of *htmw-bity* ("royal seal-bearer"), as Imhotep had before him.

Step Pyramids fell out of fashion after the time of Imhotep, with the unfinished pyramid of Sekhemkhet (Djoser's successor) being the only major example. Under the 4th Dynasty, the erection of true pyramids (without stepped sides) began, starting with Sneferu (2575–2551BCE), who built three such pyramids. The Great Pyramid at Giza built for Khufu (2552–2528BCE) was the largest construction ever, with a square base measuring 230 sq m (755 sq ft) and requiring nearly 2.6 million cu m (92 million cu ft) of stone. Hemiunu, the grandson of Sneferu, was the engineer behind this colossal building. As a prince of the royal house, his own tomb lies close to the Great Pyramid itself.

BUILDING FOR THE AFTERLIFE
This view of the Step Pyramid at Saqqara clearly shows the multi-layered structure of the building and part of the complex of buildings to its southeast, which included temples for the *sed* festival, intended to celebrate important anniversaries of the pharaoh's rule.

EUPALINOS

ANCIENT GREEK HYDRAULIC ENGINEER

GREECE 6TH CENTURY BCE

THE CONSTRUCTION OF SECURE WATER CHANNELS was one of the most difficult challenges faced by engineers in ancient times. In the 6th century BCE, Eupalinos, a Greek engineer, built a long tunnel for this purpose on the Aegean island of Samos. The structure became famous as "the tunnel of Eupalinos" and was noted for being excavated from opposite sides of a hill, with the two ends of the tunnel meeting almost precisely in the middle. It was the longest tunnel of its time and is still regarded as a major achievement in ancient hydraulic engineering.

A LIFE'S WORK

- **Theagenes**, ruler of Megara, **commissions a smaller underground channel** a few decades before Eupalinos builds the tunnel on Samos
- The Greek philosopher **Aristotle** of Stagira **compares the building programme of Polycrates on Samos** (including the Tunnel of Eupalinos) **to the pyramids of Egypt**
- **Herodotus mentions the tunnel** in his writings **100 years later**, which shows that the tunnel was still functioning then
- A **small shrine near the centre of the tunnel is built** in *c.*500–800CE, which indicates that the **structure remained intact 1,000 years after its construction**
- The tunnel is eventually forgotten until its **excavation by the French archaeologist Victor Guérin** in 1863

For many larger towns and cities in the ancient world, wells could not provide enough water. To supplement the supply, water was often supplied from a continuously flowing spring through channels, or pipes made of either clay or metal. Pipes had the advantage that they could carry the water uphill as well as down, although they were vulnerable to leaks – a single puncture could cause the whole system to run dry.

Ancient water engineers generally used an open conduit system, known as a qanat. In this system, water flowed through a channel made of stone, sealed underneath with plaster or cement to provide waterproofing. The principal disadvantage of this method was that the water needed to be channelled downhill at a gentle gradient to ensure a consistent flow. The most ancient surviving qanats (up to 2,000 years old) are the water channels of the Middle East and North Africa, which run through open ground. The earliest known example of a water channel dug through a hillside is the Tunnel of Hezekiah, built in the 8th century BCE outside Jerusalem, and even here the conduit may have followed an underground watercourse.

DESERT IRRIGATION
Qanats such as this one in Morocco sustained agriculture in areas with little rainfall. In the Near East and much of North Africa underground qanats were accessed by vertical shafts.

MAKING A BREAKTHROUGH

It was on the Greek island of Samos in the 6th century BCE that the technical challenge of driving a water channel through a mountain was solved.

Under the rule of Polycrates (570–522BCE), the Samians controlled the Strait of Mycale, a crucial trade route that enabled them to dominate naval traffic in the eastern Aegean. The revenue from this allowed Polycrates to become a patron of architects, mathematicians, and poets from all over Greece. It also paid for a building programme, which included, according to the historian Herodotus, three of the greatest construction feats of the age. The first was an enormous temple to the goddess Hera, constructed by the Samian architect Rhoecus, and the second was an artificial harbour sheltered by a breakwater about 125m (410ft) long. The third and, in engineering terms, the most significant of Polycrates's commissions was the channel driven through Mt Kastro to supply water from the spring of Agiades to the town of Samos in the southeast. The engineer behind this feat was Eupalinos.

In ancient times, one way to construct a water channel when a hillside obstructed the route was to build it around the edge of the slope and support it on a low wall (called a *substructio* by the Romans). However, this was expensive and vulnerable to enemy attack. An alternative method was to dig the tunnel underground with vertical shafts being sunk down to allow small connecting sections about 30m (100ft) long to be built in succession. Eupalinos adopted the second approach. Two of his excavators dug through solid limestone for 1,035m (3,390ft) using only simple hammers and picks, carving out a tunnel about 2.4m (8ft) in height and width, and managed to make the two tunnels meet in the middle.

There were some deviations to the route on the way – the northern tunnel started to zigzag about halfway to the junction and then turned sharp left at the junction point, a manoeuvre that may either have been designed to avoid unstable areas of soft earth or simply to minimize the risk of the southern tunnel missing its northern counterpart completely.

USING MATHEMATICS

At the point where the two tunnels joined, there was a slight drop of 60cm (2ft), which was a minor discrepancy. Apart from this, the tunnel itself is almost exactly horizontal. Inside the tunnel, the water was carried in a rectangular channel, which sloped slightly to ensure an even flow. At the northern end, this channel was about 3m (10ft) below the tunnel floor, sloping down to 9m (30ft) beneath the

ROCK TUNNEL
The interior of the tunnel was made tall enough for Eupalinos's workers to walk through. At its centre is placed a Byzantine shrine built in the 6th century CE.

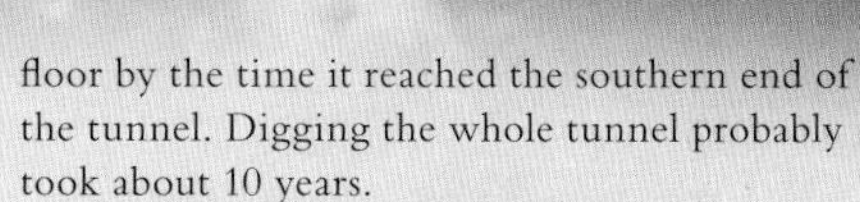

floor by the time it reached the southern end of the tunnel. Digging the whole tunnel probably took about 10 years.

Having started from the north, Eupalinos had to calculate where to begin the southern end of the tunnel, and in which direction to dig. To find the correct level, he probably sent surveyors horizontally round the side of the mountain. To find the direction, another team went over the top, keeping to a straight line above the tunnel using sighting poles, until they reached the right level. They could then check both position and angle using Pythagoras's theorem (see box, right). The tunnel was still intact a thousand years after its construction. Thereafter, however, one of the greatest marvels of ancient engineering was forgotten until its excavation by the French archaeologist Victor Guérin in 1863.

PYTHAGORAS

One of the most famous of all the ancient Greek mathematicians was Pythagoras, who was born in the town of Samos, and was probably a teenager at around the time Eupalinos was digging his tunnel. The town is now called Pythagoreio in his honour. Pythagoras went on to run a mathematical school in southern Italy, and is alleged to have invented the modern musical scale. He is remembered for what is now called Pythagoras's theorem, which states that in any right-angled triangle, the square on the hypotenuse (the long side) is equal to the sum of the squares on the other two sides. This relationship had already been known for a thousand years, but Pythagoras proved it was always true, and probably introduced the vital concept of proof into mathematics. Meanwhile the tunnel of Eupalinos turned out to be a masterpiece, and survived for 2,500 years.

THIS IS A MARBLE BUST OF THE MATHEMATICIAN PYTHAGORAS, WHO PROPOUNDED THE PYTHAGORAS THEOREM.

SOSTRATUS OF CNIDUS

BUILDER OF THE FIRST LIGHTHOUSE

GREECE 3RD CENTURY BCE

PHAROS, THE GREAT LIGHTHOUSE of Alexandria, was constructed in the Greek-ruled Ptolemaic kingdom of Egypt in the early 3rd century BCE. The lighthouse was one of the most famous buildings of its age, and Sostratus of Cnidus, the man who built it, was a part of the select band of engineers responsible for one of the Seven Wonders of the Ancient World. For many centuries, the Pharos was among the tallest man-made structures on Earth. It remains a civic symbol of the city of Alexandria, and one of the greatest engineering achievements of Sostratus.

A LIFE'S WORK

- As a military engineer, Sostratus helps with the **capture** of the **Egyptian royal capital of Memphis** in 323BCE by diverting the course of the River Nile
- He builds a **raised promenade** supported on arches in Cnidus, the first of its kind ever built
- Historians believe that Sostratus **dedicated** the Pharos **to himself rather than to Ptolemy II** as ordered
- His name appears in an **inscription in the Treasury** of the Cnidians, Delphi, but it is not clear for what service he was honoured
- Sostratus acts as a **diplomatic envoy** for Ptolemy II in 270BCE

Although the city of Alexandria was only laid out in 331BCE on the orders of Alexander the Great, the small offshore island of Pharos, from which the lighthouse takes its name, was already known to the epic poet Homer five centuries previously. He mentions "an island in the surging sea behind the Nile ... There is a harbour there with good anchorage, from which men launch shapely ships into the sea". Alexander ordered the construction of a causeway that would link the island to the mainland and make a safer harbour, and also the building of a lighthouse on the island, to guide sailors into the port.

The building was commissioned either under Ptolemy I Soter (306–282BCE) or his son, Ptolemy II Philadelphus (285–246BCE). The man responsible for building it was Sostratus, from the town of Cnidus (in Caria, on the southwest coast of modern-day Turkey).

Over the centuries, the Pharos Lighthouse was severely damaged by earthquakes, and, in 1480, the remaining stonework was reused to build Fort Qait Bey, which still stands on the eastern tip of the island. Our understanding of the appearance of the Pharos depends principally on the accounts given by Greek, Roman, and Arabic travellers, as well as depictions on coins, gems, and mosaics.

LOFTY ARCHITECTURE

The lighthouse, set on a base about 9m (30ft) square, is thought to have soared to a height of about 120m (400ft), and was crowned with a 15m (50ft) statue – probably of the god Poseidon during the Roman period. At the very top was a bronze mirror to reflect sunlight and be visible from far out to sea. This was the second tallest building in the world at the time – dwarfed only by the Great Pyramid of Khufu.

THIS TOWER ... APPEARS TO SPLIT THE SKY ... BUT AT NIGHT A SAILOR ... WILL SEE A GREAT FIRE ABLAZE ON ITS SUMMIT

POSIDIPPUS OF PELLA c.275BCE

The Pharos was meant to be dedicated to Ptolemy II, and Sostratus did indeed dedicate it to him, but according to the 2nd-century historian Lucian, Sostratus inscribed Ptolemy's dedication on a layer of shining white plaster. Over the course of a few decades this soft plaster wore off, revealing underneath a dedication to Sostratus himself, carved in much harder stone: "Sostratus of Cnidus, the son of Dexiphanes, to the Divine Saviour Gods, for the sake of those who sail on the sea", mounted near the top of the building.

The lighthouse had three stages: the bottom one was square – and accessible by a long external ramp – the middle octagonal, and the top one circular – a cylinder. The bottom storey, which was about 56m (183ft) high, was punctuated by a series of windows designed to prevent the lighthouse from being blown over in a storm. It may have contained a viewing platform for visitors. The second level was about 27.5m (90ft) high, while the topmost level was just 7.3m (24ft) in height.

The Pharos was built of white stone, which allowed it to be seen from a great distance during the day. When describing the stone, Arabic writer al-Idrisi speaks of caadzan, a type of hard white limestone, almost marble-like in quality. In about 26BCE, the historian Strabo describes the Pharos as being topped by a bronze mirror, which reflected sunlight and guided ships into the port. A mosaic from the 6th century CE even depicts such a mirror. Posidippus, a contemporary of Strabo, refers to a great fire that acted as a beacon for mariners at night. Many of the 300 internal rooms of the Pharos acted as storage space for the fuel necessary to keep this fire burning, and there is some evidence of winches being used to haul up the wood needed for the fire. According to the 1st-century historian Josephus, the light – whether it was from the mirror reflection or beacon flame – could be seen at a distance of 300 "stadia", which is perhaps 30km (20 miles).

The cost of building the Pharos was a staggering 800 talents (an ancient currency system). This was a tenth of the total amount available in the Egyptian treasury at the accession of Ptolemy I, and was significantly more than the expenditure on the great Parthenon at Athens, which had cost about 450 talents two centuries before. It is unlikely that such funds would have been available to a private individual. This increases the likelihood that Sostratus was in charge of building the Pharos, rather than simply being its patron as some theories suggest.

RECURRING DAMAGE

The Pharos, completed in 280BCE, was the last of the Seven Wonders of the World to be built, and the last to fall. It was damaged during a war in 48BCE, and then by a succession of earthquakes. In 796 the upper storey collapsed after a tremor. Then, one hundred years later, Sultan Ibn Tulun (868–884CE) built a mosque on the tower's summit. The Pharos was still in use as a lighthouse until 1183, but further collapses in 1303 and 1323 reduced it to rubble, which was then used to build Fort Qait Bey in 1480. In 1994, French archaeologist Jean-Yves Empereur discovered blocks of granite, including parts of a colossal statue, on the seabed near the Pharos. These are perhaps the last remains of Sostratus's great engineering accomplishment.

ANCIENT NAVIGATION

The development of lighthouses, which began with the Pharos, was extremely welcome to mariners, as shipwrecks were common. Most ancient vessels used the stars and landmarks to navigate, and so tended to remain in sight of the coastline. They were also slow-moving. For example, the trireme warships reached a speed of 14 knots by carrying 200 rowers, with no room left for cargo. Cargo-carrying vessels, using both sail- and oar-power, were much slower – it took nine days to travel a distance of 1,600km (1,000 miles) at an average speed of 4.6 knots.

ROWED TRIREMES SUCH AS THIS ONE CARRIED ARMED MARINERS FOR BATTLES, BUT WERE UNFIT FOR TRADE.

WONDER OF THE WORLD
This reconstruction of the Pharos shows the three levels of the building and the top lantern lit up with the fire that guided mariners into the harbour at night.

VITRUVIUS

COMPILER OF ANCIENT ARCHITECTURAL PRACTICES

ROME C.84–C.15BCE

THE EARLIEST MAJOR WORK on engineering and architecture to survive from the ancient world is the *De Architectura* ("On Architecture") written by Vitruvius in the 1st century BCE. Although Vitruvius built little himself, his book on the subject provides much of our knowledge of Roman engineering technology and architectural theory.

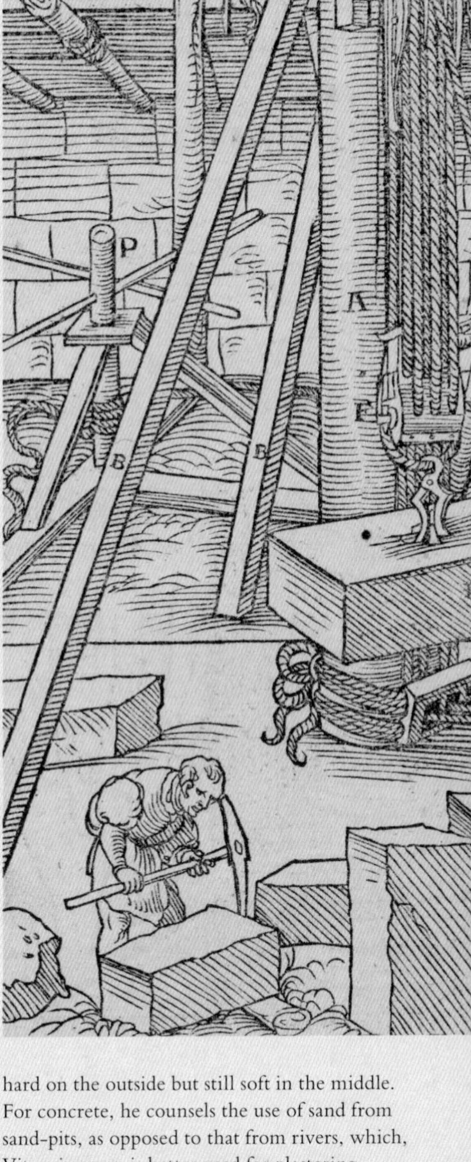

Vitruvius's book, *De Architectura libri decem* ("Ten Books about Architecture"), provides the only few details known about his life. He was born in about 84BCE and served in the Roman army under Julius Caesar as a military engineer, and was responsible for constructing siege engines. He seems to have played some role in making improvements to the water supplies around Rome under Octavian (the future Emperor Augustus), but the only building where he is known to have supervised the construction himself was the basilica, an administrative building in the provincial Italian town of Fanum Fortunae, now in the modern-day town of Fano. The Fanum basilica has long since disappeared, and even its ground plan is not known. Therefore Vitruvius's book, composed during his retirement some time after 33BCE, aided by a pension granted by Emperor Augustus, is the only surviving monument of Vitruvius.

The first book of *De Architectura* sets out the many qualities needed by a person to be a good architect – skill in writing and drafting plans, the ability to coordinate and direct building work, knowledge of mathematics, philosophy, acoustics, music, and law, in addition to the more obvious requirement of an intimate knowledge of building material and architectural styles. Vitruvius then describes what he believes to be the fundamental principles of architecture and the qualities all good buildings must possess. His list encompasses *firmitas* ("strength") – to ensure that a building is constructed well, *utilitas* ("utility") – to guarantee that it fulfils the purpose for which it was built, and *venustas* ("beauty") – to make sure it is aesthetically pleasing. His outline of these principles has formed the basis of architectural theory ever since.

A LIFE'S WORK

- Vitruvius **calls the architect a *secundus deus* ("a second god")** because of his role in shaping a building, just as God shaped the world
- He may have been responsible for **standardizing the sizes of pipes and nozzles in the public water supply** of Rome
- In the **preface to *De Architectura***, Vitruvius **refers to himself as small, old, and ugly**
- His **family name is not known**, although it is often **believed to be Pollio**, as this appears at the beginning of a supplement to his ten books in some manuscripts
- He is **sometimes considered to be the same person** as a **Roman noble named Mamurra** from the town of Formia, but this identification is far from certain

DETAILED RECORDINGS

After describing different types of construction, ranging from private dwellings to public baths, in books III to VI, Vitruvius moves on to the qualities of various building materials – a more tangible sign of his engineering expertise. Some of his recommendations are very specific. When using wood, he advises that oak is tough but warps easily, and alder is suitable for foundations in damp places because it never rots. For the best bricks, Vitruvius recommends only those made in the spring and autumn, because those made in the summer will be baked hard on the outside but still soft in the middle. For concrete, he counsels the use of sand from sand-pits, as opposed to that from rivers, which, Vitruvius says, is better used for plastering.

In the final few books of *De Architectura*, Vitruvius describes many of the inventions devised by his predecessors that could be of use to the architect and engineer. For instance, he writes about a mechanism invented by the Greek architect Cherisphron, who built the great temple of Artemis at Ephesus in the 6th century BCE. This device was used to transport massive

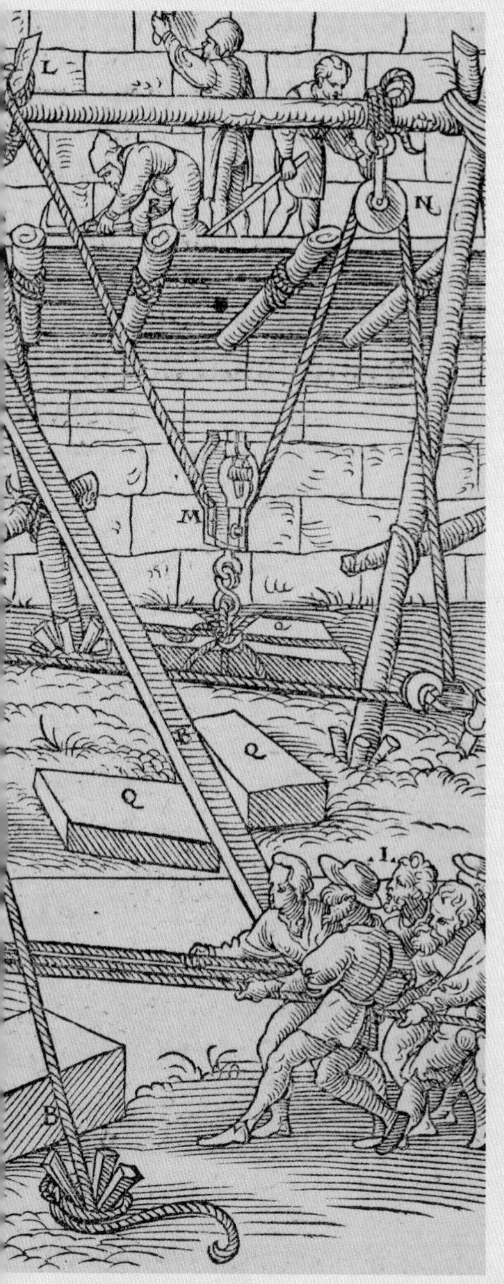

VITRUVIUS'S CRANES
A depiction of one of Vitruvius's cranes from a medieval German edition of his works clearly shows the three feet (marked B), which gave it stability, and prevented it from toppling if a heavy load swung round.

column drums (segments), which could weigh several tons. To achieve this, a wooden frame was built around the segment, and iron spigots were attached inside another wooden frame to create the "spokes" of a giant wheel with the segment in the centre, which could then be pulled by oxen. Vitruvius also details a primitive form of odometer that could be used to calculate the distance travelled by a cart.

Among the inventions by more famous engineers of antiquity, Vitruvius explains the construction of Archimedes' screw for raising water (see pp.38–39), and a type of force-pump with pistons, cylinders, and valves for pumping water – a device ascribed to the Greek engineer Ctesibius (see pp.30–31).

IMPROVISING IDEAS

Vitruvius betrays his own background as a military engineer by showing his particular interest in catapults. He discusses the method of increasing the diameter of the springs to give catapults more power. His knowledge of military engineering is also revealed by his account of various cranes, including one with a quintuple pulley system on the cable for hoisting, and others that used treadmills to provide the power needed to lift heavier weights.

However, it is in the field of hydraulics and water pumps in particular that Vitruvius made his own unique contribution. As well as the only detailed description of siphons (used for relieving hydraulic pressure in pumps) to come from an ancient author, there are also descriptions of a number of devices for pumping or lifting water, including a turning wheel that filled buckets and then emptied them into a channel as it rotated. Vitruvius refined this water wheel by altering the shape of the buckets so that less of the water spilled out as it rotated. The type of water wheel which came to bear his name – the Vitruvian mill – was an "undershot" water mill where the wheel has vanes or paddles around its circumference against which the water current pushes, forcing the wheel to move around. Attached to the main wheel is a gear with projecting pins. These pins engage with bars on a second wheel that is horizontal – at right angles to the first. The second wheel is attached to a millstone, which is used to grind corn.

Vitruvius's work was transmitted by later Roman authors who summarized or adapted *De Architectura*. Vitruvius's own writings, however, were lost, until their rediscovery in 1414 by the Italian humanist Poggio Bracciolini. They came to light just in time to have an influence on the rediscovered sciences of architecture and engineering at the start of the Renaissance.

CENTRAL HEATING

Vitruvius describes many building innovations designed to improve convenience, one of which was the hypocaust, a type of central heating system, where hot air from a fire circulated under the floor of public baths and villas. He gives instructions for maximizing fuel efficiency, for example so that the caldarium (hot bathroom) is next to the tepidarium, followed by the frigidarium. He even recommends a type of regulator to control the heat – a bronze disc set into a circular aperture in the roof, which could be opened or closed by a rope pulley to adjust ventilation.

THIS IS THE HYPOCAUST BENEATH THE CALDARIUM OF THE ROMAN BATHS IN BATH, ENGLAND. HEAT IS CIRCULATED IN THE EMPTY SPACE BENEATH THE TILED FLOOR.

SYMMETRY IS THE HARMONY WHICH ARISES OUT OF THE ASSEMBLED PARTS OF A BUILDING AND THE CORRESPONDENCE OF THOSE PARTS TO THE FORM OF THE BUILDING AS A WHOLE

VITRUVIUS, *DE ARCHITECTURA*

ENGINEERING INNOVATIONS

THE DOME

A DOME IS ESSENTIALLY a three-dimensional version of an arch in which the arch's vertical axis is rotated around to form a circular ring. It is a structural element of architecture and resembles the hollow upper half of a sphere. The form of a dome is a very ancient one, possibly dating even to before Neolithic times. Yet it was only during the early Roman Empire that the discovery of concrete, which could be cast in horizontal layers thinned out towards the top, permitted the creation of truly monumental domes.

RISE OF THE DOME

The dome has its origins in shelters built by early humans out of pliable material such as tree branches, which naturally created circular structures with curved roofs. The shape became associated early on with the "dome" of the heavens and acquired a sacred aspect. It was used for tombs in ancient Mesopotamia in the 3rd millennium BCE and in underground corbelled chambers for the tombs of the Mycenean ruling class from the 14th century BCE.

Although such primitive domes continued to be in use, the larger hemispherical domes of classical antiquity had to wait until Greek and Roman engineers had developed the mathematics to understand how the walls of buildings could be reinforced sufficiently to carry greater weights. Because of the difficulty of the engineering involved, many domes, such as that of the great pagan temple of Marnas at Gaza (130CE), were built of wood. This technique was adopted after the legalization of Christianity in the Roman Empire led to

c.15,000BCE – MAMMOTH HOMES
Round huts made of mammoth bones at Mezhirich (Ukraine) create a curved dome-like structure.

c.5,500BCE – HIDDEN VAULTS
Underground domed vaults are built at the Halafian settlement at Tell Arpachiyah (in modern-day Iraq).

c.2500BCE – QUEEN'S TOMB
The Tomb of Queen Puabi at Ur, Mesopotamia, is built using a rubble dome constructed over a timber framework.

c.1400BCE – MYCENAEAN TOMBS
Tholos tombs (tombs with underground corbelled vaults), such as the Treasury of Atreus, become widespread through the Mycenaean world.

c.300BCE – THOLOS STYLE
A Thracian royal tomb at Kazanlik (Bulgaria) is built in the tholos style, and decorated with ornate frescoes.

c.130 – PAGAN DOME
The Marneion pagan temple in Gaza is built with a freestanding bulbous dome.

c.200 – PENDENTIVES
The Tomb at Qasr an-Nuwayis (Jordan) is the first known example of a cut-stone dome using pendentives.

c.224 – SQUINCHES
The Palace of a Sassanian ruler, Ardashir, at Firuzabad (in modern-day Iran) has a brick dome with the first known use of squinches.

BCE 15,000 10,000 5000 2500 1000 500 250 CE 100 200 300

The Neolithic tomb at Maes Howe on Orkney, built in about 2800BCE, uses corbelling to create the dome of its inner chamber, which is buried under the ground. The walls, built with stone slabs, are 1.4m (4ft 8in) high and slope inwards, using smaller and smaller sections of stone until they meet at the roof, creating a dome. Excavators break the domed roof open in 1861.

2800BCE MAES HOWE

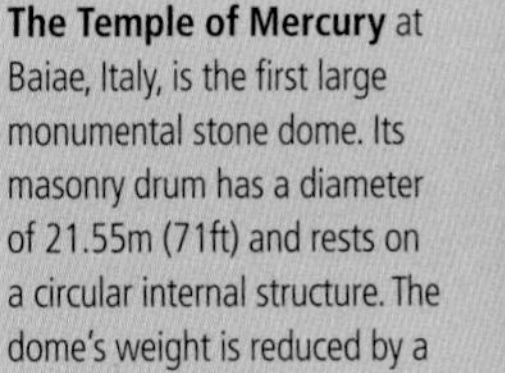

The Temple of Mercury at Baiae, Italy, is the first large monumental stone dome. Its masonry drum has a diameter of 21.55m (71ft) and rests on a circular internal structure. The dome's weight is reduced by a central oculus and four square windows, while the thickness of the drum narrows to just 60cm (2ft) near the top. It is the model for temples to Diana and Venus at Baiae and also the Pantheon.

c.80BCE TEMPLE OF MERCURY

the building of large churches, such as the domed Church of the Holy Sepulchre in Jerusalem (326–335CE).

ADVANCES IN MASONRY

The Roman discovery of the properties of concrete, a material both durable and light, allowed them to create larger domed areas, including great bath houses such as the central rooms of Nero's Domus Aurea palace in Rome (64–68CE). The domed form became the dominant one for Christian churches between the 4th and 6th centuries and masonry domes began to appear. Earlier domes, such as the Pantheon (see box, right), had relied on the dome's weight pressing down on massively thick supporting walls, but gradually engineers experimented with new techniques. In the 2nd–3rd centuries CE, pendentives (triangular sections of stone that helped transfer the pressure of the dome on to the walls) had appeared at the tomb of Qasr an-Nuwayis in modern-day Jordan and at the Church of the Holy Apostles at Constantinople (now Istanbul) in 550. Squinches – diagonal arch-like projections – were used to carry the weight of the domes. By the 6th century CE, engineering techniques had advanced sufficiently to create the great dome of the church of Hagia Sophia in Constantinople, whose vast bulk was almost freestanding. Further advances in dome technology had to wait until the Renaissance, and the width of the Pantheon's dome was not exceeded until the 19th century.

THE PANTHEON

The Pantheon in Rome was built between 118 and 128CE. Made of brick and marble, the weight of concrete in the 43.2m- (142ft-) wide dome is about 4,000 tonnes (4,500 tons). It is supported on eight barrel vaults. Other strategies to diminish the pressure of the dome include an open, unglazed space (oculus), the narrowing of the dome walls to 1.2m (4ft) near the oculus, and the coffering (or excavation of square panels) from the interior marble shell.

THE OCULUS REDUCES THE DOME'S WEIGHT AND ADMITS THE ONLY LIGHT.

c.315 – DOMED BATHS
The triple-apsed caldarium (warm room) of the Kaiserthermen baths in Trier has one of the largest domes in northern Europe.

c.326 – WOODEN DOMES
The Church of the Holy Sepulchre at Jerusalem is built with a timber dome (one of the first Christian domed churches).

c.525 – FREESTANDING DOME
The Tomb of Bizzos at Ruweha is built. It is the only extant freestanding masonry dome in Syria.

532 HAGIA SOPHIA

The Church of Hagia Sophia in Constantinople is built between 532 and 537. Its central dome is 32.5m (107ft) wide, supported by two lower half-domes that sit to the east and west, creating a longitudinal (rather than circular) space. The dome collapses in 558 and is redesigned by Isidore the Younger, who adds pendentives to spread the load on the arcades below – their first use in a major structure, creating the largest freestanding dome of antiquity.

c.550 – VAULTED DOME
The Church of the Holy Apostles at Constantinople is built in a Greek cross shape and is vaulted with five domes.

c.561 – BRICK DOME
The building of a brick dome for the Qasr ibn Wardan, a military complex in Syria, uses massive vaults of narrow span.

c.705 – MOSQUE STYLE
The Umayyad Mosque at Damascus is built. It is the first large Islamic masonry dome.

c.805 – DOMED CHURCH
The Palatine Chapel in Charlemagne's palace at Aachen has a cupola that is 31m (102ft) high and 16.5m (54ft) wide.

c.1018–37 – DOME SYMBOLS
The Church of Santa Sophia, Kiev, has one large dome representing Christ and 12 smaller domes to symbolize the apostles.

400 500 600 700 800 900 1000

1063 ST MARK'S BASILICA

The basilica of St Mark in Venice has a 42m- (138ft-) wide central dome, flanked by four smaller domes. The inner shell of the main dome is less than half that of the outer domes, which are supported on circular drums. During the 13th century, the domes are heightened by the addition of lead casing to make them fit in with the remodelling of the basilica in the Gothic style.

APOLLODORUS OF DAMASCUS

GREATEST ENGINEER OF IMPERIAL ROME

ROMAN SYRIA 2ND CENTURY CE

ONE OF THE MOST PROMINENT engineers of the Roman Empire, Apollodorus of Damascus was associated with some exceptionally high-profile building projects under the emperor Trajan. From feats of military engineering to the reshaping of a large part of central Rome, Apollodorus's career demonstrates the importance of major engineering works in Roman public life.

Very little is known about the early life of Apollodorus, except that he was from Syria, which had been ruled by the Romans since 64BCE. He was a military engineer in the early stages of his career. Emperor Trajan (53–117CE) was his greatest patron but it is unclear when they met. He accompanied Trajan during the Second Dacian War (105–106CE) against Decebalus, the ruler of the Dacian kingdom north of the River Danube (in modern-day Romania), who was threatening Roman territory. Transporting the army of 12 legions (about 60,000 men) across the Danube was achieved through the use of pontoon bridges. Trajan entrusted Apollodorus with the task of devising a permanent solution to the problem of communicating with the army and with the new province he hoped to conquer.

To achieve this, Apollodorus supervised the construction of an enormous arched bridge (see box, below). For a thousand years no one in the world built a longer arched bridge. The technical difficulty of building it over such a wide river probably meant that Apollodorus diverted the flow of the river. This required him to dam the section to be bridged, which explains how he managed to complete it in just a single building season.

Despite the effort and skill required to build the bridge, its life span was short. In about 120CE, Emperor Hadrian, Trajan's successor, decided that it made crossing the Danube southwards into the Roman Empire too easy for potential barbarian invaders, and ordered its demolition.

BRIDGE ACROSS THE DANUBE

The 1,100m- (3,600ft-) long Trajan's bridge was supported by driving 20 huge piles into the River Danube. The piers that supported the bridge were made of stone, while the upper structure was made of wood. On the north bank of the Danube, the bridgehead was at Drobeta (now in the town of Turnu-Severin, Romania), where even today, two huge stone blocks of the first supporting piers can be seen.

TRAJAN'S COLUMN DEPICTS THIS SCENE OF APOLLODORUS AND THE EMPEROR MAKING A SACRIFICE IN FRONT OF THE BRIDGE.

A LIFE'S WORK

- Apollodorus **enters the Roman imperial service**, though the date of his entry is unknown
- Writes a book on **siege-craft**, which includes a device for **pouring boiling water over defenders** manning a wall
- Is **attributed with building an odeon** (small theatre) in Rome, but no trace of it survives
- **Falls out of favour** with Trajan's successor, Hadrian, over an argument in which he **accuses the emperor of being an amateur dabbling in architecture** for his decision to build the new Temple of Venus too low and without a pedestal; some believe that **Hadrian arranges the assassination of Apollodorus**
- Owing to his fame, several buildings **constructed by others are attributed to him**, including the **Bridge of Alcantara in Spain**, built by Gaius Julius Lacer, and the **Pantheon in Rome**, built by Decrianus of Ostria during Hadrian's reign

TRAJAN'S FORUM
The monumental brick walls of Trajan's Forum still stand. The emperor funded its construction with the spoils from the conquest of Dacia which ended in 106CE. It was the last of the great imperial forums to be built.

Apollodorus accompanied Trajan back to Rome, where he received his first large civil commission, the *Thermae Traianae* ("Baths of Trajan") on the Esquiline, one of Rome's seven hills. Completed in about 109CE, the baths occupied an enormous area of 9.3 hectares (23 acres), with the central bathing block taking up almost half the area.

He included new developments such as the use of groin vaults (made by intersecting two barrel vaults) that gave a sense of spaciousness to the interior. There was also an extensive use of windows made possible by the increasing availability of glass by the early 2nd century CE. The baths were largely constructed in concrete, a material that was commonly used for the largest structures in Rome from the late 1st century BCE.

FORUM AND MARKETS OF TRAJAN

The complex for which Apollodorus has become best known was even more ambitious in scope. The construction of the Forum of Trajan and the adjacent Markets of Trajan required the excavation and levelling of the area between the Esquiline and Quirinal hills to create a flat space for the new buildings. The Forum, inaugurated in 112CE, measured about 300m (980ft) long and 185m (600ft) wide. It featured two porticoes adorned with statues of Dacian prisoners, a large entrance arch with a statue of Trajan driving a triumphal chariot, two libraries, and the Basilica Ulpia, an administrative building, which, at 170m × 60m (560ft × 200ft), was the largest in Rome.

The Markets of Trajan were intended to provide a new commercial quarter for Rome. The complexity of the terrain was such that it needed terracing at different levels and a departure from the normal symmetry of such Roman buildings, regaining considerable engineering and architectural skills to conceive and complete the project. Apollodorus laid out the complex like a military camp, with elements such as the pathways that divided up Roman forts mirrored in the layout of the Markets.

The only monument that has survived intact from the Forum is the Column of Trajan (see pp.26–27), a 30m- (100ft-) high structure made of Luna marble. A golden urn, which contained Trajan's ashes, was interred inside the base of the column.

AN ODE TO TRAJAN

All around the exterior of the column are a series of friezes, spiralling upwards in 190m (625ft) of cartoon-like strips, which provide a visual narrative of the two Dacian wars. The carvings of events in the conflicts, including the crossings of the Danube, are the richest source of data for what the Roman army looked like on campaign. As a monument for Trajan, the frieze depicts the emperor about 60 times. Building such a tall column was technically difficult, and would have pushed Roman cranes and lifting apparatus to their limit. In total, about 1,100 tonnes (1,232 tons) of stone were needed, with the stone used as the capital at the top of the column – which had to be lifted more than 30m (100ft) – weighing 53 tonnes (59 tons) alone.

The column, one of the most striking Roman structures to have survived, is a fitting epitaph to Apollodorus, Rome's greatest engineer.

A CELEBRATION OF EMPEROR TRAJAN'S VICTORY

TRAJAN'S COLUMN

Built by Apollodorus of Damascus in about 110CE, Trajan's Column formed the centrepiece of the Forum of Trajan in Rome. At 30m (100ft) in height, the column was built of 20 large blocks of Luna marble. The 190m- (625ft-) long frieze that spirals around the column illustrates scenes from the two Dacian wars fought by Emperor Trajan. These panels were almost certainly carved in a workshop, as carving them once the column had been erected would have been difficult. As well as an engineering feat in itself, the carvings reveal valuable information about the Roman army and its military engineering practices.

► **MOBILE CATAPULT**
Two Roman artillerymen are shown here manning a carroballista. This was a light mobile catapult with two vertical springs and two arms that sprung horizontally to launch an arrow or bolt against the enemy. The carroballista's light weight made it a highly mobile and effective weapon.

▼ **FIELD ARTILLERY**
In this major battle against the Dacians, Roman engineers have set up carroballista emplacements behind the ramparts of a fort (top left) and in the open, where they are protected by timber breastworks (bottom centre). The purpose was to provide covering fire to the legionaries assaulting a Dacian fortress.

▲ **WAR MACHINE**
This is a detail of a three-wheeled war machine, possibly an onager, a type of catapult.

◄◄ BUILDING A CAMP
Here, Roman legionaries are building a fortified camp at the end of the day's march. It is a relatively sophisticated structure, with strong ramparts. The "bricks" that form the outer walls are probably made of turf.

◄ ENGINEERING TOOLS
The legionaries carried a variety of entrenching and digging tools in their packs. Here, the fort is nearing completion as one soldier makes finishing touches to the wooden frame on which the turf bricks will be set. The others are using buckets to remove excess earth.

▲ IMPERIAL SHIP
Trajan is shown here re-crossing the River Danube by boat, a move forced on him by a Dacian counter-invasion of Roman territory. The boats depicted are biremes, with two banks of oars.

◄ RIVER CROSSING
Trajan's army is shown here crossing into Dacia on a pontoon bridge made up of small boats. Apollodorus of Damascus would later design a more permanent bridge across the River Danube, but the pontoons were easily and rapidly assembled, lending an element of surprise to the invasion.

ANCIENT INNOVATORS

THE ENGINEERING ACHIEVEMENTS OF THE ANCIENT WORLD WERE UNDERPINNED BY GROWING TECHNICAL KNOWLEDGE. DISCOVERIES IN MATHEMATICS, ASTRONOMY, SURVEYING, HYDRAULICS, AND MECHANICS LAID THE BASIS FOR BOTH PRACTICAL AND THEORETICAL ENGINEERING.

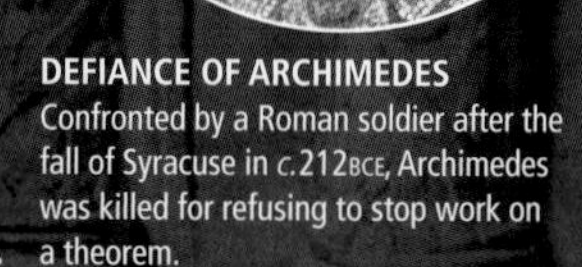

DEFIANCE OF ARCHIMEDES
Confronted by a Roman soldier after the fall of Syracuse in *c.*212BCE, Archimedes was killed for refusing to stop work on a theorem.

CHINESE WATER WHEEL
This Chinese wheel uses the flow of the river to lift water to an irrigation channel. The use of such water wheels, which were also used to power bellows, was first recorded in China in about 31CE.

The fundamental challenge for ancient engineers was to increase the level of power available to them beyond 36kg (80lb) – the weight a man could haul easily. Achieving this meant understanding the processes by which natural materials, such as water and air, could be harnessed to power devices.

NATURAL FORCES PUT TO USE

It was not until the 6th century BCE in Greece that philosophers such as Thales (624–546BCE), Anaximenes (*c.*585–25BCE), and Democritus of Abdera (*c.*460–370BCE) began to lay the foundations of theoretical engineering. By considering the matter with which the universe is composed – Thales thought it was water, for Anaximenes it was air, and for Democritus, indivisible particles called "atoms" – they moved away from the ancient belief that natural phenomena were caused by divine forces.

The establishment of the great Library in Alexandria under Ptolemy I (323–283BCE) acted as a further spur to advances in theoretical engineering, and many of the greatest innovators of the Greek world attended the Mouseion, the school that was attached to it. These included Aristarchus of Samos (*c.*310–230BCE) – who first proposed the idea that the Sun is at the centre of the solar system, and Eratosthenes of Cyrene (275–194BCE) – who made an accurate calculation

ANCIENT GEAR SYSTEM
The Antikythera mechanism, dating from about 100BCE, employed a complex system of dozens of gears to make astronomical calculations.

EXPLOSIVE DEVICES
By the 12th century, Chinese engineers had developed gunpowder and incendiary weapons, notably a flamethrower that used a piston pump first described by Pi Li in 186CE.

SETTING THE SCENE

- The **bow and arrow is the first "machine" to store energy**. The bow is probably first invented in about 25,000BCE.
- **The potter's wheel**, invented in about 3000BCE, is the **first device to use rotational movement in a machine**.
- The Babylonians are the **first to formalize the study of the stars as astronomy**, making star maps and **establishing the sexagesimal system**, with the **circle divided into 360 degrees**, and the **year into 360 days**.
- The **earliest form of piston is part of a syringe**, which the **ancient Egyptians use for mummification** in the late 3rd millennium BCE.
- The Greek scholar Thales of Miletus (*c.*625–548BCE) devises the **Theorem of Thales**, which states that **an angle inscribed within a semi-circle will always be a right angle**.
- Chinese engineers develop **devices for lifting water** in the 3rd century BCE. Before that, water had to be hauled manually out of wells using a rope and bucket, gourd, or pottery jar.

of the circumference of the Earth. Its most famous alumni were Euclid (*c.*300BCE), who established the basis of almost all subsequent geometry, and Archimedes, who founded the field of hydraulics (see pp.38–41). Later on, the Alexandrian school produced innovators such as Hero of Alexandria (see pp.32–33), who furthered the scientific understanding of water power. Without the Alexandrian engineers, ancient engineering would have been quite impoverished.

Greek philosophers paved the way for theoretical engineering. By considering what makes up the universe (water, air, or indivisible particles called "atoms") they moved away from the belief that natural phenomena were caused by divine forces.

Even so, it took until the 1st century BCE for the power of water to be harnessed for use in water mills. The earliest known mention is by the geographer Strabo in the 1st century BCE, who describes one in Pontus (now in present-day Turkey), which was probably of the "vertical shaft" type in which water is channelled by a wooden trough to a 30m (10ft) drop to drive the stone. However, the Romans never adapted their engineering innovations to large-scale industrial use.

More dedication was applied to the development of deadly war machines and siege engines, and many ancient engineers applied themselves to the science of "poliorcetics" (or siege-craft). Catapults, in particular, attracted the attention of engineers, becoming even more complicated. They were fitted with windlasses, pulleys, and torsion springs of hair or sinew coated in oil – until 190BCE when Isidore of Abydos designed a catapult that was 5m (16ft) long, and could hurl a shot weighing 18kg (40lb).

In many ways Chinese engineers were far ahead of their Western counterparts. From about 230BCE, gear wheels were found in Han tombs, ratcheted with 16 slanting teeth to allow different speeds, using techniques that were not described in Western Europe until the 4th century CE. Devices for lifting water, complicated automata, levers, pulleys, and gears were all invented and described by Chinese engineers under the Han Dynasty (206BCE–220CE) and its immediate successors. They also devised machines such as seismographs, while Ko Hung (*c.*320CE) described the principles of the rotating blade of the helicopter – a thousand years before Leonardo da Vinci (see pp.68–69) did so in Italy.

LOST IDEAS

Although the work of the early Chinese, Greek, and Roman innovators was truly clever, much of it was lost or forgotten. Hero's inventions, such as pistons, steam power, and windmills, prefigured many of the innovations that would help launch the 19th-century Industrial Revolution in Europe. Yet at that time, his machines found only small-scale use in agriculture, in homes, and in temples. It would take many centuries for engineering once again to reach the heights achieved by the ancient engineers.

ANCIENT TOOLKIT
This stone tablet depicts the tools of a Roman surveyor, including a suspended "groma" – a sighting instrument with plumb lines and right angles.

CTESIBIUS

FATHER OF PNEUMATICS

GREECE 285–222BCE

Many early Greek and Roman machines depended on compressed air for their operation. The man who first understood the theoretical basis of its use was Ctesibius, a 3rd-century BCE engineer from Alexandria in Egypt, who founded the science of pneumatics and whose air-operated devices influenced other writers until Roman times.

Ctesibius was a mathematician and an inventor. Few details of his life and career are known, but he probably lived and worked in the Egyptian city of Alexandria, during the reigns of Ptolemy I (304–283BCE) and Ptolemy II (285–246BCE).

THE CTESIBIAN DEVICE

Ctesibius's first invention was an adjustable-height mirror for his father's barber shop. The mirror was suspended on a rope that ran over a pulley, and was balanced by a lead counterweight. The weight was kept out of harm's way in a tube, and Ctesibius noticed that when the weight slid up or down in the tube there was a hooting noise, as the compressed air escaped from a small hole. He was inspired to invent the water organ, and to develop the science of pneumatics.

He went on to produce a force pump, a mechanism for raising water by means of compressed air, which could be used in wells or other places where water needed to be pumped to a higher level. The pump was made up of two cylinders standing in a water tank and operating alternately. When the piston in each cylinder was pushed down, it compressed the air underneath, and the compressed air pushed water up a central pipe between the cylinders and into a higher tank. When each piston was pulled up again, it sucked air in through a valve. This was like a precursor of Hero's force pump (see pp.32–33). The device had the advantage that it could be used to lift water to any height, as long as the piston and cylinder (which Ctesibius specified should be made of bronze) were able to withstand the pressure of the water.

A LIFE'S WORK

- Ctesibius is probably **the first head of the Mouseion**, the centre of scholarship attached to the Library of Alexandria
- **Designs an acoustic cornucopia** (horn of plenty) that emits sounds, which is **mounted on a statue of Arsinoe**, the wife of Ptolemy II, some time between 274 and 270BCE
- **Invents the hydraulis, a water organ** that uses the action of compressed air to make different sounds when the organ player pressed the keys
- Devises the **principle of the siphon**

PROPERTIES OF AIR DISCOVERED

Several such pumps have been found by archaeologists, proving that the Ctesibian pump was a practical device that was actually put to use. One discovered at Silchester (in southern England), in 1895, had an oak block 56cm (22in) long to act as the water tank, with the pipes lined with lead to prevent leakages. Another found at Bolsena (in Italy) probably formed part of the pump for a Roman fire engine.

Ctesibius also devised an improvement to the traditional clepsydra (or water clock), which had been used to measure the time available for speakers in the law-courts. In Ctesibius's version, the basic clepsydra vessel was fitted with an overflow pipe, and had an extra reservoir feeding water in from above, so that it was always full. This meant that the flow rate

WATER CLOCK
This woodcut of 1567 gives some idea of Ctesibius's improved clepsydra, or water clock. These remained the most precise clocks for several hundred years.

CTESIBIUS WAS THE FIRST WHO FOUND OUT THE PROPERTIES OF THE WIND, AND OF PNEUMATIC POWER

MARCUS VITRUVIUS POLLIO

of water out of the vessel remained constant. This outflow of water ran into a cylinder that contained a cork float. As the water ran steadily in, the float rose, and the pointer attached to it moved up a scale of hours. In the ancient world, a day was considered to have 12 hours of daylight no matter what the time of the year (so that in winter the "hours" were shorter). Ctesibius made a set of refinements, by varying the width of the "hours" on the indicator board, and then altering the valve on the water tank so that it was irregular, and letting a varying amount of water out each day, depending on the season.

IMPROVED CATAPULTS

Ctesibius also turned his understanding of the properties of compressed air to military use. The catapults that the Egyptian army had used during the reign of Ptolemy I were expensive, and the cords that provided tension for the arms of the bow tended to get damp or perished with use. Ctesibius devised a new catapult, in which the tension was achieved by using compressed air in a bronze cylinder, with piston rods linked to each side of the catapult. He also invented a catapult where a series of bronze springs, placed on top of each other, provided the torsion to shoot the bolts. Unfortunately, both of these systems were technically too difficult to reproduce on a large scale.

Ctesibius set down the details of his machines in a compilation called the *Mechanica*, but this has been lost. It was, however, known to Vitruvius (see pp.20–21), who recorded most of them. It ensures that his inventions, particularly the Ctesibian pump, were among the most renowned engineering achievements of the ancient world.

AGRICULTURAL DEVICES

Agricultural techniques in the ancient world were conservative and slave labour hindered the progress of improved technology. Pliny the Elder (23–79CE) describes an automated harvesting machine, in which the cutting shafts could be moved according to the height of the crop. The cut stalks fell into a container for easy collection; the edges of the machine's frame were splayed out to prevent its wheels from becoming entangled in the stalks. Other advances, associated with wine and olive growing, included lever-operated presses for crushing, and a wine press, which combined a lever with a screw for more crushing power.

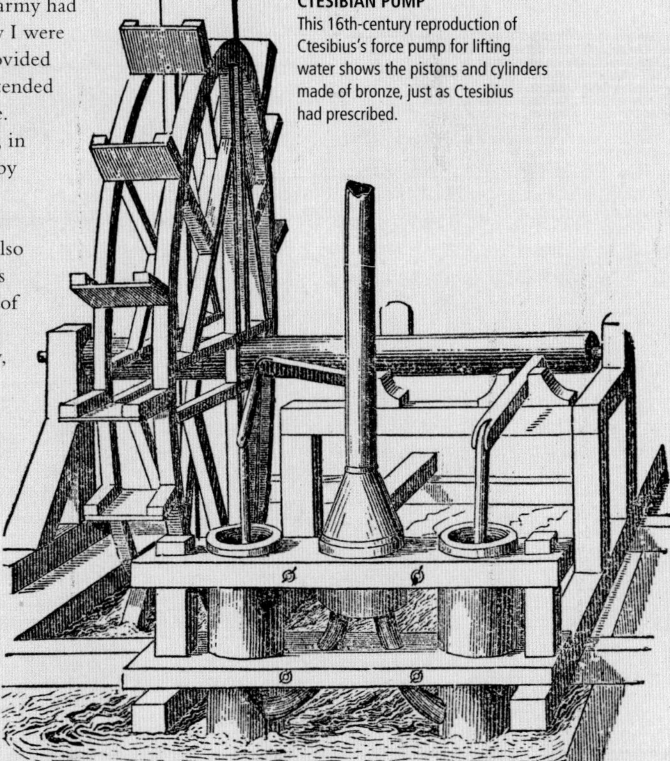

CTESIBIAN PUMP
This 16th-century reproduction of Ctesibius's force pump for lifting water shows the pistons and cylinders made of bronze, just as Ctesibius had prescribed.

HERO OF ALEXANDRIA

AN EARLY INVENTOR OF MANY DEVICES

GREECE C.10–70CE

THE 1ST-CENTURY MATHEMATICIAN and engineer Hero of Alexandria became famous for his invention and description of a wide variety of machines, including the first steam turbine, and a large number of spectacular devices for use in temples or theatres, all of which depended on carefully worked out principles of physics.

With a career spanning the third quarter of the 1st century CE, Hero probably taught at the Mouseion, the scientific academy at Alexandria, and may even have been its director. His surviving writings are in the form of lecture notes for courses in mathematics, physics, mechanics, and pneumatics, which he may have taught there. They comprise explanations and experimentation, and descriptions of mechanical devices and the principles behind them – those he invented himself and also those developed by his predecessors.

In his book *Mechanics*, Hero discusses the theory of motion and how to make a pantograph – a device with toothed wheels used for re-drawing a diagram at a larger scale.

Of more practical importance in the ancient world, he explains the basic principles of the lever, pulley, wedge, and screws, and describes a number of applications of these to cranes and hoists. The most sophisticated of these was the *barulkos* – a lifting device operated by gears – attached with a worm-gear that could not go into reverse (and allow the load to slip) and four parallel axles, all of which meant that a large weight could be lifted using little force, an invaluable aid to lifting heavy marble column drums. Conscious of the need to provide the means to construct these devices, Hero includes the first surviving description of a lathe for cutting the screws that were needed for many of his machines.

A LIFE'S WORK

- Hero **describes a lunar eclipse** that happened on 13 March 62CE in his **book *Dioptra***
- In his **book *Automatopoietica***, he **describes two miniature puppet theatres** where the **figures are powered by millet seeds** running into cylinders, which in turn power pistons
- **Proposes theories on the properties of geometrical solids**, and discusses other mathematical problems
- Invents mechanical devices such as a **drinking vessel** from which either **water or wine can flow** and **a fountain** that is operated by the **action of the Sun's rays**

FIRE AND WIND TECHNOLOGY

In his *Pneumatica*, Hero concentrates on the properties of air, steam, and introduces two of his most famous devices. The aeolipile is perhaps the most important, being the world's first steam engine (see picture, top right). In this, a hollow sphere can spin freely on two supports at opposite ends of its diameter. One support is hollow, and supplies steam to the sphere from the boiler below. The steam escapes tangentially from twin jets, and the force exerted by the escaping steam spins the sphere at up to 1,500 revolutions per minute. Although it provided a spectacular demonstration of the properties of steam and the possibility of converting this into rotary motion, the device was never adapted for practical use; calculations have shown that if it was to do the work of one human, three or four would be needed to operate it; so there was no real incentive to develop it further.

Hero also describes a type of fire engine, using a Ctesibian force pump (see pp.30–31), which has a rocker arm that drives water through two cylinders alternately, using automatic flap valves, and then out through a nozzle. The remains of such a pump were found in the Sotiel Coronado mine in Spain in 1897.

STRATO OF LAMPSACUS

GREECE 335–268BCE

Strato was the third head of the Lyceum, the school of philosophy founded by Aristotle. His interest in science and empirical observation earned him the title *Physicus* ("the physicist").

Strato's work to demonstrate how gravity acts on liquids, including on how water falling from a spout breaks up into individual droplets, was developed by Hero in his book *Pneumatica*. Strato was a materialist, believing that the universe was made up solely of matter and energy and that the "gods" were just a manifestation of natural forces. His interests included works on human physiology, animals, theology, and ethics, although very little of it has survived.

Another device, which again was never harnessed for large-scale use, was the windmill. It was used to power a *hydraulis* ("water organ") and consisted of an axle with two discs, one of which had projecting rods that lifted alternately making a piston fall as the disc rotated. The other disc had vanes attached to it, which were blown around in the wind and provided the rotational power to operate the first. The operation of the piston forced air into the pipes of the organ, allowing the user to create musical notes. It was not, however, until the Middle Ages that wind-powered mills were used widely for milling flour.

HERO'S AEOLIPILE
This modern reconstruction of the aeolipile shows the metal sphere that was spun around by steam from the cauldron below.

THEORIES PUT TO USE

One device that was certainly used was the *dioptre*, an instrument for surveying land, containing a table with a series of vertical rods, which had discs that could be slid up and down to measure the relative heights of points in the distance. Hero used it to determine the relative slope of a water course and the distance between two points not visible from one another.

His cheirobalistra, a catapult that hurled iron-tipped bolts with great accuracy over distances of 300m (980ft), saw military service in 105CE, and is depicted on Trajan's column (see pp.26–27).

Hero also invented "temple devices" to show physical theories in use. These included temple doors that opened when a fire was lit on the altar (using expansion of air caused by the fire to displace water from a bucket that then opened the doors), and a statue that poured out a libation when a devotee placed a coin in a slot. The level of power these devices produced could not be reproduced on an industrial scale. Even so, Hero's place as one of the most influential of ancient engineers is assured.

GIANT CATAPULT
Roman soldiers are shown here preparing a torsion-powered, spear-throwing catapult for shooting. It is similar in design to the cheirobalistra invented by Hero.

ZHANG HENG

FIRST ENGINEER TO MEASURE AN EARTHQUAKE

CHINA 78–139CE

THE HAN-DYNASTY astronomer and engineer Zhang Heng combined the illustrious career of a bureaucrat in imperial service with practical engineering talent. It enabled him to invent two "firsts" in Chinese history – a mechanical celestial globe that demonstrated the rotation of planets, and a seismograph that could detect an earth tremor and indicate its direction.

A LIFE'S WORK

- Zhang **opposes a calendar reform in 123CE** over its use of texts outside the standard Chinese canon, and consequently is **demoted from his official positions**
- **Compares the Earth to the yolk of an egg**, lying at the centre of the celestial sphere
- **Estimates that there are 11,250 small stars** in the universe
- Earns a salary of **600 bushels of grain** per year during his first official position at the Han court; however, **after his demonstration of the seismograph**, it is **increased to 2,000 bushels**
- His **armillary sphere survives** for 300 years, and is found **rusty and neglected by the army of Emperor Wu** in 418CE when the army recaptures Chang'an from northern barbarians

Zhang Heng was born into a family that had held important civil service positions in Nanyang, Henan Province, during the early 1st century CE under the Eastern Han Dynasty (25–220CE) of China. His grandfather was the governor of a commandery (province), and the young Zhang was sent to study Chinese classics and philosophy at the imperial capitals of Chang'an and Luoyang. When Zhang returned to Nanyang in 101CE, he gained a position in the local commandery's chancery, before being summoned to court at the age of 34.

Although he achieved the prestigious position of Chief Astronomer – first from 115–120CE and then from 126–132CE – his career was marred by rivalries with other scholars and the influential eunuch faction at court. This led to his exile from the imperial capital for several years before his recall in the last year of his life.

Zhang was a polymath, and his interests ranged widely, from civil service reform to literature, mathematics, and engineering. His lyric poetry was highly regarded in subsequent times, including works such as the *Western Metropolis Rhapsody*, which warned against the dangers that governmental corruption and decadence posed to the Eastern Han Dynasty – the same problems that had brought down their Western Han predecessors in 8CE.

Zhang's most notable mathematical achievement was the approximation of the mathematical constant "pi". One of Zhang's predecessors, Chinese astronomer Liu Zin (*c.*50BCE–23CE), had estimated it to be 3.154 but it is not known how. Zhang made calculations based on the area of a circle inside a square and the volume of a sphere inside a cube, and arrived at the figure of 3.162, which would not be improved upon in China for nearly 200 years.

CELESTIAL EXPLORATION

One of Zhang's key interests was astronomy. In his studies, he combined acute observational skills with engineering aptitude. He belonged to an astronomical school called Hun Thien ("Celestial Sphere"), which regarded the universe as a sphere, the centre of which was the Earth. The Sun, Moon, and planets were situated on this sphere, and according to Zhang, they rotated at different rates. In the star catalogue that he compiled from his observations, he identified 124 constellations, and named 320 of the brightest stars, stating that there were 2,500 large stars in the heavens. Among his astronomical advances was understanding how lunar eclipses occur, and that the Moon does not have its own independent light, but instead reflects the light of the Sun.

To illustrate his concept of the heavens, Zhang constructed an armillary sphere – a celestial globe. Zhang's was not the earliest armillary sphere in China, but it was unique in being the first that actually moved to show the rotation of the planets and the relative movement of the stars. To operate it, he used a gearing system that was linked to a clepsydra, a water clock, whose water tank released liquid into a series of pans. Eventually the weight forced the pans down, driving the gears. Zhang realized that when the inflow tank of the clepsydra was filled, the device slowed down, causing the clock to lose time. To resolve this problem, he added an extra compensating tank between the outflow and inflow tanks. The craftsmanship needed to construct the armillary sphere is indicated by the precise instructions for its assembly. For instance, Zhang specifies, "The equatorial ring goes around the belly of the armillary sphere 91 and 5⁄19 degrees away from the pole".

ZHANG'S SEISMOGRAPH

The world's first seismograph, invented by Zhang, was in the shape of an urn that had eight tubular projections shaped like dragon's heads, each containing a bronze ball. Below each of these heads was a bronze frog, mouth agape. When a tremor occurred, a pendulum placed inside the urn moved in the direction of one of the dragon's heads. A crank and lever would raise that dragon's head and release a ball, which then dropped into the mouth of the frog below. As the heads were set in the eight principal directions of the compass, the earthquake's direction could be determined. The pendulum also acted as a stabilizing device, and had a pin that engaged a toothed wheel, which halted the device after the tremor.

WHEN THE BALL DROPPED INTO A FROG'S MOUTH IN THE SEISMOGRAPH, THE APPARATUS EMITTED A NOISE, INDICATING THAT AN EARTHQUAKE HAD OCCURRED.

THE SUN IS LIKE FIRE AND THE MOON LIKE WATER; THE FIRE GIVES OUT LIGHT AND THE WATER REFLECTS IT

ZHANG HENG IN *THE SPIRITUAL CONSTITUTION OF THE UNIVERSE*, c.120CE

For his armillary sphere Zhang had applied hydraulic engineering, the use of which had been initiated in about the 6th century BCE. The earliest water wheels in China were not used for turning millstones, as was the case in Europe, but instead for operating bellows on blast furnaces, used for smelting iron. The first recorded use of this system was in 38CE by Du Shi, who had become the prefect of Nanyang in 31CE. The amount of labour that was saved as a result – previously hand-bellows had been used – meant that the innovation initially spread widely. The invention was, however, then forgotten and was only rediscovered in about 238CE. By the 5th century, the practice was so widespread that an artificial lake was built at Pei Phi for the ironworks there – an industrialized use of the water wheel that Europe had yet to discover.

GROUND-BREAKING DEVICE

Towards the end of his career, Zhang demonstrated before the Han court another revolutionary device – the world's first seismograph. It was in the form of a bronze urn with a pendulum inside (see box, opposite).

When he first demonstrated the device in about 132 or 138CE, the pendulum clanged, indicating that an earthquake had struck. Zhang's critics, however, pointed out that no tremor had been felt. They were duly confounded when, a few days later, a messenger arrived to say that there had been an earthquake in Gansu Province, some 500km (300 miles) to the northwest – precisely the direction the seismograph had indicated.

DISTANCE AND DIRECTION

There were a number of other inventions attributed to Zhang, one of which was an odometer, used for measuring the distance travelled by a cart. The device employed a mechanical wooden figure linked by a gear to the cart's wheel. The figure would strike a drum after every *li*, a distance of 500m (1,640ft) travelled, and a gong after every 10 *li*.

Zhang is also said to have re-invented the "South Pointing Chariot" – a non-magnet compass made up of a two-wheeled chariot that had differential gears, which meant that the compass constantly pointed to the south, no matter how far or in which direction it travelled.

Although many of Zhang's inventions were forgotten after his death, he was held in high esteem among his successors. He remains one of the most highly regarded early Chinese engineers.

CLOCK TOWER
This reproduction shows the water-driven clock tower built by the engineer Su Song in 1088 that incorporated an armillary sphere at the top – similar to the one constructed by Zhang.

BREAKTHROUGH INVENTION

ROMAN AQUEDUCT

The problem of delivering water to cities from distant springs was a difficult one for ancient engineers, especially when hills and gorges interrupted the direct line of approach. Although large-scale water channels or aqueducts were constructed by various ancient peoples, including the Nazca of Peru in the 6th century CE, the most spectacular examples, with the water channel carried on piers or arches, were built by Roman engineers.

By the 4th century BCE, the city of Rome had exhausted the ready supply of nearby water, and engineers turned to exploiting the network of rain-fed streams and springs in the Alban Hills. This led to the building of Rome's earliest aqueduct, the Aqua Appia, in 312BCE, the first of 11 aqueducts in a 475-km (300-mile) network of concrete-lined water channels, much of them laid out underground.

The engineers who built the aqueducts were conscious of the need to establish a gentle, near-constant gradient (normally between 1 in 200 and 1 in 350) to ensure a smooth and constant flow of water. Problems arose when the water channel had to cross a steeply sloping valley or gorge.

One solution was to construct a "siphon", a pipe that would be watertight even under outward pressure, and allow gravity to push water down one side of the valley and up the other. This solution was adopted near Lyons, France. Another method was to even out the gradient by raising the entire water channel on to a series of arches to create the most stunning examples of Roman aqueducts. The one at Nemausus (Nîmes, France), completed in about 16BCE, had three tiers of arches, as its 270-m (900-ft) length crossed the gorge of the River Gardon, while that at Segovia in Spain is nearly 825m (2,700ft) long, supported on unusually slender arches.

Probably built in the second half of the 1st century CE, under the reign of Emperors Vespasian or Nerva, the Segovia Aqueduct is made up of granite blocks, with little or no use of mortar, arranged in two-tiered arches. One of the best-preserved works of Roman civil engineering, it continues to supply water 16km (10 miles) from the Fuente Fría River to the city of Segovia. It was included in the UNESCO World Heritage list in 1985.

AS FOR WATER, THE AQUEDUCTS DELIVER SUCH QUANTITIES THAT RIVERS OF IT FLOW THROUGH THE CITY AND ITS SEWERS

STRABO *GEOGRAPHY 5.3.8*

ARCHES OF SIMPLICITY
The graceful aqueduct at Segovia has 36 semi-circular arches organized in two levels. On the upper, the arches have a width of 5.1m (16ft) and hold the channel through which the water flows. The channel's gradient adjusts to the base height of the topography below.

ARCHIMEDES

GREAT MATHEMATICIAN AND ENGINEER

GREECE c.287–212BCE

ARCHIMEDES ACHIEVED FAME in ancient Greece both as an exceptional theoretician and for his use of geometrical theorems to solve routine, practical problems. Towards the end of his life, he became even more renowned for the machines he built, which were used as war engines to defend his home city, Syracuse, when it came under siege by the Roman army.

Archimedes is believed to have been born in the Greek city of Syracuse, on the east coast of modern-day Sicily, in about 287BCE. His father was an astronomer named Phidias, whom Archimedes names in his work *The Sand Reckoner*. Archimedes is said to have spent some time in Egypt, studying at the Mouseion of Alexandria, where he may have met the scholars whom he addresses in his works, including Eratosthenes of Cyrene (*c*.276–219BCE), an astronomer who made estimates of the circumference of the Earth, and of the distance from the Earth to the Sun and the Moon. After completing his studies, Archimedes returned to Syracuse and spent most of his career in the service of the Syracusan ruler Hiero II.

MATHEMATICAL ODYSSEYS

The extant treatises of Archimedes are a virtual treasure trove and give an idea of the breadth of his work. In *On the Measurement of a Circle*, for instance, he deploys the idea of "infinitesimals" – in a manner similar to modern integral calculus – to calculate an approximation to the mathematical constant "pi", and he also proves that the area of a circle is equal to pi multiplied by the square of its radius.

The work of Greek mathematician Euclid (see p.40) provided the theoretical underpinning to Archimedes's research in geometry. In fact, such was his love for the subject that he even regarded his scientific inventions as trivial in relation to his achievements in geometry. In his books *On Conoids* and *On Spheroids*, for instance, he presents methods to calculate the volumes of geometrical solids, including a proof that the volume of a sphere that just fits inside a cylinder is two-thirds that of the cylinder. He regarded this as his greatest discovery, and requested that his tombstone be adorned with a sphere enclosed in the smallest possible cylinder.

EUREKA! EUREKA!

It was Archimedes's book *On Floating Bodies* that established him as the father of the science of hydrostatics and led to the anecdote for which he is most famous. His treatise begins with the premise that liquid particles on one level displace other particles on the same level, but under less pressure. At the same time, objects pressing down liquid particles displace them. More importantly, he shows that if an object is placed in water, it will sink to the bottom if it is denser than water, and if weighed when submerged it will seem to be lighter than its actual weight by an amount equivalent to the weight of the water that it displaces.

Archimedes is said to have used these principles to solve a problem presented to him by his patron, Hiero. The Syracusan ruler had ordered a quantity of gold to be melted and made into a votive wreath to be dedicated to the gods as a token of his gratitude for a recent military success. However, the king suspected that the craftsmen entrusted with the task had cheated him by adulterating the gold with some quantity of silver. He thus wanted Archimedes to determine how much silver had been mixed. The problem was that the wreath was irregularly shaped, and so its volume could not be easily determined. One day, as Archimedes stepped into a tub at the public baths, he saw that the water overflowed. Suddenly, he realized that the amount of water that spilled out was equal to his own volume, and that the same method could be applied to the wreath, by submerging it.

A LIFE'S WORK

- Archimedes devises a method to measure the **volume of an intricate object**
- **Constructs a huge ship**, *Syracusia*, and helps launch it into the sea
- Builds machines that are used as **defensive war engines** in the siege of Syracuse by the Roman army
- Discovers that the **volume of a sphere set in a cylinder** is two-thirds the volume of the cylinder
- Works out the **approximate value of pi** (π)
- In his work *The Sand Reckoner*, he calculates that **10^{63} grains of sand would fill up the universe**
- A crater on the Moon and a **lunar mountain range are named after him**

GIVE ME A PLACE TO STAND ON, AND I SHALL MOVE THE EARTH

ARCHIMEDES

CLAW OF ARCHIMEDES
This 18th-century impression of the Siege of Syracuse shows the "Claw" embedded in one of the besieging Romans' ships. The device is just about to pull the vessel out of the water.

In his excitement, he is said to have sprinted home naked, crying out "*Eureka! Eureka!*" ("I have found it! I have found it!").

Archimedes lowered 0.5kg (1lb) of pure gold into a bucket filled with water up to the brim, and measured how much water had spilled out. The volume of water displaced must equal the volume of the gold; so he knew the volume of 0.5kg (1lb) of gold. He then lowered in the fancy wreath, and found that its volume was greater than it should have been for pure gold – the king had been cheated.

Archimedes further developed his work on floating bodies to examine the way they tend to return to a particular position even after being tilted. The shape that he primarily looked at was the paraboloid, perhaps because it is similar in shape to an ancient Greek ship's hull. A hull's stability became important to Archimedes after Hiero commissioned him to construct a massive ship named *Syracusia*, which was large enough to accommodate 600 sailors. The vessel was so heavy (more than 3,629 tonnes, or 4,000 tons) that no one could work out how to launch it into the sea. But Archimedes, while developing work that he had carried out on increasing force by the use of levers, is said to have designed a system of levers and pulleys that required only the lightest of touches to propel the naval leviathan into the waters of Syracuse Bay. When Hiero exclaimed how impressed he was at this feat, Archimedes is said to have retorted, "Give me a place to stand on, and I shall move the Earth!"

It was *Syracusia* that led to another of Archimedes's famous inventions, the Archimedes screw or screw pump. It consists of a screw that fits neatly inside a tube, and can be turned on its shaft either by a handle or, for example, by wind power. Each time it turns, the bottom-end of the screw scoops up some water, which is then carried up the tube and out of the tap. The Archimedes screw has been used ever since for irrigation, for pumping water from a stream up to a higher level, for pumping water out of the bilges of ships, and for pumping sewage.

EUCLID'S *THE ELEMENTS*

Euclid lived at around the turn of the 4th century BCE, a few generations before Archimedes. His best-known work, *The Elements* is a compilation of geometrical propositions, some of them his own, others borrowed from early Greek mathematicians. All subsequent works on the subject for almost 2,000 years were based on the foundations he had laid. Archimedes must also have benefited from Euclid's compilations, but his genius lay in the fact that he undertook original researches to solve both old and new problems in the field of geometry.

THE GEOMETRICIAN EUCLID (SEEN CROUCHING DOWN) IS SHOWN HERE EXPLAINING ONE OF HIS THEOREMS.

THE POWER OF LEVERS
Archimedes is shown here demonstrating the principle of levers and their magnification of force, so great that in theory they might even move the Earth.

THE SIEGE AND THE HERO

A year after Hiero's death in 215BCE, Syracuse came under attack by the Roman army led by Marcus Claudius Marcellus. The siege lasted two years, and Archimedes, by then about 75 years old, turned all his engineering skill towards defending his home town against the invaders. Among the devices he is believed to have invented is the Claw of Archimedes, a mechanical hook on a chain, which would fix into the prow of an approaching ship and then pull it up, so that the ship's deck was tilted at a steep angle, which would lead to most of the sailors falling into the sea. The hook would then be released abruptly, making the vessel break up or capsize. He also deployed cranes to drop huge stones or lead weights on enemy ships to make them sink. Another device he designed, of less-certain

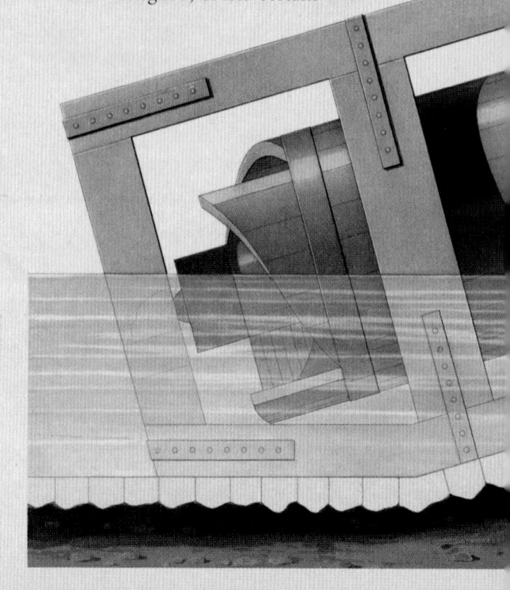

ENGINEERING TIMELINE

287–276BCE	275–264BCE	263–245BCE	c.244–241BCE	c.240–216BCE
	Hiero II becomes ruler of Syracuse and enlarges Syracusan influence in Sicily		Eratosthenes of Cyrene arrives in Alexandria; Archimedes may have met him in Egypt	

effectiveness, was an array of mirrors in a concave shape, intended to focus the Sun's rays and set fire to the Roman ships. Modern attempts to replicate the feat have concluded that it is very unlikely to have worked.

However, all of these stratagems were of no avail, and the Romans finally stormed into the city of Syracuse. Marcellus ordered that the great engineer be brought to him, probably hoping to use his talents for the advancement of Rome. But this was not to be. A Roman soldier found Archimedes examining a diagram that he had drawn in the sand, and told him to accompany him. Archimedes took no notice of the orders, and is instead said to have shouted, "Don't disturb my circles". The irate soldier stabbed him to death. Marcellus is said to have been furious, but he honoured Archimedes's memory by refusing to take any plunder from Syracuse, except for two orreries, or mechanical models of the solar system, that the scientist had constructed. With their moving planets and differential gears, the orreries, whose specifications Archimedes had described in his book *On Sphere Construction*, combined practical engineering skills with physical and astronomical theory, and were a fitting epitaph to his glorious scientific career. In recent times, Archimedes's achievements have been honoured in the form of Montes Archimedes, a mountain range on the Moon that is named after him.

PHILO OF BYZANTIUM

GREECE 350–320BCE

The 3rd-century BCE engineer Philo (also known as Philo Mechanicus) wrote an encyclopaedic work on general mechanics, including sections on pneumatics and automata, but only the parts on pneumatics and those concerned with siege engines, artillery, and fortresses have survived.

Philo listed a variety of military catapults, with improvements such as bronze plates in place of sinews to increase the tension, and therefore the range of the bolts. He also described an automatic catapult, with arrows fired automatically by a winching device. In mathematics, Philo is believed to have solved one of the three most famous geometric problems unsolvable by compass and straight-edge construction – doubling the cube. As well as his mechanical treatise, Philo was reputed to be the author of the earliest list of the Seven Wonders of the World.

THIS REPRODUCTION OF A CROSSBOW, SAID TO HAVE BEEN INVENTED BY PHILO, USED A ROCKER ARM RATHER THAN A TRIGGER TO RELEASE THE BOLT INTO FLIGHT.

ARCHIMEDES SCREW
This cross-section of an Archimedes screw in operation, shows how the "screw" rotates within the wooden tube when operated by a crank handle. The rotational force of the screw then pulls up the water.

215–214BCE	213BCE	212BCE–390CE	391–499	c.500
Archimedes's patron, Hiero II of Syracuse, dies, leading to the breakdown of the Roman-Syracusan alliance	Romans set siege to Syracuse. Archimedes plays a key role in the defence of the city, and may have employed a giant Claw to attack Roman ships	The Roman army under Marcellus captures Syracuse; Archimedes is killed by a Roman soldier for refusing to stop working on a theorem	Most copies of Archimedes's work are destroyed in a fire at the Temple of Serapis in Alexandria	Anthemius of Tralles, a Roman architect, describes Archimedes's fire-creating mirrors as an offensive weapon

ROMAN AMPHITHEATRE
Built in the 3rd century CE, the amphitheatre of Thysdrus (in El-Djem, Tunisia) had the capacity to accommodate more than 30,000 spectators. The weight of the building was supported by a series of tiered arches, which also provided a complex entrance and exit system for the crowds. It remained mostly intact until the Ottoman governor bombarded it with cannons during an uprising in 1850.

MA JUN

CREATOR OF THE SOUTH-POINTING CHARIOT

CHINA c.220–265CE

MA JUN WAS THE MOST DISTINGUISHED ENGINEER in China during the early 3rd century CE. Like most Chinese scientists and engineers of antiquity, he had a bureaucratic career and his interests were widespread. They ranged from the creation of advanced mechanical puppet theatres to a compass-type device known as the south-pointing chariot, as well as developing improved devices for irrigation and silk-weaving. Over time, however, many of his inventions were lost and had to be reinvented by later generations of engineers.

A LIFE'S WORK

- Ma Jun **supervises the construction of a palace** for Emperor Ming of Wei
- **Designs a number of new weapons**
- **Develops a new type of silk loom** that allows the use of more intricate and detailed patterns
- **His puppet theatre is so sophisticated** that it includes figures dancing on balls and hurling swords
- He is said to have **devised the first Chinese odometer** (for measuring distances travelled)
- Is believed to have **invented the first historically valid south-pointing chariot**, which used differential gears instead of magnetism to work
- **Earns the respect** of later Chinese writers

Ma Jun was born into relatively humble circumstances in Fufeng, a city on the River Wei. He did, however, manage to obtain some literary qualifications – a passport to obtaining a position in the lower rungs of the Chinese imperial bureaucracy. As a result, he became a *Ji Shih Zhong* ("Advisor on the Review of Policy") at the court of the state of Wei. China was at that time divided into several kingdoms. Under the patronage of Emperor Ming of Wei, Ma supervised the construction of a palace and is said to have designed several new weapons, including a ballista catapult operated by a rotary device.

ENGINEERING FOR PROFITS

Ma's relative poverty, however, encouraged him to turn his talents towards developing money-making devices. These included a new variant of the silk loom, which reduced the number of treadles from 50 in traditional looms to only 12 in Ma's improved version. This meant that the loom could produce more intricate patterns that could sell at higher prices than those woven with the old version.

IRRIGATION DEVICES

Ma is also said to have developed a new form of square-pallet chain pump for irrigation, in which a chain hauled up through a tube trapped water on the pallets, which were then pulled up, and discharged their load of water at the top. The contraption was apparently used to water the extensive palace grounds of Emperor Ming. However, Ma was probably not the first to discover the device, for it has been described earlier by philosopher Wang Chong in about 80CE, while a set of chain pumps had been constructed under the orders of an official named Zhang Rang, under the later Eastern Han Dynasty.

PUMPING WATER
This illustration shows Ma Jun's chain pump, also known as "dragon backbones", being operated by foot. Ma Jun probably designed it to water gardens.

ADVANCED PUPPETRY

Ma Jun also invented a moving puppet theatre, similar in conception to that of Hero of Alexandria's (see pp.32–33). A primitive form of this theatre has been found in the tomb of Qin Shi Huang Di, the first emperor of unified China. Qin Shi Huang Di's version has 12 bronze-cast musicians, which were moved in unison by pulling a rope laid beneath a mat, while the group seemed to make music when the puppeteer blew into a tube.

Ma was challenged to create an automated version of this, and he achieved it by using a large piece of wood carved into a wheel that was rotated in a horizontal position by water. The effect was astonishing, with statuettes singing, playing, and dancing, while wooden manikins beat drums and blew flutes.

THE RIGHT DIRECTION

Ma is best remembered, however, for his invention (or rather, reinvention) of the south-pointing carriage. This two-wheeled vehicle, said to have been created several times in Chinese history, is first mentioned as having been built by the Duke of Zhou at the start of the 1st millennium BCE as a means of allowing foreign envoys from far beyond the Chinese frontiers to return home safely. It had the singular property of a movable pointer to indicate south, no matter how far it travelled.

EMPTY ARGUMENTS WITH WORDS CANNOT COMPARE [IN ANY WAY] WITH A TEST THAT WILL YIELD PRACTICAL RESULTS

MA JUN

As all previous versions had been lost, Emperor Ming commissioned Ma to build another one in the face of scepticism from senior officials who did not believe the engineer's claim that such a carriage was possible and that it had, indeed, been built in the past. Ma is recorded to have said to them, "Empty arguments with words cannot compare [in any way] with a test that will yield practical results".

The chariot did not rely on magnetism for its operation, but probably used differential gears, and was one of the first mechanical devices to deploy them. These gears allowed a single input of rotation to have two different outputs, allowing wheels to go in different directions to correct the carriage, so that it always pointed to the south. However, although an ingenious invention, the south-pointing chariot was lost over time.

A "REINVENTED" LEGACY

Although it seems that he was famous and respected in his day, Ma is believed to have had weak rhetorical skills. In fact, his ideas were often spurned and, for most of the time, he was not able to persuade others to allow him to apply his ideas.

Unfortunately, few of Ma's inventions survived long after his death. In fact, several of his greatest works – in particular, the south-pointing chariot – had to be reinvented by engineers who came after him. Two centuries later, in 478, for example, Chinese engineer Zu Chongzhi (see box, right) reinvented the south-pointing chariot using a system of differential gears that was similar to the version developed by Ma.

FINDING SOUTH
This model of a south-pointing chariot shows the system of differential gears that allowed south to be correctly determined. The statue placed on top of the chariot always pointed in the right direction.

ZU CHONGZHI

CHINA C.430–510CE

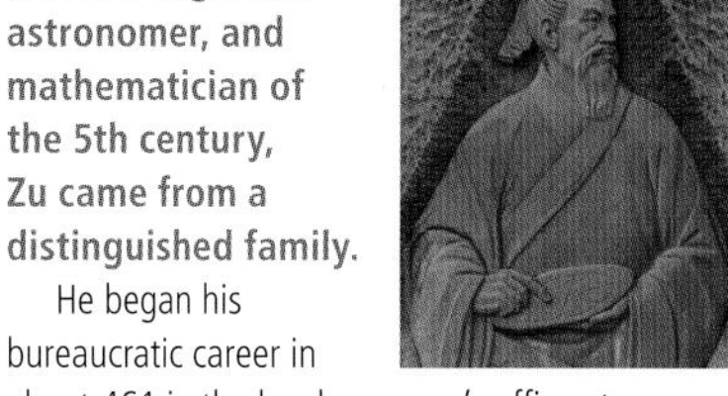

The most renowned Chinese engineer, astronomer, and mathematician of the 5th century, Zu came from a distinguished family.

He began his bureaucratic career in about 461 in the local governor's office at Zhenjiang. His major work, *Zhui Shu* (Method of Interpolation), contained formulas for calculating the volume of a sphere and also an extremely precise calculation of the value of pi (between 3.1415926 and 3.1415927), which was within a fraction of the actual value (and was not improved upon until the 16th century). His astronomical work included calculating the length of the phases of the Moon with great accuracy, and distinguishing between the sidereal year (the length of time the Earth takes to rotate around the Sun with respect to fixed stars) and the tropical year (the length of time in which the Sun makes a full rotation of the seasons).

h2
l
k
L

RENAISSANCE AND ENLIGHTENMENT

RENAISSANCE AND ENLIGHTENMENT

1200

1206
Islamic scholar **Al-Jazari** writes the *Book of Knowledge of Ingenious Mechanical Devices*, which includes designs for water-powered clocks and automata (see pp.52–53).

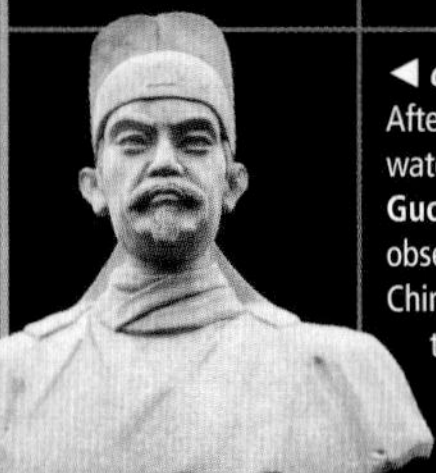

◄ *c.*1280
After designing the waterways in Beijing, **Guo Shoujing** builds observatories across China and equips them with astronomical instruments (see pp.62–63).

▼ 1420
Italian artist and architect **Filippo Brunelleschi** wins a commission to build the dome of Florence cathedral, and designs site-specific lifting gear for the project (see pp.66–67).

1421
The first industrial patent is granted to Italian engineer **Filippo Brunelleschi** for his invention of a barge with lifting gear for moving building materials.

1431
Italian military engineer **Taccola** publishes his first treatise on engineering, *De Ingeneis*, which is followed by a second, *De Machinis*, in 1449 (see pp.64–65).

1500

▲ *c.*1500
Italian polymath **Leonardo da Vinci** fills his notebooks with designs and inventions, including a double hull, a pound lock, weapons, and flying machines (see pp.68–69).

1500s
The *Taschenuhr,* a spring-driven pocket watch, is built by **Peter Henlein** (*c.*1479–1542) in Nuremberg, Germany.

1543
Polish astronomer **Nikolaus Copernicus**'s epochal book, *On the Revolutions of the Celestial Spheres*, is published just before his death in 1543 (see pp.60–61).

► 1550s
A book by Ottoman engineer **Taqi al-Din** (see pp.56–57), gives descriptions and diagrams of a programmable astronomical clock with alarm, and a 6-cylinder water pump.

1550s
The theodolite, an instrument for measuring angles when surveying, is first described by English mathematician and surveyor **Leonard Digges** (1520–99).

1570

► 1570s
Persian engineer **Fathullah Shirazi** introduces Persian technology to the Mughal Empire in India, and as adviser to Akbar designs irrigation facilities at Fatehpur Sikri (see pp.58–59).

1590
The first compound (multi-lensed) microscope is designed and built by Dutch spectacle-makers, father and son **Hans and Zacharias Janssen**.

***c.*1600**
A land yacht carrying up to 24 passengers, designed and built by Dutch military engineer **Simon Stevin** (see p.75), is demonstrated on a beach near The Hague, Holland.

▼ 1609
Italian **Galileo Galilei** (see p.61) makes his first telescope based on designs by lens-makers Hans Lippershey (1570–1619), Zacharias Janssen (1580–1638), and Jacob Metius (*c.*1571–1628).

1620

1620
Cornelis Drebbel, a Dutch engineer working in London, builds an oar-driven submarine to a design by William Bourne and demonstrates it on the River Thames (see pp.74–75).

1629
In his book *Le Machine*, **Giovanni Branca** describes a paddle wheel driven by a jet of steam, similar to the one described by Islamic engineer Taqi al-Din.

1640s
French mathematician **Blaise Pascal** (1623–62) invents the first mechanical calculator, which can add, subtract, multiply, and divide.

▲ 1642
The Canal de Briare, designed and built by **Hugues Cosniers** (1573–1629) and **Guillaume Boutheroue** (d.1648) is completed. It connects the Loire and Seine valleys in France.

1656
The pendulum clock is built by Dutch scientist and clockmaker **Christiaan Huygens** (see pp.80–81), who later designs a spiral-balance spring.

◄ PP.46–47 A drawing of 1717 shows Thomas Newcomen's steam engine operating a pump.

During the Middle Ages, the centre of philosophical nd scientific activity moved from the Mediterranean ountries and China to the Middle East. Following an slamic "Golden Age", a new movement, the Renaissance, ained momentum in Europe, bringing a resurgence of nterest in science and technology increasingly free from he control of the Church. Discoveries and inventions during the next 200 years laid the foundations for the so-called Age of Reason, or Enlightenment, and a "Scientific Revolution" in which engineering and technology flourished. New technology and improved transport drove the pace of change, especially in Britain, where the mechanization of mills and the invention of steam engines led to the beginning of the Industrial Revolution.

1660

c.1660
Englishman **Robert Hooke** (see pp.78–79) invents a balance-spring mechanism to improve the accuracy of clocks and watches, and the anchor escapement for pendulum clocks.

► 1668
In order to make detailed astronomical observations, English astronomer and physicist **Issac Newton** (1642–1727) designs and builds the first reflecting telescope.

1679
Huygens's assistant **Denis Papin** discovers that the pressure of steam in a closed vessel increases the boiling point of water, and designs his "steam digester", or pressure cooker (see pp.106–07).

◄ 1690
Denis Papin designs the first of his steam engines using the pressure of steam from water heated in a cylinder to drive a piston, which returns when the heat is removed.

1698
English engineer **Thomas Savery** (see pp.108–09) is granted a patent for his "fire engine", a steam device for raising water in mines and the first steam engine to be used commercially.

1699
French inventor **Guillaume Amontons** (1663–1705) designs a rotary steam engine that he calls a "fire wheel". It is a type of water wheel driven by steam power.

1700

► 1701
British agriculture is revolutionized by the invention of a horse-drawn seed drill by **Jethro Tull** (1674–1741), which allows seeds to be sown quickly and efficiently in rows.

1705
Denis Papin builds a second steam engine based on Savery's design. In the same year, he fits a river boat with one of his steam engines to power a paddle wheel.

c.1710
Incorporating ideas from the steam engines of both Papin and Savery, **Thomas Newcomen** and his partner **John Calley** build the first practical steam engine (see pp.110–11).

▼ 1755–59
The wooden lighthouse on the Eddystone Rocks in Devon, England, is replaced by a new granite building, designed and built by civil engineer **John Smeaton** (see pp.100–01).

1760

1761
John Harrison's fourth marine chronometer (H4), which he dubs the "Sea Watch", is tested on a voyage to Jamaica and found to be accurate to within five seconds (see pp.82–83).

1761
The **Bridgewater Canal**, designed and overseen by James Brindley (see pp.92–93), is completed. It marks the beginning of a canal network linking British cities and ports.

◄ 1772
Smeaton's assistant **William Jessop** is appointed the principal engineer of the Grand Canal of Ireland, and makes his name as an innovative civil engineer (see pp.94–95).

1774
A steamboat built by marquis **Claude de Jouffroy d'Abbans** (1751–1832), the *Palmipède*, is demonstrated on the River Doubs in France. Its oars are powered by a steam pump.

1790

1790
Jessop becomes a partner in **Butterley Iron Works** and pioneers the use of iron as a construction material, including replacing the wooden wagonways in mines with iron rails.

1801
Construction begins on Jessop's tramway for horse-drawn vehicles, the **Surrey Iron Railway** south of London, which is arguably the world's first public railway.

▲ 1810
Work begins on the nine-arched Waterloo Bridge, the first of Scotsman **John Rennie**'s three bridges across the River Thames in London (see pp.96–97). It is opened in 1817.

1811
John Rennie oversees the construction of a massive breakwater designed by him to protect the harbour at Plymouth Sound in Devon. The project takes 30 years to complete.

▼ 1825
Benjamin Wright's Erie Canal opens and allows cargo to be shipped from the Great Lakes to the Hudson River and the Atlantic seaboard (see pp.98–99).

ISLAMIC ENGINEERS

AFTER THE DEATH OF MOHAMMED IN 632, ISLAM SPREAD ACROSS THE MIDDLE EAST INTO AFRICA, ASIA, AND EUROPE. A SUCCESSION OF ISLAMIC EMPIRES FOSTERED A CULTURE OF SCHOLARSHIP IN WHAT HAS BECOME KNOWN AS A "GOLDEN AGE" OF PHILOSOPHY, SCIENCE, AND ENGINEERING.

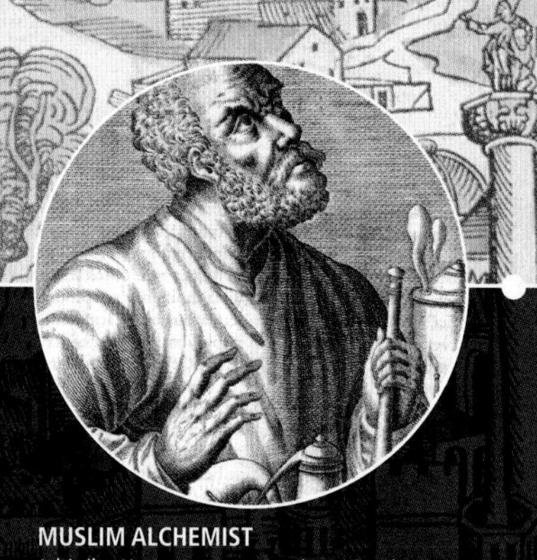

MUSLIM ALCHEMIST
Jabir ibn Hayyan was a prominent Persian alchemist of the 8th century. He was also an accomplished Koranic scholar and mathematician.

The Islamic empires had great respect for learning, drawing on the cultures of the Babylonians and Egyptians, which were themselves enriched by invading empires including the ancient Greeks and Romans. Science, in particular, was already established in many Middle Eastern countries before Islam. There was a considerable body of knowledge of mathematics, astronomy, medicine and alchemy, enhanced by contact with Chinese and Indian scholarship, as well as with the Mediterranean empires.

While medieval Christianity regarded science as a challenge to its dogmas, early Islam tolerated and even embraced the idea of philosophical and scientific research, and respected the scholarship of other civilizations. As a result, science flourished in the Islamic "Golden Age", bringing advances in technology and engineering not seen in Europe until the Renaissance.

During the 8th and 9th centuries, Islamic countries started leading the world in scientific discovery and research. In the major cities, centres of learning were established, such as the Bayt al-Hikma (House of Wisdom) in Baghdad, to encourage learning by preserving and translating ancient texts. These attracted scholars from all over the Islamic world. Many of them became respected *hakim* – polymath "wise men" – and they included some of the greatest scientific thinkers of the medieval period: mathematicians such as al-Khwarizmi (*c.*790–850); the alchemist Jabir ibn Hayyan (*c.*721–815); and physicians such as Abu Bakr Zakariya al-Razi (*c.*854–935); Ibrahim ibn Sina, known in the West as Avicenna (*c.*980–1037); and al-Zahrawi (*c.*936–1013).

NAVIGATIONAL AID
A triumph of Islamic engineering, the astrolabe was used for navigation, determining the position of celestial bodies, and finding the direction of Mecca.

PROMOTERS OF ENGINEERING

Medieval Islam did not just value scholarship for its own sake, however, and scientific research was recognized as having important practical applications. Advances in technology, which had arisen originally out of necessity in the ancient world, had led to a tradition of engineering achievements, and as Islamic cities prospered, their rulers encouraged the *hakim* in scientific research – in contrast to the way in which the Christian church in Europe was attempting to stifle what it saw as "heretical learning".

> **In contrast to the way in which the church in Europe tried to stifle what it viewed as "heretical learning", medieval Islamic rulers hailed technological advances and encouraged the *hakim* ("wise man") in scientific research.**

Under the patronage of the Islamic rulers, discoveries in experimental sciences such as alchemy, for example, led to advances in metal- and glass-working, while those in mathematics provided a base for making calculations in engineering technology.

Inevitably, as the empires vied for territory, military engineering flourished, but so, too, did architecture and civil engineering to provide infrastructure for the increasingly prosperous cities. A driving force for innovation was the management of water, a scarce resource in most of the region. This lead to advances in hydraulic engineering and the invention of water-raising devices that incorporated innovations such as camshafts and suction pumps. The engineers realized the potential of water to make machines work, and developed sophisticated water-driven devices – some, such as clocks and agricultural machines, for specific purposes; while others, such as automata and musical machines, simply for the amusement of their patrons.

Another important stimulus to scientific research came from Islam itself – practice of the religion relied on accurate timekeeping and calendars, and it was essential for worshippers to know the direction of Mecca. Astronomy was, therefore, revered as one of the most important sciences, and it, in turn, encouraged progress in other fields. These

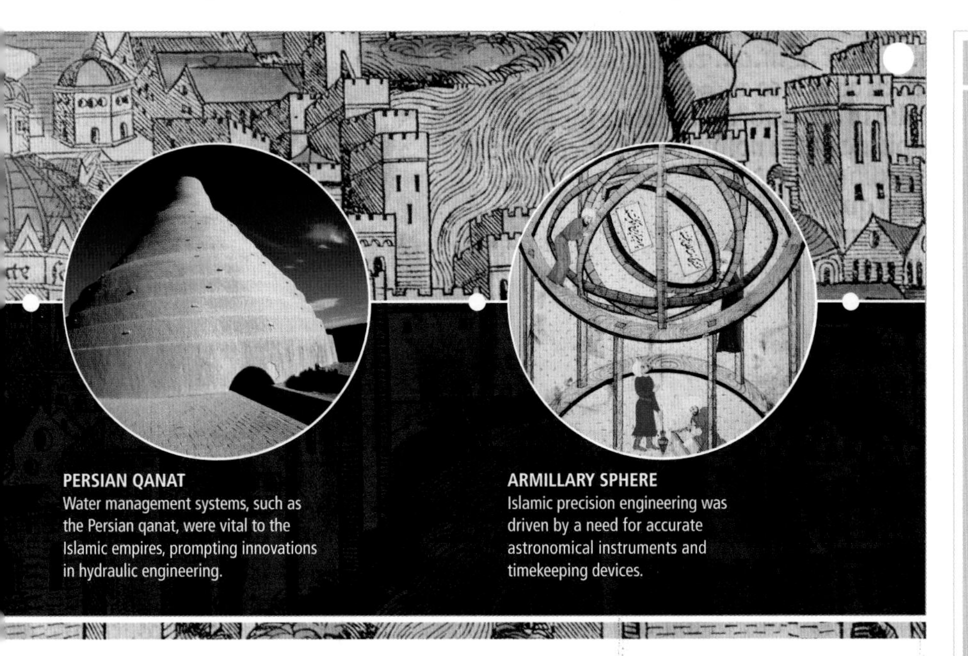

PERSIAN QANAT
Water management systems, such as the Persian qanat, were vital to the Islamic empires, prompting innovations in hydraulic engineering.

ARMILLARY SPHERE
Islamic precision engineering was driven by a need for accurate astronomical instruments and timekeeping devices.

SETTING THE SCENE

- The **Bayt al-Hikma is set up in Baghdad** in 762 as a **centre for scholarship**. Apart from encouraging research and innovation, it collects **Greek and Indian treatises**, which are translated into Arabic.
- Islamic scholars study both the humanities and the sciences. **Even those with expertise in a particular field are polymaths.**
- Islamic countries inherit a long **history of engineering**. Scientists **put their theories into practice**, especially in the areas of clockmaking, the design of astronomical instruments, water pumping devices, and **civil engineering projects to manage water in arid countries**.
- The prosperity of Islamic empires is reflected in the wealth of the courts, and **engineers design "ingenious devices"**, such as musical machines and drink-serving automata.
- **Islamic science and engineering begins to wane** at the **end of the 13th century** as the Ottoman Empire concentrates on political power. However, the culture of **learning is transferred to Europe via translations of Islamic works**.

included mathematics; the technology for making precision astronomical instruments, such as the astrolabe; and the development of clocks capable of accurate timekeeping.

Astronomers including al-Zarqali (1028–87), the designer of an astrolabe, showed their skills as engineers. Other polymaths, such as al-Jazari (see pp.52–53) and Taqi al-Din (see pp.56–57), became better known for their work in engineering rather than astronomy.

INTERMINGLING OF IDEAS

Between the death of Mohammed in 632 and the beginning of the 14th century, various Islamic empires had spread the religion and its culture of science from Arabia, across North Africa, into the Iberian Peninsula, through Persia across Asia, and into the Indian subcontinent and Mongolia. Islamic engineers and scientists continued to flourish, and there was a productive cross-fertilization with the other cultures they encountered.

Unfortunately, many creations of medieval Islamic engineering have not survived. However, Islam fostered a written tradition, so we do have at least a record of those innovations. The *hakim* wrote treatises describing discoveries, and maintained Islamic works alongside translations of similar texts by Greek, Indian, and Chinese scholars. As the church began to lose its monopoly on learning in Europe, Western scholars turned to Classical learning. The libraries of works kept by scholars proved to be a valuable resource. As well as preserving the ancient Greek texts, Islamic scholars contributed a wealth of scientific knowledge of their own, whose importance to the development of European technology has only recently been acknowledged.

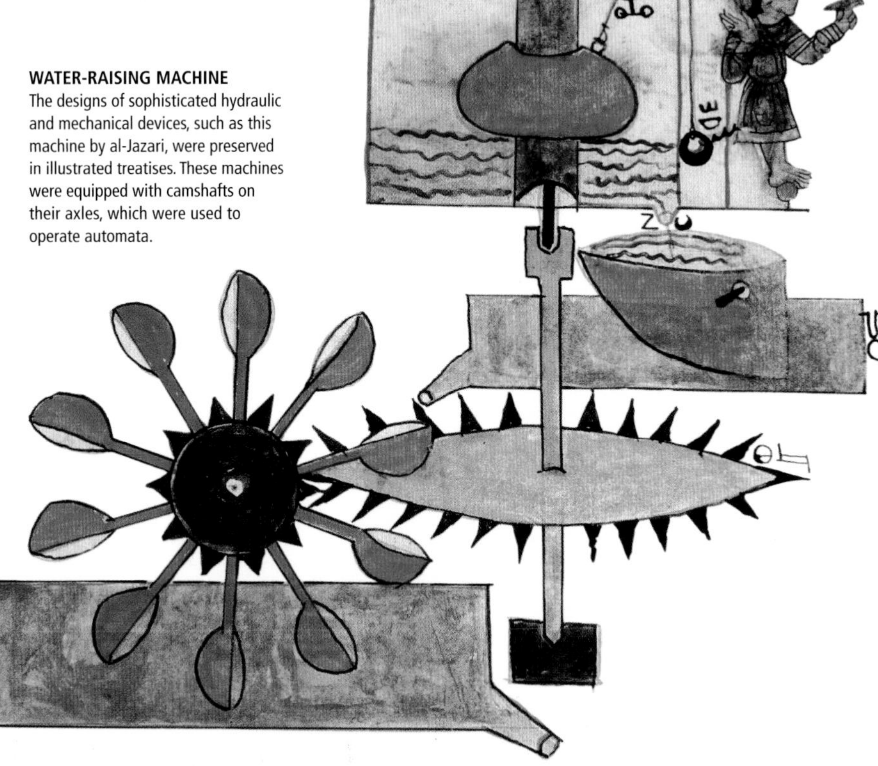

WATER-RAISING MACHINE
The designs of sophisticated hydraulic and mechanical devices, such as this machine by al-Jazari, were preserved in illustrated treatises. These machines were equipped with camshafts on their axles, which were used to operate automata.

AL-JAZARI

PROLIFIC ENGINEER OF THE ISLAMIC GOLDEN AGE

MESOPOTAMIA 1136–1206

A SCHOLARLY POLYMATH AND AN INNOVATIVE ENGINEER, Abu al-Iz Ibn Ismail ibn al-Razaz al-Jazari was the epitome of the *hakim* ("wise man") of Islamic learning and invention. Al-Jazari lived and worked at the height of the Islamic Golden Age, following in his father's footsteps as chief engineer in the Artuklu Palace, Turkey. He was an accomplished mathematician, astronomer, and artist, but is best known as an engineer and craftsman, whose "ingenious devices" included water-raising machines, clocks, and automata – all meticulously designed and often involving innovative mechanisms.

A LIFE'S WORK

- Al-Jazari's ***"The Book of Knowledge of Ingenious Mechanical Devices"*** describes in detail 50 of his engineering projects – how they were built and the mechanisms behind them
- He **designs and builds automata**, including a **"robot waitress"** that serves tea and drinks, and a **band of automaton musicians** using arguably the first programmable mechanisms
- As chief engineer to the Artuklu Palace under three successive rulers, he **designs machines for raising and pumping water for irrigation**
- He **introduces various inventions in his many different machines,** including a camshaft, an improved crankshaft modelled on the Banu Musas' design, segmental gears, and a combination lock

Al-Jazari was born in Al-Jazira, the northern part of the area between the Rivers Tigris and Euphrates. This area, known as Mesopotamia, now covers the borders of modern-day Iraq, Syria, and Turkey, and was once the centre of the Fertile Crescent – a crescent-shaped region containing the comparatively moist and fertile land of otherwise arid and semi-arid Western Asia – that is often described as the "cradle of civilization". By the mid-12th century, Mesopotamia was at the heart of the Turko-Persian Seljuq Empire, which stretched across the Middle East into Asia and North Africa, and was steeped in the Islamic traditions of craftsmanship and scholarship.

The rule of the Artukid Dynasty in eastern Anatolia, in the south of modern-day Turkey, was typical of the autonomy within the empire, and its courts were major sponsors of the arts and sciences. Al-Jazari's father was an engineer in the Artuklu Palace at Diyarbakir, and it is likely that al-Jazari learnt from him as a young man before himself entering service at the Artukid court in 1174. He studied the many treatises and texts on engineering and science translated by Islamic scholars from Greek and Asian sources, as well as the descriptions of Muslim achievements in the field, such as the Banu Musa brothers' *Book of Ingenious Devices* written some 300 years earlier. Al-Jazari, however, always described himself as a craftsman rather than a theoretician, and devoted most of his life to practical engineering projects rather than to academic study.

Having risen to the rank of chief engineer in the Artukid court under three successive rulers, al-Jazari was persuaded by Prince Nasir al-Din Mahmud to record his achievements for posterity. The result was an encyclopedia of mechanical engineering entitled *al-Jami bayn al-ilm wa'amal, al-nafi fi sina'at al-hiyal* ("*A Compendium on the Theory and Practice of the Mechanical Arts*", later known in translation as "*The Book of Knowledge of Ingenious Mechanical Devices*").

DOUBLE-PISTON PUMP
This is a reconstruction of al-Jazari's two-cylinder pump, the first known suction pump, which converts rotary motion of a wheel to reciprocating (back and forth) motion of double-acting pistons.

AN ILLUSTRATED MANUAL

This compilation was the most comprehensive treatise on the subject that had yet appeared. Its most striking aspect is its practicality – al-Jazari does not merely describe his machines, nor does he detail the scientific theories behind their mechanisms, but gives detailed drawings and instructions on how they should be constructed and operated – comprehensive enough for modern researchers to build reconstructions of his machines that work exactly as described. His manual reveals the wisdom of many years of study combined with the experience of putting his ideas into practice. The simple language and explanatory diagrams give it the feel of an instruction manual more than a learned treatise. Al-Jazari obviously considered himself an engineer, not a scientist.

The 50 devices he chose to illustrate in his work range from the useful to the simply decorative, but each shows an aspect of the elegant design and innovative technology for which he was known. He divided the book into six sections, which depict ten water and candle clocks; ten vessels and figures for drinking parties; ten jugs and basins for blood-letting and washing before prayer; ten fountains and designs for a "perpetual flute"; five water-raising machines; and five miscellaneous devices.

It seems likely that the bias towards gadgetry, rather than usefulness, was pandering to his sponsor's tastes, but al-Jazari chose a representative selection of his work, not for their applications, but for the design of the mechanisms and method of construction. The most stunning devices, and the most sumptuously illustrated, are the clocks, ranging from simple candle clocks to complex water- and weight-driven astronomical clocks.

A feature of many of these were the automata that marked the passing of time, such as those on the life-sized elephant clock (see below): the mahout beats a cymbal and the bird chirrups at regular intervals, while the scribe rotates, marking out the minutes with his pen. Even more striking is al-Jazari's castle clock, ingeniously programmed to display the position of the zodiac and the solar and lunar orbits, as well as marking the hours with figures appearing through doors – all accompanied by a band of five automata playing musical instruments.

AMAZING AUTOMATA

It was automata such as these that show al-Jazari at his imaginative best. They are also believed to have appealed most to his employer, judging by the number and variety included in his book. Some were relatively simple machines, using hydropower to dispense drinks or fill and drain a basin for hand-washing, with the mechanism disguised as a peacock or servant girl, but others included more sophisticated machinery. Perhaps the most significant of these was a floating music machine, with four automaton musicians whose actions and even facial expressions were controlled by rotating shafts studded with pegs, operating a system of levers – an early form of camshaft that was in effect programmable by altering the arrangement of the pegs.

Al-Jazari did, however, find room to include less glamorous devices, in particular a number of water-raising machines. He also included some of his most important contributions to engineering. His was the first description of a true suction pump. It was double-acting: it had a crankshaft-connecting rod to convert the rotary motion of a water-driven wheel to the desired reciprocating (back and forth) motion of two pumping pistons (see opposite).

The treatise was a mammoth task, and he may have considered it an imposition rather than a labour of love. It took him years to complete and he finished it just months before his death in 1206. It remains, however, the most comprehensive document detailing the techniques of engineering in the Islamic Golden Age, and a fitting monument to a remarkably inventive engineer.

> ... A COMPARABLE WEALTH OF INSTRUCTIONS FOR THE DESIGN, MANUFACTURE, AND ASSEMBLY OF MACHINES
>
> **DONALD R HILL**, ENGINEER AND HISTORIAN, ON AL-JAZARI'S MANUAL

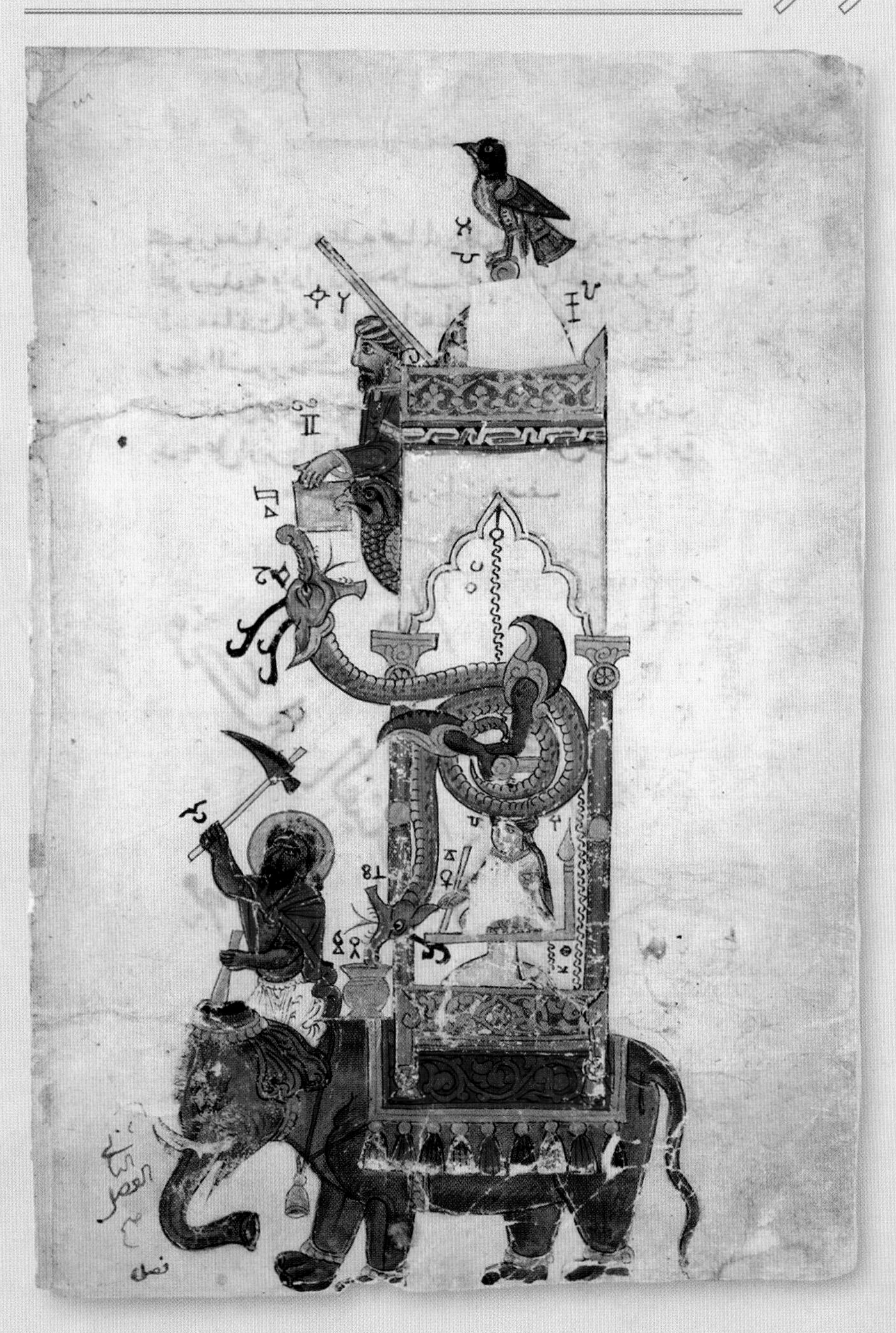

ELEPHANT CLOCK
In this contemporary illustration, the mechanism of the clock, driven by a bowl sinking into a tank of water inside the elephant, is hidden within the tower on its back. This is a beautiful example of al-Jazari's use of automata.

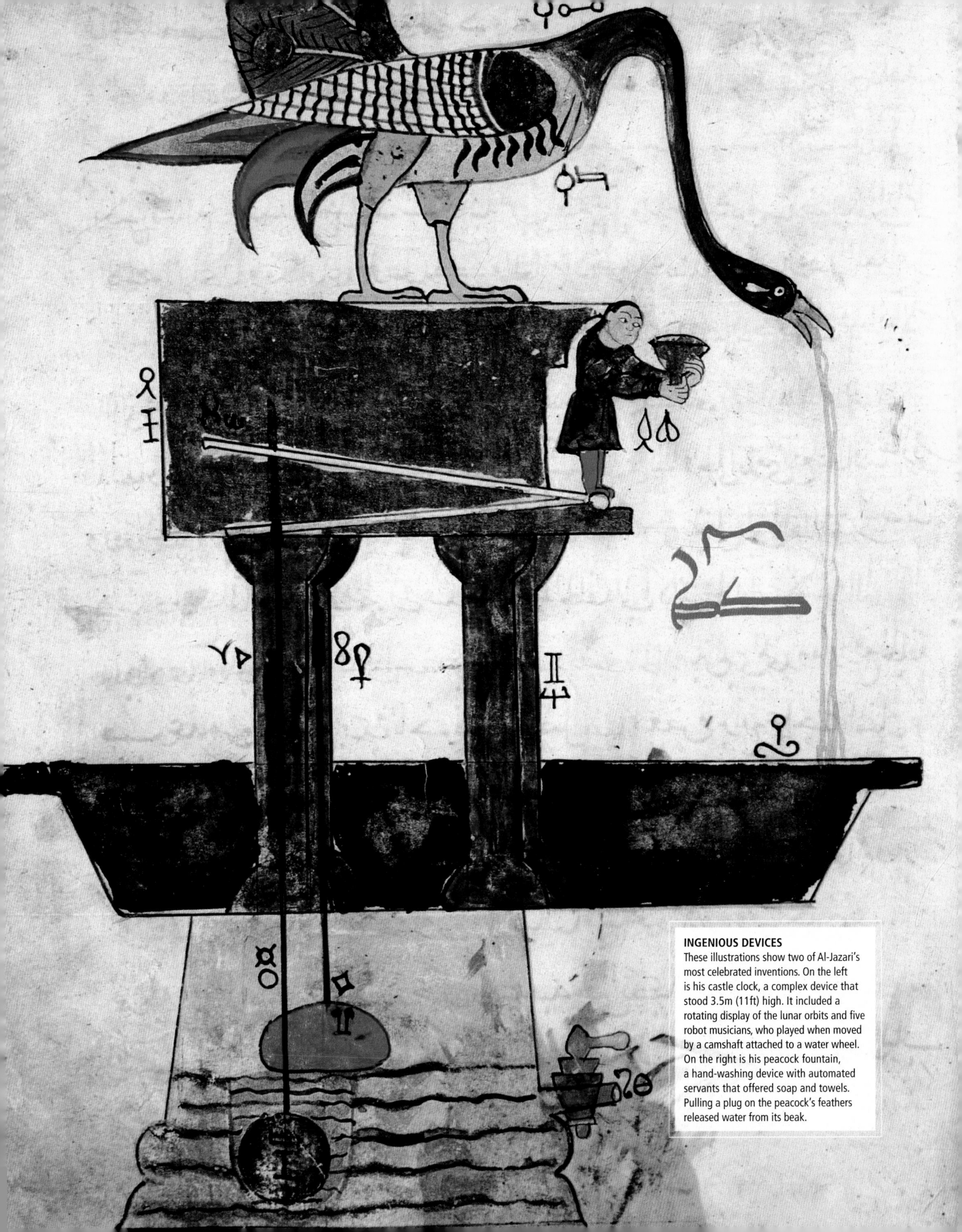

INGENIOUS DEVICES
These illustrations show two of Al-Jazari's most celebrated inventions. On the left is his castle clock, a complex device that stood 3.5m (11ft) high. It included a rotating display of the lunar orbits and five robot musicians, who played when moved by a camshaft attached to a water wheel. On the right is his peacock fountain, a hand-washing device with automated servants that offered soap and towels. Pulling a plug on the peacock's feathers released water from its beak.

TAQI AL-DIN

ISLAMIC ENGINEER, CLOCKMAKER, AND ASTRONOMER

TURKEY 1526–85

Regarded as the greatest scientist and engineer of the Ottoman Empire, and one of the last of the great Islamic polymaths, Taqi al-Din made significant contributions to areas as diverse as optics, horology, hydraulics, and steam power. He wrote more than 90 books on a wide variety of subjects, ranging from astronomy and natural philosophy to engineering, clocks, mathematics, and mechanics, which contain detailed descriptions and discussions of his work. Many of these treatises were translated into European languages and reached the West.

Little is known of Taqi al-Din's early life. He was born in Damascus, Syria, in the 1520s, shortly after it had become a part of the expanding Ottoman Empire. At some stage, he moved to Cairo (also recently under Ottoman rule), where he received most of his education. Like many Muslim scholars, he studied a number of subjects, including mathematics, astronomy, and medicine, alongside the core curriculum of theology and Islamic law.

After his studies, he taught at a madrasah in Cairo, and as a respected theologian at a local mosque was appointed a *qadi* ("Islamic judge") and religious timekeeper. He managed, however, to find time away from his duties to continue with his other interests, and published several books on scientific subjects. His position as a timekeeper at the mosque naturally involved a thorough knowledge of clocks, calendars, and astronomy, so it is not surprising that he should have become an acknowledged authority in those fields, but his expertise went far beyond the merely theoretical. As an astronomer, he became interested in optics, and even described an early form of telescope in one of his treatises. But it was as a clockmaker that he showed his genius for mechanical devices. Building on the work of al-Jazari (see pp.52–53), which he had come across as a student, he designed a weight-driven astronomical clock with an alarm activated by a movable peg in one of the dials, which could be programmed to ring at a specified time.

A LIFE'S WORK

- Before taking up science and engineering full-time, al-Din **works as an Islamic judge**, theologian, and religious teacher in Cairo
- Is **invited to work as a chief astronomer in the court of Sultan Selim II** of the Ottoman Empire, and moves to Istanbul in 1571
- Designs and constructs a **weight-driven astronomical clock with an alarm**
- Writes about **90 books on scientific and engineering subjects**, including *Al-Turuq al-samiyya fi al-alat al-ruhaniyya* ("*The Sublime Methods of Spiritual Machines*"), in which he describes several of his inventions
- Persuades Sultan Murad III to build a modern **observatory in Istanbul** in 1575, but after his incorrect astrological prediction of an Ottoman victory, the Sultan orders its demolition

BORROWING FROM THE PAST

Al-Din's interest in mechanical devices was not confined to the requirements of his job. Hydraulic engineering, an important field in a desert country such as Egypt, fascinated him, too. He had obviously researched the subject in great detail, gaining a thorough understanding of the principles discovered by Greek and earlier Islamic engineers. He designed several devices for irrigation, and some "ingenious devices", such as water-driven clocks and automata in the tradition of the Banu Musa brothers and al-Jazari, but the most important of his hydraulic machines was a water-raising machine that he invented

THE OBSERVATORY AT GALATA, ISTANBUL
In this Turkish painting of 1557, Taqi al-Din and his astronomers are shown using a variety of instruments to make their predictions, including a sextant and an astrolabe.

in 1559. This incorporated many design features from al-Jazari's water pump, but in a more sophisticated form – the camshaft, driven by a water wheel, operated pistons in a six-cylinder "monobloc" equipped with non-return valves to create a vacuum, which pumped water into the delivery pipes.

Another significant invention was also derived from his knowledge of historical machines. Using the principle discovered by Hero of Alexandria (see pp.32–33) in his mechanical curiosity known as the "aeolipile", a device that could be made to rotate by the power of steam escaping from nozzles, al-Din hit upon the idea of harnessing the power of steam to drive a wheel with vanes similar to a windmill's sails, attached to a rotating spit. This was, in essence, a steam turbine, and the first example of a steam-driven machine.

Through his books, al-Din's reputation as an engineer and scientist spread across the Ottoman Empire, which was now at its height and extended across North Africa, into Asia, and large parts of Central Europe. Selim II, who had succeeded Suleiman the Magnificent as Sultan in 1566, was anxious to make his mark and invited al-Din to become the chief astronomer at his court in Istanbul. Taking up the post in 1571, he found that the official patronage gave him the time and resources to devote himself full-time to science and engineering. He continued to publish books, mainly on subjects connected to his work as court astronomer, but also began planning a new observatory in Istanbul.

AN INCREDIBLE OBSERVATORY

Selim II died in 1574, and al-Din convinced his successor, Murad III, to fund the project. As well as the observatory itself, which when completed in 1577, was one of the finest in the world, al-Din put all his energies into equipping it with the latest technology. He set to work designing and making precision instruments, such as a sextant similar to the one at the observatory of Danish astronomer Tycho Brahe (1546–1601), and a rudimentary telescope to improve the accuracy of his astronomical observations; and, perhaps his greatest achievement, he also built a spring-driven clock, unrivalled in its accuracy at the time, with dials showing the hours, minutes, and seconds (although, as he explains in his description of the clock, "We divided each minute into five seconds").

Unfortunately, al-Din was less proficient as an astrologer (which was part of his duties as a court astronomer), and a botched prediction about Ottoman victories so enraged the Sultan that he destroyed the observatory in 1580. Fortunately for posterity, al-Din kept records of his discoveries and inventions, and published them along with his treatises on many different branches of science. In all, he wrote about 90 books, of which 24 have survived and been translated into many languages.

Al-Din died in 1585, at the height of the Ottoman Empire's military power, but at a time when Islamic scientific achievements were being eclipsed by European advances. Europe's ascendancy in engineering and sciences was, however, due largely to the tradition of discovery and invention in the Classical world, which had continued through medieval Islam and culminated in the work of al-Din, only to be rediscovered in Europe during the Renaissance.

STEAM TURBINE

A steam-driven rotating spit for roasting food was one of the most innovative of the machines described in al-Din's book *The Sublime Methods of Spiritual Machines*. Attached to one end of the spit was a wheel with vanes, which was made to turn by directing steam onto it through the nozzle of a copper vessel of water heated over the fire. Similar in concept to the "aeolipile" invented by Hero of Alexandria in the 1st century CE, al-Din's invention was no mere novelty and harnessed the power to drive a mechanical device. His rudimentary steam turbine predated Giovanni Branca's almost identical machine, often considered the first of its kind, by 78 years.

> TAQI AL-DIN ... COMPLETES THE ISLAMIC ERA'S MOST CRUCIAL PHASE IN MECHANICAL ENGINEERING
>
> **RAGHEB EL-SERGANY**, ISLAMIC HISTORIAN

THE COURT OF AKBAR THE GREAT
Here, Akbar is surrounded by the magnificence of his court at Fatehpur Sikri. He had a tolerant attitude towards other faiths and embraced culture and science. Fathullah Shirazi, a scholar and engineer, was his closest adviser.

FATHULLAH SHIRAZI

SCHOLAR WHO BROUGHT PERSIAN SCIENCE TO INDIA

PERSIA 1526–89

WITH THE GROWTH of the Persian Empire, Islamic culture and influence had spread across Asia into the Indian subcontinent. Mughal rulers, including Akbar the Great, were keen to build on India's long history of scholarship in mathematics and science, and scholars such as Fathullah Shirazi were welcomed into the courts to share their knowledge of technology from Persia.

A LIFE'S WORK

- **Shirazi originally studies medicine and science**, but is better known in Shiraz as a theologian, teacher, and administrator
- He is **invited to Bijapur** to **serve in the court of Sultan Ali Adil Shah**, but after the Sultan's assassination is unhappy with his successor, and takes up **Akbar the Great's invitation** to work in Fatehpur Sikri
- He is **commissioned by Akbar** to reform the calendar, and also **introduces changes to the syllabus** of madrasahs in the Mughal Empire to include **sciences and engineering**
- While in India, he invents numerous war machines, but also **designs carriages and agricultural machines**, and installs a **water supply system** to Fatehpur Sikri, using it to drive mills and other machinery
- Akbar admits him into the circle of **nine close advisers – the *Navaratnas*** – at Fatehpur Sikri

Born in the early 16th century, Fathullah Shirazi was educated in his home town of Shiraz, Persia, and proved to be an outstanding student. In the tradition of the Islamic scholars, he studied medicine and sciences alongside theology and philosophy, establishing himself as one of Persia's foremost intellectuals, and earning the honorific title of Hakim Amir Fathullah Shirazi. For some years, he followed an academic career, teaching and writing commentaries on the Koran, but also working as an administrator for the government finance department.

ALWAYS AN ASSET

Shirazi's reputation as an all-round scholar was not restricted to Persia, and news of his ability spread to the newly established Mughal Empire in India. This was an Islamic empire, anxious to maintain links with the Persian culture brought to India by the Mongols, so advisers were often invited to the Indian courts from Persia. Sultan Ali Adil Shah (1558–80), ruler of the Bijapur region in southern India, offered Shirazi a place as an administrator and theologian to complement his mainly literary advisers in Deccan, but discovered that he had an even more versatile and able counsellor at hand.

Shirazi became his right-hand man and, finding that in some respects Indian technology lagged behind its culture, he gave the Sultan the benefit of his scientific education, too, becoming in effect a court engineer. Unfortunately, Ali Adil Shah was murdered in 1581, and his successor, Ibrahim Adil Shah (r.1580–1627), was not so appreciative. Shirazi did not approve of the new Sultan and his frivolous, un-Islamic lifestyle, and made it plain that he wanted to leave the court. Akbar the Great (r.1556–1605), who had established himself as ruler of central and northern India, heard of Shirazi's discontent and invited him to his court at the palace in Fatehpur Sikri. It was here that Shirazi gave free rein to his talents, and where he spent the rest of his life. Akbar recognized that Shirazi had more to offer than most Persian scholars in India, and his scientific knowledge was more valuable than that of a literary or theological commentator. Initially brought in as an astronomer to reform the calendar, Akbar consulted him on all sorts of other matters, finding him more practical in his advice than many of his other, more academic ministers – Shirazi's medical training also made him an ideal personal physician.

THE SEAMLESS CELESTIAL GLOBE

While Shirazi brought many techniques of civil and mechanical engineering from Persia, there was already a long tradition of fine metal-working in India. As early as 1590, Kashmiri metallurgist Ali Kashmiri ibn Luqman had perfected a method of lost-wax casting that enabled him to produce a completely seamless globe showing the positions of celestial bodies – a remarkable feat, even if one used modern casting technology. His pioneering techniques were taken up across the Mughal Empire, making India a leader in the field of metal-casting. Another globe was produced by Muhammad Salih Tahtawi (1659–60), with Arabic and Sanskrit inscriptions, and the last was produced in Lahore by a Hindu metallurgist, Lala Balhumal Lahuri, in 1842. Twenty-one such globes were produced, and these remain the only examples of seamless metal globes.

AN INVENTIVE MIND

He designed and built military machines, including a 17-barrelled cannon and a device for cleaning several gun barrels simultaneously, and improved the efficiency of agricultural production by introducing grinding mills and other machinery. It is also probable that Shirazi designed and oversaw the installation of the water supply system in the palace at Fatehpur Sikri. As an ex-teacher, he also advised Akbar to broaden the curriculum of the madrasahs to include sciences and engineering, as well as theology and literature. Akbar greatly valued his expertise and appointed him to his council of nine advisers, the *Navaratnas*, which included Hindu and Islamic scholars, generals, and administrators.

Shirazi travelled widely in Akbar's realm, often accompanying the emperor on tours of inspection. On a trip to Kashmir, he was taken ill, and he died shortly after. Akbar's grief at his death went beyond that of an emperor for his minister, and showed the close relationship the two had built; he is reported as saying "Amir Fathullah was my advocate, counsellor, physician, and my astronomer. No one can gauge our sorrow. Had he been arrested by foreigners and had they asked all my treasure for his release, I would have considered it as a profitable transaction."

RENAISSANCE POLYMATHS

AT THE END OF THE 14TH CENTURY, THE POWER OF THE CHURCH IN EUROPE BEGAN TO WANE. HUMANIST IDEAS CHALLENGED MEDIEVAL DOGMA, SPURRED ON BY RENEWED INTEREST IN CLASSICAL LEARNING. THIS BEGAN THE RENAISSANCE, IN WHICH BOTH ARTS AND SCIENCES FLOURISHED.

COPERNICAN UNIVERSE
Using mathematics and astronomical observations, Nicolaus Copernicus's heliocentric cosmology showed that the Earth was not at the centre of the universe.

The first stirrings of a change in outlook became apparent in the Middle Ages, when ancient Greek texts preserved by Islamic scholars became available in Europe for the first time. Although only grudgingly acknowledged by the church, which had a monopoly on the preservation and dissemination of learning, demand for these philosophical and scientific writings increased. This reached a turning point in about 1400, when the rediscovery of Classical ideals, known as the Renaissance ("rebirth"), brought about a fundamental shift in thinking, placing humanity rather than God at the centre of things, with an emphasis on philosophical and scientific enquiry rather than religious dogma. These ideas were enthusiastically adopted by the independent city-states of Italy, and Florence in particular.

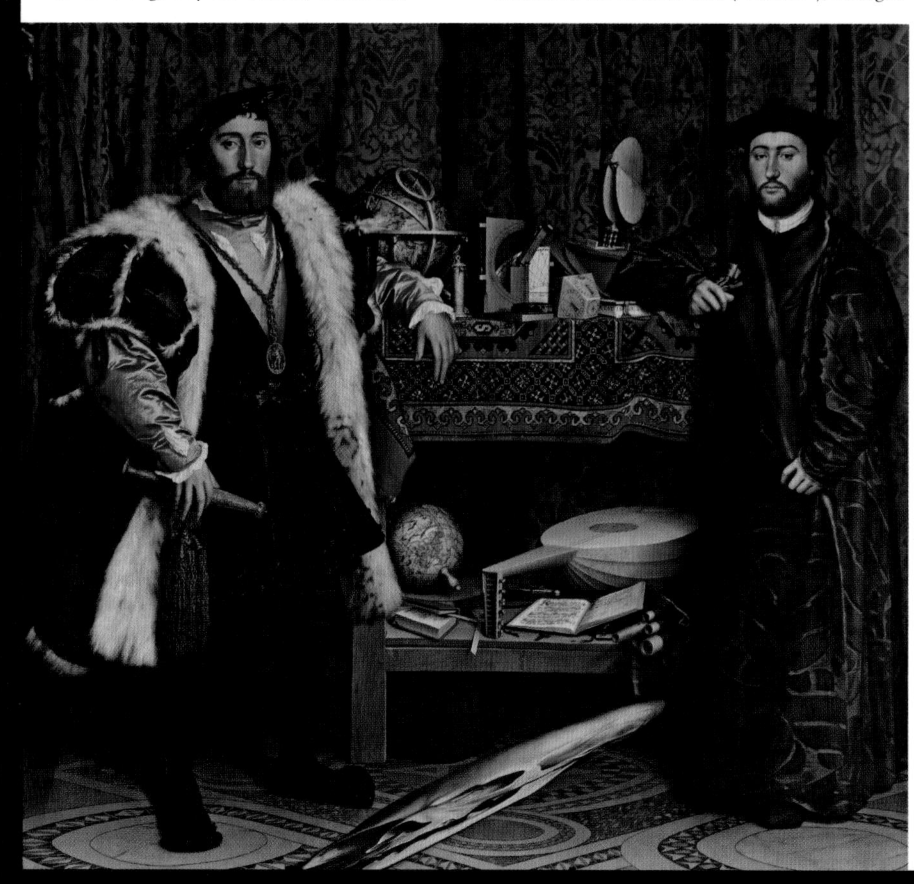

THE "RENAISSANCE MAN"

Encouraged by patrons such as the Medici family, a new breed of intellectual, the "Renaissance Man", skilled in many different areas, began to emerge – undoubtedly also influenced by the *hakim*, or polymath Islamic scholar.

As might be expected at a time when states were vying for territory and trading partners, engineering was centred on building defences, and designing and constructing military equipment. Some engineers, however, used their expertise and the new-found knowledge of Islamic engineering to produce a wealth of machine designs. Following in the footsteps of inventor Guido da Vigevano (*c.*1280–*c.*1349), Taccola (see pp.64–65) produced two books of engineering designs, which included a number of hydraulic devices, mills, and sluices. His work established a tradition of Italian artist-engineers, which included Roberto Valturio (1405–1475) and Agostino Ramelli (*c.*1531–*c.*1600).

THE AMBASSADORS
The two figures in Hans Holbein the Younger's picture, and the objects on the shelves between them, represent the combination of the sacred and the secular, the scientific and the artistic.

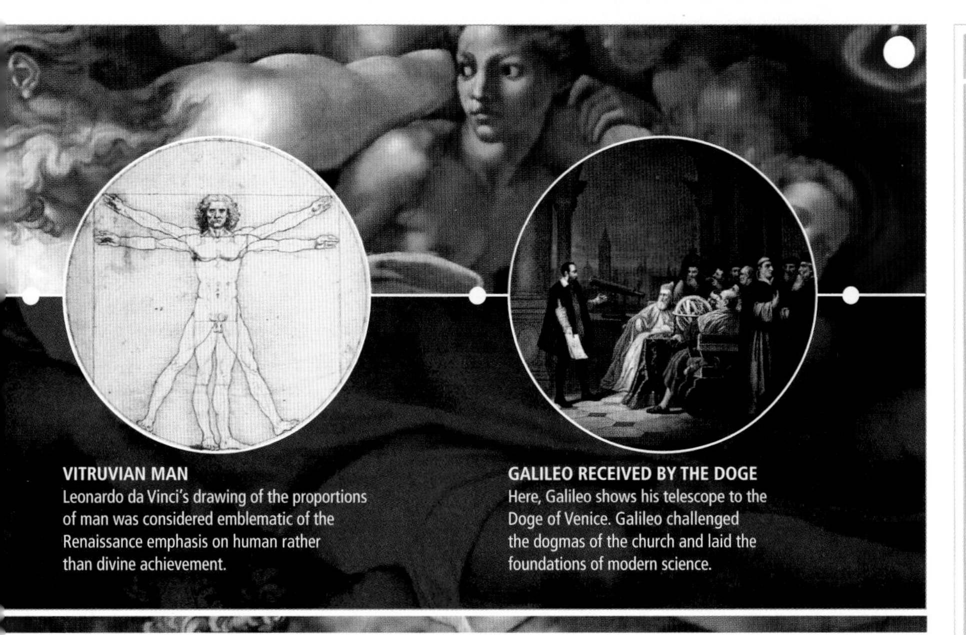

VITRUVIAN MAN
Leonardo da Vinci's drawing of the proportions of man was considered emblematic of the Renaissance emphasis on human rather than divine achievement.

GALILEO RECEIVED BY THE DOGE
Here, Galileo shows his telescope to the Doge of Venice. Galileo challenged the dogmas of the church and laid the foundations of modern science.

SETTING THE SCENE

- In the cultural climate of the Renaissance, there is **no inconsistency in being both a scientist and a poet or an artist**.
- The **development of the three- or four-masted carrack** (sailing ship) enables explorers to make longer ocean voyages and discover new territories and trading routes.
- **Copernicus's demonstration of a heliocentric cosmology** becomes symbolic of the triumph of scientific discovery over religious dogma.
- **Brunelleschi's discovery of the laws of linear perspective** inspires a new, more expressive form of drawing and painting, but also makes three-dimensional representation possible in technical and engineering diagrams.
- **Gutenberg's printing press is the single most important invention** of the Renaissance period, helping to publicize the work of scientists and engineers, and promote a culture of learning and reasoning.
- The emphasis on humanism and the influence of Islamic surgeons stimulates **an interest in human anatomy**, leading to advances in medicine and the development of the microscope.

At the same time, the work of great Florentine polymaths, such as Filippo Brunelleschi (see pp.66–67), resulted in a period of unprecedented technological inventiveness. The most noticeable advances were in the fields of architecture and building technology, pioneered by Brunelleschi in the addition of a massive dome to Florence Cathedral.

The Renaissance was a period of geographical as well as scientific exploration and discovery, epitomized by Christopher Columbus's voyage across the Atlantic, which was driven by trade and encouraged by technological advances.

The greatest of all the Renaissance polymaths was perhaps Leonardo da Vinci (see pp.68–69). As much in demand for his engineering expertise as for his skill as a painter, da Vinci showed his ingenuity in numerous military and civil engineering projects, but his curiosity about all things mechanical is best seen in his notebooks, which are filled with designs for everything from military engines to flying machines, as well as anatomical drawings that explore the movements of living things.

The subtleties of motion also occupied the astronomer, physicist, and mathematician Galileo Galilei (1564–1642), whose work on pendulums, falling bodies, inertia, and friction proved vital to the development of many new technologies in the following century.

From Italy, Renaissance ideas had spread rapidly north, sparking massive cultural and political changes. At the forefront were scientists whose discoveries called into question the infallibility of the church: Nicolaus Copernicus (1473–1543), for example, showed that the Earth was not at the centre of the universe, an idea that was considered heretical.

THE "SCIENTIFIC REVOLUTION"

This erosion of the authority of the Catholic Church was accelerated by the invention of a printing press with movable type by Johannes Gutenberg (*c.*1400–68), which allowed information to be disseminated widely and quickly. This new technology was outside the control of the church, and enabled the circulation of documents such as Martin Luther's 95 Theses, which played a large part in the protestant Reformation. It also made possible the publication of scientific treatises, and sparked the beginnings of the so-called "Scientific Revolution". The cultural climate created by the Renaissance in the 15th and 16th centuries was ideal for scientific discovery, and for the first time in Europe since Classical times, it was science and trade rather than religion that drove the pace of technological progress. The telescope, for example, developed alongside advances in astronomy and the increase in seagoing trading vessels. During the Renaissance, scientific and navigational instruments became increasingly sophisticated. This in turn allowed further discoveries in astronomical observation, and enabled voyages such as Christopher Columbus's crossing of the Atlantic Ocean in 1492.

The Renaissance, which had started as a rediscovery of Classical learning, quickly turned into a period of discovery and invention, but perhaps more importantly, it re-established science and technology as the basis for progress. Respect for the great polymaths of the Renaissance laid the foundations for a modern "Age of Reason", in which engineers were to play a vital role, finding practical applications for scientific discoveries and developing new technologies.

GALILEO'S PENDULUM
This replica clock is based on a drawing made by Galileo. It is the first known attempt to use a pendulum to regulate the movement of a clock.

GUO SHOUJING

HYDRAULIC ENGINEER AND OBSERVATORY DESIGNER

CHINA 1231–1316

Under the patronage of Kublai Khan, Chinese astronomer and engineer Guo Shoujing helped establish Dadu (now Beijing) as Khan's capital. He did so by designing the Kunming Lake, a system of waterways to provide water for the city, and a link to the Grand Canal. He also built an observatory equipped with his own innovative astronomical instruments.

By the mid-13th century, the Mongol Empire had spread southwards across China, ousting the rulers of the Song Dynasty. Kublai Khan, the grandson of Genghis Khan, founded the Yuan Dynasty (1271–1368), introducing a period of stability and prosperity and encouraging an atmosphere of science and scholarship, influenced by his contact with the Islamic world. It was in this period that Shoujing rose to prominence.

EARLY LIFE

Guo was born in Xingtai in the Hebei province of northern China shortly before Kublai Khan's brother, Ogodei, invaded the area, so Guo grew up under Mongol rule. Although his family was relatively poor, his grandfather Guo Yong was well known as a scientist and engineer, and young Guo received his early education from him, learning mathematics, astronomy, and hydraulics – for which he showed a special talent. He was particularly interested in the construction of water clocks, and designed a "lotus clepsydra" – a water clock with a lotus-shaped bowl into which water drips at a controlled rate – when he was about 14.

His education was unorthodox and stood him in good stead, and, by the time he was 20, he found work as a government official in charge of civil engineering works. Over the next 10 years, he oversaw numerous projects, including the repair of bridges, and gained a reputation as a skilled hydraulic engineer.

GRAND DESIGNS

By this time, Kublai Khan had succeeded his brother Möngke and declared himself Khan and ruler of China, and was setting up his capital, Dadu. Conscious of the need for a reliable water supply and transport links in the new city, he summoned Guo to survey the area. Guo's scheme was comprehensive, and on an unprecedented scale. He had found a source of water in the Shenshan Mountain, and designed a channel to carry the water the 30km (19 miles) to Dadu. The water was stored there in a reservoir, the Kunming Lake, that supplied the city through a network of irrigation channels, but also fed canals that linked the city via the Grand Canal to the Yangtze River. The project, which was completed in the 1290s, provided the infrastructure needed to establish Dadu as a major trading city, and Kublai Khan commissioned Guo

A LIFE'S WORK

- Guo **learns mathematics, astronomy, and hydraulics** from his grandfather, and works on improving the design of water clocks
- Aged 20, he works as a hydraulic engineer, **improving the municipal water supply** and repairing and maintaining bridges and waterways
- His colleague Zhang Wenqian **recommends him to the new emperor, Kublai Khan**, for work on the irrigation system in Dadu, the new capital
- Is commissioned by Kublai Khan to **reform the calendar and proposes building a number of new observatories** with newly designed instruments to gather astronomical data
- In 1283, he is **appointed head of the observatory he had designed**, and in 1292, is made head of the Water Works Bureau

to design similar schemes in other parts of his empire, which was continuing to expand across all of China. Guo travelled widely, surveying the land and advising the emperor on ways to improve irrigation systems, and soon became one of his most trusted advisers.

Kublai Khan was keen to establish himself as the ruler of a completely new regime, bringing in political and cultural reforms. One of the ways he envisaged doing this was by the introduction of a new calendar. Although Guo had made his name as a hydraulic engineer, Kublai Khan knew that he was competent in mathematics and astronomy, and appointed him and a colleague, Wang Xun, to a new bureau to investigate the possibilities of calendar reform. Although the need for calendar reform had been felt for many years, Kublai Khan saw political benefits in bringing in a new calendar to demonstrate that the new regime was replacing the old. Realizing that the accuracy of a calendar is dependent on reliable astronomical data, Guo set to work by establishing observatories in various locations to collect the information necessary to make the calculations. In all, he and Wang Xun built 27 observatories, including the great Gaocheng Observatory in Dadu, which Guo designed. Using the mechanical skills he had learnt as a teenager making water clocks, Guo also made astronomical instruments for these observatories. Building on the inventions of previous Chinese engineers, such as Zhang Heng (see pp.34–35), and incorporating some ideas that had come from recent contact with Islamic countries, Guo furnished the new observatories with equipment whose precision was unrivalled. Although credited with inventing the square table, which is used to measure the angles of celestial bodies, many of the devices he built were elegant improvements of existing instruments. For example, his redesign of the gnomon, the post that casts a shadow on a sundial, made it possible to tell the time and measure the angle of the Sun more accurately.

> THE FUNDAMENTAL METHOD IS TO CARRY OUT OBSERVATION AND TESTS IN THE AREA OF ASTRONOMICAL CHANGES
>
> **GUO SHOUJING**

KEEPING DATES

But perhaps his most impressive designs were of armillary spheres, models showing the movements of the celestial bodies. Guo constructed both an abridged armilla and a complex, water-powered armillary sphere known as the "Ling Long Yi". Guo and Wang completed the new Shoushi Calendar in 1280, using historical data and observations taken with Guo's improved instruments. A measure of the precision of his equipment is that they calculated a year to be 365.2425 days – with only 26 seconds difference from the modern measurement. In the process of working on the new calendar, Guo also proved his worth as a mathematician – once again combining Chinese knowledge of the subject with ideas brought by the Mongols from Islamic scholars. He came up with new formulae in spherical trigonometry to convert astronomical observations into usable data.

Kublai Khan showed his gratitude to Guo by honouring him with the titles of director of the Gaocheng Observatory and head of the national Water Works Bureau. When the emperor died in 1294, Guo was in his 60s, but he continued to work as an advisor to his successors until his death, aged 85, in 1316.

GATHERING STAR DATA
Commissioned by Kublai Khan, the observatory designed by Guo Shoujing was built in 1276 in Gaocheng, near Dengfeng in Henan Province, on a site used for astronomical observation for many centuries. It is now a World Heritage Site.

THE ARMILLARY SPHERE

A model of the celestial sphere had been invented by the ancient Greeks, but the armillary sphere (or armilla) reached a peak of sophistication in China. Zhang Heng built a water-driven armillary sphere, and it was this that formed the basis for Guo's improved precision designs, and a more complex, water-driven version, which enabled the accurate calculation of a reformed calendar.

USING THE ARMILLARY SPHERE HELPED GUO SHOUJING TO CALCULATE A MORE ACCURATE CALENDAR.

TACCOLA

THE ENGINEER WITH AN ARTIST'S MIND

REPUBLIC OF SIENA 1381–C.1453

BEST KNOWN BY HIS NICKNAME "IL TACCOLA" ("The Crow") or, simply, Taccola, Mariano di Jacopo was a Sienese artist and engineer whose treatises *De Ingeneis* ("On Engines") and *De Machinis* ("On Machines") mark the transition from medieval to Renaissance times. These beautifully illustrated treatises are packed with detailed descriptions and annotated drawings of machinery of all kinds, from weapons and mills to hydraulic engines and waterways. They reflect the beginning of an Italian Renaissance fascination with machines and technology.

A LIFE'S WORK

- Taccola is responsible for **large construction projects**, including a **harbour in Genoa**, a **bridge across the Tiber River in Rome**, and the machines described in his treatises
- Gains **the patronage** of the future **Holy Roman Emperor Sigismund** after his visit to Siena in 1432, and becomes one of his *nobiles familiares* ("noble friends")
- Inspired by Sigismund, he travels to Hungary to **fight against the Turks** in the 1430s
- From 1435 until his death, he earns his living as a sculptor while working on the massive ***De Machinis***, which he eventually finishes in 1449
- Includes in his treatises original designs for a **compound crank with connecting rod**, which converts rotary motion to reciprocal motion, and a **chain transmission system**

Taccola was born in Italy in 1381, precisely the place and time of Europe's transition from medieval religious dogmatism to the Renaissance ideas of humanism and the beginnings of the so-called scientific revolution. Little is known about his early life, apart from the fact that he was brought up on the family vineyard in Siena, Tuscany, and that he inherited his father's nickname, which is believed to refer to their skills as woodcarvers. Taccola did not follow in his father's footsteps and become a winegrower, but instead sought work in the city.

His talent for carving allowed him to work as a sculptor, and he was involved in major projects, such as the decoration of Siena Cathedral. He also became increasingly involved in civic life in the city, working as a public administrator and studying to become a notary. In 1424, Taccola was appointed chamberlain of the Casa della Sapienza, a residential centre for scholars in Siena. It was while working in this academic atmosphere that Taccola found his true calling.

As a young sculptor working in the construction industry, Taccola began to take an interest in some of the engineering technology around him, and made sketches of some of the machinery. Soon, he was mixing with scholars from the Italian universities, many of them scientists and progressive humanist thinkers. He was introduced to classical scientific texts and the ideas that had begun making their way into Europe from the Islamic world. Taccola developed an understanding of technology and engineering and, while still running the Casa della Sapienza, embarked on a project that culminated in his four-volume treatise, *De Ingeneis*.

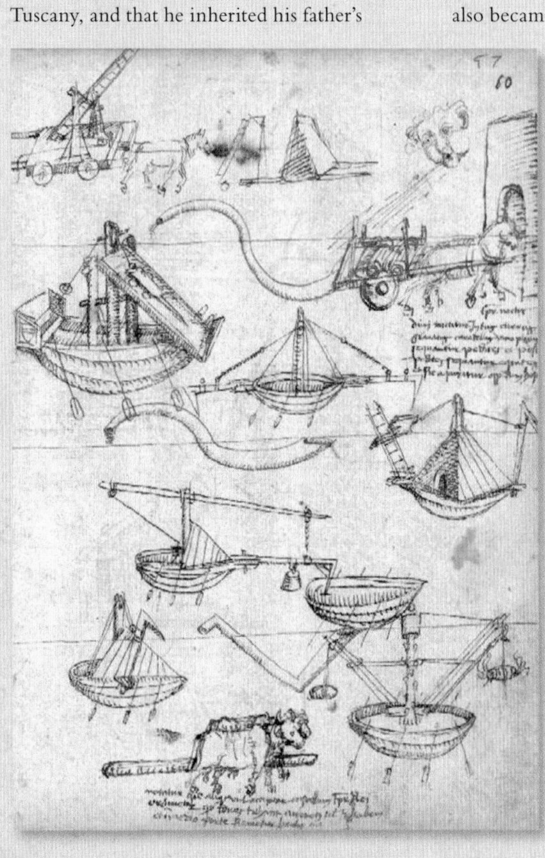

DRAWINGS FROM *DE INGENEIS*
Taccola's treatise had pen-and-ink drawings with explanatory captions, showing the construction and working of machines.

GROWING REPUTATION

Taccola left the Casa della Sapienza in 1431 and devoted the next two years of his life to completing his treatise. The book provided a comprehensive description of all aspects of engineering of the time. He included annotated pen-and-ink drawings of machines of all types in a style similar to the "books of ingenious devices" by Islamic engineers, such as al-Jazari (see pp.52–53) and the Banu Musa brothers (9th century). Not content with merely recording, Taccola also described machines he had invented, and built several of his ideas to test their practicability. News of his ambitious project spread through the academic community and he built a reputation as a talented engineer, acquiring the nickname of the "Sienese Archimedes".

His reputation won him friends in high places, including Prince Sigismund, who was soon to become the Holy Roman Emperor and a very useful patron. It also brought him into contact with like-minded engineers and craftsmen – he struck up a friendship with Filippo Brunelleschi (see pp.66–67), the engineer behind the construction of the dome of Florence Cathedral. Taccola's notebooks refer to discussions between them on engineering and its connection with architecture and construction. He later included in his treatises designs of the lifting equipment used in building Brunelleschi's dome, suggesting that there was a degree of collaboration between them. Brunelleschi's development of perspective drawing, however, does not seem to be one of the subjects they discussed. Taccola's drawings, elegant and detailed though they are, retain a two-dimensional quality and even contain certain errors resulting from his lack of feel for perspective.

LASTING INFLUENCE

After a short period in Hungary fighting the Turks as a part of the emperor's forces, Taccola returned to Siena in around 1435 and embarked on his second treatise, *De Machinis*. He spent the next 14 years working on this, partly sponsored by Sigismund, and continuing to work as a sculptor. It was similar in style and scope to his earlier work, and included many of the same ideas, reworked and improved. It also included descriptions of devices he had discovered since his first treatise, several of which were of his own inventions.

Today, it is impossible to confirm how many of the devices he described were his original designs, as his treatises contain the earliest written references to many mechanisms. Among the inventions credited to him are sluice gates, box-caisson bridges, and a suction pump, but these may well have been drawings of existing devices. However, it is certain that it was he who designed the first trebuchet – a military catapult (see box, right), a chain transmission system, and also a compound crank to convert rotary (circular) to reciprocal (back and forth or up and down) motion.

Whether or not Taccola was a great innovator, his importance in the history of engineering lies in the encyclopaedic scope of his two treatises, presented with the precision of a draughtsman and the understanding of a scientist. His work was a milestone in the establishing of an Italian Renaissance school of engineering, and influenced generations of engineers and artists, including Leonardo da Vinci (see pp.68–69). Today, his treatises provide a comprehensive record of engineering practices at the start of the Renaissance.

WAR MACHINES

Although the Renaissance began during a period of peace in Italy, the 15th and 16th centuries were far from conflict-free. The Roman Empire was continually faced with invasions, and individual states were often at war. Engineers were therefore asked not only to design cathedrals and labour-saving devices, but also to create weapons. Taccola's trebuchet is one example. He was followed by others, including Roberto Valturio (1405–75), who specialized in military engineering.

THE TREBUCHET WAS USED TO HURL STONES AND OTHER HEAVY MISSILES AT WALLS.

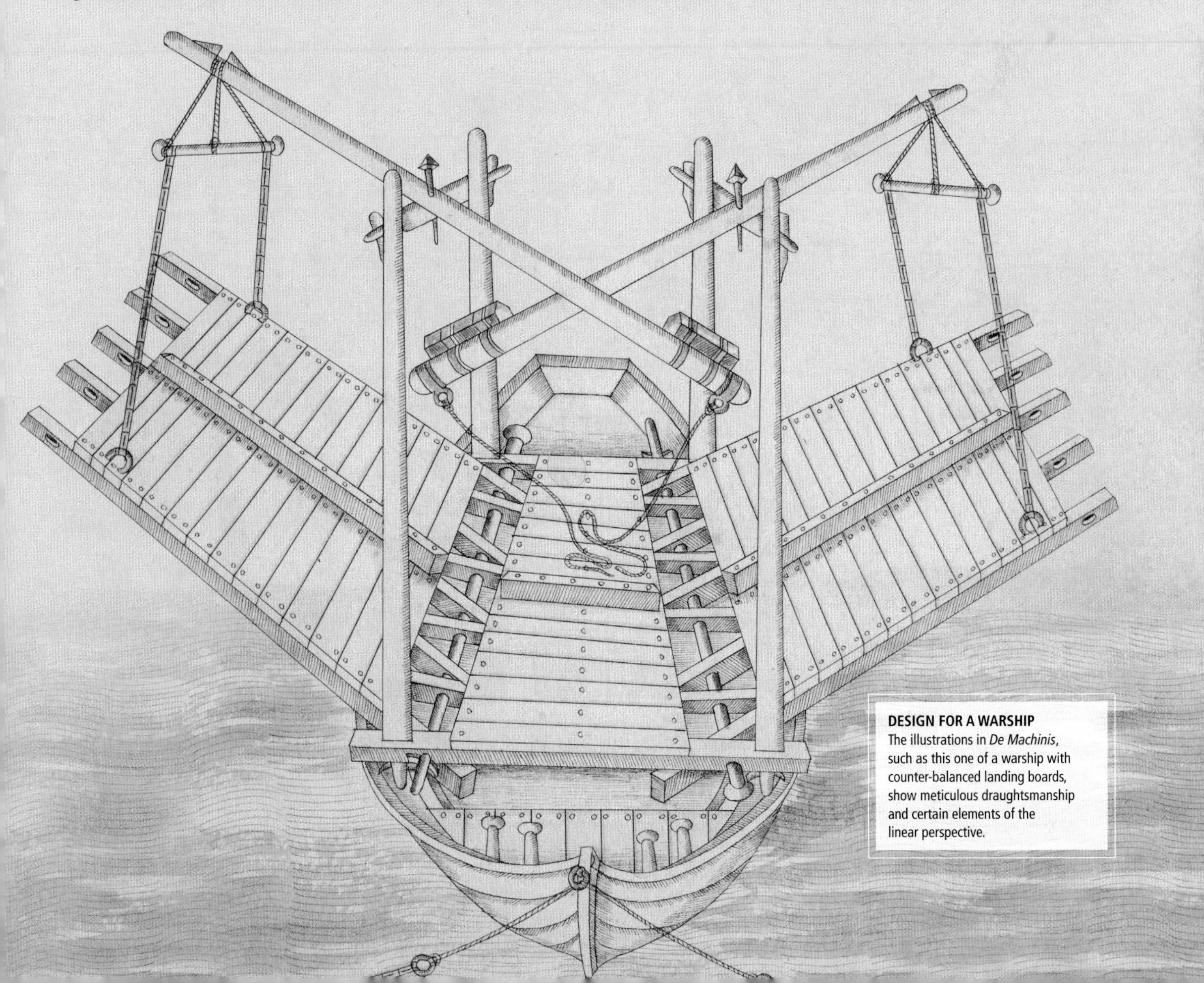

DESIGN FOR A WARSHIP
The illustrations in *De Machinis*, such as this one of a warship with counter-balanced landing boards, show meticulous draughtsmanship and certain elements of the linear perspective.

FILIPPO BRUNELLESCHI

FOREMOST ENGINEER OF THE ITALIAN RENAISSANCE

REPUBLIC OF FLORENCE 1377–1446

A GOLDSMITH, SCULPTOR, engineer, and architect, Filippo Brunelleschi introduced innovations in construction and mechanical design that set the standard for civil engineering. An ambitious and secretive man, he competed for commissions of major building projects in Florence, designing machinery for each project, and was one of the first to obtain patents for his inventions.

A LIFE'S WORK

- Brunelleschi originally **trains as a goldsmith and sculptor**, but **turns to architecture** after losing a competition to Lorenzo Ghiberti
- His teacher, Paolo dal Pozzo Toscanelli instills in him an **interest in technology and mechanics**
- Introduces the **geometrical method of perspective drawing** in about 1413
- In 1420, he beats his **old rival Ghiberti** to the commission for construction of the dome, the **largest octagonal dome in history** and, to this day, the **largest brick and mortar dome** in the world
- Is granted the **first ever industrial patent** in 1421 for the manufacture of a barge with lifting gear to transport marble from Carrara to Florence
- Construction of the dome is almost completed at the time of his death; **the project eventually is finished to his specifications**, including installation of the massive lantern 15 years later

Filippo Brunelleschi was born into a well-respected family in Florence. His mother, Giuliana, came from the aristocratic Spini family, and his father was a notary. Young Filippo, known as "Pippo", was probably expected to follow in his father's footsteps, but he showed little interest in law and civic affairs. His talents were for more practical and creative subjects and, as a boy, he was fascinated by machinery, especially clocks.

He never completed the formal education that was normal for a young man in his situation, but instead persuaded his father to let him take up an apprenticeship with the Arte della Seta, the silk merchants' guild, where he trained as a goldsmith. This was just the kind of education he was interested in, as he not only learned the craft of working metal, but also studied mathematics and mechanics, and discovered a talent for sculpture. He qualified as a master goldsmith in 1398 and soon found work on many construction projects in the city. The turning point in his career came in 1401, when he entered a competition, open to sculptors and goldsmiths, to design a pair of doors for the church of St Giovanni. Although he came second to the sculptor Lorenzo Ghiberti (1378–1455), his design earned him the recognition he wanted and fuelled his ambition to become an architect.

HOISTING MACHINE
The "block" from a block-and-tackle device, which was designed to hoist building materials into place. Machines of this kind were key to the construction of the dome of the Florence Cathedral.

With his friend Donato di Niccolò di Betto Bardi (*c.*1386–1466), an early Renaissance artist and sculptor popularly known as Donatello, Brunelleschi travelled to Rome to study architecture. Together, they spent two years examining the many styles of building there, analyzing Classical Roman architecture in particular. Brunelleschi returned to Florence in about 1407, but continued to spend much time in Rome over the next ten years, building up the expertise he needed to take on a large construction project.

A PRODIGIOUS TASK

The project he was most interested in was the jewel in the crown of Florence – the completion of the cathedral of Santa Maria del Fiore. He had been involved in discussions about the possibility of erecting a dome on the cathedral as early as 1407, but such a large structure seemed impracticable with the technology of the time. By 1418, the cathedral was still unfinished and so an open competition for the design of the dome was announced. Brunelleschi once again found himself pitted against Ghiberti, but this time won the commission and set to work on the most ambitious project of his career.

What he had in mind was audacious in both its size and method of construction. He proposed a massive brick-built dome, bigger than anything built before that would be constructed without the traditional wooden scaffolding, and would not be supported by buttresses. The dome was in the planning stage for two years, during which time he built a scale model to prove its feasibility and as a template for the builders – but left it incomplete so that nobody could steal his ideas.

Work began on the cathedral in 1420, and would continue for the rest of his life. To ensure the correct construction of the dome, he designed specific lifting apparatus to be used on the site to carry bricks and structural beams into position, and quickly obtained patents to protect his designs. On an inner dome with an octagonal framework, more than 4 million bricks were laid in a herringbone pattern – a broken zigzag pattern – which held them in place as the outer dome took shape. Brunelleschi had meticulously calculated the various stresses and had spread the load so that there would be no danger of the structure collapsing – even with such breathtaking proportions. The building work continued over the next 25 years, with Brunelleschi keeping control over every aspect of its construction. Ghiberti assisted him in the early stages of the project, but soon left his rival to work on his own.

OTHER INNOVATIONS

With responsibility for such an ambitious undertaking, Brunelleschi could well have devoted himself full-time to the cathedral, but instead chose to work on several other buildings. He had already been commissioned to design and build the Ospedale degli Innocenti ("Foundling Hospital") and was overseeing its construction even as he continued to take on architectural commissions, including defensive fortifications for the city. In addition to the workload that this must have entailed, he also found time to design and build hydraulic and clockwork machinery for the theatrical performances common in the churches of the time. Brunelleschi is also reported to have made some clocks. A less impressive but perhaps more influential legacy than the cathedral dome was his formulation of linear perspective – an idea that proved to be a breakthrough in the stylized art of the medieval period, unleashing the realism of Renaissance.

Unfortunately, Brunelleschi did not live to see the completion of his masterpiece. When he died in 1446, the brickwork of the dome was almost complete, but the finishing touch, a huge suspended lantern, had yet to be lifted into position. As with many of his designs, this was not mere ornament, but an integral part of the structure, and luckily he had left sufficient specifications to enable a team of engineers, led by his friend, the famous architect and sculptor Michelozzo di Bartolomeo Michelozzi (1396–1472), eventually to install the lantern in 1461. Eight years later, the sculptor Andrea del Verrocchio added a gilded copper ball to the very top of the small spire on the dome. Verrocchio was assisted by his apprentice – one Leonardo da Vinci (see pp.68–69) – who took a keen interest in Brunelleschi's specially designed machinery, and went on to become a great artist and engineer himself.

PERSPECTIVE DRAWING

Brunelleschi used the principles of geometry and the notion of a vanishing point to describe how an artist could achieve a three-dimensional effect in two-dimensional drawing. To demonstrate this, he produced a painting using his perspective technique and then drilled a small hole in the middle of it. By peering through the hole in the painting from the back, and viewing the painting's reflection in a mirror, the viewer received the full benefit of the three-dimensional illusion.

BRUNELLESCHI'S MASTERPIECE
The dome of the Santa Maria del Fiore has dominated Florence's skyline since 1461, and remains the largest brick-built dome in the world. A cultural as well as physical landmark, it inspired the style and construction of subsequent Italian Renaissance architecture.

LEONARDO DA VINCI

THE ARCHETYPAL "RENAISSANCE MAN"

REPUBLIC OF FLORENCE 1452–1519

Famous as the artist responsible for some of the best-known paintings and drawings of the period, Leonardo da Vinci was revered not only for his artistic talent, but also for his skill in other fields. His notebooks were filled with copious writings, sketches for his paintings and sculptures, anatomical drawings of humans and animals, and detailed designs of machines and mechanical devices.

A LIFE'S WORK

- Da Vinci's **"Vitruvian Man"** of *c.*1487 shows the proportions of the human body and becomes symbolic of Renaissance ideals, **putting man, rather than God, at the centre of the universe** and **linking humanity with both nature and mathematics**
- Although already established as an artist, he **emphasizes his engineering expertise** when applying to the Duke of Milan for a post in his court, and later works as a **military engineer** for the authorities in Venice and for Cesare Borgia in Rome
- Among the many **unrealized projects** in his notebooks are **designs for flying machines, parachutes, tanks, bridges,** and **hydraulic machinery**
- At the height of his career as an artist in the decades around 1500, he paints the masterpieces ***The Last Supper*** and ***La Gioconda***, better known as the ***Mona Lisa***

The illegitimate son of Florentine notary Ser Piero da Vinci and a peasant woman, Leonardo da Vinci was born just outside the town of Vinci, 40km (25 miles) west of Florence. Until the age of five, he lived with his mother on a farm in nearby Anchiano, but was then brought up in his father's household in Vinci. There he received a conventional education appropriate for the son of a gentleman. Almost nothing is known for certain of da Vinci's boyhood, but he must have shown a talent for art, as his father arranged an apprenticeship for him when he was about 14 or 15 in the studio of one of the best-known artists in Florence, Andrea di Cione, known as Verrocchio. Under Verrocchio, da Vinci learnt to paint and sculpt. He was also taught metalwork and given some instruction in mathematics and sciences.

Da Vinci assisted his master in making and installing the gilded ball on the top of the dome of Florence cathedral, using lifting machinery designed by the architect Filippo Brunelleschi (see pp.66–67). Da Vinci found such mechanical devices fascinating. After his apprenticeship, he stayed on as an assistant in Verrocchio's studio until he began to receive some independent commissions in the late 1470s. On the strength of one of these, an "Adoration of the Magi" for a local church's altarpiece, he set up his own studio in 1481, but the project was left unfinished when he accepted a post at the court of the Duke Ludovico Sforza of Milan.

CREATIVE YEARS

In his letter of application for the position of court artist, da Vinci stressed his expertise as a military and civil engineer, and only as an afterthought mentioned that he could paint – an indication of the importance he placed on his versatility. He worked in Milan until the French invasion of the city in 1499, and during his time there produced some of his best-known paintings, including the *Virgin of the Rocks* (1483) and *The Last Supper* (1495–97). The project that seemed to occupy him the most, however, was a massive equestrian statue. The technical problems of designing a free-standing statue of a leaping horse and then casting it in bronze appealed to da Vinci's inquisitive mind but, despite numerous different designs, the project remained unrealized. The Duke of Milan also secured him plenty of work designing the military machines and apparatus used in the regular pageants and performances of the court. Da Vinci struck up a friendship with the court architect, Donato Bramante (1444–1514), collaborating with him on many building projects around the city.

After Milan fell to the French, da Vinci moved to Venice, where he worked briefly as a military engineer, and then, after spending a short time in Mantua, returned to Florence. With Florence as his base, he embarked on the most productive period of his career, not only creating many of his finest paintings, including the *Mona Lisa*, but also making frequent visits to Rome to work for the powerful Roman nobleman Cesare Borgia as a military engineer on defence and civil projects.

Now in his fifties, with his reputation secured, da Vinci could afford to be more choosy about the work he took on. He accepted an invitation from the French governor of Milan to return to the city in 1506, but after only a year returned

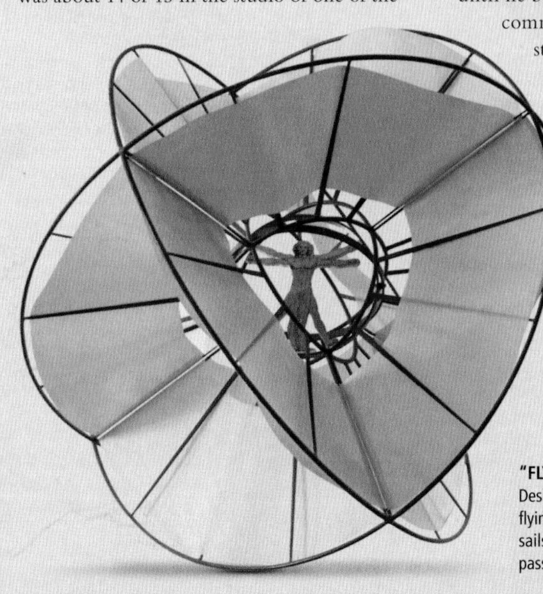

"FLYING SPHERE"
Designed to catch the wind like thistledown, the flying sphere consists of three intersecting, circular sails. The gimbal at its centre ensures that the passenger remains upright as the sphere rolls.

> # WHERE THE SPIRIT DOES NOT WORK WITH THE HAND, THERE IS NO ART
>
> **LEONARDO DA VINCI**

to Florence to sort out his late father's estate. Finding himself financially secure and free of commissioned work, he took the opportunity to follow up on projects of his own. He devoted his energies to his notebooks, building up an impressive collection of anatomical, scientific, and engineering drawings, with notes written in mirror writing, perhaps as a simple code.

ILLUSTRIOUS POSITIONS

In 1513, the newly installed Pope Leo X – a Medici from Florence – invited da Vinci to Rome again, where he lived at the Vatican. Da Vinci was not required to produce paintings or oversee projects, and the stay gave him the opportunity to mix with other Renaissance artists, such as Raphael and Michelangelo, who were working there at the time and who were grateful for the contact with the great master.

In 1516, Francis I of France visited Rome, and offered da Vinci a nominal post as first painter, architect, and mechanic to the king, with residence at the court of Fontainebleau. Da Vinci was also provided with living quarters at the Château de Cloux, now known as Clos Lucé, and a substantial allowance, amounting to a comfortable retirement. It was a feather in Francis's cap, having such a revered figure on his staff. Da Vinci moved to Fontainebleau with his companion Count Francesco Melzi, and lived there, more as a guest than as an employee, until his death in May 1519 at the age of 67.

Da Vinci's most obvious legacy lies in his paintings, which had an immediate influence on artists of the time and remain among the most famous in history. His skill as an engineer was also well known during his lifetime, and he was as much in demand for that as for his painting and drawing. Less recognized, until recently, are the many unfinished and unrealized projects, designs, and sketches that fill his numerous notebooks – with many of these designs not appreciated until several years after his death.

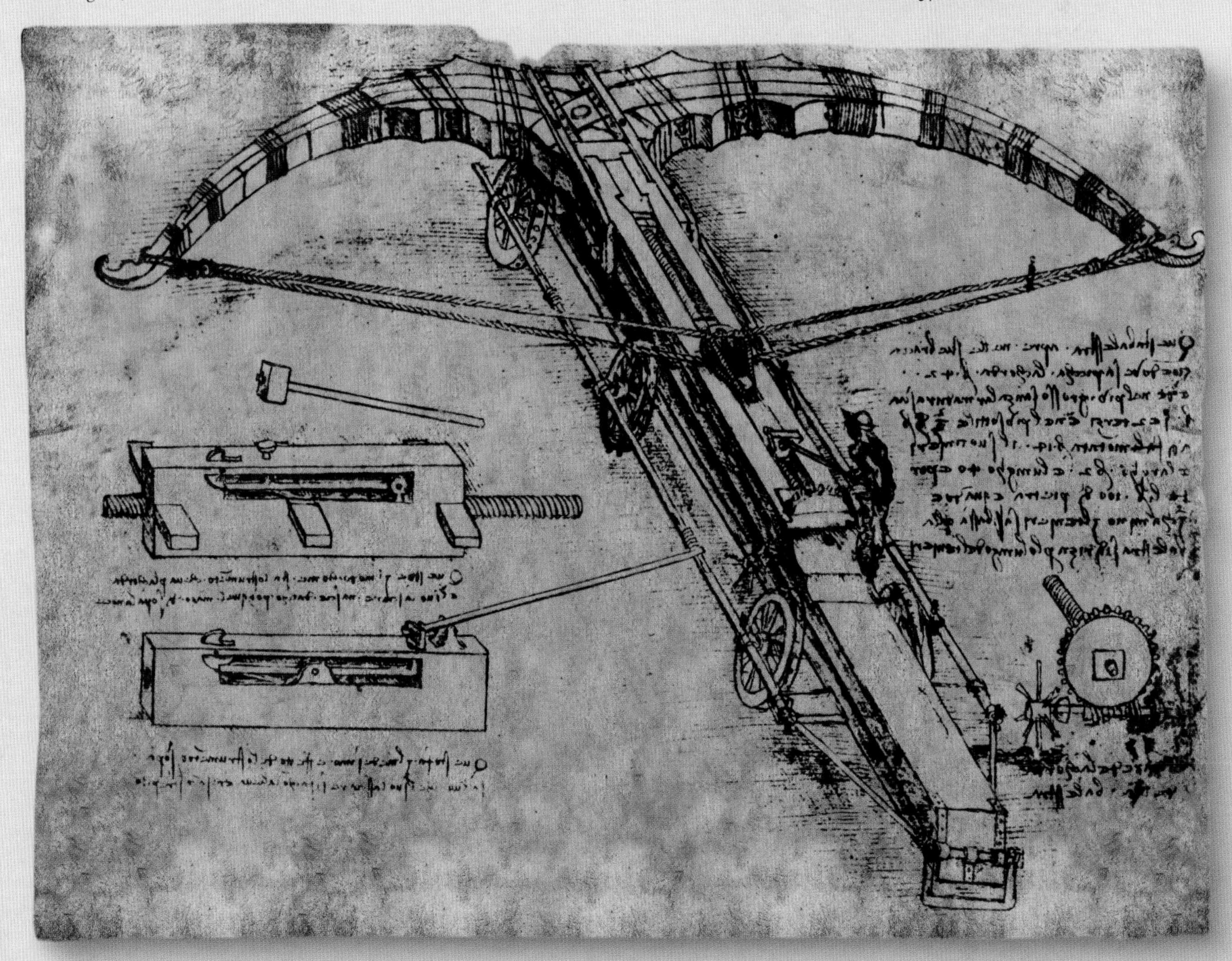

GIANT CATAPULT
This mobile catapult, similar to an outsized crossbow, was designed to hurl stones, and, possibly, flaming bombs. The string is pulled by rotating a crank, and released by knocking out the holding pin with a mallet.

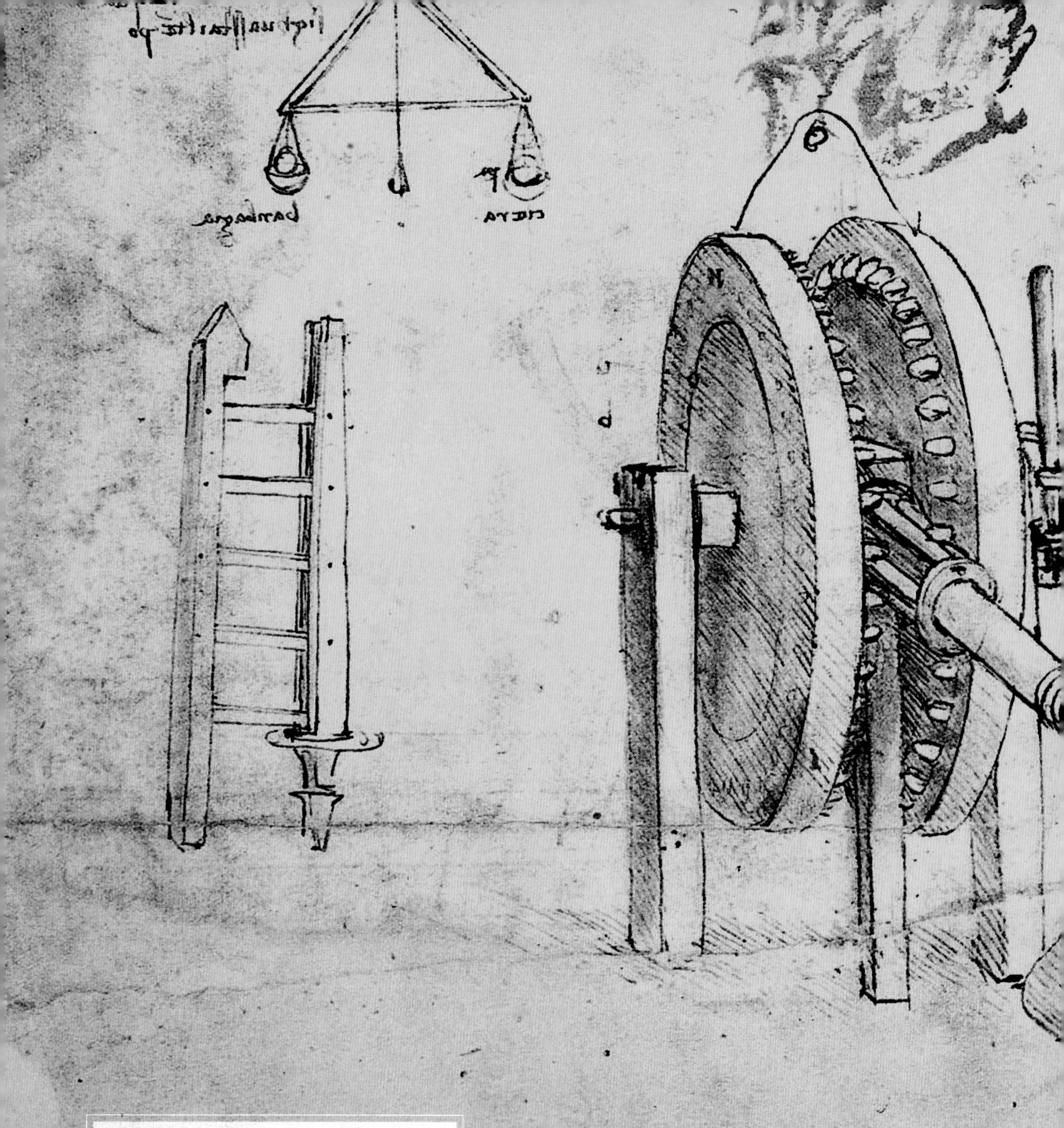

RENAISSANCE ENGINEER
Leonardo da Vinci's notebooks are packed with designs for a vast number of inventions, ranging from musical instruments and flying machines to steam-cannon and hydraulic pumps. Some are practical, and have been proved to work; others fared less well when built and tested by modern-day engineers. The idea illustrated here is highly practical: it is a two-wheeled hoist for lifting heavy objects, with a caged gear that would mechanically assist the lift and stop the rope from slipping.

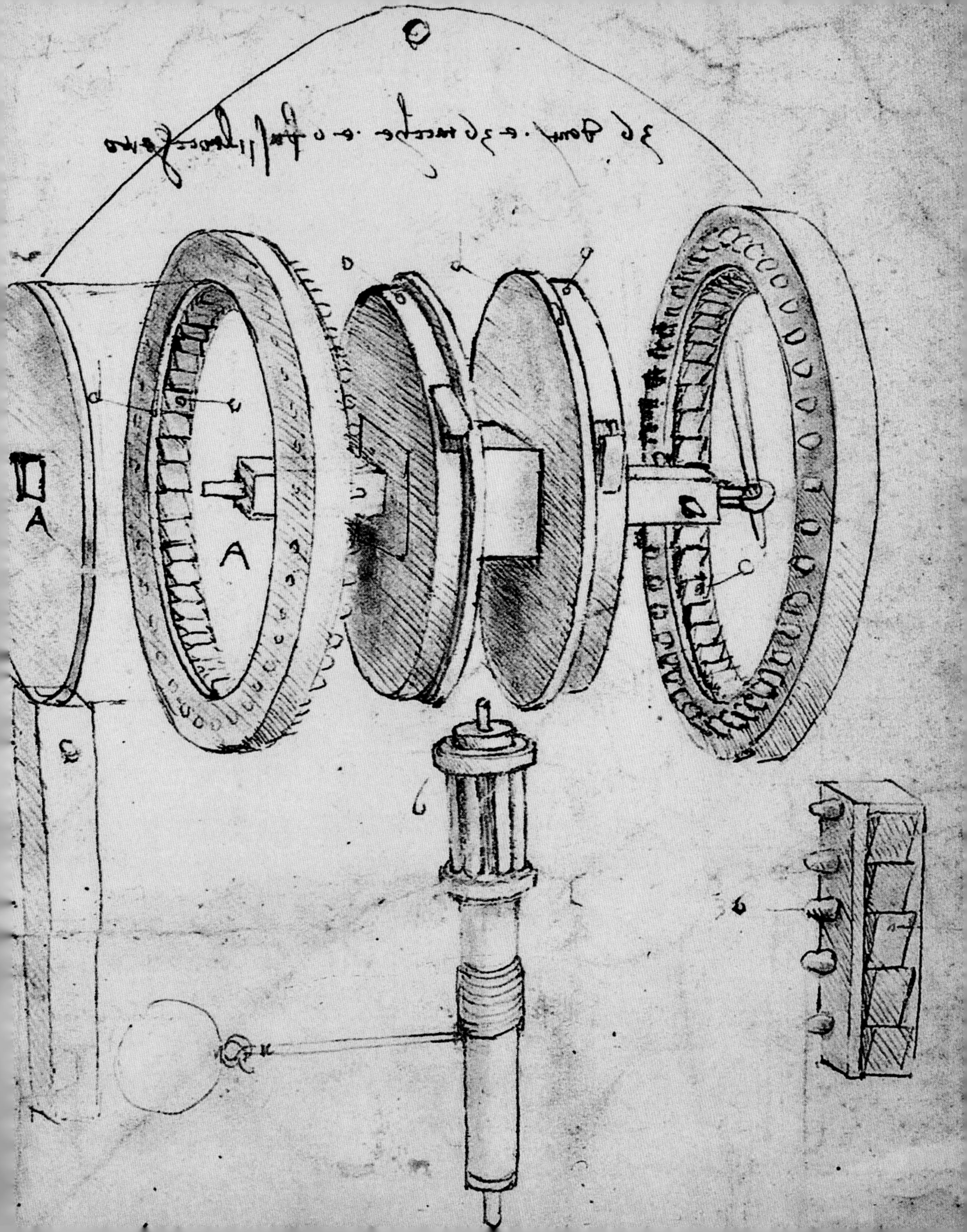
A
A

A NEW TECHNOLOGY

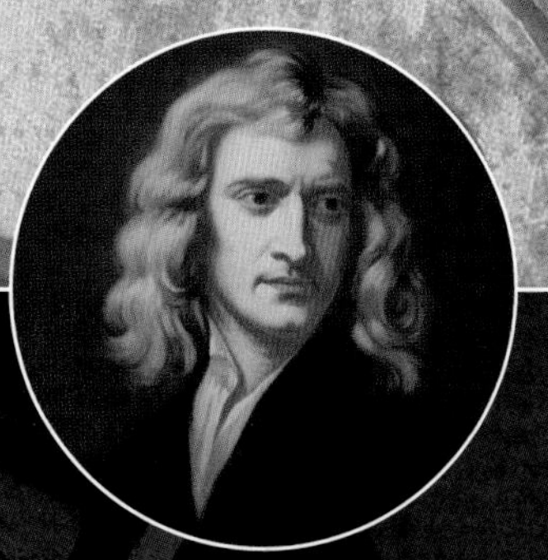

NEWTON'S GENIUS
The Scientific Revolution reached its peak in the 17th and 18th centuries with the aid of brilliant work by scientists such as Isaac Newton (1642–1727).

THE PACE OF INNOVATION INCREASED THROUGH THE RENAISSANCE, AND THE 17TH CENTURY SAW NEW DISCOVERIES IN SCIENCE. EUROPE HAD BECOME PROSPEROUS THROUGH TRADE AND COLONIZATION, AND THERE WAS AN INCREASING DEMAND FOR NEW ENGINEERING TECHNOLOGY.

The "Scientific Revolution" that provided the impetus for this new technology was a natural progression from the enormous cultural changes brought about during the Renaissance. While the Renaissance had been a primarily cultural and artistic movement, it had produced some remarkable scientists and engineers, and created a climate in which science could flourish. With the religious certainties replaced by a belief in scientific enquiry and observation, Europe entered what became known as the Age of Reason.

SCIENTIFIC ADVANCES

This new era was marked by a move towards a division between the arts and sciences. The notion of the Renaissance polymath was gradually changing to an emphasis on a more specialized approach. For the first time, science was being regarded as a separate discipline. Inspired by philosophers such as Francis Bacon (1561–1626) and René Descartes (1596–1650), scientists progressed from observation to rigorous examination involving experimentation, and from this developed the scientific method. The scientists of this period not only discovered new scientific principles, but were also quick to find useful applications for them. Also, the new science created a demand for more and better instruments and equipment. Consequently, many scientists developed engineering skills, or collaborated with craftsmen, to make new devices and machines during this period.

Leading the way were the astronomers, including Galileo Galilei (1564–1642), who heard about the invention of the telescope by two Dutchmen, built his own much more powerful version, and used it to discover the moons of Jupiter. Another example of this symbiotic relationship between science and technology in the 17th century was the invention of the mercury barometer by Galileo's pupil Evangelista Torricelli (1608–47). Galileo had heard that pumps could not suck water up more than 10m (33ft), and Torricelli investigated this, using mercury in narrow glass tubes; he found the pressure of the atmosphere would support a mercury column about 76cm (30in) high, higher in fine weather and lower in storms. Mathematicians also made their contribution to the new technology. Blaise Pascal (1623–62) built an ingenious mechanical calculating machine, and the invention of the vernier scale increased the precision of instruments for measuring.

PASCAL'S CALCULATOR
Pascal devised this "Arithmetic Machine" – a mechanical calculator capable of addition, subtraction, multiplication, and division.

Inventions such as the telescope and sextant enabled cartographers to make more accurate maps. Navigators, however, demanded increasingly accurate instruments to find their way reliably along established trade routes.

Another measuring device, the clock, also saw considerable advances. Christiaan Huygens (see pp.80–81) built the first pendulum-regulated clock in 1657, and these weight-powered clocks proved to be the first that were more precise than Ctesibius's water clocks (see pp.30–31). Spring-driven clocks and watches, regulated by balance wheels, were also making their first appearance, but did not keep good time.

Some advances, however, were driven by economics and trade as much as science. The maritime nations of Europe, in particular England and Spain, had colonized much of the world, and built up their navies to protect their territories. This led to improvements in ship design, both to increase capacity on cargo ships, and to carry more cannons on the warships, resulting in the faster and more manoeuvrable galleon used both by the British navy and the Spanish Armada. At the same time, the threat of war inspired the constructions of the first navigable submarine.

Following the age of exploration and colonization, Europe became increasingly reliant on shipping to transport goods across the oceans. Improvements to instruments such as the sextant and telescope had helped to make navigation more reliable, but the problem of determining the longitudinal position of a

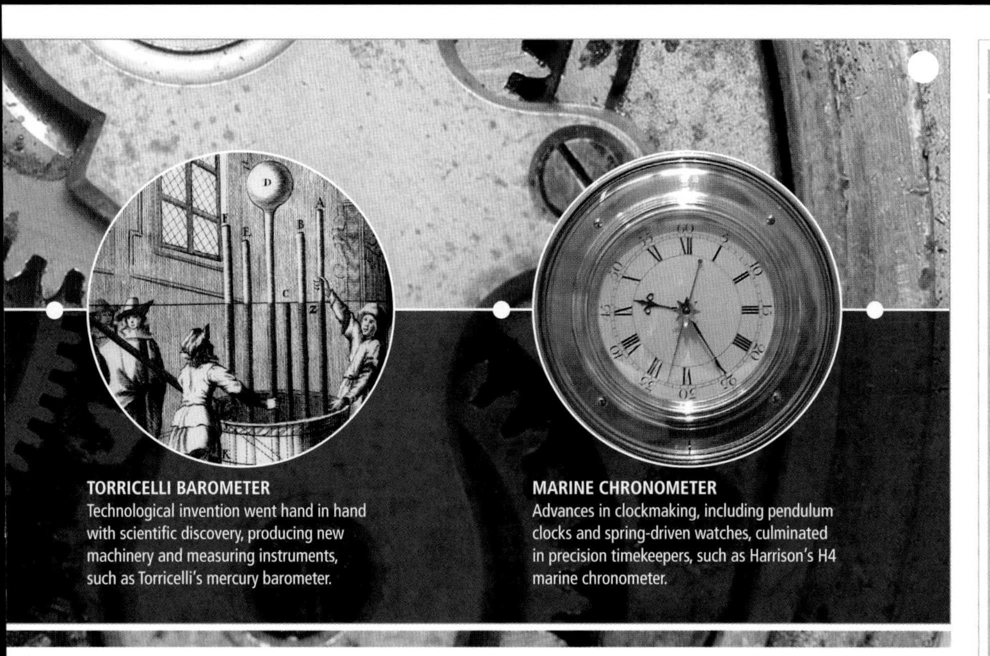

TORRICELLI BAROMETER
Technological invention went hand in hand with scientific discovery, producing new machinery and measuring instruments, such as Torricelli's mercury barometer.

MARINE CHRONOMETER
Advances in clockmaking, including pendulum clocks and spring-driven watches, culminated in precision timekeepers, such as Harrison's H4 marine chronometer.

SETTING THE SCENE

- The **Scientific Revolution** becomes centred in northern Europe in the 17th-century Enlightenment, producing thinkers such as **Bacon, Descartes, and Pascal**, who pioneer the scientific method, and institutions such as the **Royal Society of London and the German Academy of Sciences**, established to promote scientific research.
- The age of exploration and colonization of the world leads to **advances in ship design** and the development of more **sophisticated weapons**.
- The **discovery of the logarithm** by John Napier (1550–1617) leads to the **invention of the slide rule** for complex calculations.
- Much of the **new technology** of this period is **developed for scientific instruments**.
- The need for accurate timekeeping, particularly in solving the **"longitude problem"**, leads to **innovations in clockmaking** including the pendulum clock, balance spring, spring-driven pocket watch, and marine chronometers.
- By the end of the 17th century, **engineering** begins to emerge as **a distinct profession** from science, concerning itself with devising practical applications for the new scientific discoveries.

ship still remained. The solution lay in the development of an accurate chronometer capable of withstanding sea voyages and changes of climate. This was achieved by John Harrison (see pp.82–83) after devoting a lifetime to the task.

ENGINEERING AS A WAY OF LIFE

The combination of scientific and technological innovation during the Age of Reason led to a new way of thinking about the world. Natural phenomena were being explained in mechanistic terms. Descartes described humans as a kind of machine with a soul, and Isaac Newton derived his laws of motion and gravity from a model of the universe in which the planets revolved like wheels in a clockwork machine.

The 17th century saw the establishment of learned societies devoted to science, and an increased specialization in branches of science that has continued to the present day. The scientists of this age were also inventors, engineers, and craftsmen. The new technology that they produced created a more mechanized society that had become ever more reliant on machines. From this, engineering emerged as a separate discipline, to devise and implement practical applications for scientific discovery. And, to a large extent, it was engineers who were beginning to dictate the pace of progress.

LAND YACHT
Inventions such as Simon Stevin's land yacht and Cornelis Drebbel's submarine (see pp.74–75) heralded the beginning of a period of advances in transport technology.

CORNELIS DREBBEL

BUILDER OF THE FIRST NAVIGABLE SUBMARINE

THE NETHERLANDS 1572–1633

EPITOMIZING THE ROLE of the eccentric inventor, Cornelis Drebbel is best known today for building the first submarine, but during his lifetime he was respected as a cartographer, a maker of telescopes and microscopes, a hydraulic engineer, and an alchemist. However, through a combination of bad luck and mismanagement, he died neglected and in poverty.

A LIFE'S WORK

- Drebbel **trains as an engraver** under artist and alchemist Hendrik Goltzius
- He builds a name for himself as an **etcher, cartographer, and instrument-maker**
- He is granted patents for a **water-raising fountain** and a **self-winding clock**
- He is credited with the **invention of the thermometer**
- Works as an **engineer in the court of James I of England**, and accepts the post of hydraulic engineer for Emperor Rudolf II in Prague
- In 1620, he **demonstrates a submarine** built in accordance with a design by William Bourne
- **His sons-in-law** take over the dyeworks he founded after his death and **make their fortune** from his **scarlet-dying technique**

Drebbel was born in Alkmaar in the Netherlands, and initially trained as an engraver in Haarlem under artist and alchemist Hendrik Goltzius (1558–1617). He soon established his reputation as a fine etcher and cartographer, but also took an interest in Goltzius's scientific experiments, dabbling in alchemy and becoming proficient at instrument-making, through which he learnt the principles of mechanical engineering.

He returned to Alkmaar in 1595 and set up business as an engraver and cartographer, but found himself in financial difficulties (largely due to his marriage to Goltzius's sister, who expected a standard of living beyond their means). He then turned to engineering as a way of supplementing his income. There was a market for skilled instrument-makers at the time, and Drebbel established a reputation for his scientific equipment – in particular for precision lens-grinding for telescopes, magic lanterns, and cameras obscura. Instrument-making quickly replaced engraving as his main source of income as demand for his work increased, and he began to extend his expertise into other branches of engineering.

GROWING FAME

It was at this time that he discovered a talent for invention. Having mastered the rudiments of hydraulic engineering, he took out a patent on a water-raising pump, and was commissioned to design a fountain for the town of Middelburg and a water supply system in Alkmaar. He was also granted patents for an improved design for a chimney, and for the invention that made his name, a self-winding clock that he claimed was a perpetual-motion machine, powered by changes in temperature and atmospheric pressure. On the strength of his growing reputation as an inventor, he and his family moved to London in 1604, where he had secured a job as an engineer in the court of James I. His role there, however, was not quite what he had expected. He was put in charge of firework displays and the mechanics of theatrical productions for the young Henry, Prince of Wales, and although he gained further publicity for his perpetual-motion machine, he had little time to spare for new inventions.

IMPRISONMENT AND FREEDOM

Frustrated by his job in London, Drebbel accepted an invitation to work for Emperor Rudolf II in Prague as a mining engineer, but this ended in disaster when Rudolf's brother, Matthias, conquered the city a year later. Drebbel was imprisoned and had all his possessions confiscated, and only after appealing to Prince Henry was he finally freed and returned to London in 1613.

Reattached to James I's court, but with less onerous duties, Drebbel resumed instrument-making, embarking on a period of renewed innovation and invention. He developed and improved lens-grinding techniques, building telescopes and a compound microscope with two convex lenses, which became much sought after by scientists across Europe. Working from principles he had learnt while designing his perpetual-motion machine, he also designed thermostats for use in ovens and furnaces, which led to his invention of an incubator for hen and duck eggs; and he is often credited with the invention of the thermometer. But the invention that was to make his name was actually not his own. English mathematician William Bourne (1535–82) included in his 1578 book *Inventions and Devices* a plan for a rudimentary submarine, basically an upturned boat working on the same principle as the diving bell, and powered by oarsmen. Drebbel's submarine was simply a realization of this design, and his own contribution was in the process of removing carbon dioxide and supplying oxygen for the chamber by heating saltpetre. Nevertheless, Drebbel's sons-in-law, who were angling to become his business partners, publicized the submarine as his invention, and were responsible for the hype that surrounded its launch on the River Thames in 1620.

Drebell had opened a dyeworks in Stratford-le-Bow in the east of London some years before, and it was in this firm that his sons-in-law, the brothers Abraham and Johannes Kuffler, were keen to become involved. Although the firm was struggling financially, Drebbel had made several important chemical discoveries, including a

WORKING SUBMARINE
Drebbel incorporated some modifications to Bourne's design in his submarines, which helped in improving the vessel's watertightness and provided breathable air for the crew.

THERE IS SOMETHING IN THAT BADLY DRESSED MAN ... WHICH WOULD MAKE ANY OTHER MAN [SEEM] RIDICULOUS

PETER PAUL RUBENS, FLEMISH BAROQUE PAINTER

process of fixing scarlet dye using tin chloride, which was marketed by the brothers after his death as "Colour Kufflerianus".

NAVAL ENGINEER

Drebbel went on to build two further submarines, successively larger and with slight modifications, which were also successfully demonstrated on the Thames. It was these that brought him to the attention of the Royal Navy, and although they could not see any military use for the vessels, they were impressed enough by Drebbel to employ him as a military and naval engineer. He spent some years designing torpedoes, firebombs, and fireships, which were used in the expeditions to end the siege of La Rochelle, France, in 1628. However, he was made a scapegoat when the campaign ended in failure, and was dismissed from service.

Once again he found himself out of work and in need of money, but his expulsion from the navy had severely damaged his reputation and made it difficult for him to find employment or sell his expertise. In desperation, he took on the management of an alehouse in a less-than-respectable area of London near London Bridge. In the early 1630s, he found some work with a fellow Dutchman, Cornelius Vermuyden, who was an engineer in charge of a drainage scheme in the marshes north of London, but Drebbel's career as an engineer and inventor was over and he died in poverty and obscurity in 1633.

SIMON STEVIN

THE NETHERLANDS 1548–1620

The illegitimate son of a wealthy couple from Bruges, Antheunis Stevin and Cathelijne van de Poort, Simon Stevin worked for some years as a merchant's clerk in his native Flanders. However, he went on to become an accomplished military and hydraulic engineer, employed as an adviser to Prince Maurice of Orange.

His advice was often solicited on matters of defence and navigation, and he set up a school for engineers at Leiden. His work stems from the scientific revival in the 16th century that followed the increasing prosperity of countries such as the Netherlands.

Stevin designed a number of drainage and pumping machines, and a system of sluices for flooding low-lying areas as a defence against invasion, but the invention for which he was best known was a sail-driven carriage or land yacht capable of carrying up to 26 passengers. He also published several treatises on mathematics, and was one of the early champions of the decimal system.

ENGINEERING INNOVATIONS

SHIP DESIGN

From simple, single-sailed open boats used in ancient Egypt to the supertankers, nuclear submarines, and luxury liners of today, ship design has reflected and shaped the society we live in. As nations and empires grew, maritime powers emerged and new designs evolved. Maritime engineers played an increasingly important role, developing new technology that included the use of steam engines, and later diesel and turbine engines, as well as iron for building ships – all innovations made in response to the demands of trade, exploration, and warfare.

VESSELS FOR TRADE AND WAR

The first sea-going vessels evolved from the river boats used in ancient Egypt and China. Fitted with a mast and sail and propelled by oars, early Egyptian ships were built of planks tied together and sealed with pitch. This simple design of an open boat, with paired rows of oars and a single square sail, continued into the time of the Phoenicians in the 8th century BCE, whose larger ships, known as galleys, established trade routes across the Mediterranean. The galley also proved useful as a warship, and was later used by the Greeks and Romans. With only slight alteration, it became the longship used by the Vikings in the Middle Ages for long voyages of exploration.

During the Renaissance, the emphasis moved from oar-propelled ships to sailing ships, such as the three- or four-masted carracks and the caravels developed in Portugal. More suited than previous ships for ocean voyages, they ushered in an age of exploration, and established the trade routes to Asia and the Americas. From these there followed a new

BCE 3000 2000 1000 500 CE 1300 1400 1500 1600 1700 1800

c.3000BCE – RIVER BOATS
Ancient Egyptians build boats with a single sail and hulls made of planks lashed together and sealed with papyrus and pitch.

c.1000BCE – SIMPLE SAIL
Sailboats have a steering oar at the stern that later evolves into the stern-mounted rudder.

c.1170 – NAVIGATION
In China, versions of a magnetic compass are used in navigation.

c.1300 – WATER-TIGHT
Chinese ships are built with water-tight double hulls.

LATE 14TH CENTURY – CARRACK
A three- or four-masted sailing ship, the carrack (or nau), with towers at the stern and bow, is used by early Spanish and Portuguese explorers.

c.1500 – DOUBLE HULL
In Venice, Leonardo da Vinci suggests a design for a double hull as a safety measure for ships in the event of leaking.

c.1620 – FIRST SUB
Cornelis Drebbel's underwater boat (see pp.74–75) is the first successfully working submarine.

c.1776 – STEAM OARS
In France, Claude-François-Dorothée (1751–1832), shows his *Palmipède*, a boat propelled by rotating oars powered by steam.

c.1800 – CLIPPERS
Fast sailing ships, known as clippers, appear and become the dominant trading vessels of the 19th century.

The rowing ships known as galleys, used by the Phoenicians and Greeks since the 8th century BCE, become larger and are fitted with a second and even third row of oars, sometimes with a total of more than 50 oars. Known as biremes and triremes, these are used as trading vessels but also, thanks to their increased speed and manoeuvrability, as warships capable of ramming and damaging other vessels.

c.500BCE OARED GALLEYS

The caravel, a streamlined version of the carrack, has a narrower keel and smaller sterncastle and forecastle instead of towers, making it more seaworthy than the carrack. Early caravels have triangular lateen sails that make them fast and manoeuvrable, but are later rigged with square sails. Christopher Columbus made his voyage across the Atlantic with a large carrack, the *Santa Maria*, accompanied by the caravels *Pinta* and *Niña*.

c.1485 CARAVEL

type of warship – the galleon, capable of carrying cannons – and a period of European naval warfare that later produced the ship-of-the-line with its characteristic broadside arrangement of guns. Another product of continual conflict was the development of submarines, which became a practical possibility during the Napoleonic wars.

BIGGER AND BETTER

Although steam engines were being fitted into boats as early as the 1770s, it was a while before steamships were commercially viable and, in the meantime, the clipper, a narrow-hulled sailship, dominated the trade routes. However, with the introduction of the screw propeller, steam gradually replaced sail, and changes in ship design accelerated. Iron, and later steel, which had previously only been used to reinforce and protect wooden-hulled vessels, was now being increasingly used as a primary construction material. The 19th century saw an increase in the number of passenger ships being built, leading to even bigger and more luxurious liners.

THE JOURNEY OF SHIPS

Throughout history, sea vessels have had a number of uses. Small boats have been used for fishing since earliest times. As civilizations grew, so did trade, and ships became an important means of transporting goods. Military vessels were developed not only to protect trade routes but also to expand empires. Ships capable of long ocean voyages were increasingly used to explore and colonize the world. Soon, passenger ships also appeared.

HERE, SHIPS OF THE LINE FORM "BATTLE LINES" TO FIRE THEIR CANNONS.

In the 20th century, ship design was led by the search for different means of propulsion. Diesel engines, which in the beginning had been costly and inefficient, were improved and they slowly replaced steam as the dominant form of power. However, despite the successful demonstration of steam turbine engines, advances made during World War II led to the installation of gas turbine engines and nuclear engines in many naval vessels.

1802 – PADDLE STEAMER
Scottish engineer William Symington (1764–1831) builds the *Charlotte Dundas*, the first practical paddle steamboat.

1807 – RIVER SERVICE
Robert Fulton's *North River Steam Boat* (see pp.188–89) starts a scheduled service on Hudson River.

1819 – STEAM SHIP
The steamship SS *Savannah* becomes arguably the first steam-powered ship to cross the Atlantic.

1839 – SCREW PROPELLER
SS *Archimedes*, constructed by shipbuilder Henry Wimshurst (1804–84), makes its maiden voyage in London; it is the first steamship driven by a screw propeller.

1843 – LARGEST VESSEL
The SS *Great Britain*, designed by I K Brunel (see pp.194–97), is launched; it is the largest vessel of its time, and the first to be built entirely of iron.

1850

1894 – FASTEST VESSEL
The *Turbinia*, powered by a steam turbine patented by Sir Charles Parsons (1854–1931), is demonstrated at the Royal Navy Review, and is shown to be the fastest vessel of its time.

1900

1903 – DIESEL ENGINE
A Russian tanker, the *Vandal*, is the first to be fitted with a diesel engine, which will gradually replace steam as the most usual form of power for ships.

1947 – GAS POWER
A gas turbine engine is fitted to British Royal Navy motor gun boat MGB 509, and is relaunched as MGB 2009, the first gas turbine powered naval vessel.

1950

Although described by James Watt (see pp.118–21) and Richard Trevithick (see pp.128–29), the screw propeller is first patented by Austrian engineer Josef Ressel (1793–1857), and soon similar designs by John Ericsson (1803–89) and Francis Pettit Smith (1808–74) are given patents. Significantly more efficient than the paddle wheel and capable of working in heavier seas, the propeller enables ocean-going vessels to be fitted with steam engines.

c.1830 SCREW PROPELLER

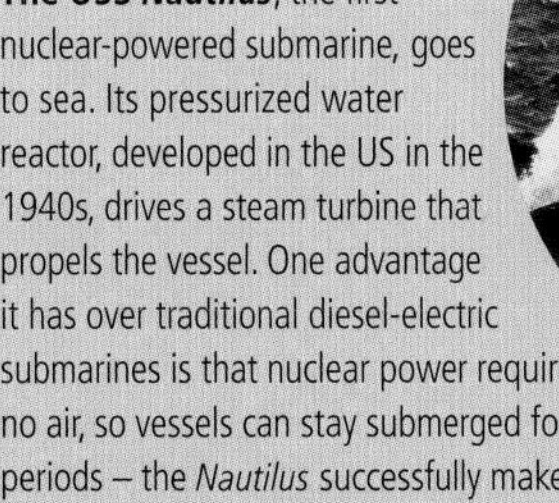

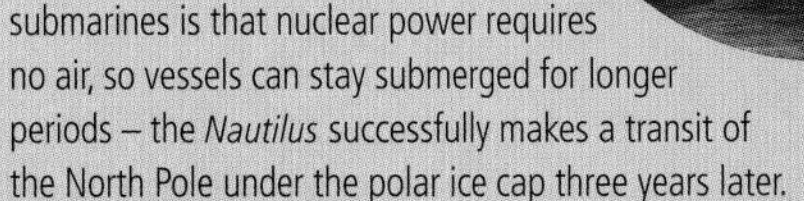

The USS *Nautilus*, the first nuclear-powered submarine, goes to sea. Its pressurized water reactor, developed in the US in the 1940s, drives a steam turbine that propels the vessel. One advantage it has over traditional diesel-electric submarines is that nuclear power requires no air, so vessels can stay submerged for longer periods – the *Nautilus* successfully makes a transit of the North Pole under the polar ice cap three years later.

1955 NUCLEAR SUBMARINE

ROBERT HOOKE

DISCOVERER OF THE LAW OF ELASTICITY

ENGLAND **1635–1703**

Robert Hooke was a scientific polymath at the forefront of the Scientific Revolution. Having built his own microscope, he wrote a remarkable book, describing what he saw. He made significant contributions to engineering, and may have invented both the anchor escapement and a balance spring mechanism for clocks and watches, as well as the law of elasticity that bears his name.

A LIFE'S WORK

- While studying at Oxford University, Hooke meets Robert Boyle and becomes his **assistant in experiments with air pumps**; he begins his work in astronomy and mechanics, **inventing the balance spring** and discovering what is now known as **Hooke's Law of Elasticity**
- Is **appointed "curator of experiments"** at the newly founded Royal Society, and later **elected a Fellow**
- After the Great Fire of London in 1666, he is made a **Surveyor of the City of London**, and works alongside Christopher Wren in **planning the reconstruction** of the city
- In later life, he feels that he has been **poorly treated by his peers**, and becomes involved in disputes about the **authorship of his inventions and discoveries**

Born in 1635 to a Church of England minister in the village of Freshwater, Isle of Wight, Robert Hooke was a sickly child, and had only a sporadic education from his father at home. Frequently left to study on his own, Hooke amused himself by drawing, painting, model-making, and tinkering with mechanical devices – where he showed a real talent, and progressed so far as to dismantle a brass clock and make a replica in wood that he said worked "well enough". He even learned to draw, making his own materials from coal, chalk, and ruddle (iron ore). He was just 13 when his father died in 1648, leaving him a small inheritance to continue his education.

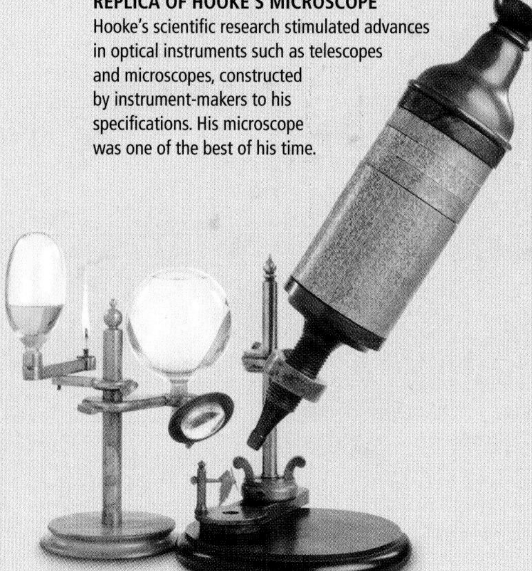

REPLICA OF HOOKE'S MICROSCOPE
Hooke's scientific research stimulated advances in optical instruments such as telescopes and microscopes, constructed by instrument-makers to his specifications. His microscope was one of the best of his time.

Hooke was sent to London to become an apprentice to the artist Sir Peter Lely, but he soon realized that he wanted a more academic education. He used the remainder of his inheritance to study at Westminster School, living with the master Dr Richard Busby. Although Hooke was not a full-time student and was taught separately from the rest of the school by Busby, he gained a grounding in the classics that enabled him to earn a place as a chorister at Christ Church College, Oxford. It was here that Hooke's scientific education began in earnest.

WATCHES AND ELASTICITY

At Oxford, Hooke found work as an assistant to the chemist Thomas Willis, who recognized the young man's enthusiasm and flair, and gave him both instruction and encouragement. Hooke also met Robert Boyle (see 1627–91), who was at that time looking for an assistant to help with the construction of his "pneumatic engine". Hooke eagerly accepted Boyle's invitation, built Boyle's air pump, and became his assistant in investigating the vacuum. As a result he met a circle of "natural philosophers" and academics under whom he was able to complete his scientific education, even though he never actually graduated from the university.

While working for Boyle, Hooke undertook some research of his own – related to their experiments – with pumps and compression of gases. This led Hooke to the work on elasticity and springs for which he later became famous. He applied himself to the improvement of the pendulum and its mechanisms, and in about 1657 went on to study both gravitation and the mechanics of timekeeping.

Combining his childhood talent for clockmaking and the discoveries he was making while working with Boyle, Hooke invented a device that would considerably improve the precision of spring-driven clocks and watches. By adding a balance spring to regulate the motion of the balance wheel, he ensured that a clock could run at a consistent rate rather than wind down slowly. This invention was a significant step forward in clock design, but as Hooke was a humble assistant and not one of the scientific elite, he failed to find backing to develop the idea, and it went unnoticed for several years. As did his discovery in 1660 of the law of elasticity, according to which the extension of a spring is in direct proportion to the load applied to it. Hooke expressed it as *Ut tensio, sic vis* ("as the extension, so the force"), and it later became known as "Hooke's Law" in his honour.

"CURATOR OF EXPERIMENTS"

Although on friendly terms with Boyle and the other Oxford scientists, Hooke must have felt somewhat excluded because of his status, and maybe a little patronized by their attitude towards him. When in 1660 a group of them formed the Royal Society, the first of the national learned societies for science, they agreed to employ a "curator of experiments", and Hooke was a natural choice for the job. Although he was flattered by the offer and was glad of the opportunity to carry on his own research while drawing a salary, his talents were still not being properly acknowledged.

Nevertheless, he proved his worth as their employee, setting up experiments and demonstrations – of his own methods or those

THE SCIENCE OF NATURE ... SHOULD RETURN TO THE ... OBSERVATIONS ON MATERIAL AND OBVIOUS THINGS

ROBERT HOOKE

PETER HENLEIN

NUREMBERG C.1479–1542

A German locksmith and clockmaker, Peter Henlein designed and built the *Taschenuhr* ("pocket clock"), one of the earliest examples of a portable watch.

Henlein's small, drum-shaped watches incorporated his own design of balance wheel, and were capable of running for 40 hours without winding. Often wrongly credited with the invention of a spring-driven movement, he did, however, perfect the design of a "clock without weights" that used a spiral mainspring. The idea of using a spring as the driving force for a clock first occurred in the 15th century, and was adopted by Henlein in the design of the first portable timekeepers.

suggested by the members – and giving lectures on mechanics. Among Hooke's earliest demonstrations were discussions on the nature of air and the implosion of glass bubbles sealed with hot air. There were also experiments on the subject of gravity, the falling of objects, the weighing of bodies, and measuring of barometric pressure at different heights. Hooke finally received the recognition he sought when he was elected a Fellow of the Society in 1663, the same year he was granted an MA degree from Oxford University.

With his position now secure, he devoted himself to scientific research, but also found the time to work on improvements to the scientific equipment he needed. He built telescopes and microscopes, and is believed to have invented the iris aperture and the cross-hair sighting device for use in telescopes, which in turn enabled him to make important discoveries in astronomy and biology. He also coined the term "cell" for the basic building block of plants and animals. He detailed these discoveries in his book *Micrographia* in 1665.

REBUILDING LONDON

The following year, the great Fire of London destroyed a large part of the city, and English architect Christopher Wren (1632–1723) was put in charge of reconstruction. He appointed Hooke, a friend since their days at Oxford, as his chief assistant, and the two collaborated on many of the major architectural projects involved in restoring the city.

Hooke continued working as a lecturer at Gresham College for the rest of his life, and diligently carried out his duties as Secretary of the Royal Society. Hooke met his friends Christopher Wren and astronomer Edmond Halley in coffee shops almost every day, but also gained a reputation for being irascible and argumentative. He had a long-running dispute with Isaac Newton, for example; Newton never forgave Hooke for criticizing his first scientific paper. There were also disputes over discoveries and inventions, notably Christiaan Huygens's claim to have invented the balance spring (see pp.80–81) before Hooke.

Contrary to his reputation, Hooke maintained a close circle of friends, including scientists, horologists, and engineers, and although he never married, he is rumoured to have had a lively love life. His funeral in London in 1703 was attended by members of the Royal Society, but his body was buried in an unmarked grave.

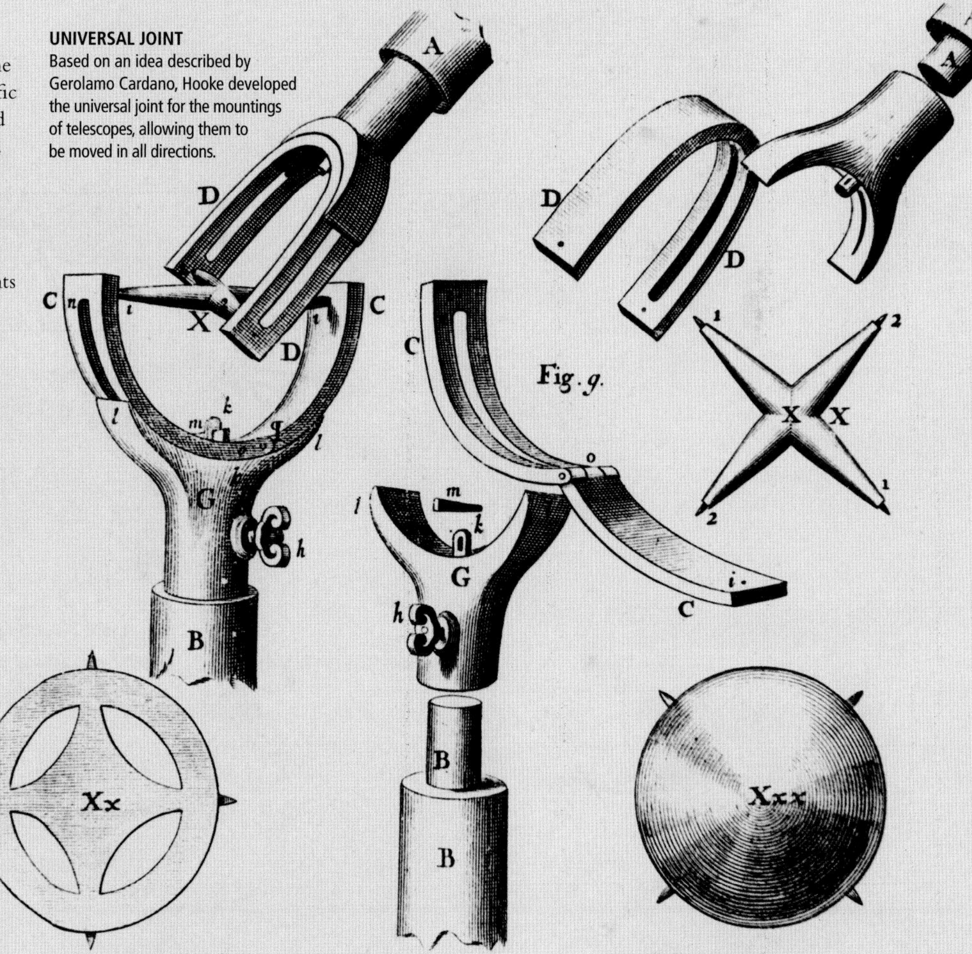

UNIVERSAL JOINT
Based on an idea described by Gerolamo Cardano, Hooke developed the universal joint for the mountings of telescopes, allowing them to be moved in all directions.

CHRISTIAAN HUYGENS

INVENTOR OF THE PENDULUM CLOCK

THE NETHERLANDS 1629–95

A SCIENTIST WITH an immensely practical turn of mind, Christiaan Huygens ranks among the most influential thinkers of the 17th-century's Scientific Revolution. Primarily an astronomer and physicist, he also designed optical instruments. In his search for a solution to the "longitude problem", he contributed several advances in clockmaking, including the invention of the pendulum clock.

A LIFE'S WORK

- **Encouraged by the French mathematician Blaise Pascal**, Huygens writes the first ever **book on probability, *De Ratiociniis in Ludo Aleae*** ("On Reasoning in Games of Chance")
- After **patenting the first pendulum clock**, he continues his experiments with pendulums and, in 1673, publishes an analysis of the mathematics of pendulums in ***Horologium Oscillatorium sive de motu pendulorum***
- In his ***Treatise on Light* (1678)**, he argues that light consists of waves and, in experiments with optics, he develops an **early form of magic lantern**
- He discovers a method of **dividing the musical scale into 31 equally spaced intervals**, and builds a keyboard to play in this 31-note temperament
- His final book, ***Cosmotheoros***, completed shortly before his death, presents the evidence for **his belief in life on other planets**

Son of a well known Dutch diplomat named Constantijn Huygens – whose friends included the French philosopher and writer René Descartes – Huygens grew up with all the advantages of a cosmopolitan family in The Hague. He was educated at home by private tutors, with an emphasis on mathematics, sciences, and, unusually, mechanical engineering. At the age of 16, he went on to study law and mathematics at the University of Leiden and the College of Orange at Breda.

After graduating in 1649, he returned to The Hague, and seemed set to follow a diplomatic career. But a correspondence with the French mathematician Mersenne whetted his appetite for mathematics and he spent the next few years working on a series of books on the subject. Huygens also became increasingly interested in astronomy, and, as was often the case with him, began to experiment. In order to make more accurate astronomical observations, he designed and built his own improved telescopes, even devising a new method of grinding the lenses, which in turn introduced him to the subject of optics.

MEASURING TIME

Most significantly, astronomy – with its intimate connection with clocks and calendars and a need for accurate timekeeping – also took him into the realm of horology, the science of measuring time. A hot topic at the time was the difficulty of determining longitude at sea (see p.83). Without this information, navigation was very imprecise. The search was on for a truly accurate timekeeping device that would enable mariners to calculate longitude. Huygens rose to the challenge, and, in 1656, patented a design for the first pendulum clock – more accurate than any other clock of the time, but, unfortunately, impractical for use on a moving ship at sea. Undeterred, he continued to make improvements to the design throughout his life, and used his experience as the basis for a mathematical analysis of the motion of pendulums.

Huygens was no dilettante, however, and his work in such diverse but interconnecting fields was often ground-breaking. During the 1650s, he gained a reputation across Europe for his discoveries and innovations, and it brought him into contact with other leading scientists and thinkers. He travelled to Paris to discuss his astronomical findings, particularly relating to the moons and rings of Saturn. In the 1660s, he spent a couple of years in London, where he was elected to the newly established Royal Society, before settling in France as director of the Académie des Sciences in 1666.

PENDULUM TIME
This reconstruction of the first pendulum clock shows the movement and, at the rear, the adjustable pendulum.

PERSISTENT RESEARCH

While much of Huygens's time in Paris was taken up with astronomical observation, he continued to pursue his many other interests. Working at the Paris Observatory, he built several telescopes, refining and improving the design over time, and exploring the science of optics more deeply. This culminated in the *Treatise on Light* in 1678, which included an explanation of his theory that light was composed of waves.

The search for an accurate clock still occupied him, and, as well as making improvements to his pendulum clock, he worked on a chronometer with a totally different kind of mechanism, using a balance wheel and a spiral spring. There is some controversy about whether he was in fact the inventor of the balance spring, but it is certain that Huygens's was not an entirely original idea. Robert Hooke (see pp.78–79)

HUYGENS IN COSTER'S CLOCKMAKING SHOP
This painting shows Huygens (centre) examining his finished pendulum clock in the workshop of horologist Salomon Coster (right) in The Hague.

had made a clock with a balance spring some years before, but Huygens's use of a spiral spring – rather than a coil spring, as used by Hooke – was a true innovation. It formed the basis of watch mechanisms until the advent of quartz crystal technology in the 20th century.

Never one to miss an opportunity to experiment and research, Huygens continued to develop other ideas. He had seen the vacuum pump made by the Anglo-Irish physicist Robert Boyle (1627–91), and became fascinated with the idea of a pump that could be used to raise water from mines. Reproducing Boyle's experiments at first, and then developing his own, he designed an apparatus in which a piston was driven by gunpowder exploding in a cylinder – a rudimentary combustion engine. This was never built. However, his assistant in these experiments was the young Denis Papin (see pp.106–07) who went on to work with Robert Boyle, and eventually pioneer the atmospheric steam engine.

Huygens's health troubled him throughout his life, and a serious illness prompted him to return to the Netherlands on more than one occasion. In 1676, he went back to The Hague for a two-year stay, intending to divide his time between there and Paris, but after a couple of visits to France, found it a great strain on his health. After 1681, he left The Hague only on one occasion – to visit the English polymath Isaac Newton (1642–1727) and Boyle at the Royal Society in London in 1689. Although Huygens now felt isolated from the centres of science, he continued working, in particular on refining both his pendulum and spring-regulated clocks, and on the improvement of lenses. He took a keen interest in scientific theories and discoveries until his death in 1695.

Huygens's influence on science was immense, paving the way for many scientific discoveries of the 18th century. However, his contribution to the world of engineering was sometimes overlooked. Undoubtedly, his greatest innovation was the invention of the pendulum clock, and, more indirectly, his experiments with a combustion-driven machine. But it is in his improvement of lens technology and the refinement to Hooke's balance spring mechanism that Huygens left a lasting legacy.

CLOCKS AND WATCHES

The need to measure time prompted some of the earliest human inventions. Sundials and candles were eventually superseded by more complex devices, the first true clocks, which were powered by flowing water. These were replaced by weight-driven clocks in the late medieval period, and the first spring-driven clocks in the 15th century. However, these clocks were inaccurate. Huygens's invention of a pendulum clock, and Robert Hooke's development of a balance spring mechanism around the same time, produced the first reliable chronometers.

JOHN HARRISON

SOLVER OF THE "LONGITUDE PROBLEM"

ENGLAND 1693–1776

WHAT STARTED AS A HOBBY for John Harrison turned into his life's work – designing and making an accurate timepiece that would maintain its reliability in extremes of weather and temperature, unaffected by the motion of sea transport. In his quest for this ideal chronometer, Harrison introduced many innovations to the science of clockmaking.

A LIFE'S WORK

- In 1713, Harrison builds his first clock, a longcase **pendulum clock made entirely of wood**
- In his twenties, he invents the almost frictionless **"grasshopper" escapement** and the **"gridiron" pendulum**, which compensated for changes in temperature
- Determined to win the **generous prize** offered by the **Board of Longitude**, he starts work on his **first marine chronometer** in 1730, and devotes the rest of his life to the task
- He finds an ally in **King George III** when the Board refuses to award him the prize money for a version of his **H4 marine chronometer**
- As **choirmaster at Barrow parish church,** he develops an interest in **bellringing and acoustical theory**, and goes on to write a paper on the manufacture and tuning of bells

Born in the village of Foulby, Yorkshire, John Harrison learnt his trade as a carpenter from his father, but, as a boy, discovered an interest in clocks. His family moved to Barrow in Lincolnshire when he was seven, and, during his childhood there, he earned some pocket money repairing clocks and watches.

Discovering an aptitude for the work, he taught himself the necessary skills, and, by the age of 20, felt confident enough to start building clocks. Using his skill as a joiner, he made the first of these entirely of wood, carving the intricate parts from oak and lignum vitae – a hard wood that is so oily that it does not need any lubrication when used for bearings. Working with his brother James, he built several pendulum clocks – some of the most accurate of the period. Three of these have survived, and are still in working order.

What marked Harrison out as an engineer, rather than a skilled craftsman, was his understanding of the principles of clockmaking. This allowed him to make improvements to the traditional mechanism. The first of these, which he would later incorporate into his precision marine chronometers, was the ingenious "grasshopper" escapement for controlling the rate of movement of the clock's gears. He also solved the problem of the effects of temperature on the length of the pendulum with his "gridiron" pendulum, made from brass and iron rods whose different rates of expansion cancelled each other out – another concept that was to prove useful in his later designs.

TRIAL AND ERROR

In the 1720s, Harrison became increasingly interested in the "Longitude Problem". A surge in worldwide trade and exploration had highlighted the need for an accurate method of determining a ship's position. It was relatively easy to measure latitude, but a reliable way of establishing longitude was yet to be found (see box, opposite). The British government set up a Board of Longitude, which offered a prize of up to £20,000 to anyone who could provide a solution to the problem.

One school of thought was that this could be achieved by astronomical observations – the lunar distance method. This method was based on the fact that the moon's position changes predictably in relation to stars. Thus, the moon's motion could be regarded as a clock to calculate Greenwich time. A seafarer could calculate his longitude by measuring the position of the moon and a nearby bright star, and ascertain the time that this would correspond to in Greenwich. He would calculate his longitude by determining the difference between the Greenwich time and the local time. However, this method could not be used in cloudy weather and Harrison believed that a more accurate measurement could be made with a reliable chronometer that would keep the time of a reference location, which could then be compared to local time. The problem, of course, was designing and building a timepiece that was accurate and would also be able to withstand the stresses of a long sea voyage.

By 1730, Harrison was confident that he could make a marine chronometer that was up to the task, and started work on the design. Seeking some financial assistance for the project, he took his ideas to the Astronomer Royal, Edmond Halley (1656–1742), who introduced him to George Graham (1673–1751), Master of the Clockmakers Company. Graham was so impressed that he personally lent Harrison money and became a champion of his work.

The first of Harrison's sea clocks – now known simply as H1 – was completed in 1735. Previous attempts by Christiaan Huygens (see pp.80–81)

FIRST MARINE TIMEKEEPER
This is a front view of Harrison's spring-driven chronometer, known as H1, which won the promised prize of £23,065. It has four dials, and, at the top, a motion-compensating mechanism.

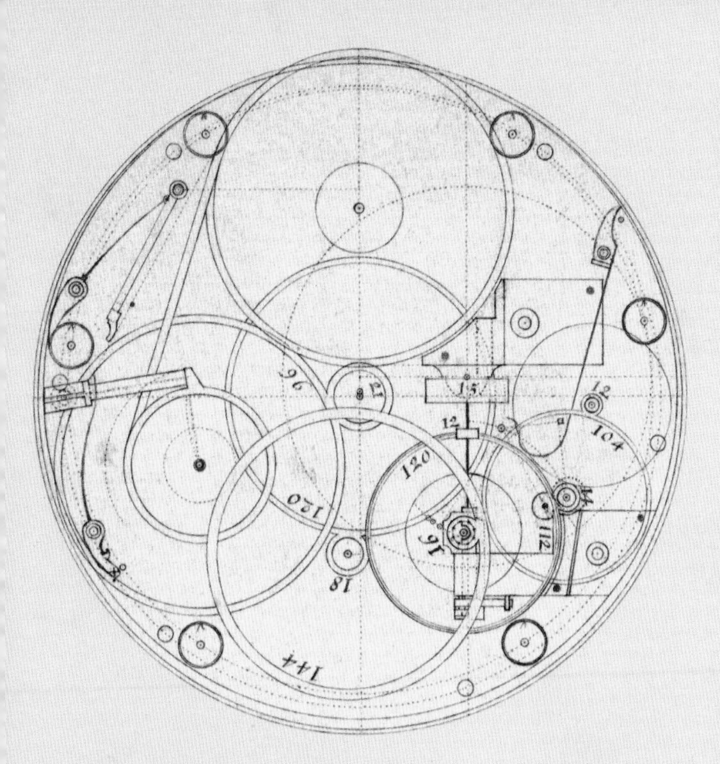

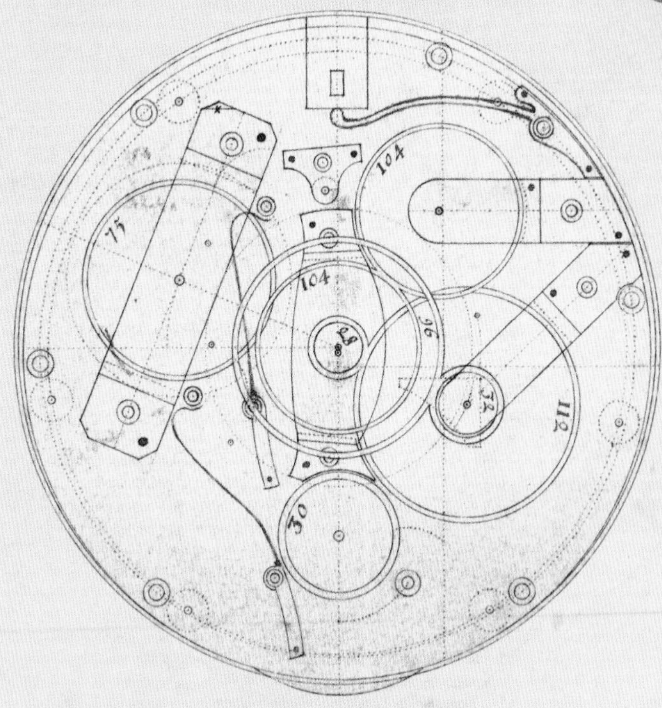

DESIGN DRAWINGS FOR THE "SEA WATCH" H4
The fourth of Harrison's marine chronometers, H4 was based on a pocket-watch design. It incorporated many innovative features that made it both robust and accurate enough to be used by mariners to calculate longitude.

and English clockmaker Henry Sully (1680–1729) had shown the shortcomings of pendulum clocks on board a ship. Harrison therefore opted for a linked balance mechanism to compensate for the effects of motion. H1 performed remarkably well when Harrison carried it on a trial voyage to Lisbon, but this journey did not meet the Board of Longitude's specifications of a transatlantic trial. Rather than ask for a second voyage, Harrison persuaded the Board to grant him £500 to build an improved version of the clock.

"LONGITUDE PROBLEM"

Crossing the oceans in the 18th century was a rather hit-and-miss affair, as navigating was done largely by the stars. A ship's latitude could be easily determined from the altitude of the sun using a sextant, but longitude was not so simple. The relationship between time and longitude offered a solution – as the Earth rotates 360 degrees each day, 15 degrees of longitude corresponds to a time difference of one hour. If a means of maintaining a fixed reference time to compare with local time could be found, the calculation would be a simple matter. The only obstacle was devising a sufficiently accurate marine chronometer. Harrison's H4 provided a solution.

He worked on the new version, named H2, from 1737 to 1740, but, as he progressed, he realized that there were fundamental flaws in the design. He negotiated a further £500 payment from the Board, and spent the next 19 years resolving the problems and introducing innovations. These included a bimetallic strip to compensate for changes of temperature on the balance spring, and an improved anti-friction device – the caged roller bearing. The new model, H3, however, proved to be a disappointment, and failed to meet the requirements of the Board.

BATTLE FOR RECOGNITION

Even before finishing H3, Harrison knew that he had been on the wrong track. He discovered that Thomas Mudge (1715–94), who had succeeded Graham as Master of the Clockmakers Company, was making precision pocket watches that rivalled the accuracy of his clocks. Harrison realized that a "sea watch" was the answer. He had already designed a pocket watch for his own use, and set about applying the same principles to a slightly larger, more robust version of it.

The resulting marine watch, now known as H4, was Harrison's masterpiece. The design included a completely new escapement mechanism with diamond pallets, and incorporated devices to compensate for changes of temperature and to keep the watch running while being wound. His son William took it on a trial voyage to Jamaica in 1761, where the watch was found to be only five seconds slow. This represents a navigational error of only 1.25 minutes of longitude, or one nautical mile. The Board, however, was not convinced, and demanded a second trial. This infuriated Harrison so much that he took the matter to Parliament. Eventually, however, he agreed to another transatlantic trial.

Once again, the watch proved its worth, being only 39 seconds slow on arrival in Barbados. But a rival claim was being made by Nevil Maskelyne (1732–1811), using the lunar distance method with some success. As Maskelyne was appointed Astronomer Royal, and an ex officio member of the Board, soon after Maskelyne's voyage, the awarding of the prize became a controversial political topic. Harrison felt, with justification, that he had earned the £20,000, but Maskelyne ruled against him. The argument raged on for years, during which time Harrison designed a second marine watch. Frustrated by the lack of recognition, he took the design of the second watch to King George III, and earned himself a powerful ally in persuading Parliament to award him a further payment of £8,750 in 1773.

Harrison was now in his eighties, and had lived comfortably on the income from his marine chronometers for much of his life – he recieved a total of £23,065 for his work on chronometers – but had been denied the honour of winning the Board's prize. Captain James Cook took a copy of H4 on his second (1772–75) and third (1776–79) voyages, and came back delighted with its reliability, but this endorsement came too late for Harrison, who died in 1776.

BREAKTHROUGH INVENTION

MARINE CHRONOMETER

The English horologist John Harrison devoted much of his life to the development of an accurate marine chronometer that would solve the "longitude problem". To this end, he made a series of seagoing clocks (now known as H1, H2, and H3) between 1730 and 1760. The breakthrough came, however, when he hit upon the idea of a pocket watch rather than a clock, and, in 1759, he completed his masterpiece, the "sea watch" H4.

It was while working on H3 that John Harrison (see pp.82–83) decided to change tack. Inspired by the work of Thomas Mudge (1715–94), Master of the Clockmakers, he had designed a pocket watch for his own use (made by John Jeffreys in 1753 to his specifications), and soon realized that this was the way forward. He approached the Board of Longitude in 1755 with the idea of a marine watch, "having good reason to think from the performance of one already executed ... that such small machines may be render'd capable of being of great service with respect to the Longitude at Sea".

The sea watch represented a fundamental shift in his approach to the construction of marine chronometers – not only because of its size, but also in terms of the design of its movement. It incorporated features such as a bimetallic strip to compensate for changes in temperature, and a "maintaining power" mechanism to keep it going true during winding. But because of its size, he had to dispense with the innovations of his marine clocks – the anti-friction wheels, caged bearings, and grasshopper escapement. He discovered that diamond had the hard-wearing and low friction properties he was looking for, and made pallets of diamond, cut with a precision that was remarkable for the time, in a completely new design of escapement, with a remontoire (secondary spring). The movement, housed in a paired silver case, and protected by a decorative engraved plate, ran at five beats per second, and when fully wound would run for 30 hours. At 13cm (5in) in diameter, and weighing 1.45kg (3lb), it was larger than a normal pocket watch, but smaller than his previous marine timekeepers. On its first sea trial, to Jamaica, H4 lost a mere five seconds, and later proved its worth when Captain James Cook took a copy of it on his second and third voyages – but this still failed to win Harrison the Board's £20,000 prize earmarked for anyone who could find a reliable way of establishing longitude.

THE WATCH HAS BEEN OUR FAITHFUL GUIDE THROUGH ALL THE VICISSITUDES OF CLIMATE

CAPTAIN JAMES COOK

THE GREAT CLOCKMAKER
Harrison posed for this famous portrait in 1767, holding the watch made for him 14 years earlier by John Jeffreys. This watch inspired Harrison to build his world-beating "timekeeper for longitude" H4.

WATERWAYS AND HARBOURS

GROWTH IN TRADE CREATED THE NEED FOR IMPROVED TRANSPORT LINKS TO CITIES AND PORTS IN EUROPE. WATERWAYS PROVED TO BE ONE SOLUTION, AND, IN THE 18TH CENTURY, AN INTENSE PERIOD OF CANAL-BUILDING BEGAN TO MEET THE DEMANDS OF INDUSTRIALIZATION.

GRAND CANAL OF CHINA
Work began on the first major man-made waterway connecting the Yangtze and Huai Rivers in 486BCE. Its final, total length is 1,776km (1,103 miles).

For centuries, goods had been transported by barges on rivers, and cities and ports tended to grow on the banks of these navigable natural waterways. The notion of man-made waterways, or canals, was nothing new – the Grand Canal in China, for example, dates back to the 5th century BCE. In Europe, however, canals were constructed primarily as a means of irrigation, and only began to be considered for transport in the late Middle Ages. Canal systems in the Netherlands and in Milan, Italy, were found to be useful for boats and developed into waterways, but perhaps the first to be specifically designed for transport was the Stecknitz Canal in Germany, built in 1398 to connect the River Elbe to the Baltic Sea. The scope of these early canals was limited, however, and they were incapable of sustaining the volume of traffic necessary for trade in the 17th century. They were also restricted to low-lying areas.

THE FIRST ADVANCED CANALS

The breakthrough came in France, with the construction of the Canal de Briare to link the Loire Valley with Paris on the River Seine. Its designer, Hugues Cosnier (1573–1629), pioneered the use of pound locks to take the waterway across elevated land between the two river valleys, and inspired a new era of European canal-building. Following the success of this

BARGE DEVELOPMENT
As traffic on the canals increased, specially designed barges for carrying cargo evolved. At first sail-driven or horse-drawn, they were later fitted with steam engines.

SMEATON'S EDDYSTONE LIGHTHOUSE
John Smeaton pioneered modern stone-built lighthouse design in the 1750s to replace the old wooden structures. The previous structure, by Henry Winstanley, was destroyed in a storm.

INDUSTRIAL BRIDGE
Canal-building in the 17th and 18th centuries brought innovation in engineering techniques, such as bridge and aqueduct building and complex systems of locks.

project, Pierre-Paul Riquet (see pp.88–89) constructed the Canal du Midi in southwest France to connect the Mediterranean coast with the Atlantic, bypassing the long sea route around the Iberian peninsula.

> **As the canal system grew, specialized boats for carrying goods evolved. Long boats or barges were designed to maximize the load that they could carry on narrow waterways, and these were sometimes towed together in "trains".**

"CIVIL ENGINEERS"

The advantages of canal transport are obvious. A horse can carry about half a tonne, or pull 10 tonnes (9.8 tons) in a cart, but it can tow 100 tonnes (98 tons) in a barge. Merchants realized the commercial opportunity and commissioned engineers to build waterways across Europe. In Britain, where industrialization was progressing faster than elsewhere, this led to a massive programme of canal-building, which lasted from the 1750s until the advent of the railway in the 19th century.

Because canals were a new venture, there were no engineers with experience of building them, and the first were designed using the combined expertise of surveyors, hydraulic engineers, and millwrights, such as James Brindley (see pp.92–93). However, they quickly learnt the techniques of building bridges, tunnels, aqueducts, and embankments, and began introducing their own innovations. From these early canal builders, a new elite breed of engineer emerged, including John Smeaton (see pp.100–01), William Jessop (see pp.94–95), John Rennie (see pp.96–97), and Thomas Telford (see pp.160–63), for which Smeaton coined the term "civil engineer". Eventually, the success of British canals inspired other countries to follow suit, and in the newly independent United States the development of a canal system, spearheaded by the work of civil engineer Benjamin Wright (see pp.98–99), was seen as a step towards becoming a major trading nation. By the end of the 18th century, canals had become arteries for carrying goods and raw materials to and from cities and ports, and remained an important means of transporting non-perishable goods, even after the arrival of the railway. A second era of canal-building was heralded in the second half of the 19th century with some larger-scale projects for seagoing vessels, such as the Suez Canal (completed in 1869) and the Kiel Canal (built 1887–95), culminating in the Panama Canal in 1914.

BEACONS FOR BUILDINGS

The enormous growth in trade that was made possible by canals put increased pressure on docks in many ports, and civil engineers were in demand to help improve and expand their facilities. It was also important to maintain military facilities, especially in Britain, as it faced the threat of a Napoleonic invasion. Many of the techniques of waterway engineering were also applicable to dock and harbour design, and canal builders, such as Jessop and Rennie, were often called upon to oversee the work.

Smeaton was involved with harbour engineering, too, but he also specialized in another area vital to maritime trading – lighthouse design and construction. The concept of a beacon marking the entrance to a harbour or warning of hazards to shipping had existed since Classical times, but the construction of purpose-built buildings only began in the 16th century. Smeaton made his name by replacing the second Eddystone Lighthouse with a granite structure, marking the beginning of the modern age of lighthouse design.

The establishment of civil engineering coincided with the development of the steam engine, and the techniques of building bridges, tunnels, embankments, and cuttings were to provide the infrastructure for the next revolutionary form of transport – the railway.

CANAL BUILDER
This statue of Pierre-Paul Riquet stands in Béziers, his home town. Riquet's great work, the Canal du Midi, connects the Mediterranean Sea with the Bay of Biscay.

SETTING THE SCENE

- The **technology of irrigation and water management**, including sluices and locks, is **adopted by canal builders** in the 17th century, making navigable waterways a viable proposition.
- The **first canals in Europe** are regarded as a **commercial opportunity** by entrepreneurs to transport goods to and from cities and ports.
- **Ground-breaking projects**, such as the Canal de Briare and the Canal du Midi in France, show the potential of waterway transport, and **spark a period of canal-building across Europe**.
- **New techniques** of constructing aqueducts, tunnels, cuttings, and embankments are **developed to take canals across difficult terrain**, and the profession of **"civil engineering" emerges**.
- As the Industrial Revolution gets under way in Britain, **engineers build a system of waterways** to cater for **the demand for improved transport links** and expand dock facilities at ports.
- Trinity House is founded in 1514 to oversee British lighthouses, but **modern lighthouse-building begins in the 18th century** with the construction of stone lighthouses.

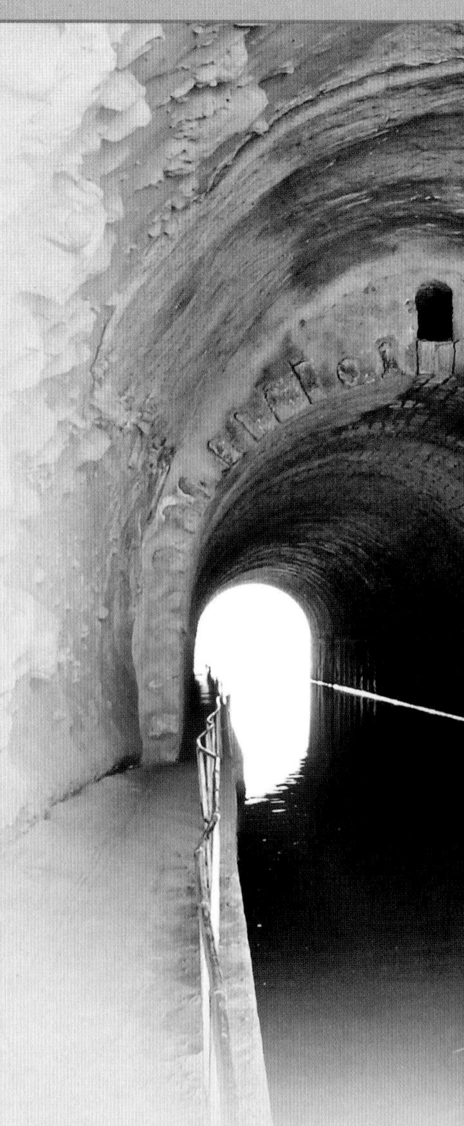

PIERRE-PAUL RIQUET

FIRST OF THE GREAT EUROPEAN CANAL BUILDERS

FRANCE C.1609–80

A WEALTHY ARISTOCRAT and astute businessman, Pierre-Paul Riquet boldly took up the challenge of building a canal across the southwest of France. Unlike others who had considered the project before, however, he had money, influence, and the know-how to make the dream a reality. He was responsible for the Canal du Midi connecting the Mediterranean Sea with the Bay of Biscay.

Pierre-Paul Riquet was born in Béziers, in the heart of the Languedoc region of France. His family had its roots in the Italian and Provençal *haute-bourgeoisie* ("upper middle class") in Provence, and his father was a well-respected local notary, public prosecutor, and businessman. Pierre-Paul, however, showed little interest in his conventional middle-class education, preferring instead mathematics and engineering, and even speaking the local *langue d'oc* – a dialect of French – in preference to French. However, he eventually gave in to his father's wishes and went into an administrative job as a *fermier général* ("tax-farmer") for the area, a position secured for him by his godfather. This involved assessing and collecting the salt taxes, a duty he carried out with typical efficiency, and which gained him some standing in the local community, but more importantly, got him noticed by the central authorities. His diligence was rewarded by royal permission to levy and administer his own taxes, which helped to make him a wealthy man.

In the 1630s, he married and started a family, and it seemed as if he would settle down to life as a country gentleman at his home in Mirepoix. Despite his wealth and various business interests, however, he was constantly on the lookout for a new project to occupy his mind. In 1648, he moved to Revel, and, four years later, acquired the Château de Bonrepos, where he planned the project that was to become his life's work.

A LIFE'S WORK

- As ***fermier général*** ("tax-farmer") of **Languedoc-Roussillon**, Riquet is appointed inspector of the *gabelle* ("salt tax") in 1630, and is later given **royal permission to levy his own taxes**
- From 1648 onwards, Riquet explores the area around Revel with an engineer, Pierre Campmas, and **formulates the idea of a canal linking the waterway in Toulouse with the Mediterranean Sea**
- He **becomes Baron de Bonrepos** in 1662, and the owner of large areas of farmland and forest, with residences in Revel and Toulouse, where he is **also the Royal Judge**
- The official **inauguration of the Canal Royal de Languedoc takes place** in May 1681, only seven months after Riquet's death

A GREAT IDEA CONCEIVED

Situated as he was in the southwest of France, midway between the Mediterranean and the Atlantic coasts, he was aware of the unrealized plans that had been made to connect the two with a navigable waterway. Toulouse, not far from where he lived, was already linked to the Bay of Biscay through the River Garonne, and it seemed to him that it should be possible to construct a waterway from there to the Mediterranean coast. From this germ of an idea, Riquet began to devote his time to assessing the feasibility of the project. He enlisted the assistance of a surveyor and engineer, Pierre Campmas, to study the area, and the two spent much of the 1650s planning the route and, most importantly, searching the Montagne Noire area northwest of Toulouse for a source of water to feed the proposed waterway.

In 1662, he felt confident enough to approach Louis XIV's finance minister, Jean-Baptiste Colbert, for state backing for the scheme. Having recently been ennobled as Baron de Bonrepos, Riquet was in a good position to gain support from the king, who saw the advantages of bypassing the long sea route round Spain and Portugal. In 1665, the project got official approval. A commission was also appointed, which approved a healthy budget for the project the following year.

Although now in his sixties, Riquet took charge of the building of the canal, overseeing a team of engineers. The first major work was to dam the River Laudot at a point on the

CHALLENGE OF THE MALPAS TUNNEL
The first of the great European canals, the Canal du Midi presented Pierre Paul Riquet with completely new engineering problems, such as the construction of a tunnel to avoid crossing the River Aude.

Montagne Noire that he and Campmas had chosen, about 20km (12 miles) from the highest point of the proposed route of the canal. A major feat of engineering in its own right, the dam was the largest in Europe, and created a reservoir – the massive Bassin de St Ferreol – that could feed the waterway both westwards to Toulouse and southeastwards to the coast, as well as supply water to the town of Revel via a specially constructed channel.

CONSTRUCTION BEGINS

Work then began on the canal itself, following a meandering course through the often mountainous terrain, in order to avoid building too many locks. In all, 91 long, wide, oval locks were used, and the staircase of eight locks (today only six are in use) at Fonserannes, just outside Riquet's home town of Béziers, is an impressive achievement. Unsurprisingly, the project ran hugely over budget, but he had found an ally in Chevalier de Cerville, a military engineer who persuaded the commission to provide the extra funding. Riquet also invested a great deal of his own money in the canal, which, despite his wealth, put him into severe debt.

The completion of the canal was never in doubt, as it had become a source of national pride. From the new canal port of Revel, the canal made its way up to Toulouse, and down to the Mediterranean town of Sète on the south coast of France. Work on the 240-km (150-mile) waterway was all but complete when Riquet died in October 1680. The Canal Royal de Languedoc, as it was then called, was officially opened on 15 May 1681, and hailed as a triumph for France. The first of the great European canal builders, Riquet became a national hero.

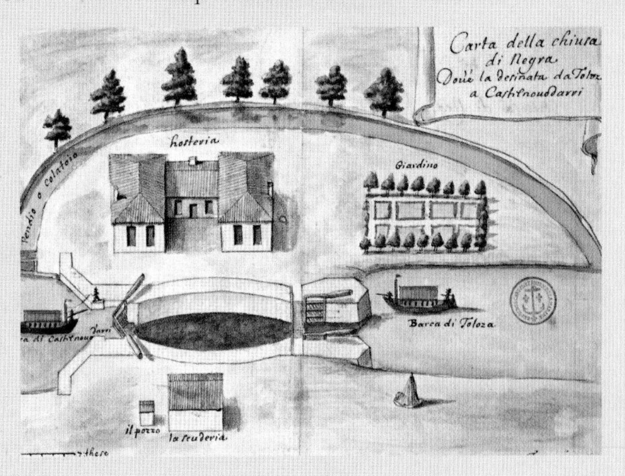

LOCK ON THE CANAL DU MIDI
This drawing shows the distinctive oval-shaped locks designed by Riquet, which prevented the side walls from collapsing.

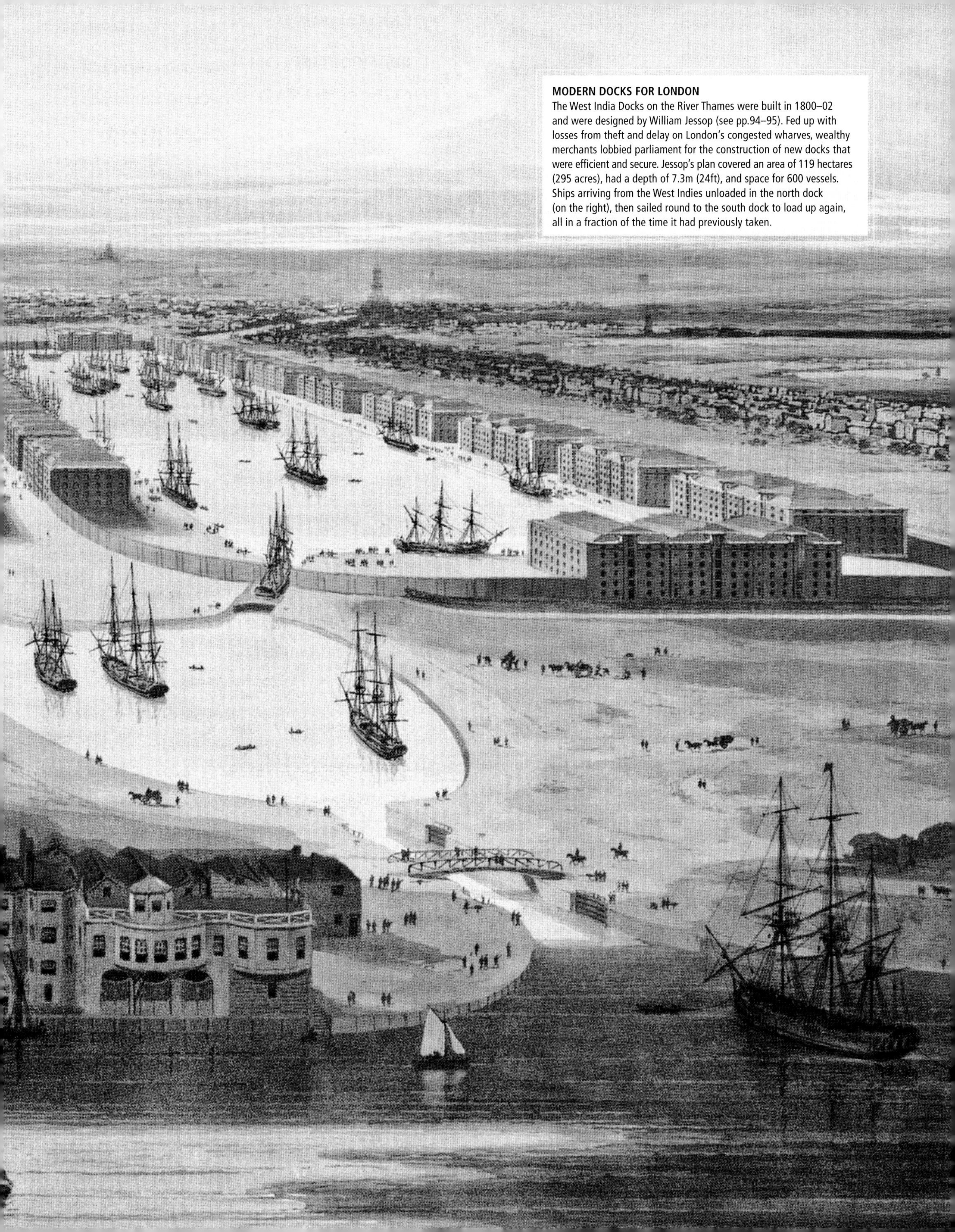

MODERN DOCKS FOR LONDON
The West India Docks on the River Thames were built in 1800–02 and were designed by William Jessop (see pp.94–95). Fed up with losses from theft and delay on London's congested wharves, wealthy merchants lobbied parliament for the construction of new docks that were efficient and secure. Jessop's plan covered an area of 119 hectares (295 acres), had a depth of 7.3m (24ft), and space for 600 vessels. Ships arriving from the West Indies unloaded in the north dock (on the right), then sailed round to the south dock to load up again, all in a fraction of the time it had previously taken.

JAMES BRINDLEY

DESIGNER OF THE ENGLISH CANAL NETWORK

ENGLAND 1716–72

Using the engineering skills he learnt as a millwright in the English midlands at the beginning of the Industrial Revolution, James Brindley went on to become one of the first of the great British canal-builders. He designed a network of waterways that would link the growing industrial heartland with the coastal ports, and invented the process of clay puddling.

A LIFE'S WORK

- Brindley gains **the nickname "the Schemer" for his tinkering with machinery** in his business as a millwright, and **designs a pump for draining coal mines**
- He **installs several atmospheric engines**, and obtains **a patent for an improved boiler design**
- In 1758, he is hired to **survey the area** between Liverpool and the potteries of the River Trent
- Makes his name as the **engineer of the Bridgewater Canal**, completed in 1761, and as the visionary of the **"Grand Cross", linking the rivers Mersey, Trent, Severn, and Thames** with a network of canals
- After getting **soaked while surveying** near Etruria, Staffordshire, in 1772, Brindley **falls ill and dies** shortly afterwards

The eldest of seven children, James Brindley was born in Tunstead, Derbyshire. When he was about ten years old, the family moved to a farm that his father had inherited near Leek in Staffordshire. As a boy, Brindley showed an aptitude for working with machines, and rather than stay on the farm, soon became apprenticed to Abraham Bennett, a wheelwright and millwright, near the mill town of Macclesfield in Cheshire.

Brindley learnt his trade well, working in local silk and paper mills, and after his apprenticeship stayed on with Bennett, virtually running the business for him. After Bennett's death in 1742, Brindley moved back to Leek, and set up on his own as a millwright. Thanks to his skill in installing, maintaining, and, above all, improving machinery in the mills, he soon had a thriving business, and expanded into a second workshop in nearby Burslem, Staffordshire.

This was an exciting period for Brindley's trade. The steam engines designed by Thomas Newcomen (see pp.110–11) were enormously increasing coal production, and in the increasingly industrial midlands, Brindley was ideally placed to take advantage of the times. He had a good understanding of the new machines, and, during the 1750s, he made his name installing atmospheric engines in the mills, and suggesting modifications to improve their performance.

NETWORK OF CANALS

A turning point came when Brindley designed a scheme for draining a colliery near Manchester, proving his skill as a hydraulic engineer. The project involved building a pump that was powered by a water wheel, driven by water diverted through a tunnel from the River Irwell. On the strength of this and similar projects, in 1758, Brindley was commissioned to work on a proposed canal between the potteries on the River Trent and Liverpool, via the River Mersey. His survey was impressive, and came to the attention of the Duke of Bridgewater, who was working on plans for a canal to transport coal from his mines to the growing industrial town of Salford.

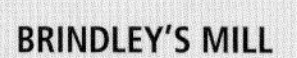

BRINDLEY'S MILL
This water-driven corn mill built by Brindley in Leek, Staffordshire, has been restored to working condition, and now houses the James Brindley Museum.

In 1759, Brindley was hired to help with the Duke's scheme for the Bridgewater Canal, and came up with a more ambitious route, taking it into Manchester, where it would link up with his proposed Trent and Mersey Canal. Brindley was now thoroughly immersed in his work on canals, and had the foresight to see beyond the individual schemes to an integrated system of waterways.

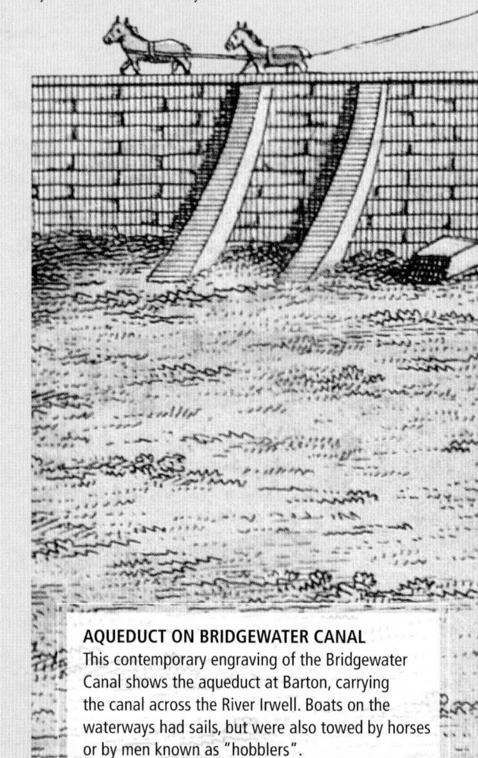

AQUEDUCT ON BRIDGEWATER CANAL
This contemporary engraving of the Bridgewater Canal shows the aqueduct at Barton, carrying the canal across the River Irwell. Boats on the waterways had sails, but were also towed by horses or by men known as "hobblers".

The Bridgewater Canal was completed in 1761 to Brindley's design, and was later extended to meet up with the River Mersey at Preston Brook. Although work on the Trent and Mersey Canal was delayed until 1766, Brindley was soon regarded as the foremost canal engineer in the country. He had enough expertise to realize his vision of a "Grand Cross" of canals, connecting the industrial midlands with the seaports via the Rivers Mersey, Trent, Severn, and Thames. Companies were set up to finance and administer a number of different canal projects, and industrialists were keen to invest in these exciting schemes. Appointed chief engineer on most of them, and as a consultant on many of the others, Brindley was able to coordinate and integrate the projects.

Despite the many administrative responsibilities, Brindley remained very much a hands-on engineer. As well as surveying and designing ten of the new canals – including the Coventry Canal, the Oxford Canal, and the Staffordshire and Worcestershire Canal – and overseeing their construction, he was actively involved in the building process, and personally trained a team of engineers to assist him. Canal-building on such a scale had not been seen in Britain – or anywhere in the world – before, and Brindley was pioneering new construction techniques. Partly because of financial constraints, he had opted to cut comparatively narrow channels, following contour lines (lines that join points of equal elevation) to avoid having to dig cuttings and embankments. But when necessary, his designs were bold and ambitious. He built several impressive aqueducts, including one carrying the Bridgewater Canal across the River Irwell at Barton, and a tunnel – the first of its kind in Britain – from the mines at the head of the Trent and Mersey Canal at Harecastle. Among his innovations was the process of clay puddling (see box, right) to make the channels watertight.

DIVERSE INTERESTS

At the same time as his involvement with the burgeoning canal network, he continued to run his millwright business, installing and maintaining atmospheric engines. He also expanded his interests to become a partner in his brother's pottery, and, with other members of the family, bought the Turnhurst estate and ran its colliery, building a system of tunnels to carry coal to the Trent and Mersey Canal.

CLAY PUDDLING

Along with the problem of finding a suitable route for a canal, and providing a constant supply of water to feed it, canal engineers had to overcome the problem of leakage and seepage from the sides and bottoms of the channels. The locks, tunnels, and aqueducts could be made watertight by lining them with bricks or metal, but this was impracticable for long open stretches. Brindley's solution was simple and remarkably effective. He lined the channels with successive layers of a clay and water mix, compacted with flat tools called punners, in a process known as clay puddling.

Inevitably, the strain of work began to take its toll. Towards the end of the 1760s, Brindley's health began to suffer. To make matters worse, he had developed diabetes, but he continued to work as he had always done. He was reluctant to delegate any of his tasks, and set himself a punishing schedule, sometimes travelling long distances between his various places of work. In 1772, after getting drenched while surveying a proposed canal in Straffordshire, Brindley succumbed to an illness from which he did not recover. Although he lived only until his fifties, he left an enormous legacy in the shape of a network of canals that helped accelerate the pace of the Industrial Revolution, and lay the foundation for transport systems in the following centuries.

WILLIAM JESSOP

MASTER-BUILDER OF CANALS AND TRAMWAYS

ENGLAND 1745–1814

CIVIL ENGINEER WILLIAM JESSOP emerged as the foremost canal-builder in a period of "canal mania" in England in the second half of the 18th century. This period not only witnessed some of Jessop's most important canal schemes, but also his expansion of the docks in London and Bristol, and his pioneering use of iron rails for tramways, the precursors of the steam railway.

A LIFE'S WORK

- Jessop learns his trade by **assisting Smeaton** in the building of the Calder and Hebble and the Aire and Calder canals in Yorkshire
- He **makes his name as a canal engineer** with the Grand Canal of Ireland, which features an aqueduct across the River Liffey and an ambitious embankment crossing the Bog of Allen
- Joins Outram and others as a **co-founder of Butterley Iron Works**, which manufactures ironwork for canal and railway projects
- Builds a **tramway for horse-drawn vehicles** from Wandsworth in London to Croydon as part of the Surrey Iron Railway, arguably the **first public railway**
- Serves as an alderman in Newark-on-Trent, and is **twice elected mayor** of the town in 1790–91 and 1803–04

Jessop was born in Devonport, Devon, to a shipwright, Josias Jessop, whose duties included the maintenance of the wooden lighthouse on the nearby Eddystone Rock. In 1755, engineer John Smeaton (see pp.100–01) was commissioned to design a granite lighthouse for the site, and, when construction began the following year, Josias assisted in overseeing the construction of the building. Smeaton became a family friend, and, when Josias died in 1761, he took young William under his wing.

Jessop moved in with Smeaton's family near Leeds and became apprenticed to his engineering business. Jessop could not have had a better mentor, as Smeaton was one of the most respected civil engineers of the period, whose work included bridge- and canal-building, as well as lighthouses and harbour works.

Once he had completed his apprenticeship, he continued to work as Smeaton's assistant, taking an active part in the design and construction of canals in Yorkshire and Scotland. However, by 1772, Smeaton felt that it was time for his protégé to take on a project of his own.

MASTER CANAL-BUILDER

Smeaton had been approached by a company in Ireland to complete the Grand Canal of Ireland, which had run into difficulties and needed a skilled engineer. He sent Jessop to oversee the project. This was a chance for the young man to prove his worth, and he rose to the challenge. After surveying the area, he suggested an ambitious change to the proposed route, which called for an aqueduct at Leinster to take the canal over the River Liffey, and an embankment across the Bog of Allen. The plan was approved, and, under his supervision, work on the canal began. He remained involved in the project for a number of years, and even after settling in Newark-on-Trent, Nottinghamshire, in about 1784, made frequent visits to Ireland until he was certain the project could be completed successfully. The Irish canal secured his reputation. By the end of the 1780s Smeaton had retired and Jessop was acknowledged as the most able canal engineer in the country.

At this time, England was entering a period of what became known as "canal mania", as hundreds of companies were set up to build canals and cash in on the "canal boom". Jessop was, naturally, much in demand, and could afford to pick and choose his projects. The first of these was as chief engineer of the Cromford Canal, for which (in contrast to his design for Ireland's Grand Canal) he earned a reputation for simplicity of design and cost awareness. In fact, when part of the Derwent Aqueduct collapsed, he was so concerned for its financial viability that he paid for the repair himself.

Jessop's honesty and frugality made him even more popular, and, throughout the 1790s, he was involved in several important canal-building schemes, such as the prestigious Grand Junction Canal linking James Brindley's Oxford Canal (see pp.92–93) to the River Thames at Brentford, a distance of about 145km (90 miles). It was a huge challenge for Jessop, as it involved several river crossings requiring aqueducts, and the digging of two tunnels. Other schemes included the Grantham Canal, fed entirely from reservoirs, and the Ellesmere Canal, for which he employed a young and inexperienced Thomas Telford (see pp.160–63) as assistant engineer – in fact trusting him with much of the design.

ANOTHER DIRECTION

Towards the end of the century, Jessop began to take an interest in projects other than canals. He was appointed chief engineer for the expansion and improvement of the West India Docks in

SURREY IRON RAILWAY

With the improved rails developed by Jessop and companies such as Butterley Iron Works, commercial tramways became a viable alternative to canals. The Surrey Iron Railway in England was part of a large project to connect London to the south coast, and it opened to goods traffic in 1803. Hauliers paid for the use of the standard 1.5-m (5-ft) gauge tracks, but had to provide their own wagons and horses; the advent of the steam locomotive prompted railway companies to provide both the railway and the train.

HORSE-DRAWN COAL WAGONS ON THE SURREY IRON RAILWAY MAKE THEIR WAY OVER A ROAD.

MODERN HARBOUR
To improve facilities at Bristol docks, Jessop designed a larger harbour that included the Cumberland Basin, connected by locks to the tidal River Avon, and a non-tidal area, which became known as the Floating Harbour (shown here). It opened in 1809.

East London, and later worked on designs for the harbours in Bristol and Shoreham-by-Sea. But perhaps his most radical departure from canal-building was yet to come; it was also the one with the most significance for the future.

RAILWAY PIONEER

While working on the Cromford Canal, Jessop had befriended land surveyor Benjamin Outram (1764–1805), who was planning to set up in business as an engineer himself. In partnership with a local businessman, Outram had acquired a site at Butterley, and he invited Jessop to join him in founding an ironworks company. Jessop jumped at the opportunity, as he was an enthusiastic advocate of iron as a construction material, and this arrangement would give him a ready supply of ironwork made to his specifications. Many of the canals Jessop worked on were meant for carrying coal from mining areas to industrial cities, and, often, the coal was transported from the mines to the canal by horse-drawn wagons on a wooden track. Jessop suggested replacing these tracks with iron rails. The idea caught on, and soon Butterley Iron Works was busy making rails for these new "railways".

By the turn of the century, the idea of a commercial public tramway was being discussed, and Jessop was consulted on the viability of a scheme to build one from London to the south coast at Portsmouth. He built the first section of this tramway, the Surrey Iron Railway, from Wandsworth to Croydon, in 1801–02. This was extended the following year to Merstham, making a total distance of 29km (18 miles). In spite of these achievements, however, Butterley Iron Works was left in a state of confusion when Outram, a competent engineer but an inadequate administrator, died suddenly aged 41, and intestate, in 1805.

At the very end of his career, Jessop was engineer for the Kilmarnock and Troon Railway, which opened in 1812 as a horse-drawn tramway. But, in 1817, just four years after his death, it became the first railway in Scotland to carry a steam locomotive.

CREDIT DUE

Modest and unassuming, Jessop's contributions to civil engineering and the development of the early railway are sometimes overlooked today. In fact, many of his achievements are credited to those who served as his assistants. In his time, however, Jessop was highly respected, particularly by his fellow engineers, and twice served as town mayor of Newark-on-Trent, Nottinghamshire.

JOHN RENNIE

BUILDER OF BRIDGES ACROSS THE RIVER THAMES

SCOTLAND 1761–1821

A MILLWRIGHT WHO MOVED from Scotland to London to work as a mechanical engineer, John Rennie was to become one of the greatest civil engineers of his day. Following a successful career designing machinery, he started building canals, and later became famous for his bridges across the River Thames and for his work on docks, including the Plymouth Breakwater scheme.

FLIGHT OF LOCKS
The flight of 29 locks between Rowde and Devizes, Wiltshire, takes the Kennet and Avon Canal up a total of 72m (237ft) in 3km (2 miles) from the Avon Valley to the Vale of Pewsey.

The youngest of nine children, John Rennie was born on a farm named "Phantassie" in what is now East Lothian, Scotland. His father died when John was only five, leaving his eldest son, George, as head of the family and running the farm. John was educated at the local school, but learnt his early engineering skills as a boy working with the millwright and inventor of the threshing machine, Andrew Meikle (1719–1811), who lived nearby. At the age of 14, he went to high school in Dunbar, East Lothian, where he performed so well with his studies in sciences and mathematics that he was asked to stay there as a faculty member.

A LIFE'S WORK

- Rennie's interest in machinery is fostered when, aged 12, he **works with Andrew Meikle, windmill designer and inventor of the threshing machine**
- After graduating from Edinburgh University, he **travels through England to look at sites of modern engineering**, and meets **James Watt**
- At Watt's invitation, he moves to London to **install machinery in the Albion Mills**
- During the 1790s, he **shifts his attention to civil engineering**, working on canals, docks, and bridges
- The high point of his career is **designing the three bridges across the River Thames** – Waterloo Bridge, Southwark Bridge, and London Bridge (which is completed by the firm set up by his sons John and George)

However, Rennie preferred a more practical career, and sought Meikle's help in learning the necessary skills to become a millwright, which he did in 1779. He did not abandon academic studies, however, and while setting up his business he also found time to study at Edinburgh University. There, he received a thorough grounding in chemistry, physics, and theoretical engineering that complemented his apprenticeship under Meikle.

TWO GREAT MINDS MEET

After completing his studies at Edinburgh in 1783, Rennie decided to take a working holiday in England, visiting places in the newly industrialized areas to see for himself how the techniques of modern engineering worked in practice. As well as the machinery that already fascinated him, he saw the growing canal network in the English midlands, and developed an interest in bridges and aqueducts. It was on this trip that Rennie was introduced to the Scottish inventor and mechanical engineer, James Watt (see pp.118–21), in what proved to be a life-changing meeting.

The two got on well together and, impressed with the young man's knowledge and abilities as a millwright, Watt invited Rennie to work with him on installing and improving his steam engines. Very soon, Rennie was put in charge of the London branch of Watt and Boulton's engineering business, with the specific task of supplying engines for the Albion Mills. This was a prestigious commission, as the new flour mill was designed with all the latest technology. Rennie rose to the challenge and redesigned some of the machinery, creating an uneasy relationship with the original designer, Samuel Wyatt, by replacing timber structures with iron.

The Albion Mills established Rennie's name as an able millwright. He was soon being commissioned to supply machinery not only for flour mills, but for other industries too, including the rolling mills for the new Royal Mint, and was even getting orders from mainland Europe. Throughout the 1780s, he gained a reputation for ingenuity and innovation, continually suggesting improvements to machines and novel solutions to engineering problems, and expanded his scope to include dredging and construction work. It was this that led to another chapter in his career, when he was asked to survey the proposed Kennet and Avon Canal in southern England in 1790.

DEVELOPING WATERWAYS

From then on, although Rennie maintained an interest in his mechanical engineering business, he devoted the majority of his time to civil engineering projects. The work fascinated him and gave him the opportunity to combine his expertise in managing water with his love of bridges, aqueducts, and tunnels. Over the next 20 years or so, he became involved with the design and construction of several canals, including the Rochdale Canal, the Lancaster Canal, the Aberdeen Canal, and the Royal Canal of Ireland.

At the same time, the threat of an invasion of England by Napoleon was growing, and Rennie was approached to help improve defences. As well as the

[RENNIE] HELD THAT THE ENGINEER HAD NOT MERELY TO CONSIDER THE PRESENT BUT THE FUTURE

SAMUEL SMILES, SCOTTISH REFORMER AND BIOGRAPHER OF RENNIE

Royal Military Canal along the south coast, he was called to assist with the design of the London Docks and both the East India and West India Docks. The government also appointed him in charge of the expansion of the naval dockyards in Woolwich, which was followed by the most ambitious project of his career, a massive breakwater to protect the naval port of Plymouth. This was begun in 1811, but completed only in 1848 by his sons' engineering company.

It was in the designing and construction of bridges, however, that Rennie found most satisfaction, and in which he made his greatest mark. Having developed a love of them as a student, and having gained some experience with bridges and aqueducts for the canals he built in the 1790s, he had become a master of bridge design, combining elegance with sturdiness.

THE LOST BRIDGES

Three bridges spanning the River Thames were the apotheosis of his engineering career, but sadly none of them remain today – Waterloo Bridge, considered to be his greatest achievement, was demolished in the 1930s; Southwark Bridge, with Britain's widest cast-iron span, was rebuilt to a different design in 1921; and London Bridge, which he designed shortly before his death in 1821 and which was completed by his sons in 1831, was famously moved to Arizona in 1968 with the design spoilt by being shortened.

Without these lasting monuments to his engineering genius, Rennie has been a somewhat neglected hero of civil engineering, overshadowed by his friendly rival Thomas Telford (see pp.160–63). Rennie's contribution to advances in mechanical engineering is similarly often overlooked. He had significant influence on the following generation of engineers, many of whom he trained personally, in particular his sons George and John, who carried on the family name with the engineering company J & G Rennie.

CONSTRUCTING WATERLOO BRIDGE
The first of Rennie's Thames bridges, Waterloo Bridge was supported by nine arches 36.6m (120ft) wide, and took seven years to complete.

BENJAMIN WRIGHT

THE FATHER OF AMERICAN CIVIL ENGINEERING

UNITED STATES 1770–1842

From humble beginnings as a country lawyer and surveyor, Benjamin Wright taught himself, and hundreds of others, the skills of civil engineering. In so doing, he became one of the major figures in the expansion of transport systems in the United States during the Industrial Revolution. Wright was the mastermind behind the construction of the Erie Canal.

Born in Wethersfield, Connecticut, Wright's childhood was disrupted by the American Revolution. His father, Ebenezer, a lieutenant in Washington's Continentals, suffered financially in the struggle for independence, and, as a result, Benjamin received little formal education. He was, however, hard-working and quick to learn, and with the help of his uncle, Joseph, taught himself the basics of law and surveying.

A LIFE'S WORK

- With the help and encouragement of his uncle, Wright **teaches himself law and surveying**
- While working as a surveyor on the Mohawk River project, he **learns the principles of canal-building and civil engineering**
- He is **elected a county judge in 1813**, and serves for many years in the state legislature
- The **Paw Paw Tunnel on the Chesapeake and Ohio Canal**, one of the **longest canal tunnels in the world**, takes nearly 14 years to complete
- His versatility and leadership qualities gain him respect as **the foremost civil engineer of the period**, even when making the transition from canal- to railroad-building
- Dies in 1842, aged 71, with his **five sons following in his footsteps as civil engineers**
- He is **declared the "Father of American Civil Engineering"** by the American Society of Civil Engineers in 1969

The family moved to Rome, in rural upstate New York, when he was 19, and there he established himself as a surveyor. Working mainly for farmers in the neighbouring counties, he built a reputation for diligence and honesty, and became a respected member of the local community. With his knowledge of the law, he was also involved in legislature and local politics and held several posts, including that of county judge.

BIRTH OF AN ENGINEER

By the end of the century, Wright was much in demand as a surveyor, and his work soon attracted the attention of the firm building sections of canal and locks on the Mohawk River, a tributary of the Hudson. Originally employed to survey parts of the proposed route for a navigable link to the Hudson and then down to New York City, Wright developed a taste for engineering work and assisted in some of the construction. Inspired by the success of the expanding canal networks in Europe, plans were being made in America to connect the Great Lakes southwards to the coast, bypassing the St Lawrence River, and Wright found himself caught up in the excitement of the project.

Perhaps because of his unconventional early education, he learnt the skills of canal-building quickly, and by 1811 he was proficient enough to plan the route of a section of the new Erie Canal, connecting Rome and the Hudson River at Waterford. The project as a whole was a huge one, from Buffalo on Lake Erie to Albany on the Hudson, a distance of some 584km (363 miles). After years of planning, fundraising, and seeking government approval, the New York State Canal Commission was ready to start work in 1817 and, due to the expertise he had shown in the design stages, Wright was appointed as one of the three engineers in charge of its construction. He oversaw the building of the middle and eastern sections – over some of the most difficult terrain on the course of the waterway – and proved to be an inspiring leader, teaching his workmen his new-found skills, as well as being an accomplished engineer. He also showed considerable diplomatic skill in dealing with the authorities, and, as the project neared completion, he was promoted to Chief Engineer.

The Erie Canal was opened for cargo traffic in 1825, bringing prosperity to New York state and enhancing New York's status as a major Atlantic port. Wright emerged as very much the hero of the hour, with the first boat to sail on the canal being named *Chief Engineer* in his honour. On the strength of his success with the Erie Canal, he went on to work on many of

CANALS OF AMERICA

At the end of the 18th century, adopting the new technologies of the Industrial Revolution became a matter of pride for the newly independent United States. An essential component in establishing the nation as a competitive industrial trader was a modern transport system. Taking Europe as its model, work began on a network of canals to carry goods and produce from inland via the Great Lakes to ports on the eastern seaboard. By the mid-19th century, the American canal network was the envy of the world, and complemented by the ever-expanding railroad system, transformed the United States into a prosperous trading nation.

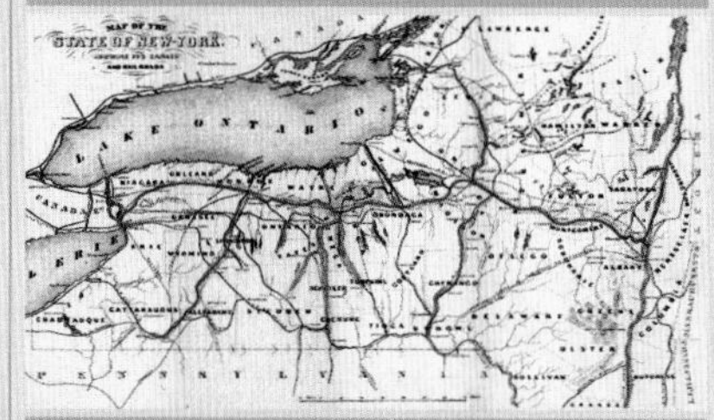

THIS MAP OF THE ERIE CANAL, 1858, SHOWS THE ROUTE OF THE WATERWAY FROM LAKE ERIE TO THE RIVER HUDSON.

FLOW OF WEALTH
Towns along the Erie Canal, such as Lockport, New York (shown here in the 1830s), were transformed from rural communities into prosperous ports. The canal not only carried goods, but also transported people. In its first year of opening, more than 40,000 people used the waterway, reaching destinations as far apart as the Atlantic Ocean and the Great Lakes.

the canals being built to form a network of navigable waterways in the northeastern US and Canada. His surveying skills also put him in an ideal position to plan the route and design of what later became the Delaware and Hudson Canal, which was completed in 1828.

GROWING STATURE

Wright's reputation as a versatile and inspiring engineer spread through the world of transport, and he was headhunted by the canal company planning the Chesapeake and Ohio Canal, just two weeks before construction was due to begin. This was to stretch his abilities to the utmost, as the route was planned through a mountainous area, and had to cross a new railroad and several rivers. As Chief Engineer of the "Grand Old Ditch", as the new canal was nicknamed, he rose to the challenge, incorporating into its 298-km (185-mile) length 74 locks, 11 aqueducts (including the impressive seven-arch aqueduct across the Monocacy River), and a 950-m (3,118-ft) tunnel through the mountain at Paw Paw. At the peak of canal-building in the US, rail lines were also being built, not just in competition with the waterways, but often to complement them, particularly in areas where water transport was impracticable, or to provide portage between sections of the canal network with no navigable connection.

The New York and Erie Rail Road company was set up in 1832 to build a rail line from the Hudson River port of Piermont to Lake Erie. As a result of his experience in surveying and working in the area, Wright was approached for the project, and was appointed the first Chief Engineer when construction began in 1837.

This was the last of Wright's considerable achievements as an engineer. He then went into semi-retirement in New York City, where he was the chairman of a committee to establish the American Society of Civil Engineers – the organization that later officially honoured him with the title "Father of American Civil Engineering".

> ... THEY HAVE BUILT THE LONGEST CANAL ... FOR THE LEAST MONEY AND THE GREATEST PUBLIC PROFIT
>
> **JESSE HAWLEY**, EARLY ADVOCATE OF THE ERIE CANAL, ON WRIGHT AND HIS TEAM

WINSTANLEY AND SMEATON

BUILDERS OF THE LIGHTHOUSE ON EDDYSTONE ROCKS

ENGLAND 1644–1703; 1724–92

HENRY WINSTANLEY

JOHN SMEATON

ALTHOUGH THEY WERE BORN NEARLY a century apart, Henry Winstanley and John Smeaton pursued a common goal – the construction of a lighthouse on the Eddystone rocks. However, building and maintaining an open-sea lighthouse was fraught with engineering challenges.

A LIFE'S WORK

- Winstanley dazzles the public with the **mechanical inventions in his Wonder-House**, in Littlebury, Essex
- After losing a ship on Eddystone reef, Winstanley "does the impossible" and **builds a lighthouse there**
- During the construction of Eddystone, **he is kidnapped by French privateers** and imprisoned
- He **dies in the Great Storm** of 1703
- Smeaton **builds the third lighthouse at Eddystone** early in his career as a civil engineer
- He is **involved in the design and construction** of more than 50 water- and windmills
- The **River Calder navigation system** and the **bridges of Coldstream, Perth, and Banff** are some of his most accomplished, ingenious, and famous projects

Henry Winstanley was born miles away from the sea, in Saffron Walden, Essex, and baptised there on 31 March 1644. He became a porter and then a secretary at the nearby Audley End House, which had been sold to Charles II. His natural aptitude caught the king's attention, and he recommended that Winstanley undertake a grand tour of Europe. So in the 1670s, Winstanley set off for France, Germany, and Italy, where he was impressed by the local architecture and the engravings in which it was portrayed. Upon his return, he began to dabble in engraving and soon issued a set of "geographical" playing cards, which became very popular and sold well. He also began to make mechanical devices. When he moved to Littlebury, Essex, he built a home – known as the Wonder-House – with enough space for all his inventions, and was well known there for his fascination with gadgets, both mechanical and hydraulic. The exterior included a windmill in the garden and a lantern on the roof, and the inside was a treasure trove of contraptions. His royal patron was said to be a frequent visitor. Winstanley was appointed Clerk of Works at Audley End in 1679 and held the post until 1701.

"MORNING AFTER A STORM"
An engraving of 1789 shows Smeaton's lighthouse engulfed in a spume of seawater. The building was strong enough to withstand the immense weight of crashing waves.

EDDYSTONE ROCKS

These dangerous rocks have led to the destruction of many ships – 50 a year before the construction of the lighthouse. The rocks lie about 22km (14 miles) southwest of Plymouth, in the English Channel. Eddystone consists of three ridges, which at low tide peek out, but at high tide only a tiny bit of the largest rocks can be seen. Another danger lies in the eddies – counter-current swirls of water – after which the rocks were named. The sea-bed formation creates dangerous currents that pulls ships onto the rocks. The rocks have played host to four versions of the Eddystone Lighthouse, and are still home to the current lighthouse and the stub of its predecessor.

SUCCESS AND FAILURE

However, his royal patronage underwent a significant change with the arrival of William III. The new king sailed a fleet of 400 ships into Plymouth harbour, and while they were unscathed by the Eddystone rocks, William wanted to do something about this reef. Shipowners, too, had been lobbying the lighthouse authority, Trinity House, to take action, but putting a beacon on the spot was considered impossible. One night, when Winstanley was in a pub in London, he overheard news of the latest Eddystone casualty – the *Constant*. He owned shares in the unfortunate ship, and immediately decided to go to Devon and investigate the possibility of building a lighthouse. What he found was an almost insurmountable situation – near-permanent bad weather, dangerous currents, and only one suitable bit of barnacle-encrusted rock where he could build.

He was not deterred, however, and it took three summers to build the lighthouse. The first summer, in 1696, saw the initial phase of gouging 12 holes in the rock and fixing iron bars into each of them. The following summer he began to build the stone base. Due to ongoing hostilities with the French, a Royal Navy ship, the *Terrible*, was sent to guard the project. But its crew sailed off one day, allowing a French privateer to mount an attack, destroying some of the stone and kidnapping Winstanley. He was imprisoned in France, although Louis XIV allowed him to return to England.

The next summer was less eventful, and the lighthouse was completed. From base to top, it was 24m (80ft) high and, on 14 November 1698, it went into service with Winstanley lighting 60 tallow candles in the lantern. He made many modifications over the following years, increasing the base thickness to 7.3m (24ft) and the height by 6m (20ft). On 25 November 1703, he rowed out to undertake more work, ignoring warnings about the uncertain weather. That night in the lighthouse, he was caught in the Great Storm of 1703. The next day, there was no sign of the lighthouse, or of Winstanley. The storm had destroyed them both. Two days later, on 28 November, the *Winchelsea* struck the reef and sank – the first ship to do so in five years.

REBUILDING ON THE ROCKS

After the wreck of the *Winchelsea*, the need to replace the lighthouse on the dangerous rocks was more urgent than ever. Another engineer was employed – John Rudyerd (*c.*1650–1718), who built the tower around an 18.5m (61ft) timber mast, with heavy oak as an encasement. It opened in 1709 and lasted until 1755, when its wooden structure was destroyed by fire.

John Smeaton was invited soon after to build the lighthouse for the third time. Smeaton was born near Leeds, and although he was initially pushed into law he gave it up because he was far more interested in mechanical pursuits. After he accepted the offer in 1756 to work on Eddystone, he started on a new design, determined to learn from the mistakes of his predecessors.

He praised Winstanley and Rudyerd's work, but when it came to his own plan, he moved far beyond the previous two, using stone blocks that were dovetailed to fit tightly together. This provided a sound structure while having none of the fire risk of wood. The lighthouse was completed in October 1759 and it stood nearly 25m (80ft) above the mean sea level. The structure required 1,000 tonnes (1,100 tons) of granite carved into 1,493 stone blocks. It remained in use until 1877 – this time it was cracks in the rock that led to its dismantling. It was reassembled in Plymouth, where it stands today.

Smeaton eventually became one of Britain's leading civil engineers. He built bridges, water- and windmills, river canals, fen drainage systems, and harbours.

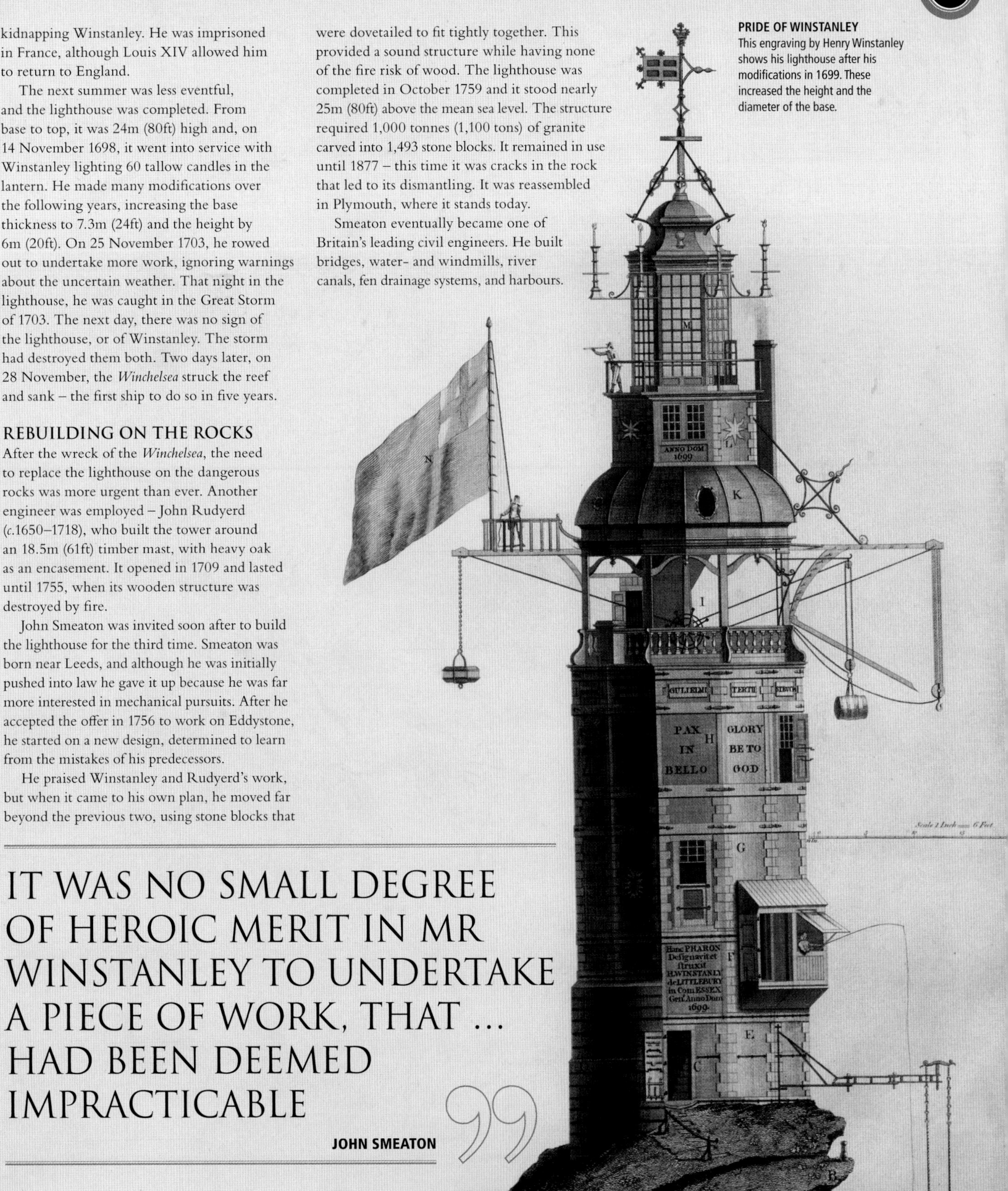

PRIDE OF WINSTANLEY
This engraving by Henry Winstanley shows his lighthouse after his modifications in 1699. These increased the height and the diameter of the base.

IT WAS NO SMALL DEGREE OF HEROIC MERIT IN MR WINSTANLEY TO UNDERTAKE A PIECE OF WORK, THAT … HAD BEEN DEEMED IMPRACTICABLE

JOHN SMEATON

ENGINEERING INNOVATIONS

LIGHTHOUSES

SAILING, FOR ALL OF ITS HISTORY, has been a difficult and dangerous undertaking. One of the most consistent threats lurks beneath the water – lethal rock formations that can destroy the hull of a ship. In the past, the safety of ships was improved by the construction of lofty towers that cast out warning beams of light, but their development took place over thousands of years. Today, lighthouses continue to stand tall, fulfilling their age-old role, although many manned lighthouses are now being replaced with automatic machines, along with radar and GPS.

FLAMES OF WARNING

Before the advent of lighthouse towers, beacon fires were burnt on hilltops to warn approaching ships of dangerous waters. Such practices have been recorded since the 4th century BCE – and, indeed, they even make an appearance in the ancient Greek epics of the *Illiad* and the *Odyssey*. Fire illuminated early lighthouses, and the use of wood and coal persisted well into the 19th century, although oil lamps were later installed. However, all these methods produced smoke, which dirtied the tower's panes and made the light more difficult to see.

It was an innovation by Gustav Dalén (see timeline, below) that allowed many lighthouses to switch to the use of acetylene-gas flames for a few decades until electricity became more widespread. The increasing use and understanding of optics and glass in the 19th century was also crucial with respect to the visibility of light, which improved radically in the 20th century. This meant that lighthouses

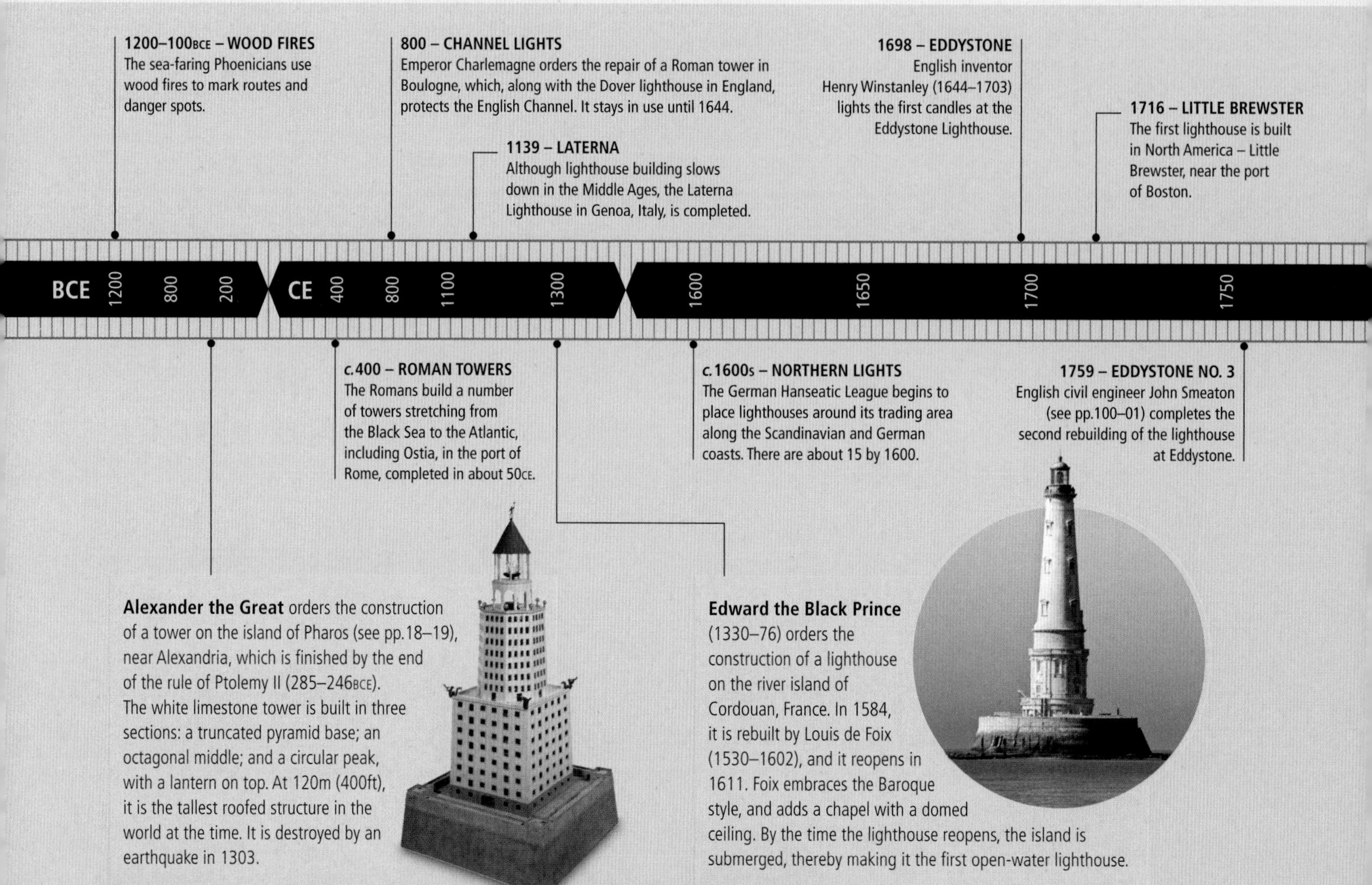

1200–100BCE – WOOD FIRES
The sea-faring Phoenicians use wood fires to mark routes and danger spots.

800 – CHANNEL LIGHTS
Emperor Charlemagne orders the repair of a Roman tower in Boulogne, which, along with the Dover lighthouse in England, protects the English Channel. It stays in use until 1644.

1139 – LATERNA
Although lighthouse building slows down in the Middle Ages, the Laterna Lighthouse in Genoa, Italy, is completed.

1698 – EDDYSTONE
English inventor Henry Winstanley (1644–1703) lights the first candles at the Eddystone Lighthouse.

1716 – LITTLE BREWSTER
The first lighthouse is built in North America – Little Brewster, near the port of Boston.

c.400 – ROMAN TOWERS
The Romans build a number of towers stretching from the Black Sea to the Atlantic, including Ostia, in the port of Rome, completed in about 50CE.

c.1600s – NORTHERN LIGHTS
The German Hanseatic League begins to place lighthouses around its trading area along the Scandinavian and German coasts. There are about 15 by 1600.

1759 – EDDYSTONE NO. 3
English civil engineer John Smeaton (see pp.100–01) completes the second rebuilding of the lighthouse at Eddystone.

Alexander the Great orders the construction of a tower on the island of Pharos (see pp.18–19), near Alexandria, which is finished by the end of the rule of Ptolemy II (285–246BCE). The white limestone tower is built in three sections: a truncated pyramid base; an octagonal middle; and a circular peak, with a lantern on top. At 120m (400ft), it is the tallest roofed structure in the world at the time. It is destroyed by an earthquake in 1303.

332BCE EGYPTIAN PHAROS

Edward the Black Prince (1330–76) orders the construction of a lighthouse on the river island of Cordouan, France. In 1584, it is rebuilt by Louis de Foix (1530–1602), and it reopens in 1611. Foix embraces the Baroque style, and adds a chapel with a domed ceiling. By the time the lighthouse reopens, the island is submerged, thereby making it the first open-water lighthouse.

c.1300 CORDOUAN'S CHAPEL

could send out beams equivalent to those of millions of candles into the dark and often dangerous night.

DESIGNS FOR THE FUTURE

External lighthouse design also went through many modifications. From open-air beacons to lavish Baroque French creations, to simple yet elegant modern models, the engineering of these structures became an important way of testing the relationship of buildings to wind and water. This was especially the case with the Eddystone Lighthouse (see pp.100–01), off the coast of England. As building materials improved and the knowledge of engineering increased, the number of lighthouses quickly multiplied around the coastal areas of the world.

Later, technological developments such as the electric light and timers, as well as more recent developments such as radar and GPS, meant that these structures no longer required anyone to live in them, and so the job of the lonely lighthouse-keeper was phased out over the course of the 20th century. In addition, other forms of warning technology have been developed. Automatic floating buoys proliferated as a result of Dalén's inventions, while lighthouses themselves can be monitored remotely through radio and satellite links. Notwithstanding these developments, however, modern, state-of-the-art navigational systems are yet to consign the solitary, steady beam of the lighthouse to history.

FRESNEL LENS

Developed by French inventor and lighthouse engineer Augustin-Jean Fresnel (1788–1827), the first Fresnel lens was installed in the lighthouse at Cordouan in France. The refracting – or dioptric – lens collects light at an angle and bends it. When the lamp rotates, it appears to flash, and the overall effect is of a far stronger light that can be seen for much longer distances – about 32km (20 miles) out at sea.

THIS WORKING FRESNEL LENS IS INSTALLED AT THE ROCHES-DOUVRES LIGHTHOUSE IN THE ENGLISH CHANNEL.

1811 – STEVENSON'S ROCKS
Scottish engineer Robert Stevenson (1772–1850) finishes a lighthouse on Bell Rock, which is part of a submerged reef, about 22.5km (14 miles) off the coast of St Andrews.

1820 – GLOBAL CONSTRUCTION
By this time, there are some 250 lighthouses scattered around the world.

1823 – FRESNEL LENS
The first-ever Fresnel lens (see box, above) is installed in the lighthouse at Cordouan.

1882 – EDDYSTONE FINALE
The Eddystone Rocks receive their third and final rebuilding of the lighthouse tower, this time designed by James Douglass (1826–98).

1901 – STRONG WICKS
British engineer Arthur Kitson (1861–1937) develops the vaporized oil burner, which uses kerosene and air, and is six times stronger than oil wick lights.

1913 – LIGHT SWITCH
In the North Sea, the German Helgoland lighthouse switches to electric-carbon arc lamps and mirrors, producing the equivalent of 38 million candles of light.

1920s – SWITCH OVER
Electric filament lamps come into use in lighthouses.

1998 – AUTOMATIC BEACONS
The last manned lighthouse in Britain, North Forland, converts to automatic operation.

1800 1850 1900 1910 1920 1990 2000

Swiss inventor Aimé Argand (1750–1803) constructs an oil lamp incorporating a hollow cylinder within the circular wick. This helps provide a steady, as well as smokeless, light. The design is patented in 1784, and is the first design change to lamps for thousands of years. The light of Argand's lamp is about 10 times brighter than the light of earlier lamps of the same size, but it consumes a greater amount of oil.

1782 OIL LAMP

Swedish inventor Gustav Dalén (1869–1937) wins the Nobel Prize in physics for his work on the storage and use of acetylene. Dalén paved the way for the automated lighthouse with two inventions. The first was a revolving lens that emitted a bright light by using a mixture of acetylene gas and air. The second was the development of his "sun valve", which turned the gas "on" at dusk and "off" at dawn, saving fuel.

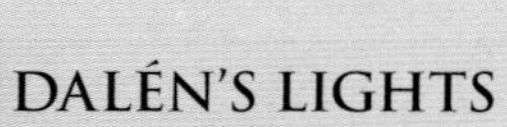

1912 DALÉN'S LIGHTS

EARLY STEAM ENGINES

ALTHOUGH FIRST DEMONSTRATED BY THE ANCIENT GREEKS, THE POWER OF STEAM WAS NOT EXPLOITED TO DRIVE MACHINES UNTIL THE 17TH CENTURY. STIMULATED BY A NEED FOR MORE POWERFUL MACHINERY, ENGINEERS EXPLORED VARIOUS WAYS TO HARNESS STEAM POWER.

SCIENTIFIC COLLABORATION
Development of early steam engines began with the collaboration between engineers such as Denis Papin (left) and experimental scientists such as Robert Boyle (right).

Machines have always been powered in a number of different ways. While small devices such as clocks were driven by weights and springs, or were hand-operated, large machines such as mills and water-raising devices were driven by wind or water power, or by animals. However, 17th-century demands for machines – particularly for pumping water from mines – called for a more reliable and efficient driving force.

Probably the first to see the possibility of using steam to power machinery was the 16th-century Islamic engineer Taqi al-Din (see pp.56–57). He was undoubtedly familiar with Hero of Alexandria's novel "aeolipile", which relied on water being heated in a closed vessel and the steam being forced through twin nozzles for its spinning action (see pp.32–33). Taqi realized that this jet of escaping steam could be used to drive a wheel, which he used to work a rotating spit. Similar devices were also described in the following century by Giovanni Branca (1571–1645) and John Wilkins (1640–72), but the development of this simple design into the steam turbine would have to wait until the 1880s.

MINE ENGINE
This engraving of the Dolcoath Copper Mine in Cornwall, England, shows the engine houses containing the Newcomen engines that pumped water from the tunnels. The end of the beam engine can be seen projecting from the upper floor of the engine house.

FROM THEORY TO PRACTICE

Instead, other ways of exploiting the power of steam began to emerge. Engineers such as Giovanni Battista della Porta (1535–1615) and

SAVERY'S PUMPING ENGINE
By heating water in a closed container and then cooling the steam to condense it, Thomas Savery created a partial vacuum that could be used to pump water.

THE "FIRE ENGINE"
Savery's steam engine was the first commercially viable steam pump, and he took out a patent on all "Fire engines", which included Thomas Newcomen's more successful design (above).

SETTING THE SCENE

- A **steam-driven rotating spit** – a rudimentary steam turbine – is described in the 16th century by Taqi al-Din.
- The **Enlightenment preoccupation with science** prompts a move towards mechanization, but mills are still powered by wind or water power, and other machines are hand- or animal-driven.
- In the early 17th century, several engineers describe **steam-powered siphonic pumps**.
- Experiments in the field of **"natural philosophy"**, notably Robert Boyle's work, open up new possibilities.
- Thomas Savery's **siphonic steam pump** controlled by valves, invented in 1698, becomes the first steam engine to be used commercially.
- At the beginning of the 18th century, Thomas Newcomen combines ideas from Papin's and Savery's designs and **builds a steam-driven piston engine**, which works a beam engine.

Salomon de Caus (1576–1626), designed pumping machines that used the pressure created by heating water in a closed container, but these were never built. Edward Somerset (see p.109), in a book of 1655, claims to have constructed a similar machine at his castle – used to raise water to the top of a tower – but no evidence remains of the machine.

Inventors in the 17th and 18th centuries were well aware of the commercial possibilities of their inventions, and were keen to stop others cashing in on their ideas. As a result, patent laws were brought into force in many countries.

Engineers and scientists were preoccupied with the idea of pumps at this time. The pioneer in the field was German scientist Otto von Guericke (1602–86), whose dramatic demonstrations helped to publicize his experiments. He showed that air could be pumped out of a container to create a vacuum, and that this had an enormous suction power. Vacuum pumps also fascinated Robert Boyle (1627–91), whose experiments on the expansion and contraction of gases led to Boyle's Law, concerning the relationship between volume and pressure of a gas.

As always, engineers quickly began to look at the practical applications of these scientific discoveries. Christiaan Huygens (see pp.80–81) copied some of Boyle's experiments, and later built a piston that was driven by gunpowder in a cylinder – a crude internal combustion engine – but did not take the idea further. His assistant, Denis Papin (see pp.106–07), however, immediately saw the potential of using expanding gas in a cylinder to drive a piston, and replaced the gunpowder with heated water to produce steam. He also realized that if the steam was allowed to cool, its contraction would draw the piston back down the cylinder. The idea was sound, but in practice the engine was painfully slow.

ENGINES FOR MINES

Meanwhile, Thomas Savery (see pp.108–09) had approached the problem from another angle. Instead of driving a piston, the steam would move water through pipes in two consecutive processes. First the steam was blown down to the level of the water, and then it was condensed. This would create a partial vacuum and suck water upwards perhaps 5m (16ft) into twin reservoirs. Then, after the lower valves were closed and the steam valves were opened, steam at high pressure would push the water further upwards. Unfortunately, the engine had to be installed half-way down the mines that Savery had hoped to pump dry. Further problems were that the high temperature and pressure were often too much for the soldered joints of the machine. Although Papin and Savery had shown that steam could be harnessed to drive machinery, their engines were rudimentary, slow, and no more efficient than the horse-powered pumps in use at the time.

The breakthrough came with Thomas Newcomen (see pp.110–11), who saw a way to incorporate the best of both their designs. To the idea of a piston in a cylinder, he added a separate boiler, and (by accident, when a leak allowed cold water into the cylinder) he invented an internal condensing device to cool the steam.By adding a system of automatic valves, the piston could be pushed up and drawn down continuously, and by connecting the piston to a beam engine, Newcomen harnessed its reciprocating action to work a mechanical pump. When the Newcomen engines were installed in coal mines, they were found to be reliable and commercially viable. They were truly ground-breaking, and remained in production for many years, long after the advent of James Watt's steam engine (see pp.118–21).

All these engines were significant milestones in the Industrial Revolution. They established steam as the main source of power until the 20th century.

STEAM POTENTIAL
Denis Papin designed this steam engine to power a paddle boat. The boat was not built, but the potential of the engine was clear.

DENIS PAPIN

EARLY PIONEER OF THE STEAM ENGINE

FRANCE 1647–C.1712

A QUALIFIED MEDICAL DOCTOR, Denis Papin spent most of his adult life away from his native France in England and Germany, and made his name as a physicist and engineer. As well as inventing the "steam digester", he designed and built the earliest steam engines, and published his ideas, although he never attempted to develop them commercially.

A LIFE'S WORK

- Denis Papin's **work with Huygens** leads to further **experiments with vacuum pumps**
- Undertakes a trip to England in 1675, where he **meets Robert Boyle** and **invents the pressure cooker**
- Is **elected as a member of the Royal Society**
- He is appointed **professor of mathematics at Marburg University in Germany** in 1687
- Drawing from his invention of the "steam digester", he **designs a simple piston steam engine** in 1690, **some years before Savery's** steam engine
- He **suggests modifications to Savery's steam pump**
- He designs a **steam-driven boat**

Denis Papin was born into a well-to-do protestant family in Blois, France. At the age of six, he was sent to study at the protestant academy in Saumur, where he stayed with his uncle, a doctor. There was a tradition of medicine in the maternal side of his family, and Denis went on to qualify as a doctor himself at the University of Angers in 1669. However, after moving to Paris to set up a medical practice, he soon realized that he was not suited to the profession.

His interests were related to the fields of mathematics and physics, and, by pulling a few strings with a family friend, he found a job as assistant to a Dutch scientist, Christiaan Huygens, in Paris (see pp.80–81). This was exactly the opportunity that Papin was looking for. Huygens was an inquisitive experimenter whose interests ranged from astronomy to optics and clockmaking, and when Papin came to work in his laboratory, he was occupied with studying the work of Anglo-Irish scientist Robert Boyle on vacuum pumps (see box, opposite). Papin even designed a rudimentary piston machine powered by gunpowder.

THE POTENTIAL OF STEAM

Papin was fascinated by Huygens's work, especially on pumps, and was soon developing his own ideas on the subject, which he presented in a treatise in 1674. Working with Huygens also brought him into close contact with other great scientists, including Gottfried Wilhelm Leibniz (1646–1716), who became a lifelong friend and a supporter of his work. Then, when he was sent to London in 1675 to present Huygens's balance-spring watch to the Royal Society, he took the opportunity to introduce himself to Boyle, who had become a hero to him.

Papin decided to remain in England, as much because of the persecution of protestants in France as his desire to work with Boyle. A Huguenot community in London took him under its wing, and Papin continued his work on vacuum pumps, among other things, with Robert Boyle.

Meanwhile, he discovered that water heated in a sealed container produces steam at high pressure, which causes the water to boil at a higher temperature than normal. This allows food to be cooked more quickly; in fact he had invented the pressure cooker, although he called it "A New Digester or Engine for Softening Bones". On 12 April 1682 he held a "philosophical dinner" for the president and fellows of the Royal Society, about which diarist John Evelyn wrote: "I went this afternoon with several of the Royal Society to a supper which was all dressed, both fish and flesh, in Monsieur Papin's digestors, by which the hardest bones of beef itself, and mutton, were made as soft as cheese."

Papin spent some time in Paris with his old mentor Huygens, then worked for a while in Venice, but returned to London in 1684. He had been elected as a member of the Royal Society on the strength of his steam digester, but the society offered him only menial work, and he was relieved when the chance of a professorship at Marburg University in Germany came along in 1687.

ACADEMIC PURSUITS

For some years, Papin led a productive life in Marburg, which, like London, had an ex-patriot Huguenot community. Academic life gave him the time to work on his own experiments, and it was here, in 1690, that he succeeded in building the world's first piston-and-cylinder steam engine model, eight years ahead of Thomas Savery (see pp.108–09) and more than 20 years

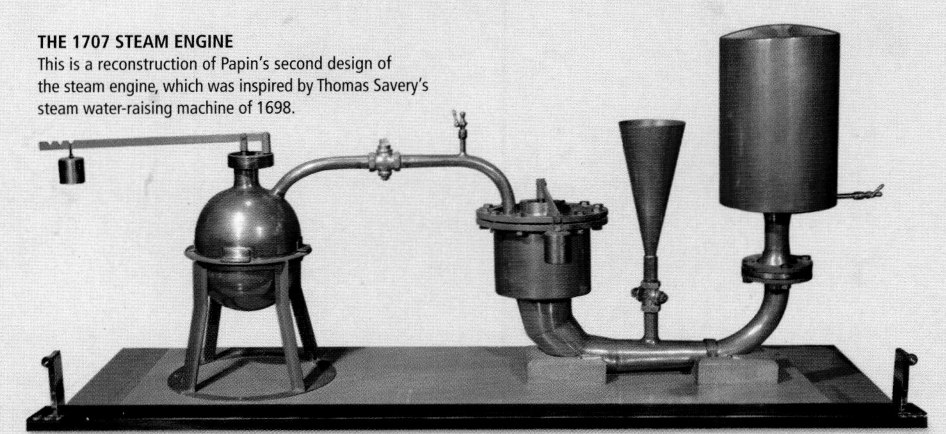

THE 1707 STEAM ENGINE
This is a reconstruction of Papin's second design of the steam engine, which was inspired by Thomas Savery's steam water-raising machine of 1698.

before Thomas Newcomen (see pp.110–11). He published his ideas, and continued to work on steam on and off for the next 15 years.

Somewhat disillusioned with university life, he took on some routine engineering work, draining water from mines. His interest in steam engines was rekindled when his friend Leibniz sent him a description of Savery's steam pump in 1705. Papin set to work suggesting improvements, and, in 1707, he had designed a second steam engine based on Savery's model. A couple of years later, he built a boat driven by a paddle-wheel rather than oars, which he envisaged as eventually being powered by a steam engine.

AN OBSCURE END

Once more enthused, Papin returned to London in 1707 to present his ideas to the Royal Society, but received a lukewarm reception. His champion, Boyle, had died several years previously, and he found no friends in the scientific community. Isaac Newton (1642–1727), now president of the Royal Society, seemed uninterested in Papin's work, but did allow his treatises to be presented. Despite the lack of response, Papin continued to send papers for the society's consideration, many demonstrating that he was still working on new ideas.

Perhaps there was some British national pride involved in snubbing Papin: the invention of the first steam engine was being credited to Savery – it is even possible that Isaac Newton, who disliked the French, was determined that an Englishman should get priority. The similarities between Newcomen's steam engine design and Papin's first piston-and-cylinder engine also suggest that Newcomen might have seen Papin's articles published by the Royal Society. Whatever the case, Papin's contribution to the development of a practical steam engine was not acknowledged until long after his death, and his final years were spent in obscurity.

BOYLE'S INFLUENCE

Inspired by Robert Boyle's work on the expansion of gases, Papin experimented with steam. He discovered that boiling or steaming food in a closed container – a steam digester – reduced the cooking time, and could easily soften bones. Observing the movement of the safety valve on his pressure cooker may have given him the idea of his first piston steam engine.

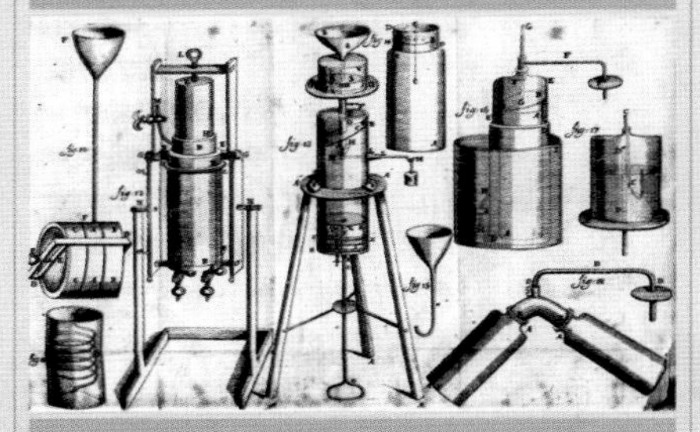

THIS ILLUSTRATION SHOWS PAPIN'S DIAGRAMS OF THE PRESSURE COOKER AND ITS SAFETY VALVE.

PAPIN'S STEAM BOAT
Here, Papin is supervising the building of his paddle-wheel boat, and demonstrating to boatmen his proposal for powering it with a steam engine.

THOMAS SAVERY

INVENTOR OF THE FIRST PRACTICAL STEAM ENGINE

ENGLAND c.1650–1715

THE PROBLEM OF DESIGNING an effective pump to raise water had long preoccupied engineers, and, as mining gained importance in the 17th century, there was an increasing need to drain mines. Working from an idea by Denis Papin, Thomas Savery developed a steam pump, which, though it could raise water and did so in a few places, was never useful for draining mines.

A LIFE'S WORK

- Savery works as a military engineer but develops an **interest in mining engineering** and, in his free time, **invents mechanical devices**
- Is granted a **patent on his steam engine in 1698**, which is **extended** by an Act of Parliament **from 14 to 21 years**, and **covers all steam-driven water-raising machines** – including Thomas Newcomen's
- He demonstrates his steam engine to the Royal Society in 1699, and is **elected a Fellow of the Society in 1706**
- **Sets up a workshop in London** to build, demonstrate, and sell his engines
- **Resumes work as a military engineer**, and applies for patents on several inventions, including a seagoing mill, and a device for measuring the distance travelled by ships

Thomas Savery was born into a wealthy family of merchants near Modbury, Devon. However, despite his comfortable upbringing, he chose not to follow a career in commerce and instead became a military engineer, eventually rising to the rank of captain. His interest in engineering was not restricted to his duties to the military, however, and he spent much of his spare time applying his skills to experiments with mechanical devices. Apart from learning the principles of mechanical engineering through this hands-on experience, he discovered that he had a talent for innovation, and, in his 40s, began to apply for patents for his inventions.

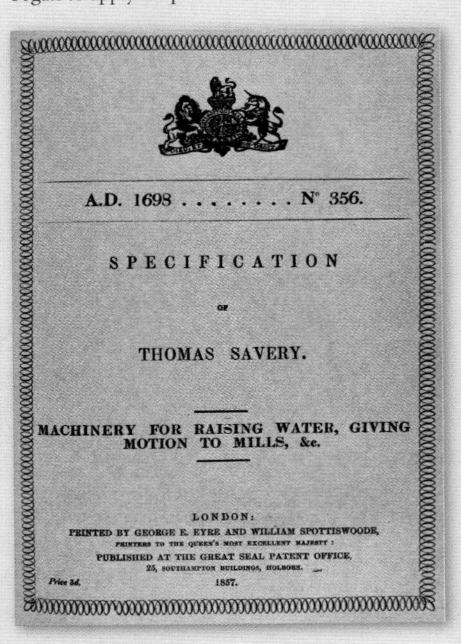

A.D. 1698 . . . . N° 356.

SPECIFICATION

OF

THOMAS SAVERY.

MACHINERY FOR RAISING WATER, GIVING MOTION TO MILLS, &c.

LONDON:

PRINTED BY GEORGE E. EYRE AND WILLIAM SPOTTISWOODE,

PRINTERS TO THE QUEEN'S MOST EXCELLENT MAJESTY:

PUBLISHED AT THE GREAT SEAL PATENT OFFICE,

25, SOUTHAMPTON BUILDINGS, HOLBORN.

Price 3d. 1857.

The first of these was for a machine for grinding and polishing glass, granted in 1696. In the same year, inspired by the naval heritage of his native Devon, he patented a device for propelling ships by means of a pair of paddle wheels. This was initially received with enthusiasm after he built a yacht equipped with the machine and demonstrated its practicality, but as he was not an officially recognized naval engineer, it was shelved by the Royal Navy – prompting an angry retort from Savery in his description of the invention in *Navigation Improved*, published in 1698.

TO THE MINERS' RESCUE

Devon was a mining area, and Savery was well aware of the problems faced by miners as their mines were flooded by rainwater. The deeper they went, the more water came in; they could not work under water, and they could not pump the water out quickly enough, even using horses. The news of Denis Papin's experiments with steam power (see pp.106–07) had spread throughout the engineering world, and this gave Savery the idea of designing a steam-driven pumping device.

By 1698, he was ready to apply for a patent for his "invention for raiseing of water and occasioning motion to all sorts of mill work by the impellent force of fire, which will be of great use and advantage for drayning mines, serveing townes with water, and for the working of all sorts of mills". This machine was not designed to drive a piston, but instead worked by introducing steam under pressure into a closed vessel, forcing water in the vessel through a pipe to a higher level. The vessel was then cooled, and, as the steam condensed, it created a vacuum that drew more water into the vessel, and the process was repeated. To facilitate a more continuous flow, he designed the machine with two vessels working in tandem. The only moving parts in the engine were the taps acting as valves on the inflow and outflow pipes of the working vessels. The machine worked, but perhaps not as well as Savery had hoped.

Heat transference to the water to be raised caused some inefficiency, and as the vacuum relied on atmospheric pressure to raise water into the vessel, the working height of the machine was limited. On a more mundane level, at that time, there was no solder available for the joints that was capable of withstanding the high temperature and pressure of the steam, so it required constant maintenance and repair.

Nevertheless, a patent was granted, initially for 14 years, which was extended to a further 21 years in 1699 by an Act of Parliament, which became known as the "Fire Engine Act". The wording of the patent was vague, with no accompanying description or diagrams of the machine, and it effectively gave Savery rights in all subsequent steam engines. It therefore ensured that he and his successors would benefit from Thomas Newcomen's improved steam engine design (see pp.110–11) 14 years later.

Savery demonstrated his steam engine to the Royal Society in 1699 to great acclaim. Having achieved some measure of fame, he

STEAM PUMP PATENT
This is the title page of Savery's pump patent, granted in 1698. It was the first practical high-pressure steam engine for pumping water from mines.

set out to make his fortune by building and demonstrating his engines for commercial use, and publishing a full description in his instruction manual *The Miner's Friend*.

AN UNFORTUNATE TRANSITION

Several mine owners bought and installed Savery's engines in the early 1700s, making him set up a workshop in London to supply steam engines. However, once the novelty wore off, the demand for his machines dwindled, and he was forced to close the business in 1705, though he did manage to sell a few pumps for domestic use later.

Somewhat disillusioned, he abandoned any further involvement in mining engineering and returned to his duties in the military. Shortly afterwards, he found work as treasurer of the Commission for Sick and Wounded Seamen, but continued dabbling in mechanical inventions and applying for patents. When Newcomen built his steam engine in 1712, he had to go into partnership with Savery because of the Fire Engine Act.

Once his work at the Commission for Seamen came to an end in 1713, Savery was appointed as surveyor of the waterworks at Hampton Court, where he worked until his death two years later. Despite his successes as an engineer and inventor, he left large debts to his widow, who later sold his steam engine patent to a committee, which administered it until its expiry in 1733.

EDWARD SOMERSET

ENGLAND C.1601–67

A staunch Royalist and Catholic, Edward Somerset fought for Charles I during the English Civil War (1642–51) and was imprisoned in the Tower of London in July 1652, but was released by order of Oliver Cromwell in 1654.

He had, however, lost the family estate in Raglan, where he had been brought up, and pursued an amateur interest in science and engineering. He claimed to have invented a steam-driven machine to raise water (predating Thomas Savery's by about 60 years), and a perpetual-motion machine. These and other inventions were described in a book that he published in 1663, when the restoration of the monarchy allowed him to resume his experiments with his former assistant, and Dutch engineer, Caspar Calthoff.

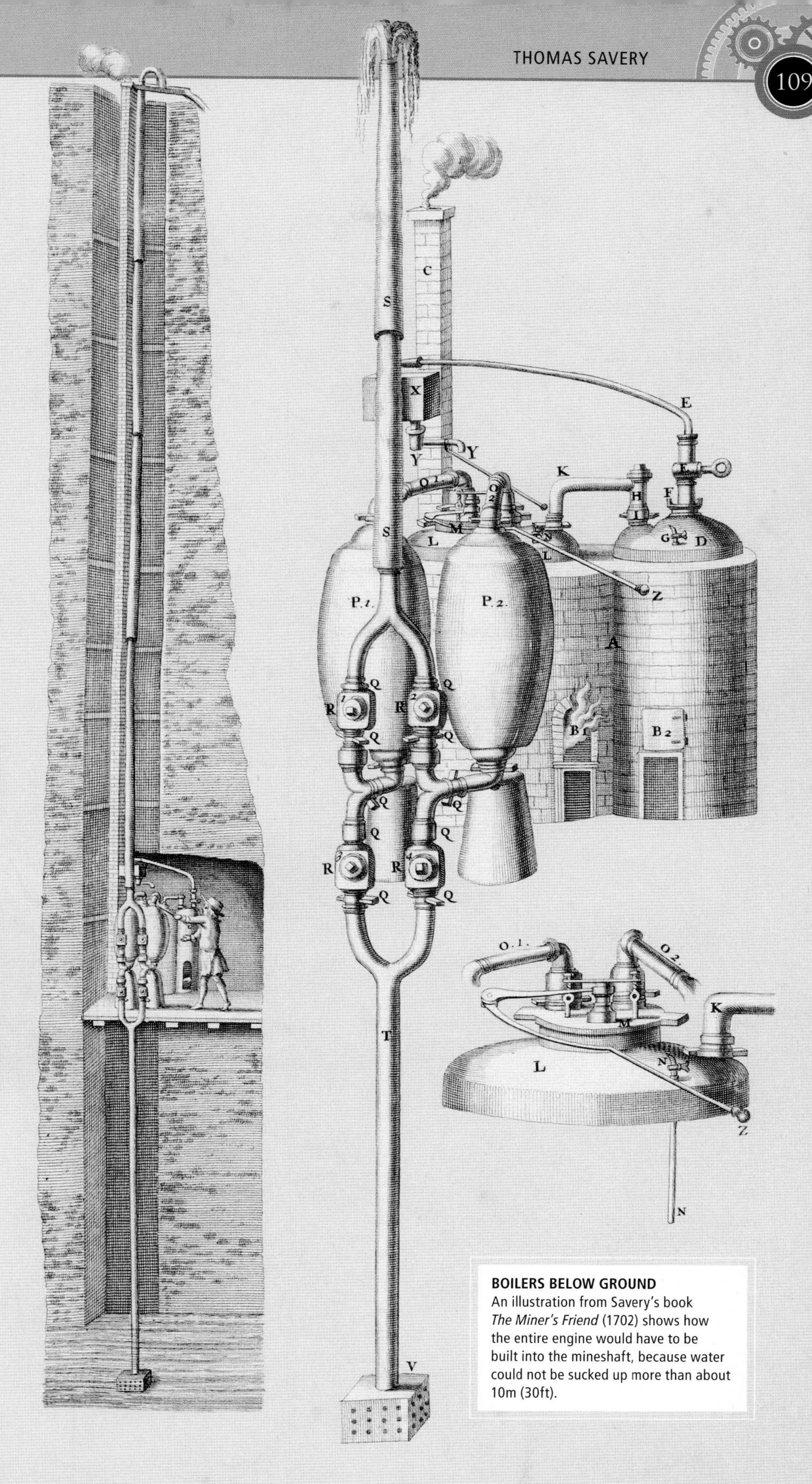

BOILERS BELOW GROUND
An illustration from Savery's book *The Miner's Friend* (1702) shows how the entire engine would have to be built into the mineshaft, because water could not be sucked up more than about 10m (30ft).

THOMAS NEWCOMEN

MASTERMIND OF THE NEWCOMEN ENGINE

ENGLAND 1664–1729

IN THE EARLY 1700S, in a seaside town in southern England, Thomas Newcomen and his partner John Calley built the world's first useful steam engine, and so kicked off the Industrial Revolution. This was arguably the most important step forward in the whole history of technology. So successful was the design that it was scarcely improved upon until Scottish engineer James Watt came along some 60 years later, and many original Newcomen engines remained in use well into the 19th century.

A LIFE'S WORK

- **A staunch Baptist**, Newcomen is well known throughout his life as a **lay preacher**, and for many years **serves as a pastor** of his local parish
- After **trading as an ironmonger** for several years, he goes into partnership with the metalworker and plumber John Calley, and sets up an engineering workshop in Dartmouth
- **Combines the principles** behind **Denis Papin's and Thomas Savery's engines**, adding improvements of his own, and creates the **Newcomen engine**
- Installs his first working steam engines to **pump water from mines** in Cornwall and the Midlands in 1712, and goes on to oversee **installation of about 100 more** throughout England and Wales
- Is **forced into partnership with Savery** by the terms of Savery's **patent on all steam-driven engines**; nevertheless, Newcomen spends the rest of his life **promoting and selling** the Newcomen engine

Newcomen was born in Dartmouth, Devon, into a family of shipowners and merchants. Little is known of his early life, but, in 1685, he set up business as an ironmonger in Dartmouth. This was a successful venture, and he built up contacts with iron foundries in the Midlands, which were later to prove useful. Living close to the tin mining areas of the west of England, Newcomen was well aware of the problems of draining water from the mines, and, in the early 1700s, he began experimenting with mechanical devices to replace the horse-powered pumps used at the time. He acquired property in Dartmouth, and went into partnership with a local trader and a fellow Baptist, John Calley, whose expertise in metalwork and plumbing complemented Newcomen's engineering skills.

It is not clear how much they knew about the work of other engineers in the same field. Thomas Savery had demonstrated his steam pump in 1698 (see pp.108–09), and published a description of it in 1702. Newcomen might have heard of the invention. It is also possible that he and Calley were aware of Denis Papin's designs for a steam-driven piston, given his fame as inventor of the "steam digester" (see pp.106–07). The engine Newcomen and Calley designed incorporates elements of both inventions, but it may well be that they came up with these ideas independently.

NEWCOMEN ENGINE

Whether or not they were influenced by Papin and Savery, Newcomen and Calley developed a more effective machine than either of their predecessors, with several innovative features. By about 1710, they had built a working engine whose design was so successful that it remained essentially unchanged for almost a century. Like Savery's engine, it used steam from a boiler to fill a chamber, which was then cooled to create a vacuum. But rather than using this to draw in water, Newcomen hit upon the idea of harnessing the pressure of the atmosphere to drive a piston in a cylinder.

NEWCOMEN'S HOUSE
The success of his ironmonger's business allowed Newcomen to buy several houses in Dartmouth, including this one in North Town where he lived above his engineering workshops.

Newcomen's genius lay in combining the two principles, and then coupling the piston to a mechanical engine.

The design also incorporated some ingenious improvements. Steam from the boiler passed into the working cylinder, and this allowed the piston to rise. Cold water was then squirted in to condense the steam and form a vacuum. Atmospheric pressure would push the piston back down to the bottom of the cylinder, containing the piston. To condense the steam to form a vacuum, the cylinder was cooled from the inside rather than the outside, by an injection of cold water. Pressure would then drive the piston into the cylinder, and the cycle would start again.

One of the advantages this had over Savery's high-pressure engine was that the Newcomen engine ran at barely over 100°C (210°F) with steam at a pressure of one atmosphere, and this was within the tolerances of the materials that were available. In place of the cumbersome taps of the previous rudimentary engines, Newcomen invented a system of automatic valves to control the operation of the condensing jet and the inlet for the steam.

The main advantage, however, was that Newcomen's was a beam engine. When the piston was pushed down, it pulled down one end of a massive wooden beam pivoted on the wall of the engine house. The other end of the beam rose, and pulled up pump rods that hung down the mine and pumped the water out. So the engine could be built beside a mine shaft to pump up water from any depth.

Initially, Newcomen and Calley had difficulty marketing their invention. Not until 1712 did they manage to arrange a demonstration of the machine in London. But because of Newcomen's northern connections, their first customer was in the Midlands: Conygree Coalworks near Dudley, Staffordshire. They bought and installed a Newcomen engine, as it was to be known. News of its effectiveness spread quickly around the mining industry, and orders came in from collieries in Warwickshire, Yorkshire, and Wales.

IMPOSED PARTNERSHIP

The partners were anxious to protect their interests in the invention. However, the patent that Savery had taken out on his steam-driven atmospheric engine was written in such vague terms that he effectively had rights over any steam engine. To continue selling their engines legally, Newcomen and Calley were obliged to include Savery as a partner.

Nevertheless, Newcomen and Calley oversaw the installation of engines in mines all over Britain. Newcomen was in his 50s when Calley died in 1717, but continued to run the business. After his death, the company continued to produce the engines. More than 2,000 were built, and, many of these were still in use more than a century after Newcomen's death.

ENGINES FOR MINING

Designing machines for pumping water from mines was a major spur to the development of the steam engine. Mechanization created demand for raw materials such as metal ores and coal, making mining a vital industry. Engineers were called upon to increase the efficiency of pits. Safety was a crucial part of this, finding ways to remove water from pits and ensuring the shafts and tunnels did not collapse.

NEWCOMEN ENGINES, SUCH AS THIS ONE, WERE STILL BEING BUILT A CENTURY AFTER THEIR INVENTION.

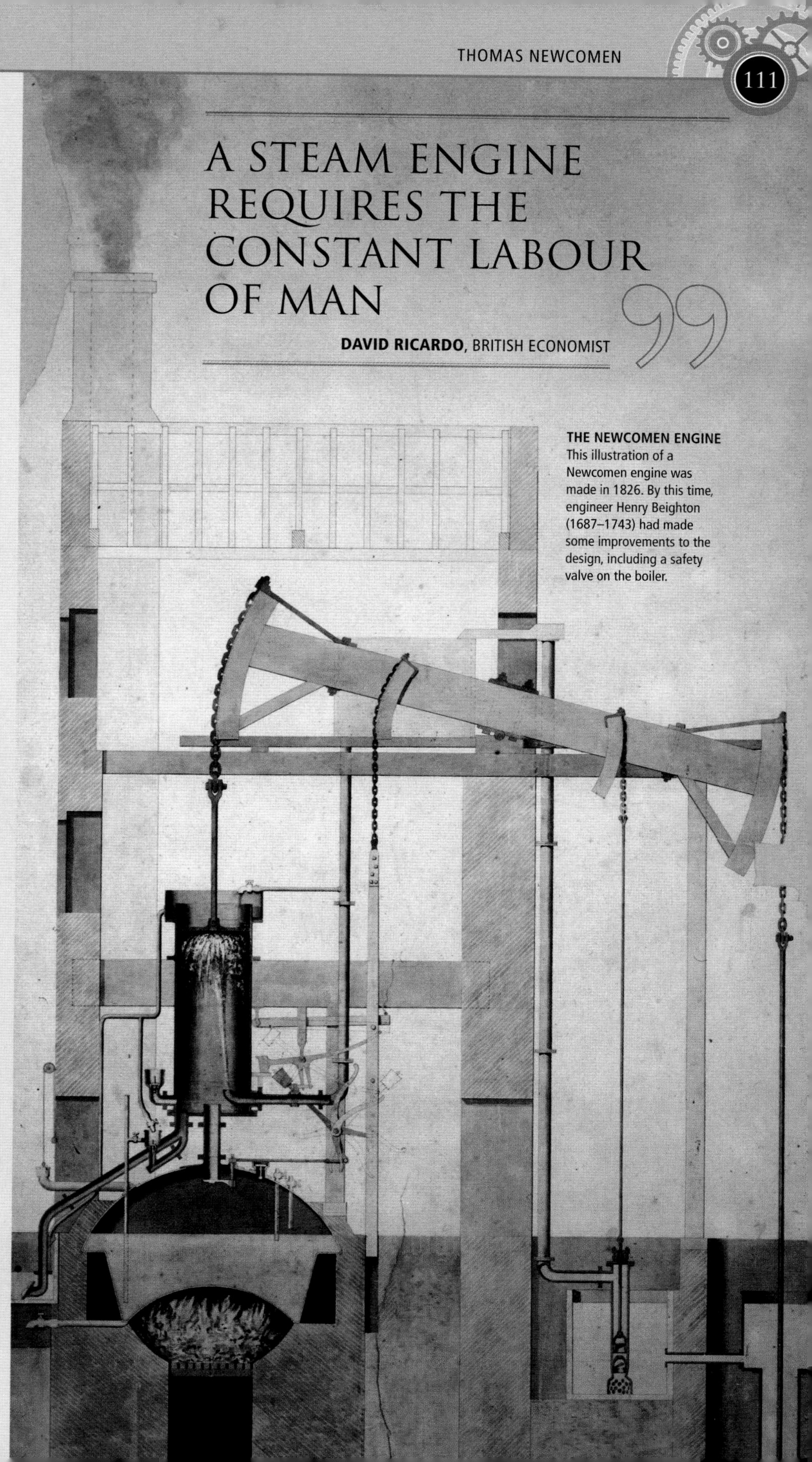

A STEAM ENGINE REQUIRES THE CONSTANT LABOUR OF MAN

DAVID RICARDO, BRITISH ECONOMIST

THE NEWCOMEN ENGINE
This illustration of a Newcomen engine was made in 1826. By this time, engineer Henry Beighton (1687–1743) had made some improvements to the design, including a safety valve on the boiler.

THE INDUSTRIAL REVOLUTION

THE INDUSTRIAL REVOLUTION

1720 | 1770 | 1780 | 1800

1720
English dyer **Stephen Gray** explores electrical conduction and transmits a charge along a 250m- (800ft-) long twine (see pp.134–35).

▲ 1730s
John Kay (see pp.144–45) creates a shuttle to weave cotton, and James Hargreaves (1720–78) invents a jenny to spin it (above), launching the cotton industry.

1752
American polymath **Benjamin Franklin** proves that lightning is electricity with his famous experiment in which he flies a kite into a thunderstorm (see pp.136–37).

1764
Scottish engineer **Robert Mylne** completes the Blackfriars Bridge over the River Thames in London (see pp.158–59).

1769
French inventor **Nicholas-Joseph Cugnot** builds a steam carriage, which is possibly the world's first self-propelled vehicle (see p.129).

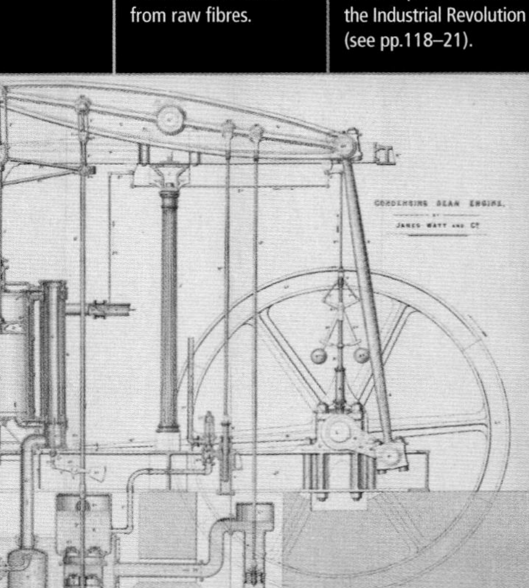

► 1771
Richard Arkwright opens his first factory at Cromford in Derbyshire, England, where a massive water wheel drives his revolutionary "water frames" to spin cotton (see pp.146–47).

1774
English engineer **Samuel Crompton** (1753–1827) invents an even faster spinning machine known as a "mule" that creates finished cotton cloth from raw fibres.

▼ 1776
Scottish engineer **James Watt** completes his steam engine, the crucial revolutionary improvement that turns the steam engine into the powerhouse of the Industrial Revolution (see pp.118–21).

◄ 1779
English architect Thomas Pritchard builds the world's first all cast-iron bridge at **Ironbridge** in Coalbrookdale in Shropshire, England (see pp.156–57).

▼ 1783
The first manned ascent is made in the **Montgolfier brothers'** hot-air balloon in Paris (see pp.172–73), followed by the first gas balloon ascent by Jacques Charles (see pp.174–75).

1788
American inventor **Oliver Evans** introduces his automated flour mill, which turns the entire process of milling flour into an operation of a single machine (see pp.124–25).

1789
US industrialist **Samuel Slater** introduces the factory system to the US as he sets up a cotton spinning factory on Rhode Island (see p.149).

1789
English locksmith **Joseph Bramah** introduces his "unpickable" lock and goes on to invent other important household items, such as the flush toilet (see pp.150–51).

► 1795
Eli Whitney invents the cotton gin, a machine for separating cotton fibres from the plant. It transforms cotton production in the American South (see pp.148–49).

1800
Alessandro Volta creates the first real electricity battery, later called the Voltaic pile, providing the first continuous supply of electricity (see pp.138–39).

► 1801
Richard Trevithick builds the first high-pressure steam road vehicle. He then builds the first steam engine at Pen-y-darren (see pp.128–29).

1801
Judge James Finlay (1762–1828) builds the first known metal suspension bridge over Jacob's Creek in Pennsylvania, US.

1802
Samuel Bentham (1757–1831), **Marc Brunel** (1769–1849), and **Henry Maudslay** (see pp.150–51) set up the Portsmouth Block Mills, establishing new standards in precision manufacture.

1805
Oliver Evans builds **America's first steam vehicle**, the amphibious *Orukter Amphibilos* in Philadelphia, US.

▲ 1805
Thomas Telford completes the Pontcysyllte aqueduct on the Welsh borders. It carries a canal high up in an iron trough that rests on 19 masonry arches (see pp.160–63).

◄ **PP.112–13** A 19th-century work by Adolph Menzel depicts an industrial iron-mill.

n 1771, British entrepreneur Richard Arkwright opened a large building filled with machines for spinning cotton on the banks of the Bonsall Brook in Derbyshire, England. Arkwright's Cromford Mill was the world's first real factory, and it marked the beginning of the Industrial Revolution, which started in Britain and spread to the rest of Europe and to North America over the next century.

The Industrial Revolution was the most profound change in human society since the creation of the first cities, and saw new factories and towns filling the landscape. There were dramatic developments in engineering on every scale, from precision cutting of screws to fine tolerances to the construction of mighty bridges. And driving it all was the new power of steam, which had finally come of age.

1820

1820
Hans Christian Ørsted (see pp.134–35) discovers the **link between electricity and magnetism** when he notes that an electric current could make the needle of a magnetic compass swivel.

1821
English physicist **Michael Faraday** discovers the principle of the electric motor, making a magnet move around an electric wire (see pp.140–41).

1825
French engineer **Marc Séguin and his brothers** build the world's first large wire suspension bridge across the Rhône River between Touron and Tain in France (see pp.164–65).

1825
Jacob Perkins demonstrates his high-pressure steam gun. The gun can fire a projectile clean through quarter-inch thick iron plates (see pp.126–27).

▲ 1826
Telford's **suspension bridge over the Menai Straits** in Anglesey, North Wales, is suspended at least 30m (100ft) above the highest possible tide.

1826
English physicist **William Sturgeon** creates the first electromagnet – capable of lifting a 3-kg (7-lb) iron weight (see p.142).

► 1827
English bridge-builder **William Tierney Clarke** builds the Hammersmith Bridge in London – the first suspension bridge over the River Thames (see p.163).

1830

► 1830–31
Faraday and Joseph Henry (see pp.142–43) discover that the movement of a magnet can induce an electric current. They pave the way for the development of electric generators.

1832
The **Séguin brothers** complete the first railway in France, which runs 56km (35 miles) from St Étienne to Lyons. The longest tunnel in Europe is dug for its construction.

▼ 1842
Scottish engineer **James Nasmyth** invents the precision steam hammer, which revolutionizes the shaping of metals (see pp.152–53).

1843
German engineer **Ernst Alban** writes a definitive book on high-pressure steam, which makes him internationally renowned (see pp.132–33).

1849
Tierney Clarke builds the **Szechenyi suspension bridge** in Budapest, Hungary, linking Buda and Pest over the River Danube.

1850

1851
English engineer **Joseph Whitworth** demonstrates his "micrometer" to measure up to one-millionth of an inch (see pp.152–53).

◄ 1852
French engineer **Henri Giffard** (1825–82) creates the world's first dirigible (a light and easy-to-steer airship) that has propellers driven by a steam engine.

1883
After **John Roebling's** death, his son **Washington** and daughter-in-law **Emily** oversee the completion of the Brooklyn Bridge, New York, the world's largest bridge at the time (see pp.166–67).

1898
David Schwarz (1852–97) builds the world's first rigid framed airship. **Ferdinand von Zeppelin** develops it to create the world's largest flying machines (see pp.178–79).

▼ 1899
Brazilian aviator **Alberto Santos-Dumont** builds the world's first successful dirigible powered by an internal-combustion engine (see pp.176–77).

DEVELOPING THE STEAM ENGINE

BY THE 1770s, THE CLANK AND PUFF OF NEWCOMEN ENGINES COULD BE HEARD IN COAL MINES ACROSS BRITAIN. BUT THE AGE OF STEAM REALLY BEGAN WHEN JAMES WATT MADE THE IMPROVEMENTS THAT STARTED DRIVING THE INDUSTRIAL REVOLUTION, FIRST IN BRITAIN, THEN THE WORLD.

CROFTON PUMPING ENGINE
Installed in 1807, this Boulton and Watt beam engine, designed to pump water on the Kennet and Avon Canal, has been working continually for more than 200 years.

Newcomen's engine (see pp.110–11) had proved that steam could be a practical and effective source of power, and by the 1770s, there were hundreds of them pumping water from mines all across the country. But these big, slow engines were underpowered, inefficient, and notoriously temperamental. To turn the steam engine into the reliable, powerful, and efficient workhorse that drove the Industrial Age needed a technological revolution. Leading this revolution was James Watt (see pp.118–21), who completely transformed the steam engine.

INGENIOUS IMPROVEMENTS

Watt's vital innovation was to introduce a separate condenser, so that the steam could be condensed outside the working cylinder, which could always remain above 100°C (210°F). This saved a huge amount of heat. He later made double-acting engines, letting steam in at both ends of the cylinder – these worked more smoothly. He was then able to convert the to-and-fro motion of the piston into a rotational motion, and so drive machines other than pumps. This simple, but key change took steam engines into the factories.

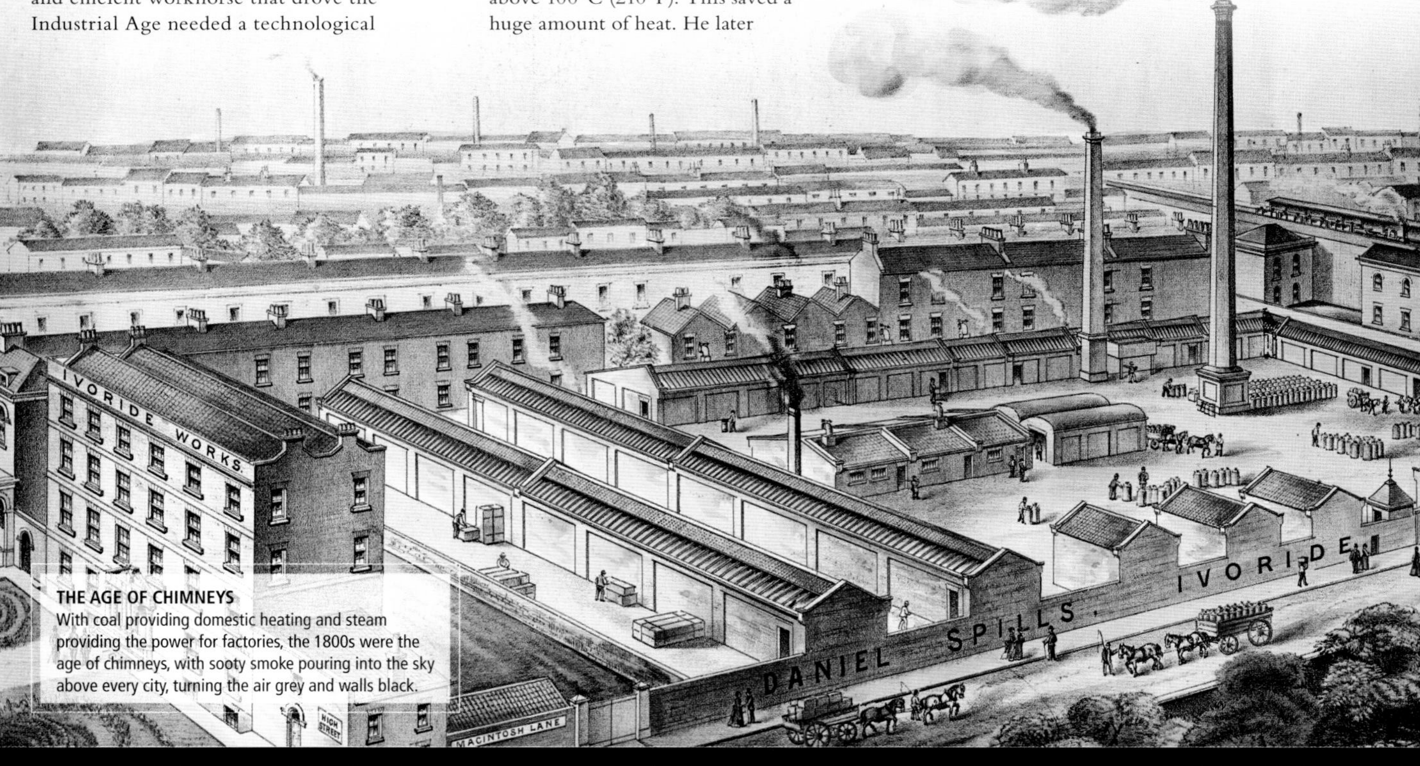

THE AGE OF CHIMNEYS
With coal providing domestic heating and steam providing the power for factories, the 1800s were the age of chimneys, with sooty smoke pouring into the sky above every city, turning the air grey and walls black.

BOULTON AND WATT
The entrepreneurial drive of Matthew Boulton and the inventive genius of James Watt combined to put the steam engine at the heart of the Industrial Revolution.

STEAM LOCOMOTION
Richard Trevithick's Pen-y-darren locomotive of 1804 marked the start of the railway age, dramatically transforming the way people and goods moved.

SETTING THE SCENE

- In 1709, Abraham Darby develops a way of **producing iron in blast furnaces with coke**, making **iron available** in the quantities needed to **build big machines like steam engines**.
- Britain's colonies in India and America provide **vast new markets**, especially in textiles, that are opened by **manufacturing products** on a hitherto unprecedent scale in factories **using machines**.
- As rival sources develop, Cornish tin miners and others **become increasingly concerned about the wasteful fuel use** of the Newcomen engine, and look for an alternative.
- James Watt develops his **steam engine**, which is **capable of running** not only devices such as **pumps but machinery as well**.
- **Steam locomotion** is ushered in with the efforts of Richard Trevithick and George and Robert Stephenson.

It is hard to overstate the impact of the steam engine. Steam gave people a previously unimaginable source of power. For many, it was quite intoxicating. By using the new machines of the age, a few men could produce in a few hours what might have taken an army a month to produce. Indeed, it created a new breed of man – the industrial magnate.

The half century from 1780 onwards saw the steam engine transformed from a local curiosity to the driving force of the industrial age, powering machines in factories across the world and providing locomotion for the new railways.

POWERING INDUSTRY

The factory age had begun with water wheels driving the belts that ran the spinning and weaving machines in the northern textile towns of Britain. But the steam engine took the factory revolution to a whole new level.

Water wheels worked well as long as the stream or river was fairly fast-flowing and consistent, but often the flow would stop completely in a cold winter. Steam engines provided much more power, reliability, and consistency, and they could be installed pretty much anywhere they were needed.

Soon they were being installed in factories and workshops, mines and mills – pumping water and driving machines of every kind. Chimneys, venting the fumes from these engines, filled the skies with smoke as factories spread across the landscape, and new towns mushroomed around them. For the first time in the history of the world, industry, powered by steam, was becoming the driving force of society.

STEAM TRAVEL

At the start of the Industrial Revolution, horse carts and waterways were quite enough to carry raw materials to the mills and finished products to their markets. But soon the mill owners began to look for a way to move things faster, farther, and in larger quantities – both to cope with rising demand, and to beat the competition, which is where a new kind of steam engine came in.

This first generation of engines were too big and heavy to be used for transport, and therefore had to be stationed in one place. But in the 1790s, American engineer Oliver Evans (see pp.124–25) and Cornishman Richard Trevithick (see pp.128–29) began to create compact, powerful engines by using high-pressure steam to force the piston down rather than using the vacuum of condensing steam to pull the piston. In 1804, Trevithick mounted his engine on a carriage and set it on a rail track to create the first steam locomotive. It was George Stephenson and his son Robert, however, that initiated the railway age with the Stockton and Darlington Railway, which opened in 1825, and the Liverpool and Manchester Railway five years later (see pp.190–91).

By the end of the 19th century the railway revolution had spread to Europe, the US, and beyond. In the 1880s, railroad pioneer Charles Francis Adams (1835–1915), referring to the arrival of steam engines in the American Midwest, said, "the young city of the West has instinctively … realized the great fact that steam has revolutionized the world, and she has bound her whole existence up in the great power of modern times".

MODEL ROTARY BEAM ENGINE
The first steam engines had a reciprocating motion suitable only for pumping. Watt introduced rotary motion so that they could be used for driving machines.

JAMES WATT

THE "POWER" BEHIND THE INDUSTRIAL REVOLUTION

SCOTLAND 1736–1819

SCOTTISH ENGINEER James Watt was one of the pivotal figures in the history of the steam engine, transforming it from a pump of limited use to a dynamic source of power driving the Industrial Revolution. His business partner, Matthew Boulton, ensured that Watt's engine was readily adopted by factories and mines across Europe and the United States.

James Watt was born in the Scottish seaport of Greenock on 19 January 1736. The Watt family was comfortably off, with Watt's father earning a living as a merchant, a maker of nautical instruments, and as a small-scale shipbuilder. In fact, it was in the workshop at his father's shipyard that young Jamie found contentment. Here he learned to make wonderful models in wood and metal – from ships' capstans to cranes.

It was no surprise, then, that at 17 he enrolled as an apprentice instrument maker in Glasgow. After a year, he travelled to London, where he worked with the instrument maker John Morgan and learnt to make precise brass rules and scales, and quadrants and barometers.

A year later, Watt returned to Glasgow, where, with the aid of influential friends, he became the Mathematical Instrument Maker to the University of Glasgow. It was a position that brought him into close contact with some of the most brilliant scientific minds of the age, including the Scottish physician Joseph Black, who discovered carbon dioxide. One of Watt's key tasks was to build models demonstrating scientific principles for lecturers. In 1764, he was asked to improve Glasgow University's model of the Newcomen engine (see pp.110–11), which ran rather splutteringly, if at all. But Watt decided to try a few scientific experiments to explore the limitations of the engine before making any adjustments.

THE "PERFECT" ENGINE

It was Watt's approach to improving the workings of a machine that made him so original. He combined a scientific turn of mind with an extraordinary mechanical inventiveness to transform Thomas Newcomen's puffing, clanking, erratic pump engine into the burnished, whirring powerhouse of the new factories. Later in life, he used the same aptitude and skills to create everything from a device to duplicate sculptures to the world's first copying machine.

At the university, he realized that Newcomen engines were extremely wasteful of heat. Steam was let into the cylinder under the piston as it rose. Then the cylinder was cooled with a spray of water to condense the steam and create a vacuum under the piston so that it was forced back down again by atmospheric pressure. But cooling the cylinder to condense the steam meant that a large amount of heat was wasted in simply heating the

THE SOHO MANUFACTORY
This building was the forerunner to the great Soho Foundry in Birmingham where Watt's engines were built. It was established by Matthew Boulton in 1761.

A LIFE'S WORK

- Watt combines **mechanical inventiveness** with a thirst for **scientific knowledge**
- **Transforms the steam engine** from an erratic pump to a factory powerhouse
- **Invents the world's first copying machine** and a new bleaching process
- Uses **horsepower** (following Thomas Savery) **as a measure of engine performance**
- **Creates the "double-acting" engine**, with both the upstroke and downstroke of the piston powered
- **Invents a mechanism** for controlling the **speed of steam engines**
- The SI (International System of Units) **unit of power – the watt –** is named after him

STUDYING THE NEWCOMEN ENGINE
Watt epitomized the genius whose hard work overturned established technology. Here he is seen studying a model of the Newcomen engine and understanding how it could be dramatically improved.

WATT [WAS NOT] ONLY A GREAT MECHANIC; HE WAS EQUALLY DISTINGUISHED AS A NATURAL PHILOSOPHER AND A CHEMIST

HUMPHRY DAVY, BRITISH CHEMIST AND INVENTOR

cylinder again when steam entered for the next stroke. This drove Watt to search for the "perfect" engine that would turn all the heat released by burning coal into mechanical work.

Then Watt had a stroke of genius – to install a separate condenser outside the cylinder. The condenser could be kept cool to ensure a rapid condensation, while the working cylinder could be kept hot constantly. As he developed his idea into a working model, he added further innovations: lubrication, for instance, was introduced to reduce friction.

Later he redesigned his steam engine to make it double-acting. When the steam from the lower end of the cylinder was condensed in this new engine, the piston was pushed down not by air but by more steam, at atmospheric pressure, let in from the top. Then the steam above the piston was condensed in turn, and more steam let in at the bottom. This meant that both up and down strokes were power strokes, and the power of the engine was doubled. This was the first real steam engine, since it did not depend on atmospheric pressure.

A REWARDING FRIENDSHIP

News of Watt's idea reached Matthew Boulton, who was not just a dynamic entrepreneur and owner of one of the country's biggest factories at Soho in Birmingham, but also a brilliant mind with an interest in the latest technological innovations. Watt's idea appealed to him, so he formed a partnership with John Roebuck, owner of Carron Iron Works near Falkirk in Scotland, to develop Watt's engine. But in 1773, Roebuck went bankrupt and Watt went into partnership with Boulton. Their first task was to protect the invention, and Boulton managed to persuade parliament to extend Watt's 1769 patent to give them a monopoly on all steam engines until 1800.

By 1776, Watt's first engine was operational, and it soon became clear that it was far superior to Newcomen's. Cornish tin mines quickly adopted the Watt engine, which not only used less coal, but could also draw water from deeper levels. Watt introduced "horsepower" as a measure of engine performance to show just how powerful his engines were.

However, the real impact of Watt's engine was yet to be felt. Steam engines at that time demonstrated only up-and-down motion, useful primarily for mining and canal pumps. The really large potential market was for factory machinery, but such machines needed rotary motion. So at Boulton's urging, Watt began to develop a rotary engine.

INGENIOUS ENGINE
In this replica of an early Watt beam engine, the balls of the "governor" can be seen in the middle, with the rods of Watt's ingenious parallel motion visible top left, below the end of the beam.

ROTARY POWER

In 1780, inventor James Pickard patented a modification to Newcomen's engine, with a crank to translate the piston's up-and-down motion into rotary motion. So if Watt used the crank for rotary motion, he might have had to pay a royalty to Pickard. Watt then patented an alternative for generating rotary motion, using a "sun-and-planet" gear. It was probably the idea of Watt's engineer William Murdoch (1754–1839), but it was Watt who made it into a viable machine.

The idea was to have two toothed wheels of the same size – one (the sun) in the centre, and the other (the planet) running around it. The planet gear was attached to a swinging rod attached to the engine's beam. As the beam see-sawed, the rod was pumped up and down, driving the planet around the sun, and dragging the sun around with it.

ENGINEERING TIMELINE

1736–54

1755–56
Watt returns to Glasgow and sets up his own business, making and repairing brass quadrants, and parts for telescopes and barometers

1757–64

1765–72
Watt creates a working model of an entirely new type of engine, but he lacks the capital to develop it further

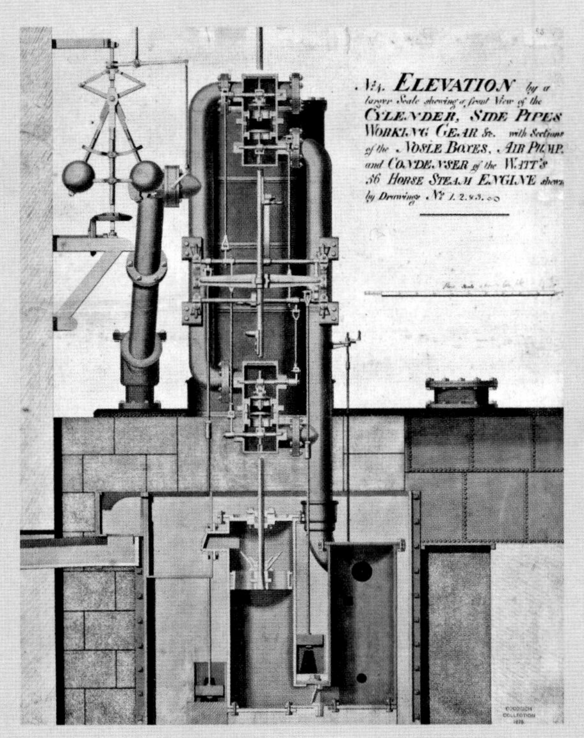

"56 HORSE" WATT PUMPING ENGINE
A contemporary drawing of a Watt engine with the power of 56 horses – based on Watt's calculation that one horse could turn a mill wheel 144 times in an hour.

One of the problems with every early engine, however, was the "dead time" in the cycle between each downward power stroke of the piston. With a pump, that was fine, since all it did was create a regular pause in the lifting of water. Such pauses would not work with spinning machines, which had to run smoothly. Even worse, the mechanism could actually reverse direction during the dead time, with disastrous effects on delicate textile machinery. Again, Watt came up with a solution, patented in 1782. Instead of just the downstroke of the piston being powered, the upstroke would be powered too, with steam let in alternately at each end of the cylinder, creating a "double-acting" engine. This had the added benefit of making the engine twice as powerful for the size.

Watt made one further major modification. The piston rod from his double-acting engine had to push the beam upwards, and moved vertically in a straight line, but the end of the beam moved in a curve, so a direct connection was impossible. Watt solved this problem by using a linking parallelogram of rods. Of all his inventions, Watt was most proud of his "parallel motion".

THE LEGACY

Watt is also credited with the invention of the speed regulating mechanism with the wonderful name "governor". This was a pair of balls that spun on hinged arms around an axle driven by the engine. The faster the engine ran, the higher the balls were flung out by centrifugal force– until they reached the point where, if the engine was running too fast, they would operate a steam valve that let out steam and slowed the engine again.

Many engineering companies wanted to capitalize on the success of the steam engine and launch their own engines in the market. With their legal protection, however, Boulton and Watt were able to maintain their monopoly. Some say that this held back the development of steam technology, as, for instance, Watt was unwilling to experiment with high-pressure steam. But the steam engine might not have become the powerhouse of the industrial age without the commitment of Boulton and Watt.

In 1800, Watt retired to his home in Heathfield, Birmingham, where he worked in his workshop. He died in 1819, and was buried in St Mary's Church, Handsworth, Birmingham.

MATTHEW BOULTON

ENGLAND 1728–1809

Without the backing and encouragement of Matthew Boulton, it is unlikely that Watt would ever have been able to make his engines.

Born in Snow Hill, Birmingham, Boulton was the son of a successful toy and buckle manufacturer. When his father died in 1759, he built the giant Soho Manufactory to bring all his employees together under one roof.

He invited Watt to join him at Soho in 1774 and gave him the backing that ensured the success of Watt's engine. At the same time, he transformed Soho into the world's foremost steam-engine factory. Between 1775 and 1800, Boulton and Watt built a third of the 1,500 steam engines installed around the world, including 300 double-acting engines that produced rotational motion. Each of these large engines could power an entire factory, and it was they, more than anything, that signalled the new industrial age. On a visit to Soho, Boulton told the diarist James Boswell, "I sell here, Sir, what all the world desires to have – POWER".

Boulton also revolutionized coinage with new minting techniques. In fact, impressed by the quality of coins manufactured at his unit, the British government commissioned him to produce 45 million new penny and two-penny coins.

At the same time, however, Boulton was much more than an entrepreneur: he was a man of science who brought together great minds as Joseph Priestley (who discovered oxygen), James Watt, Erasmus Darwin (famous English physician), and Josiah Wedgwood (potter and founder of Wedgwood company). They met as an informal club once a month, on the Sunday nearest to the full moon, and called themselves the Lunar Society.

The first of Watt's engines are installed and working in commercial enterprises, many in Cornish mines, pumping water

Further improvements are added, producing an engine that is more fuel-efficient than the Newcomen engine

1773–74	1775–80	1781	1782–87	1788–1819

att's engine is acquire
ulton, who invests in
att's ideas

CREATIVE HUB OF AN ENGINEERING HERO

WATT'S WORKSHOP

After his death in 1819, James Watt's workshop in his house in Handsworth, Birmingham, was locked and left untouched until 1924. When the house was to be demolished that year, its contents were taken to London's Science Museum and almost forgotten. Only recently has the workshop been restored and put on display. An amazing time capsule of the engineer's work, it contains more than 6,000 items, including the oldest circular saw in the world, parts of flutes, packets of chemicals, and much more. Curator Andrew Nahum recalled that one early witness of the room wrote that it felt "as if the hand and eye of the master has just crossed the threshold never more to enter in this world".

► **SCULPTURE COPIER**
The centrepiece of the workshop is Watt's equal-size sculpture copying machine. It has an arm that follows the contours of the original sculpture precisely. This arm controls the movement of a drill powered by a wheel and belt. Moved by the arm, the drill grinds away new stone in exactly the same pattern. Watt wrapped the end of the arm in rags to avoid injuring his head as he bent to check progress.

▲ **HEAD COUNT**
Perhaps one of the most surprising elements of Watt's workshop is the large number of busts. Watt devoted a lot of time in his later years to developing a machine for copying sculptures. There are more than 400 sculptures in the workshop. There are also moulds for casting sculptures, including a head of Watt himself.

◄ **HEAD SHRINKER**
The slightly smaller machine on the bench is for making copies of sculptures at a reduced size. Seen here is the tracking pointer on the arm above the large original bust, the arm that follows it and the small drill hovering above the tiny reduced version of the bust. The miniatures dotted around the workshop show just how well his machine worked.

◀ **FLUTE-MAKING TOOLS**
Watt spent a great deal of time making musical instruments even though he had no musical skill himself. Musical instruments were widely played in the 18th century, which is why he was fascinated by the idea of reproducing high-quality instruments for a wider market.

▼ **PRINT MAKER**
Watt invented a press for copying letters and drawings. Despite its worn out appearance, this press was cutting edge technology in Watt's day. It made copying quicker and reduced the chances of error when making duplicates.

▲ **CAST-IRON STOVE**
One of the homeliest objects in the workshop is the stove with its coal scuttle. Cast-iron stoves were a new form of heating in Watt's day – safe, low maintenance, and good at keeping the heat in the room. Watt's stove has a frying pan on top, allowing him to cook meals while working, without any interruption.

OLIVER EVANS

CREATOR OF THE FIRST STEAM-POWERED VEHICLE

UNITED STATES 1755–1819

The first American inventor of true genius was Oliver Evans. His automated flour-mill anticipated modern production lines, but his talent was often frustrated by a lack of investment and appreciation. Many of his brilliant ideas were left to others to develop, including the concepts of steam locomotion using a high-pressure steam engine, and steam passenger trains.

A LIFE'S WORK

- Evans introduces **automated production lines** in 1788 – more than a century before Henry Ford mass produces the Model T
- Devises the **high-pressure steam engine**, and, in 1790, **obtains a patent** for it, only the **third awarded in the US**
- Builds America's **first steam-powered vehicle, a dredger**, in 1805
- In 1805, he designs a **refrigeration machine** that runs on vapour – he is often regarded as the inventor of the refrigerator, although he never builds one; his **design is instead modified by Jacob Perkins**
- He titles his first book on steam engineering ***The Abortion of the Young Steam Engineer's Guide*** to indicate his frustration at having his ideas repeatedly shot down

Oliver Evans was born in Newport, Delaware, in 1755 – the same year, incidentally, that America's first Newcomen engine was used to pump water from a New Jersey mine. As a boy, he took a keen interest in engineering and maths, and, at the age of 14, became an apprentice wheelwright.

In the 1770s, while still a teenager, Evans started thinking about ways that carriages could be powered without horses. In his memoirs, he tells how, in 1773, his brothers told him about a blacksmith's boy who filled the blacksmith's gun with water, then rammed in wadding and put the butt end in the fire. As the water turned to steam, it shot out the wadding with a crack as loud as gunpowder. This set young Evans thinking. Atmospheric engines relied on the limited power of atmospheric pressure, brought into play by condensing steam. He wondered if the greater power of steam at high pressure could be exploited instead.

AMPHIBIOUS STEAM POWERED CARRIAGE
Evans' *Orukter Amphibolos* was America's first steam powered vehicle, with a high-pressure steam engine and wheels. It was also the first amphibious vehicle in the US.

AUTOMATED PRODUCTION

By 1777, Evans had made his first invention – a machine for making the wire-toothed leather "cards" used for combing muck from wool and cotton. Evans had the ability to see machines as interlinked steps. He created an assembly line in a box that bent the wires, cut them to length, punched holes in leather, and mounted the wire teeth – all in a sequence.

Evans soon applied his ability of being able to see processes as a whole to automate flour-milling. His mill scooped up the grains of wheat in cups on a belt, poured them onto a conveyor, carried them onto the millstone, caught the flour in a hopper, and then raked and sifted it – with all of the processes driven by a single rotating shaft. It was an automatic production line in 1788 – some 120 years before Henry Ford (see pp.282–85) mastered the process. The automated mill allowed one worker to do the work of five and speeded up flour production dramatically. However, unable to convince mill owners of the benefits of the system, he wrote *The Young Mill-wright and Miller's Guide* (1848), which became a bestseller for a generation. Then, a key sale of his system to the Ellacott brothers in Maryland saw their profits leap, and soon Evans had sold licences to 100 users, including George Washington himself.

HIGH-PRESSURE STEAM

Evans gained America's third ever patent in 1790 for his high-pressure steam engine. He realized that if the steam could be made much hotter, the extra pressure would make James Watt's condenser (see pp.118–21) redundant. The piston could be driven by the pressure of steam that could then be vented into the air. However, to get the steam this hot, he had to find a better way to heat water than the method used in either Watt's or Newcomen's (see pp.110–11) engines. In these, the boiler was little more than a giant kettle. Evans's solution was to wrap the boiler right around the furnace like a jacket. In this way, he could build engines with the same power as Watt engines, but ten times as small. In 1803, visitors to his Philadelphia workshop were impressed to see a relatively small steam engine driving 12 saws to slice through 30m (100ft) of marble in just 12 hours.

Evans had a vision for his high-pressure steam engines. Decades before the Liverpool and Manchester Railway carried the world's first railway passengers, he wrote, "the time will come when people will travel in stages moved by steam engines from one city to another almost as fast as birds can fly". It is possible that British engineer Richard Trevithick (see pp.128–29) saw the drawings. But at that time, no one in the US was very interested in steam locomotion.

A JOURNEY AHEAD OF ITS TIME

When the Philadelphia Board of Health commissioned Evans to build a steam-powered dredger for the Schuykill River in 1805, he seized the opportunity and built it in an extraordinary way. His workshop was more than a mile inland, so after building the dredger, he added wheels and linked them to its high-pressure steam engine to drive it through the streets of Philadelphia in front of the city's citizens, before launching it into the river. The *Orukter Amphibolos* ("Amphibious Digger") thus became not only America's first steam-powered vehicle, but the first amphibious vehicle, too.

Evans was by now famous across the US for his revolutionary automation of flour-milling, but his other far-sighted ideas, such as refrigeration, were ignored. So were his schemes for steam locomotion, intercity passenger trains, and steamships – schemes that sowed the seeds of most of the great advances in transport over the next century. In a book in 1805 he bitterly writes, "And it shall come to pass that the memory of those sordid and wicked wretches who opposed [my] improvements, will be execrated by every good man, as they ought to be now".

However, even if there were no takers for steam locomotion, industrialists were impressed by Evans's dredger engine. In 1811, he opened a factory in Pittsburgh to build customized high-pressure steam engines for various industries, from iron mills and paper mills to waterworks and steamboat companies. By the time Evans died in 1819, he was a rich man, and high-pressure steam was very much a part of the American industrial landscape. Meanwhile, high-pressure steam locomotion was well under way in Britain.

THE IRON CITY

The year 1811 was an important one in the history of Pittsburgh. It saw Robert Fulton (see pp.188–89) launch his steamship in the city, and Evans open his steam engine factory. From then on, Pittsburgh quickly became the industrial powerhouse of the US as it expanded westwards. Situated in the middle of a coalfield and linked by the Ohio River to the Mississippi, Pittsburgh was well placed to supply the US with manufacturers when British goods were cut off in the War of 1812. Over the next 50 years, Pittsburgh's population grew tenfold and it became known as the "Iron City", famous for its iron foundries, and, in 1835, as the first place to build steam locomotives west of the Allegheny mountains.

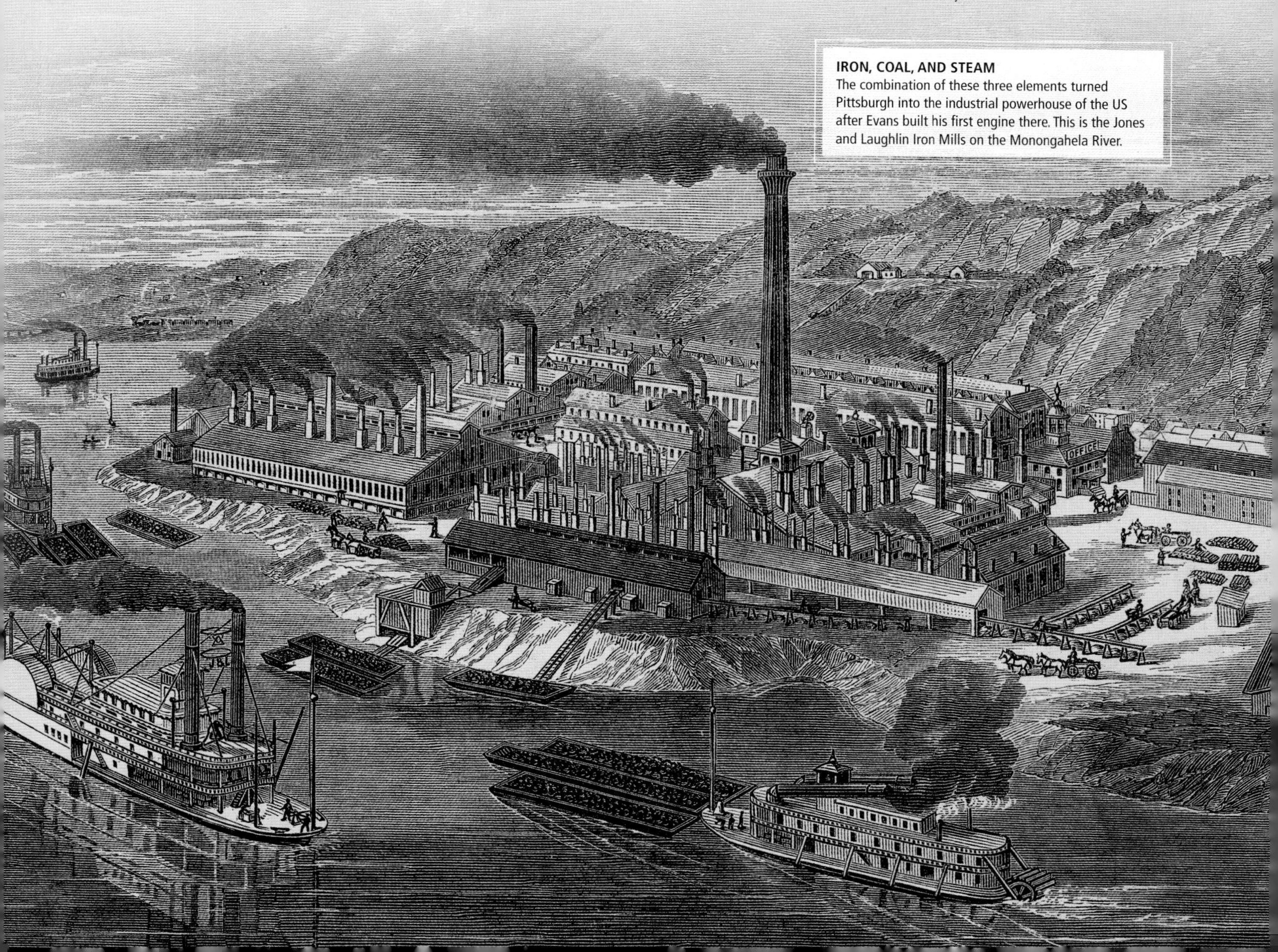

IRON, COAL, AND STEAM
The combination of these three elements turned Pittsburgh into the industrial powerhouse of the US after Evans built his first engine there. This is the Jones and Laughlin Iron Mills on the Monongahela River.

JACOB PERKINS

PROLIFIC GENIUS OF STEAM INNOVATION

UNITED STATES 1766–1849

A SKILLED ENGRAVER and an ingenious problem solver, Jacob Perkins pioneered refrigeration, central heating, and gas cooking. His fame, however, rested on taking steam pressure to unprecedented levels. Public demonstrations of his high-pressure steam gun inspired awe among spectators in London, his adopted home.

A LIFE'S WORK

- Perkins creates the **first unforgeable bank notes** in 1819, by engraving the notes on steel
- Prints **Britain's first postage stamps**, including the famous Penny Black
- Invents an **experimental high-pressure steam engine** operating at pressures of 13,790kPa (2,000psi)
- Works on a **high-pressure steam space rocket** patented in 1824
- Stuns London crowds with his awesome **steam gun** in 1825
- Works in 1839 on the **concept of the hermetic tube**, which is now used in computers and nuclear reactors
- Establishes the **National Gallery of Practical Science** at Adelaide Street, London

Jacob Perkins was born in Newburyport, Massachusetts, on 9 July 1766. At the age of 12 he was apprenticed to a goldsmith and quickly became a skilled engraver. When his master died five years later, the teenaged Perkins took over the business. Within a few years, he was commissioned to make unforgeable dies for the State of Massachusett's coinage. At about this time, he also invented a machine to cut and head nails in a single operation, which heralded a revolution in nailmaking. He went on to invent pumps for fire engines, and machines stamping coins and reconditioning cannons. In 1810, he installed one of the world's first central heating systems in Massachusetts Medical College in Boston, using large pipes to heat the whole building from a single stove.

PERKINS'S PRINTING PRESS
Perkins developed hard-to-forge printing systems for bank notes and stamps, like this press for the Penny Black stamp.

Perkins also developed a method of producing unforgeable bank notes by printing from multiple engravings on tough steel rather than copper. Impressed by his results, Charles Bagot, a British minister in the US, invited him to move to England where there had been a spate of forgeries. In May 1819, the 53-year-old Perkins took his family and key craftsmen to London to set up in business as steel plate engravers with English engraver Charles Heath (1745–1848).

Despite the high quality of their work, the Governor of the Bank of England decided against using a printer from a country Britain had recently been at war with. Yet the business thrived, and Perkins and Heath became the leading printers in London. Significantly, they printed Britain's first postage stamps, including the famous Penny Black.

SIZZLING STEAM

During this time, Perkins was also experimenting with high-pressure steam. In 1822, when pressures of even 34.5kPa (kilopascals) or 5psi (pounds per square inch) unnerved people, Perkins was achieving pressures of 3,450kPa (500psi). His engine had a revolutionary way of heating, using a boiler he called the "generator" – a cylinder of thick gunmetal entirely immersed in the furnace. Steam flashed through pipes from the generator at extremely high temperatures and pressures into the engine's main cylinder with enormous force. No one but Perkins had the skill at the time to build an engine that would work at these extremes. While inventing the engine, he also had to create a special alloy that would work without lubrication, since the extreme temperatures burned any oil off.

Over the next decade, Perkins developed his high-pressure engines, and, in the mid 1830s, introduced the innovative Uniflow engine that exhausts steam through a number of ports after the piston completes its stroke, allowing a one-way flow of steam. In 1839, at the age of 73, Perkins developed the hermetic tube or heat-pipe system, in which heat is transferred along a pipe at high speed with virtually no heat loss – an idea so far ahead of its time that it was not until the 1920s that it was used extensively in steam locomotives. Hermetic tubes are now used to cool computers and nuclear reactors, and were crucial in the Space Shuttle.

THE REVOLUTIONARY GUN

Perkins's most spectacular machine was perhaps his steam gun, which used high-pressure steam to shoot projectiles from a gun barrel.

More than 2,000 years earlier, Archimedes (see pp.38–41) had stunned people with his Architronito steam gun, which could fire a ball weighing 35kg (80lb) more than 1,000m (3,280ft). But Perkins's could punch lead balls through iron plates. The noise was like thunder, and, after a false rumour of an accident, trials at Perkins's factory near Regent's Park were curtailed. After some time, however, the streets were cordoned off for a demonstration in front of top military personnel including the Duke of Wellington.

The impressive demonstration finished spectacularly. As *The Times* reported, "[Mr Perkins] screwed on to the gun barrel, a tube filled with balls, which falling down under their own gravity into the barrel, were projected, one by one, with such extraordinary velocity, as to demonstrate, that by means of a succession of tubes filled with balls, fixed in a wheel … nearly one thousand balls per minute might be discharged." It was a machine gun, a century before its time.

But although the luminaries were impressed, and news of Perkins's terrifying weapon of mass destruction spread quickly across Europe, only the Russians – whom Perkins would not sell to – offered to buy any. In the event, despite the gun's power, the steam generator seemed far too cumbersome to take to battle, and Perkins was reduced to thrilling the public with daily demonstrations at his National (or Adelaide) Gallery of Practical Science.

By that time, Perkins was nearly 80. He died on 30 July 1849 and was buried at Kensal Green, London.

A SKILLED CRAFTSMAN

In his long career, Perkins received 21 patents in the US and 19 in England, for everything from machines for making beds to a steam-powered rocket. What they all had in common was that they usually worked well (although the rocket was never developed). Perkins was a practical man who could turn his hand to anything. He had the mechanical and craft skills to make things work, and his development of high-pressure steam pushed machine-making technology to its limits.

His investors, however, were often frustrated by the amount of time and money he spent toying with new ideas. Although ground-breaking, his steam gun and uniflow steam engine technology were never taken up in the way he had expected.

REFRIGERATION

While working in Philadelphia, Perkins met Oliver Evans (see pp.124–25) and was probably inspired by him to develop the idea of refrigeration. Perkins knew that a liquid absorbed heat when it vaporized. The key to his "refrigerator" was to circulate the vapour (the refrigerant) through metal tubes where it could be alternately compressed and condensed to a liquid, then expanded and vaporized. In 1834, Perkins got mechanic John Hague to build a working refrigerator, which produced some ice, using ether as a refrigerant. Although not developed commercially by Perkins, the vapour-compression cycle is now used for cooling in both domestic and large-scale refrigerators and air-conditioning units.

NATIONAL GALLERY OF PRACTICAL SCIENCE
In 1832, Perkins opened this gallery off London's Strand to showcase the latest scientific discoveries and innovations, including his high-pressure steam gun and a collection of rare electric eels.

RICHARD TREVITHICK

THE UNSUNG HERO OF STEAM POWER

ENGLAND 1771–1833

ONE OF THE GREATEST PIONEERS of steam power, Richard Trevithick died in poverty and relative obscurity, unlike his compatriot James Watt, who was honoured with a memorial in Westminster Abbey, London. In fact, it was Trevithick who introduced the world to high-pressure steam engines and built the first ever steam locomotive in 1802.

A LIFE'S WORK

- Trevithick **stands up for the Cornish miners** against the Boulton and Watt Company
- Develops the **world's first high-pressure steam engine**
- Builds the **first powered road vehicle in Britain**
- Devises the world's **first steam locomotive**
- His later work includes **a new cannon** and **innovative designs for paddle steamer engines**
- Designs a 305m (1,000ft) tall **iron tower**, which bears **a resemblance to the tower built by Gustav Eiffel** in Paris, half a century later
- Is **hailed as a hero in Peru** for his work in the country's silver mines
- **Barely survives a perilous journey** through the Amazon jungle, then luckily meets Robert Stephenson, who pays for his passage back to London

For his fellow Cornishmen, Richard Trevithick was a hero championing their cause. He refused to succumb to the pressure of the Boulton and Watt Company (see pp.118–21), which was unpopular among Cornish miners for the royalty charges on its pump engines and for blocking rival technical innovations. Also, the Cornish people recognized the scale of Trevithick's engineering achievements in developing high-pressure steam long before the rest of the world. Compact, powerful, and economical, high-pressure engines were a huge advance on lumbering low-pressure behemoths, such as James Watt's engines, and paved the way for steam engines to be used for a variety of tasks. This also made steam locomotion possible, since low-pressure engines had insufficient power to move their immense weight on land.

CATCH ME WHO CAN
This is a replica of the "Catch Me Who Can", Trevithick's fourth and last steam locomotive, built in 1804.

THE CORNISH ENGINEER

Trevithick was born in Illogan in Cornwall on 13 April 1771 in the shadow of copper mines, where both his father and uncle were engineers. By the age of 13, young Richard was also working as a mining engineer, and his tasks included looking after steam pumps.

Richard earned a high reputation as a mine engineer. He also endeared himself to the miners in 1796 by testifying against Boulton and Watt when the company took Jonathan Hornblower to court for patent infringement for his modification to the Watt engine design. In 1800, Boulton and Watt's patent expired, and Trevithick began to experiment with his own ideas, possibly after being shown Oliver Evans's work on high-pressure steam (see pp.124–25) by his neighbour and engineer William Murdoch (1754–1839).

Most engineers believed that there was no real power gain with high pressure – and the risk of explosion made high-pressure steam seem dangerous. But Evans showed that the gains could be significant, and physicist Davies Gilbert (1767–1839) confirmed that dispensing with Watt's condenser and venting steam straight into the air should not pose a problem.

In 1800, Trevithick built his first engine. On Christmas Eve the following year, he amazed the citizens of Camborne, Cornwall, with a carriage that moved under its own power without a horse in sight. Another surprise was the chimney, with its puffs of steam vented into the air, earning the vehicle the name "Puffing Devil".

Trevithick later designed a slicker version with his cousin Andrew Vivian (1759–1842), and demonstrated "Mr Trevithick's Dragon" in London in 1803 to much public enthusiasm. However, no investor came forward; the engine was hard to steer, and it frightened horses. To make matters worse, one of Trevithick's stationary engines exploded in Greenwich, killing four men. Boulton and Watt were quick to highlight the dangers of high-pressure steam, even though Trevithick soon came up with a brilliantly simple safety device: a lead plug in the boiler that was normally below the water line, but popped out if the boiler boiled dry.

In the meantime, Trevithick was deciding a bet between two iron foundry owners in South Wales. Samuel Homfray of the Pen-y-darren Works had asked Trevithick to build an engine for his foundry – and he suggested that it could double as a locomotive on the railway line along which horses drew trucks to the wharf,

15km (9.5 miles) away. Richard Crawshay, owner of Cyfarthfa ironworks, bet Homfray that no steam locomotive could do the job of horses. To win the bet, Trevithick needed to build an engine capable of hauling 9 tonnes (10 tons) of iron ore to the wharf. On 21 February 1804, Trevithick succeeded, making history with the world's first steam railway journey.

SOUTH AMERICAN ADVENTURE

Yet still steam locomotion had few takers, and the demands of high pressure made it seem unreliable. In fact, it was so slow to be adopted that by 1811 Trevithick was bankrupt. So, when he was invited to mechanize silver mines in Peru, he was only too willing to accept the offer. To his credit, he made the mine work. But rebellion flared up, the mine was wrecked, and Trevithick was forced to flee. After trekking through the Amazon jungle, and narrowly escaping death several times, the exhausted engineer miraculously met an old friend's son, Robert Stephenson (see pp.190–91), who lent him money to return to England. Stephenson went home to build the world's most celebrated steam locomotive, *Rocket*. Trevithick returned to find that his family had given him up for dead.

Undaunted, Trevithick moved to London and embarked on other schemes. Finally, the years of effort took their toll. He died on 22 April 1833, and was carried to his grave in Dartford, London, by a party of mechanics who revered the remarkable inventor.

CUGNOT'S STEAM CARRIAGE

The "Puffing Devil" was not the first vehicle to move under its own steam power. That honour belongs to a machine built by Frenchman Nicolas Cugnot (1725–1804). Cugnot's carriage had three wheels and a large copper boiler hanging over the front wheel. It was one of the first machines to translate the piston's up-and-down motion into rotary motion, using a ratchet. However, the *fardier a vapeur* ("steam dray"), as it was known, could only run for 20 minutes at a time and was difficult to steer.

THIS IS A MODEL OF CUGNOT'S FIRST CARRIAGE FROM 1769. IT WAS LATER SAID TO HAVE CRASHED INTO A BRICK WALL.

TREVITHICK'S STEAM CIRCUS
In 1808, Trevithick laid out an oval track for a steam locomotive, surrounded by a high fence, in London. The aim was to give people rides in a carriage hauled by an engine, known as "Catch Me Who Can", for a shilling a time. After thrilling visitors for several weeks, the track broke under the weight and the engine overturned.

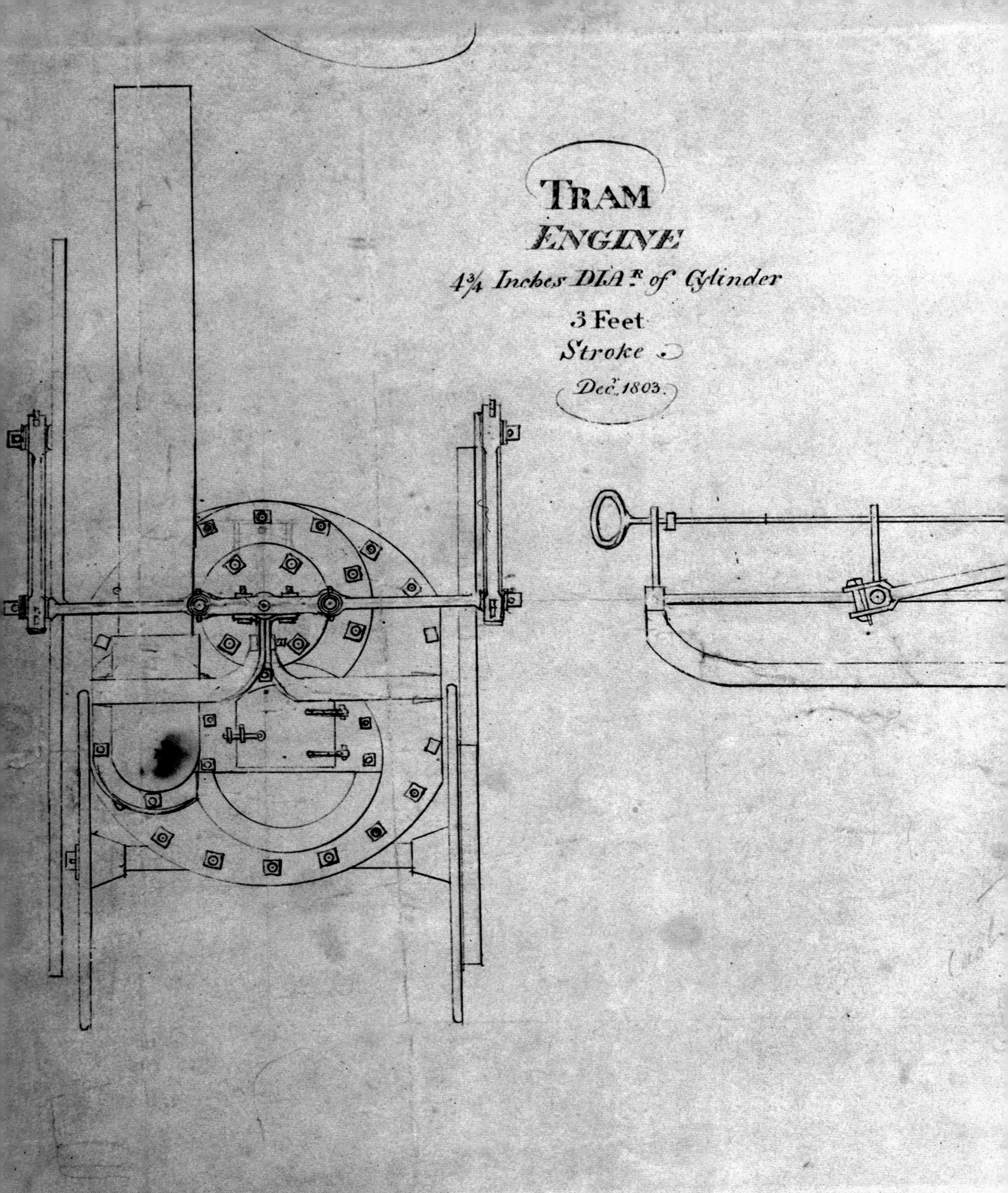
TRAM
ENGINE
4¾ Inches DIA.R of Cylinder
3 Feet
Stroke
Decr. 1803

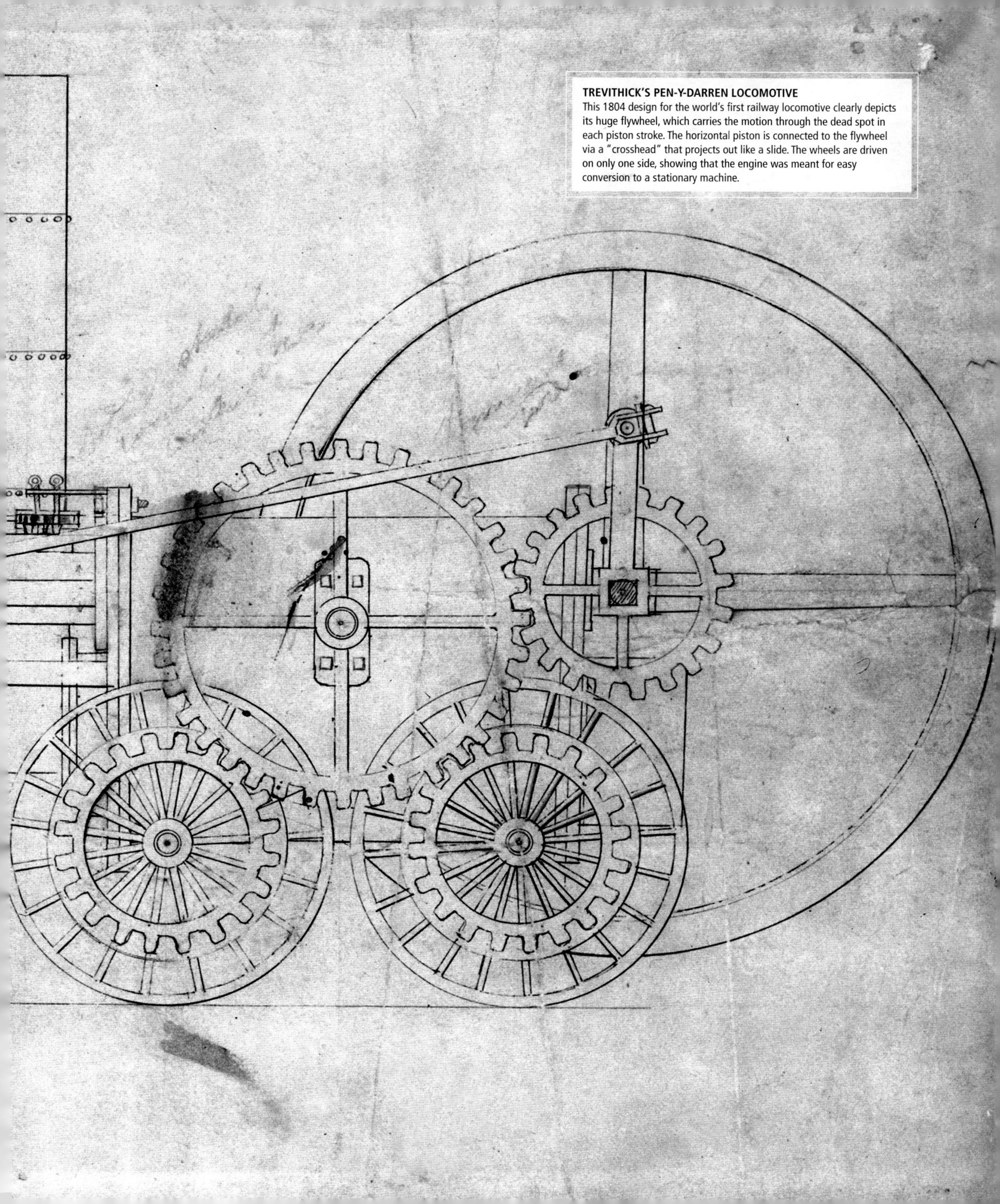

TREVITHICK'S PEN-Y-DARREN LOCOMOTIVE
This 1804 design for the world's first railway locomotive clearly depicts its huge flywheel, which carries the motion through the dead spot in each piston stroke. The horizontal piston is connected to the flywheel via a "crosshead" that projects out like a slide. The wheels are driven on only one side, showing that the engine was meant for easy conversion to a stationary machine.

ERNST ALBAN

EXPERT ON HIGH-PRESSURE STEAM

GERMAN CONFEDERATION 1791–1856

ALTHOUGH TRAINED AS AN EYE SURGEON, Ernst Alban became a master of steam-engineering. He pushed high-pressure technology to its limits, achieving pressures ten times as high as Richard Trevithick, and showed just what it was capable of doing. His textbook *The High-Pressure Steam Engine Investigated* became the definitive guide for steam engineers.

A LIFE'S WORK

- Alban becomes a **successful eye surgeon** by the age of 23
- In 1815, **creates an engine** that runs at **ten times the pressure achieved by Trevithick**
- In 1827, he **decides to build another steam engine**, in England, with the aid of British engineers
- Sets up his own workshop in England in 1840, and **establishes a foundry to make his own engines**, which become famous
- In 1843, his book, ***The High-Pressure Steam Engine Investigated***, is published and becomes the definitive textbook on steam power
- He **builds the steamship *Plauer***, in Plau, in 1845 – the first steamship to **operate on an inland lake** in Europe

In the early 1800s, Germany was a very different country from Britain, where steam engineers such as Richard Trevithick (see pp.128–29) and Jacob Perkins (see pp.126–27) were working. By 1815, Britain had thousands of steam engines, factory after factory churning out goods, and an extensive network of roads and canals to carry raw materials and goods to and fro. North Germany, where Ernst Alban lived, was entirely rural, and towns were only linked by rough trackways and natural rivers. Moreover, Germany was still fragmented into many small states ruled by feudal nobles who had never read the economist Adam Smith and knew nothing of the benefits of free trade.

It is all the more remarkable then that it was here in this largely medieval world that Alban not only became Germany's first major steam engineer, but one of the world's leading experts on high-pressure steam, the cutting edge technology of the day.

THE EYE SURGEON

Alban was born in the city of Neubrandenburg (New Brandenburg) in the southeast of Mecklenburg on 7 February 1791. His father was a pastor and, when young Ernst expressed an interest in studying mechanics, his father quickly scotched the idea. So at the age of 18 he was sent off to the University of Rostock on the northern coast of Germany to study theology. Clearly study did not increase the appeal of theology to Ernst, and after 18 months he managed to persuade his father to let him study medicine, a subject they could both agree was suitable.

In 1812, Alban went to Berlin to study medicine, but took the chance to study physics and mechanics. Two years later, Alban graduated successfully as a doctor and went to Göttingen to learn about eye surgery. Within a year, he had completed his studies, and not only set up his own practice in Rostock, but also started lecturing in anatomy, physiology, and ophthalmology at the university. He was barely 23 at the time. For ten years he built up his practice and became well-known for his highly skilled eye surgery.

During this time, he had also become increasingly fascinated by steam engines, and in particular high-pressure steam engines. He read all of the relevant literature and learned of Trevithick's advances. In 1815, he built his own working model high-pressure steam engine. There were no foundries in Rostock that could

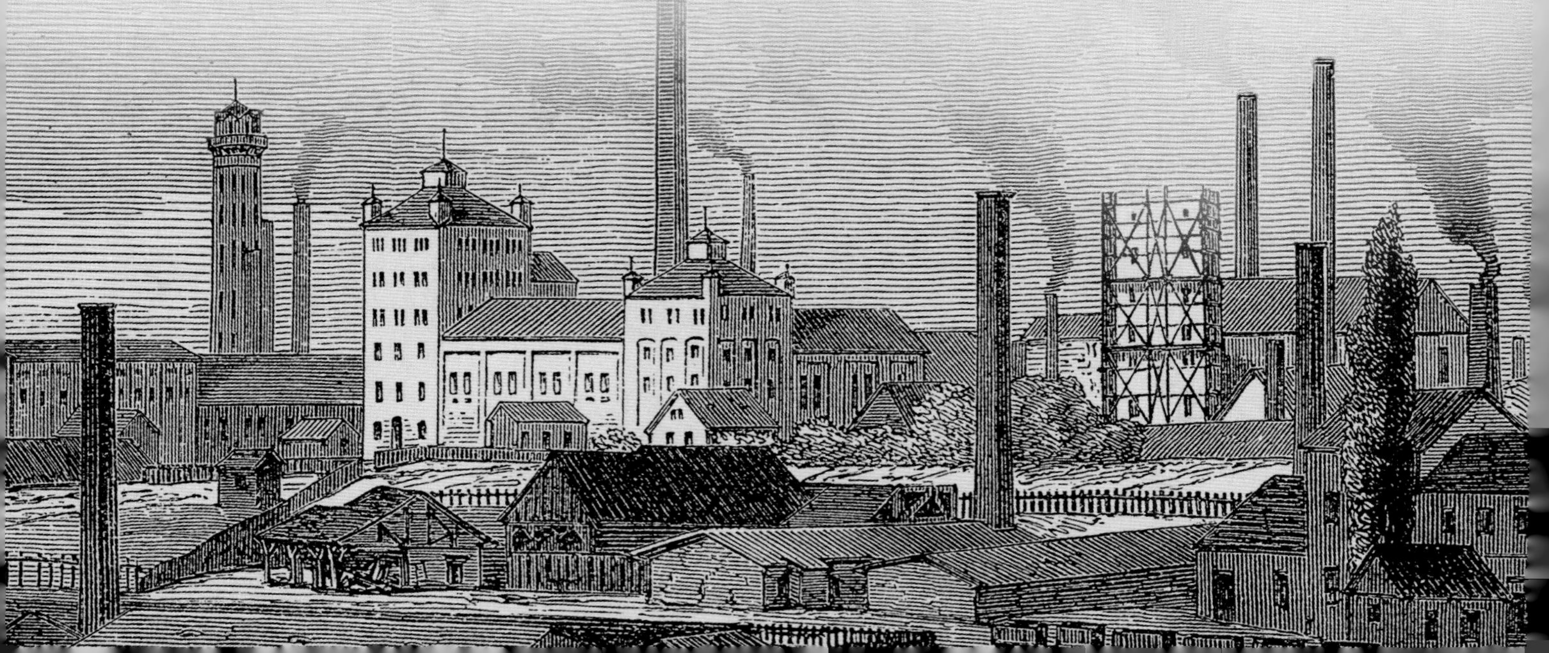

help him with his project, so he improvised, constructing his boiler from an old tin hot-water bottle. Amazingly, with this crude equipment, Alban managed to achieve pressures ten times higher than Trevithick and even higher than Perkins.

In 1825, with money from his practice, Alban decided to leave surgery and devote his life to engineering. After meeting the English consul to Mecklenburg, he decided to travel to London to learn from the skilled steam craftsmen there. He worked in several factories, was able to consider the varied steam engine designs and their operations, and developed an in-depth knowledge of the British methods. In 1827, he decided to build his own steam engine, with the aid of British engineers, but he could not get investors to take it further. So that year, he decided to return to north Germany.

MACHINES FOR THE FARM

Back in his home country, Alban met agricultural pioneer Domänenrat Carl Pogge (1763–1831) and Pogge encouraged him to use his mechanical inventiveness to develop farm machinery rather than steam technology. Alban established Mecklenburg's first engineering institution in 1829 on the farm he bought near Wehnendorf. There he taught enthusiastic recruits to build farm machines to his designs. His most lasting success was a seeding machine, still in widespread use a century later. But he faced continual opposition from Mecklenburg's very conservative government, who felt it was better for the peasants to keep working in the fields than to give up their jobs to machines.

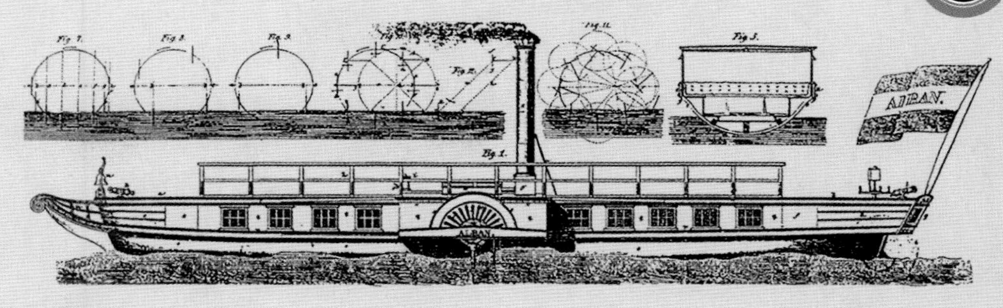

THE *PLAUER*
Ernst Alban's paddlesteamer, the *Plauer*, built to operate on Plau Lake near his home town of Neubrandenburg, was the first steamship to operate on inland waters in Germany.

After a decade, Alban decided to try steam-engineering again, and set up a factory in the town of Plau on the shores of Germany's third largest lake. The factory also made farm and textile machinery, but it was for its steam engines that it became internationally famous. Alban set the standards in high-pressure steam with a boiler that consisted of horizontal tubes, which directed the furnace's heat right into the middle of the water to achieve very high temperatures and pressures. Perkins reached pressures of about 4,000kPa (600psi), but Alban achieved more than 7,000kPa (1,015psi) – so was able to make his engines produce a lot of power while burning relatively small amounts of fuel. In 1843, Alban wrote a textbook, *The High-Pressure Steam Engine Investigated*, that was to make him an internationally renowned expert on high-pressure steam technology once it was translated into English in 1847.

In 1845, Alban built his own paddle steamer, the *Plauer*, to ply the lake. It was the first steamship to work on any inland water in Europe. Despite the fact that it was a great sensation, it was not a success. Fishermen thought it was driving away the fish and passengers were suspicious, preferring to travel by horse and cart. It was sold for scrap after just a few years. Alban's health subsequently deteriorated and he died in 1856.

> ... ONE CANNOT BE TRULY HAPPY UNLESS YOU MAKE YOURSELF AS USEFUL AS POSSIBLE
>
> **ERNST ALBAN**

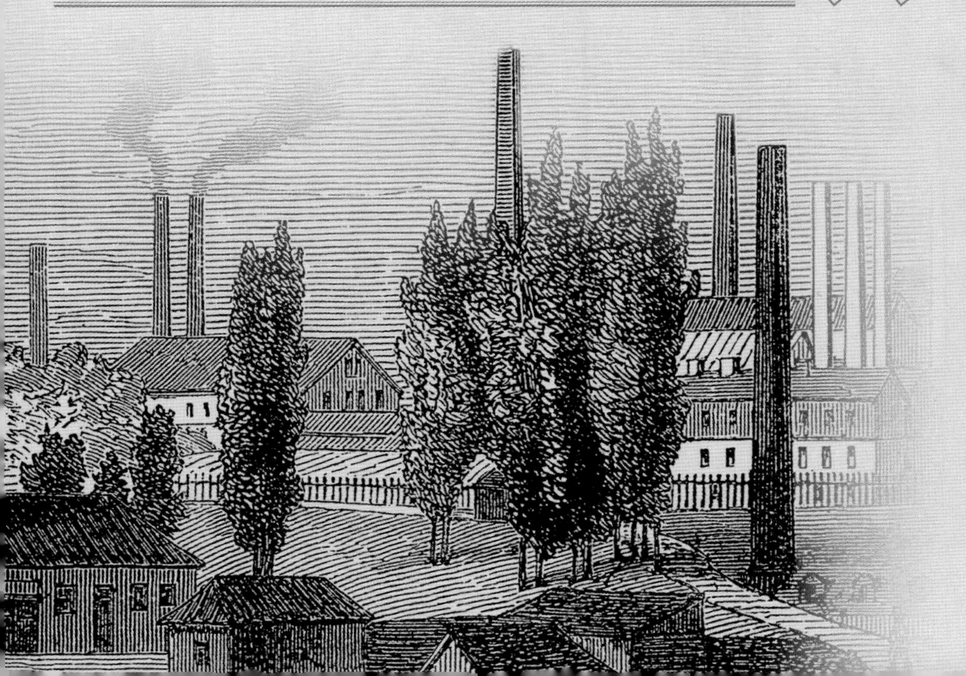

THE RUHR

Germany lagged behind Britain and France in industrializing, but from the mid-1800s began to catch up fast – most notably in the Ruhr Valley. First came water-powered textile factories, powered by the region's fast-flowing streams. As the area was rich in coal, however, it soon got its first steam railway when the Prince William horse railway near Essen was converted to steam and renamed the Steele-Vohwinkler Eisenbahn. In 1849, coke was used for steel production for the first time. By the late 1850s, the Ruhr had 300 coal mines, which turned it into one of the industrial heartlands of Europe.

INDUSTRIALIZING GERMANY
In Alban's lifetime, Germany went from a medieval land of farms to one of the world's most advanced industrial nations. The Ruhr Valley (left) where the famous Krupp steel works were based, became the powerhouse of industrial Germany.

HARNESSING ELECTRICITY

AT THE START OF THE INDUSTRIAL AGE, ELECTRICITY SEEMED TO BE LITTLE MORE THAN AN ENTERTAINING CURIOSITY. BUT AS SCIENTISTS FROM FRANKLIN TO FARADAY TRACKED DOWN THE TRUTH, IT TURNED OUT TO BE THE MOST FAR-REACHING DISCOVERY OF THE ERA.

STORM COVER
The hopeful aim of this novel umbrella was to keep off not only rain but lightning by including a lightning conductor. It was invented by Franklin in the 1750s.

For thousands of years, electricity was thought to be a strange property of materials – such as amber, which is fossilized tree resin – that made things stick together. When rubbed with a cloth, amber becomes highly attractive towards small pieces of paper.

In the 16th century, however, English scientists such as William Gilbert (1544–1603) began to investigate its true nature. In fact, it was Gilbert who coined the term "electric", which comes from "elektron", the Greek name for amber.

EARLY ELECTRIC ENCOUNTERS

Not much progress was made, however, until 1671, when German scientist Otto von Guericke (1602–86) showed how static electricity could be generated by holding one's hand against a sulphur ball that was turned quickly with a handle. Then, in 1709, Francis Hauksbee (1666–1713), a brilliant but little-known English scientist, built an electrical machine – a hand-cranked wheel, which he could use for spinning a glass ball to generate static electricity by holding wool, amber, and other substances to it. With this, Hauksbee created large sparks and electric

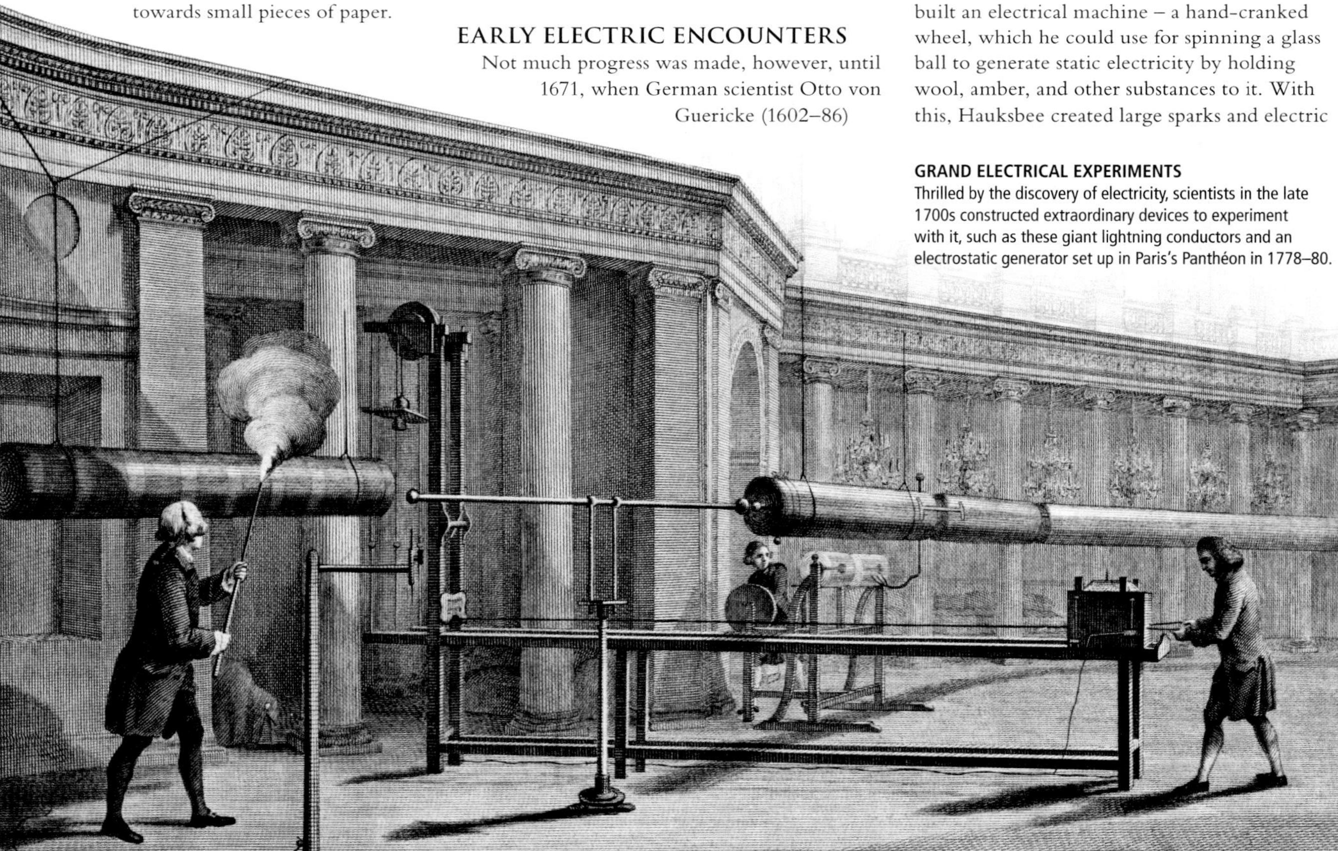

GRAND ELECTRICAL EXPERIMENTS
Thrilled by the discovery of electricity, scientists in the late 1700s constructed extraordinary devices to experiment with it, such as these giant lightning conductors and an electrostatic generator set up in Paris's Panthéon in 1778–80.

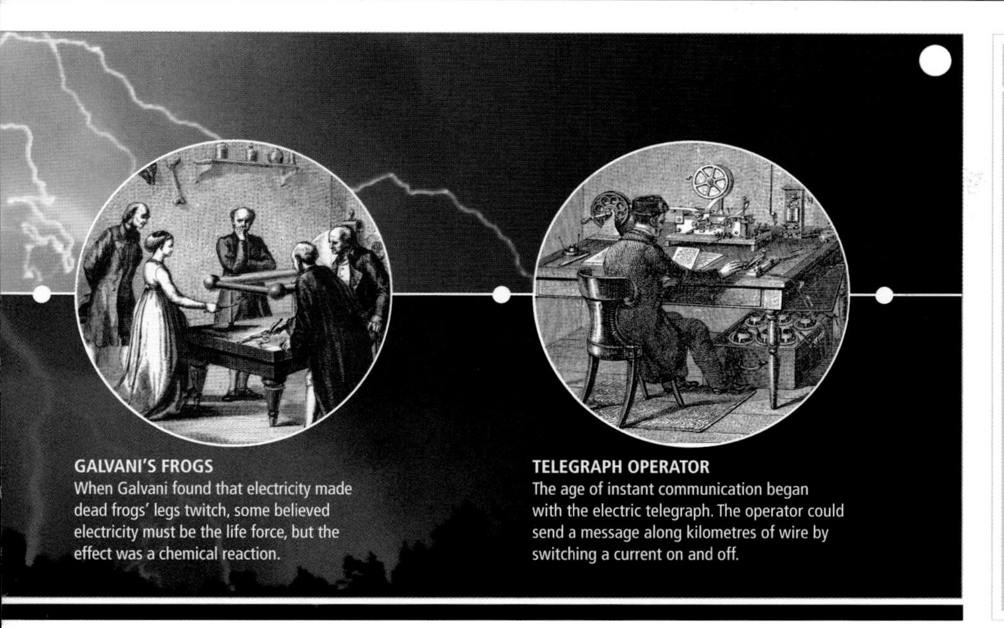

GALVANI'S FROGS
When Galvani found that electricity made dead frogs' legs twitch, some believed electricity must be the life force, but the effect was a chemical reaction.

TELEGRAPH OPERATOR
The age of instant communication began with the electric telegraph. The operator could send a message along kilometres of wire by switching a current on and off.

SETTING THE SCENE

- People from **ancient Greece** notice that **amber, when rubbed**, develops a **strange power of attraction**.
- Many **Christians believe** that **lightning is a sign of God's wrath**. When Franklin shows it is simply a **natural phenomenon** caused by the discharge of electricity, he **challenges fundamental beliefs**.
- For a while, **electricity is thought to be the spark of life**. Volta shows, however, that it **can be produced by a chemical reaction**.
- **Volta's pile** – the **first battery** – allows scientists to **investigate electric current** for the first time.
- The discovery that **magnets can generate electricity** ushers in a **new age of electrical production**, and powers the world for 150 years.

glows so bright that one could even read by them. Stephen Gray (1666–1723), an enthusiast for all things electrical, used similar machines to show how electrical effects were conducted in different ways through different materials, and transmitted electricity through a 250m (800ft) length of twine.

The first major breakthrough in the application of electricity came when Benjamin Franklin demonstrated that lightning is in fact electricity – and not a symbol of the wrath of God, as was the widely held Christian belief.

By the mid-18th century, spectacular electrical sparks, glows, and frizzes could be generated by electric machines and Leyden jars (double-walled glass jars that could store a static charge). Indeed, the effects were so impressive that electricity became a popular form of entertainment. Showmen would, for example, use an electrostatic generator to charge up an attractive young woman to a high voltage, and invite young men from the audience to give her a "shocking kiss". As their lips approached, a painful spark would jump the gap, which was usually enough to deter even the most ardent participants. Other popular forms of electrical entertainment included giving vicars "glowing halos". The human body became increasingly conspicuous in electrical demonstrations.

FORCE OF NATURE

A major breakthrough came when American scientist Benjamin Franklin (see pp.136–37) showed that lightning is electricity – and that electricity is not just an interesting novelty, but a fundamental feature of nature. The impact of Franklin's discovery has been likened to the discovery of nuclear power in the 20th century.

Yet where did electricity come from? Some scientists believed it was biological – a fluid produced by an animal's nervous system. This seemed to have been confirmed when Italian physicist Luigi Galvani (1737–98) found that the severed legs of a frog twitched in response to lightning. Some people believed that electricity was the very force of life.

MODERN EXPERIMENTS

Italian physicist Alessandro Volta (see pp.138–39), however, showed that electricity could be generated by a simple chemical reaction – and created the first battery, the voltaic pile, the world's first continuous source of electricity. A new kind of electricity was soon discovered, one that flowed steadily through a wire like a current of water, instead of discharging itself in a single spark or shock. However, the electric age really began when Danish physicist Hans Ørsted (1777–1851) and English physicist Michael Faraday (see pp.140–41) revealed the link between magnetism and electricity in 1820–21 – and Faraday and others went on to create the first electric motors.

Ten years after Ørsted's discovery, another link between electricity and magnetism was discovered. Independently, Faraday in London and American scientist Joseph Henry (see pp.142–43) in New York found that magnets can create electricity. When a magnet is moved near an electric circuit, it creates a surge of electricity. Using this principle, huge machines could be built to generate an endless amount of electricity. In time, this electricity would be widely distributed, with power plants sometimes located more than 1,600km (1,000 miles) away from consumers. In fact, it was the discovery of the link between magnetism and electricity that, more than anything else, gave us our modern age of electricity, and our lights, televisions, and computers.

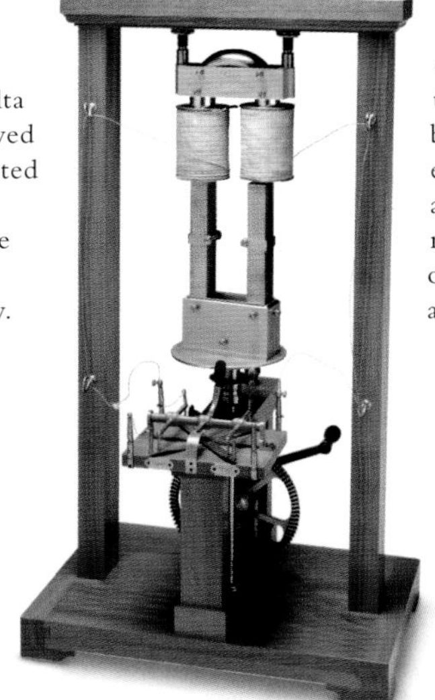

PIXII GENERATOR
Following Faraday's discovery of electromagnetic induction in 1831, young French electrician Hyppolyte Pixii made the first electrical generator. Winding the handle turned a magnet around electric coils.

BENJAMIN FRANKLIN

THE MAN WHO "CAUGHT" LIGHTNING

UNITED STATES 1706–90

ONE OF THE MOST extraordinary figures of the 18th-century Enlightenment, Benjamin Franklin is revered as a founding father of American independence. However, Franklin first became famous as a scientist, carrying out investigations into the nature of electricity and lightning, which eventually led him to invent the lightning rod.

Benjamin Franklin was born in Boston on 17 January 1706. His boldness and curiosity were evident from an early age. Not content with his pace of swimming, for example, little Ben experimented with flippers to increase his speed, and once let a kite tow him swiftly across a pond.

At 13, he joined his elder brother James's print shop as an apprentice, and soon became adept at printing. During this time, he also wrote satirical columns for James's newspaper, the *Courant*, under the pen name Silence Dogood. However, he found Boston's puritanism stifling, and, in 1723, ran away to Philadelphia – his home for the rest of his life.

A LIFE'S WORK

- Franklin establishes a name for himself as a **printer and satirical journalist**
- Makes key discoveries about the **nature of electricity**
- Proves that **lightning is electric**
- **Invents the lightning rod** and proposes its use in protecting buildings
- Devises **bifocal glasses, a heat-saving stove, and the rocking chair**
- Emerges as a key **spokesman for the American cause**
- Becomes one of the **creators of the American Constitution**

Young Franklin set up as a printer in the city, working with such diligence and enterprise that he was soon a revered figure. He was appointed postmaster, and one of his tasks was to despatch botanical samples to Peter Collinson (1694–1768) of the Royal Society in London, which began a lifelong friendship with Collinson.

ELECTRICAL ESCAPADES

In 1743, Franklin was thrilled by a demonstration of electrical effects, and Collinson sent him the equipment to start his own experiments with electricity. At that time, electricians could generate quite large static charges by turning a wheel to rub on sulphur. They could even send a charge several metres along a wet string. But electricity was still little more than a party trick for showmen to thrill the public.

Franklin soon became adept at his own tricks, including one called "The Conspirators", in which he made sparks dance around a portrait of King George II.

Observing that electricity can both attract and repel, scientists such as Charles-François du Fay (1817–78) had suggested that those were two types of electricity, "vitreous", as in glass and wool, and "resinous", as in silk and amber. But Franklin realized that there was just one type – the difference is just the charge moving from one place to another, creating a surplus here and a deficit there, a "positive" and a "negative" charge. In fact, he deduced that electricity is not unique to certain materials but present everywhere. He rightly guessed that it is

ELECTRICITY IN THE SKY
Here, Franklin performs his most famous experiment, proving that lightning is electricity by flying a kite into a thundercloud and watching the sparks fly off a brass key on the kite's line.

THE GULF STREAM

Franklin had long been intrigued by the fact that ships could travel far faster eastwards across the Atlantic to Britain than they could on the return trip. In 1768, the British Treasury commissioned him to investigate this. He rightly guessed that the key was the Gulf Stream, long known to sailors but never studied. He and veteran mariner Timothy Folger made the first map of this current, showing its origins in the Gulf of Florida and its course up the eastern seaboard of the US, before heading towards the east, across the ocean. He also made a series of measurements of deepwater temperatures to establish the location and width of the stream.

carried by "subtle" particles, which are so small that they could easily slip through all matter – an idea finally proved when J J Thomson (1856–1940) discovered electrons 150 years later. Franklin also noted that electric sparks seemed to be drawn towards needle points.

THE POWER OF LIGHTNING

Franklin was always intrigued by the similarity between thunder and lightning and the snaps and sparks in his demonstrations. He was convinced that lightning is natural electricity, and thus must be drawn towards sharp points on the Earth's surface. However, rather than leaping to snap judgments, he decided to experiment and observe.

By July 1750, he had devised his "sentry box" experiment to prove that lightning is electric. An iron rod was placed on top of a structure similar to a sentry box. The rod projected into the box, where a "sentry" would hold a wire attached to the rod via an insulated wax handle. If lightning struck the rod, he would see the spark.

In 1751, Collinson persuaded Franklin to publish his observations in the book *Experiments and Observations on Electricity*. In this small work of 86 pages, Franklin coined basic electrical terms, such as positive and negative, which was his most significant scientific contribution. The book was a great success, and Franklin's fame in Europe was secured in 1752, when French scientist Thomas d'Alibard (1709–99) tried the sentry box experiment – and found that it worked.

Meanwhile, Franklin came up with another lightning experiment. This involved flying a kite in a thunderstorm to draw electrical charge down the line to a key hanging by a silk thread. In June 1752, Franklin went out in a summer storm to fly his kite into the clouds, after insulating himself against electrocution Within minutes, sparks were streaming off the key. It was enough to show Franklin the cloud was electrified; if lightning had struck, of course, it would have killed him.

Franklin also proposed a lightning rod, a needle of iron pointing into the sky to protect buildings. He believed that lightning did its damage when its path to the Earth was blocked by insulating materials, into which lightning's charge flows faster than it can escape, making the material explode. A lightning rod would let the charge escape harmlessly through a good conductor.

INVENTOR AND STATESMAN

Franklin's scientific fame rests principally on his electrical discoveries, but he was interested in all nature's wonders, making key observations on storm movement and ocean currents (see box, left).

His fame as a scientist ensured that when he travelled to England in 1764 he was greeted enthusiastically by scientists such as Joseph Priestley (1733–1804). Increasingly, however, his ability to articulate the American cause led him to spend much of his energy on diplomacy, first in England – he lived in London for 16 years – and then in France, where he was the first American ambassador. He was in Paris in 1783 to witness the first flight of the Montgolfier brothers' hot-air balloon (see pp.172–73). By then, Franklin was almost 80, but that did not stop him from playing a key role in framing the American constitution. He died in 1790, and was buried in the presence of 20,000 mourners.

GLASS ARMONICA
After seeing wine glasses used to make music, Franklin invented a new instrument, a glass armonica, in which 24 glass bowls were mounted on one shaft so that they could all be rotated together, and then played with wet fingers.

ALESSANDRO VOLTA

INVENTOR OF THE WORLD'S FIRST BATTERY

LOMBARDY 1745–1827

ITALIAN ARISTOCRAT Count Alessandro Volta was a meticulous experimenter whose careful studies of electricity revealed that a charge can be generated chemically as well as by friction. His disagreement with fellow scientist Luigi Galvani over the concept of animal electricity led to his invention of the Voltaic pile – the first true battery and a ready supply of electricity.

A LIFE'S WORK

- Volta **discovers methane** while collecting gas from marshes
- Argues with Luigi Galvani on the **source of electricity that makes a frog's legs twitch**, when they are detached from its body
- **Rejects Galvani's "animal" electricity in favour of bimetallic electricity**, which is generated by the contact between different metals
- In 1779, he is appointed **professor of experimental physics** at the University of Pavia in Italy
- In 1800, he invents the **world's first battery, the Voltaic pile**
- Also pioneers the **remotely operated pistol**, whereby a Leyden jar is used to send an electric current from Como to Milan, a distance of 50km (30 miles), which in turn, sets off the pistol; this invention was a forerunner of the **idea of the telegraph**, which also makes **use of a current to communicate**

Alessandro Volta was born in Como in Italy on 18 February 1745 to a deeply Catholic family. Of the seven surviving Volta children, only Alessandro did not join the Church. He was educated at a Catholic school, but after making friends with the Italian physician Giulio Gattoni (1741–1801), who had his own science laboratory and natural history museum, Volta took up science.

At the age of 23 Volta wrote his first paper on electricity, and his interests were wide ranging. In 1776, he discovered methane (natural gas) and showed how it could be ignited by an electrical spark, just as sparks are now used to ignite fuel in internal combustion engines.

FASCINATION WITH ELECTRICITY

In 1774, Volta was appointed professor of physics at the University of Pavia – a position he held for 40 years – and established a reputation for his contributions to various branches of science. His focus, however, was on electricity, and his demonstrations were immensely popular.

In the late 18th century, people were captivated by the extraordinary phenomena of electricity. The sparks, flashes, tingles, and invisible attractions induced by "artificial" electricity were thrilling – whether generated by a spinning disc or released from a Leyden jar – a device that stores electricity between two electrodes on the inside and outside of a jar. Furthermore, Benjamin Franklin had shown in the 1750s that lightning was a natural form of electricty (see pp.136–37). Miraculous powers were attributed to this strange force, from curing sexual impotency to raising the dead, and electrical demonstrations were hugely popular.

Amid all the sensationalism, however, there was some serious scientific research. Italy became the focus of this research, and Italian scientists such as Giuseppe Toaldo (1719–97), Eusebio Valli (1755–1816), and Giambatista Beccaria (1716–81) all made notable contributions.

Volta is credited with inventing the electrophorus in 1776. The device provided an instant source of electricity for the experimenter, and was the equivalent of a modern-day capacitor. It consisted of a resin disc rubbed with cat fur to give it an electric charge. Each time a metal disc was placed over the resin, the charge was transferred, electrifying the

VOLTA AND NAPOLEON
In 1801, Volta demonstrated to Napoleon how electricity from his new pile could decompose water, earning himself the emperor's praise and an aristocratic title.

THE LANGUAGE OF EXPERIMENT IS MORE AUTHORITATIVE THAN ANY REASONING: FACTS CAN DESTROY OUR RATIOCINATION, NOT VICE VERSA

ALESSANDRO VOLTA

FRANKENSTEIN

In her famous book *Frankenstein* (1816), Mary Shelley (1797–1851) never specifies the means by which Frankenstein brings his monster to life. Yet it is clear she is talking about galvanism – the contraction of muscles by electric current – named after Galvani by Volta. Many people genuinely believed electricity was the vital spark that gave life to plain flesh and blood. At the time Shelley was writing, Italian physicist Giovanni Aldini (1762–1834) was conducting public shows in which he tried to bring dead bodies to life with shocks of electricity. It seems absurd, but Aldini was a serious scientist who made major contributions to lighthouse technology.

metal disc repeatedly. Swedish professor Johan Wilcke (1732–96) had invented a similar device in 1762, but it was Volta who named it and popularized it.

BATTLE OF THE FROG'S LEGS

Volta's respectful scientific rivalry with Italian physicist Luigi Galvani (1737–98) began in 1786. In April that year, Galvani took the legs of a frog – severed at the spinal cord – and put them on a table on a terrace at the Zamboni Palace near his home in Bologna. He wired up the legs to a metal clothesline, and waited for a storm. As the storm broke out, the frog's legs twitched – just as Galvani had seen them do when connected to a Leyden jar. Galvani's experiment was the proof that natural electricity from the sky is the same as artificial electricity from a Leyden jar or a generator. Both triggered the frog's nerves, also proving that nerves work electrically. All this made sense, but there was another observation Galvani made that did not.

When the frog's severed legs were touched by a metal scalpel near a spinning generator disc, the legs twitched too – even though the scalpel was not touching the generator. It was this observation that launched the experimental skirmishes between Volta and Galvani. The legs were certainly being twitched by electricity, but where was it coming from?

THE FIRST BATTERY

After a series of experiments, Galvani concluded that the source was the animal's muscles, and that "animal electricity" is inherent in all animals. At first, Volta was convinced, saying that Galvani's experiments showed that animal electricity was "among the demonstrated truths". But then Volta began to have his doubts. He believed that the electrical charge was actually bimetallic electricity coming from the contact between two different metals.

For a decade, the argument went back and forth. Every time Galvani came up with an experiment that seemed to prove the existence of animal electricity once and for all, Volta responded with a counter experiment that proved his idea of metal-contact electricity. It was Volta's work, however, that had the lasting impact.

In 1800, for his final experiment to prove the existence of bimetallic electricity, Volta took several dozen discs made of copper and zinc, stacked them alternately, and separated each one with cardboard dipped in salty water. Touching the stack gave Volta a mild shock. Adding more discs resulted in a bigger charge. Volta had invented the battery, soon known as the Voltaic pile. It became the key electrical source for the next 30 years until English physicist Michael Faraday (see pp.140–41) and American scientist Joseph Henry (see pp.142–43) discovered how to generate electricity magnetically.

EQUAL VALIDITY

After Volta's triumph in creating the battery, Galvani and his theory of animal electricity were ridiculed – even though Galvani had come up with an equally ingenious answer that proved his case, too. When Napoleon conquered northern Italy in 1797, Galvani resigned from his professorship. He died in 1798. Volta, on the other hand, was made a count by Napoleon in 1801 in recognition of his work.

Volta died on 5 March 1827 at his estate at Camnago near Como. It is now known that both Volta and Galvani were right. Volta's contact electricity is just one example of how a chemical reaction creates electricity, while Galvani's animal electricity – electricity created by body cells – is another.

VOLTA'S PILE
The chemical reactions between alternating zinc and copper discs in Volta's pile created the first true battery and the first continuous stream of electricity, leading to systematic study of the process.

MICHAEL FARADAY

DISCOVERER OF ELECTROMAGNETISM

ENGLAND 1791–1867

ONE OF THE GREATEST of all experimental scientists, Michael Faraday was a visionary who saw the underlying unity among the forces of nature. Faraday's discoveries in electromagnetism gave us the electric motor and the generator – unlocking the door to the electrical revolution and paving the way for a host of modern technologies, from television to mobile phones.

A LIFE'S WORK

- Faraday **escapes from his life as a poor bookbinder** and becomes an **assistant to British scientist Humphry Davy**
- **Discovers the principle of the electric motor**, whereby a **magnet moves in a circle around an electric wire** and vice versa
- He discovers **how to generate electricity with magnets** – electricity that would bring **telecommunications and electric lighting** to the world
- He **excites London with his scientific lectures** at the Royal Institution, where he also gives **Christmas lectures for children**, in which he **explains crucial scientific ideas**
- In 1845, he makes the remarkable revelation about the **link between electromagnetism and light**
- **Proposes the idea of "fields of force"**, a revolutionary assertion at the time that paved the way for all of modern physics

In 1791, when Faraday was born, electricity was still the domain of the showman – replete with displays of sparks and flashes, and of fanatical scientists, who were convinced that electricity could animate lifeless flesh and blood. For example, in 1818, Scottish doctor Andrew Ure gruesomely made the corpse of an executed Glasgow murderer, named Matthew Clydesdale, dance like a puppet.

But amid the electrical hysteria, serious scientific advances were being made. Alessandro Volta (see pp.138–39) showed that electricity can be created by a chemical reaction, and with the help of Volta's battery, various scientists studied electrical circuits and currents. French physicist André Ampère (1775–1836) learned about the strength of currents, and German physicist Georg Ohm (1789–1854) studied electrical "resistance".

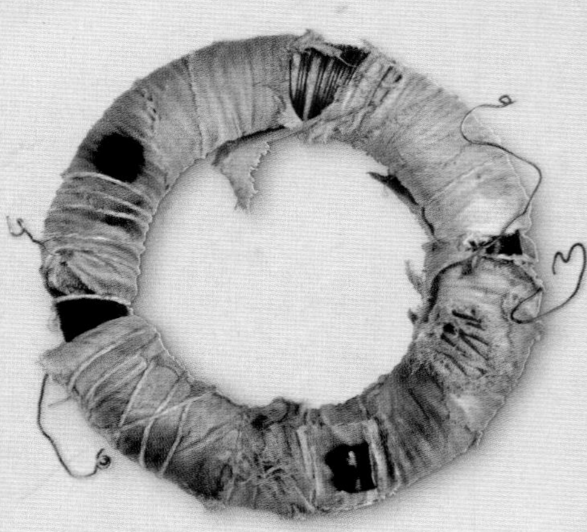

MAKING ELECTRICITY
This iron ring has two coils of wire wound through it. Faraday found that switching on a current in one coil induced a magnetic field in the ring, which in turn induced a current in the second coil.

Faraday was born in the poor district of Newington Butts in London. His father was a blacksmith, but was so disabled that he could barely work. Faraday recalled how many times the family had just one loaf of bread to sustain them for a whole week. So he considered himself lucky when he was taken on as an apprentice in George Riebau's bookshop at the age of 13.

LEARNING FROM THE BEST

In Riebau's shop, Faraday ran errands and learned how to bind books, and as he bound the books, he read them avidly – especially the science books. He became fascinated by the subject, and the generous Riebau let young Faraday set up a makeshift laboratory in the back of the shop.

By another stroke of luck, one of Riebau's customers, William Dance, a member of London's Royal Institution, heard of Faraday's interest in science and gave him tickets to the celebrated lectures of Humphry Davy (1728–1829). Faraday made avid notes, with meticulous illustrations – and sent a bound copy of it to Davy. Suitably impressed, Davy made Faraday his assistant, and, when he went on a tour of Europe the following year, took Faraday along, introducing him to scientists such as Ampère and Volta. Faraday received his scientific education at the hands of the greatest minds of the age, and within a few years, he was not just helping Davy with his experiments, but actually conducting his own.

MAGNETIC ATTRACTION

In 1820, the Danish scientist Hans Ørsted (1777–1851) found that an electric current could make the needle of a magnetic compass swivel. This startling observation, demonstrating a hitherto unsuspected link between electricity and magnetism, unleashed a torrent of scientific experimentation.

By 1821, as news of Ørsted's discovery of the link between electricity and magnetism began to spread, Davy asked Faraday to make a survey of the latest research on electricity. Faraday not only did that, but went much further – setting up an ingenious demonstration of his own to show how a magnet would move in a circle around an electric wire, and an electric wire would move in a circle around a magnet. Faraday had discovered the principle of the electric motor.

Ten years later, Faraday made an even more important discovery. He found that moving a magnetic field can create or "induce" a current of electricity. This principle of electromagnetic induction – discovered independently by Joseph Henry in the US at about the same time (see pp.142–43) – meant that machines could be built to generate huge quantities of electricity, paving the way for everything from electric lighting to telecommunications.

Yet the electric motor and the principle of electric induction are not the only great achievements. With his mentor Davy, Faraday also investigated the process of electrolysis to discover new chemical elements.

UNITY OF THE FORCES

In 1845, in one of his most brilliant experiments, Faraday showed that light is also affected by magnetism. He went on to suggest that there was ultimate unity between all forces, including electricity, magnetism, light, and gravity. He also developed the idea of "fields of force", in which a field is a measurable physical phenomenon that contains some form of energy and occupies space.

Faraday came to believe that fields were made up of lines of force – these lines demonstrated graphically by the patterns iron filings made around a magnet. He appreciated that magnets induced electric currents by creating moving lines of magnetic force that dragged the electric charge as these magnets moved. The idea of force fields is almost taken for granted today, but in Faraday's time, it was so radical that few people even understood it, let alone accepted it. Faraday's crucial insights set the stage for all of modern physics and many of today's technological achievements, from television to mobile phones.

By the 1850s, however, Faraday began to have frequent dizzy spells, headaches, and memory loss. In 1862, he finally wrote to his friend, German chemist Christian Friedrich Schönbein, "Again and again I tear up my letters, for I write nonsense. I cannot spell or write a line continuously. Whether I shall recover – this confusion – I do not know. I will not write any more". In his last days, Faraday was given a "Grace and Favour" residence at Hampton Court Palace by the Queen. He died there on 25 August 1867, at the age of 76.

Faraday was at the forefront of the tremendous advances in the science of electrochemistry and electromagnetism – advances that not only revealed it as one of the fundamental forces of the universe, but saw electricity transformed into the incredibly versatile energy source now so crucial to modern technology.

CHRISTMAS LECTURES

Faraday became famous for his exciting and carefully prepared public lectures at London's Royal Institution, where he worked from 1824. In the most spectacular of these lectures he placed himself inside a steel cage – later called a Faraday cage – while gigantic sparks of electricity were shot around the outside to demonstrate that the cage would protect him. Faraday had proved that charge resided only on the exterior of a charged conductor, and exterior charge had no influence on anything enclosed within a conductor. His inspiring Christmas lectures for children were also very popular. It included a series called The Chemical History of a Candle, in which he used a candle to introduce a wealth of scientific ideas, from chemical elements to human respiration. The tradition of science lectures for children at Christmas continues at the Royal Institution today.

AT WORK IN HIS LABORATORY
Faraday was based at the Royal Institution in London all his scientific life. He was the first Fullerian Professor of Chemistry at the institution. This fanciful illustration shows him conducting experiments there in the 1850s.

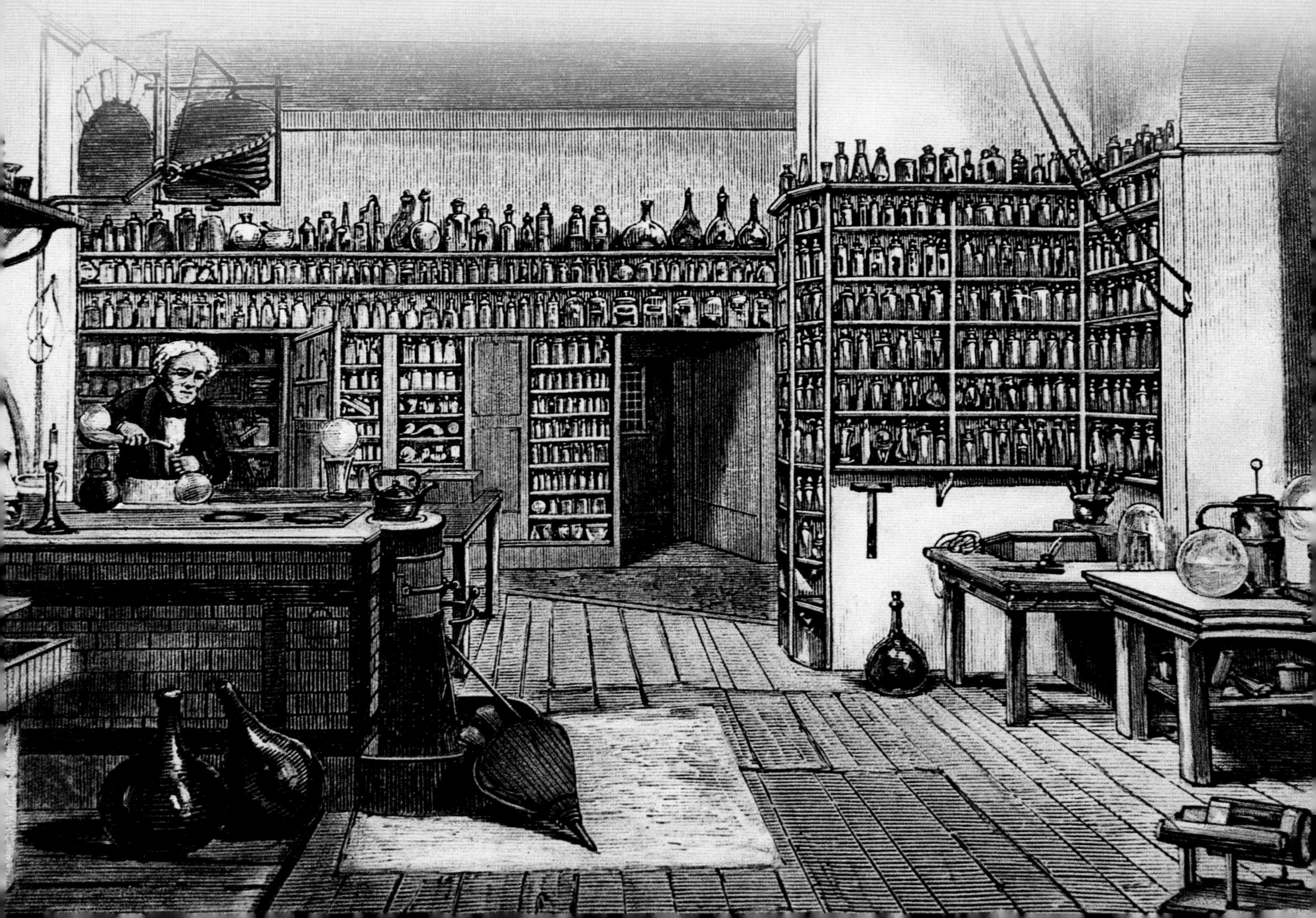

JOSEPH HENRY

DEVISER OF THE ELECTROMAGNETIC RELAY

UNITED STATES 1797–1878

Perhaps the greatest American scientist and engineer of the 19th century, Joseph Henry made immense contributions to the study of electricity. Like his contemporary Michael Faraday, he discovered the principle of electromagnetic induction, and his insights led to the development of the electric telegraph, the forerunner of modern telecommunications.

After Benjamin Franklin's discoveries of positive and negative charges and the electrical nature of lightning (see pp.136–37), the focus of research on electricity shifted to Europe, and the US became something of a science backwater. It is perhaps all the more surprising, therefore, that Joseph Henry, a poor boy from upstate New York, should manage to rival the brilliance of his contemporaries across the Atlantic in developing electromagnets, as well as deploying magnets to generate electricity.

Furthermore, like many scientists of the time, Henry's achievements spanned several disciplines. He made key contributions to the science of acoustics, the study of sunspots, balloon flights, and lighthouses.

Henry was born in Albany on 17 December 1797 to Scottish immigrants who had arrived in the US in the turbulent year of 1775, the beginning of the American Revolutionary War. His father, William, was a labourer who died when Henry was young, after which he was sent to live with his grandmother in Galway, New York. There he worked in a general store, going to school after work.

A few years later, he returned to Albany to work as an apprentice watchmaker and silversmith. In his spare time, he made a name for himself as an amateur actor, standing out with his delicate good looks and vivacious nature. At about this time, he chanced upon George Gregory's *Popular Lectures on Philosophy, Astronomy and Chemistry*. He was so captivated by the glimpse of science it afforded that he enrolled himself in the local college, Albany Academy, supporting himself by teaching maths.

A LIFE'S WORK

- Despite poverty, Henry performs exceptionally well in college and goes on to become **Professor of Mathematics at the age of 26**
- Creates the **most powerful electromagnet** in the world
- Discovers **electromagnetic induction**
- Makes **pioneering discoveries in acoustics**
- Contributes towards the **study of sunspots**
- Helps make the **Smithsonian the world's leading scientific institution**

BRUSH WITH SCIENCE

At first Henry wanted to become a doctor, and studied natural sciences. Then, one summer, he was asked to help survey the route for the new road between the Hudson River and Lake Erie. Henry found the experience so exciting that he decided to study engineering instead. By this time, he had also shown such skill as a teacher that he was appointed as Professor of Mathematics at the Albany Academy.

However, he found teaching arduous, and was instead thrilled by the experiments that he tried in his spare time. He was particularly drawn towards electromagnetism. Inspired partly by reports of the electromagnet made by William Sturgeon (see box, above), and partly by reading the works of André Ampère (1775–1836), Henry experimented with electromagnets and electrical circuits. In doing so he worked out for himself many of the principles of circuits being formulated at the same time in Europe.

MIRACULOUS MAGNETS

Aided by his new-found knowledge of electricity, Henry made a simple but important discovery about Sturgeon's electromagnets. Like Gerrit Moll (1785–1838) in the Netherlands, he found that if the wires making the coil of an electromagnet are insulated from one another, their magnetic effect is magnified. It is even said that he used strips of his wife's silk petticoats for insulation (and they worked). In 1827, he made the first genuinely powerful electromagnet, able to carry 340kg (750lb) of iron and powered by a small battery. A few years later, he made an electromagnet that could lift more than 1.36 tonnes (1.5 tons). Sturgeon was quick to acknowledge the superiority of Henry's magnets, writing, "Professor Joseph Henry has been enabled to produce a magnetic force, which totally eclipses every other in the whole annals of magnetism;

INSTANT ELECTROMAGNET
There could be no simpler electromagnet than this one, made by Henry. It was made by using two plates of copper and zinc that would be immersed in diluted acid. This made a simple Voltaic cell that generated a current in the insulated coils of wire that thus become magnetized.

WILLIAM STURGEON

ENGLAND 1783–1850

Sturgeon started out as an apprentice shoemaker and then joined the army. But he taught himself mathematics and physics because of his love for the subjects.

He was fascinated by the discoveries of the links between electricity and magnetism, and that an electric current could move a magnet. But these ideas were theoretical. It was Sturgeon who created the world's first electromagnet, able to lift a 3kg (7lb) iron weight, in 1826; and, a year later, the first solenoid. He is also credited with making the world's first electric motor in 1832.

and no parallel is to be found since the miraculous suspension of the celebrated Oriental imposter in his iron coffin".

Henry was amazed by the great potential of electromagnets, and realized that they could be used to send signals across long distances. As early as 1830, he could switch on an electrical circuit to activate an electromagnet a mile away and make a bell strike. After he moved to Princeton University (then called the College of New Jersey) in 1832, he set up a similar system to send signals from his office to home to let his wife know when he was on his way, or to order lunch. It was a small step from this to the electric telegraph. A bitter dispute later arose when Samuel Morse (1791–1872) claimed to be the sole inventor of the telegraph – even though Henry had actually encouraged Morse. Henry was also forced to testify in England that the principle of the telegraph had been known to him as well as to physicist and inventor Charles Wheatstone (1802–75).

The field of electroscience was developing so fast that there were continual disputes over priority. In 1830, Henry discovered that magnets could induce an electric current to flow – a discovery that led to the creation of electric generators. Meanwhile, in London, Michael Faraday (see pp.140–41) had established the same principle, and the credit usually goes to Faraday for the discovery, because his results were published first, in 1831. Even Faraday had to fight to ensure his priority after two Italian scientists, Leopoldo Nobili (1784–1835) and Vincenzo Antinori (1792–1865), heard of his ideas and reported their own successful experiments in induction.

A LASTING LEGACY

In 1846, Englishman James Smithson left a large bequest to set up an institution "for the increase and diffusion of knowledge among men". Henry was asked to be the Smithsonian Institution's first head, and he moved with his wife and three daughters into its fantastic multi-turreted home, The Castle, in Washington. During the next 30 years, Henry turned the institution into one of the world's leading centres of science and knowledge, providing everything from a magnificent library to route surveys for telegraphs and railroads. Henry was still at work, when, at the age of 80, he fell ill from nephritis. He died the next year, on 13 May 1878.

AS A PHYSICAL PHILOSOPHER HE HAS NO SUPERIOR IN OUR COUNTRY; CERTAINLY NOT AMONG THE YOUNG MEN ”

BENJAMIN STILLMAN ON HENRY'S APPOINTMENT AS PROFESSOR AT PRINCETON

TIMES OF PROGRESS
Communication in the 1880s was accelerated with the development of the steamboat, the steam locomotive, and, finally, the electric telegraph, made possible by Henry's discoveries.

THE FACTORY FLOOR

THE INDUSTRIAL REVOLUTION WAS THE AGE OF MACHINES. FOR THE VERY FIRST TIME, MACHINES BEGAN TO REPLACE HUMAN HANDS, TOOLS, AND ANIMAL MUSCLE. THEY CARRIED OUT A HUGE RANGE OF TASKS, FROM WEAVING TEXTILES TO MAKING RIFLES.

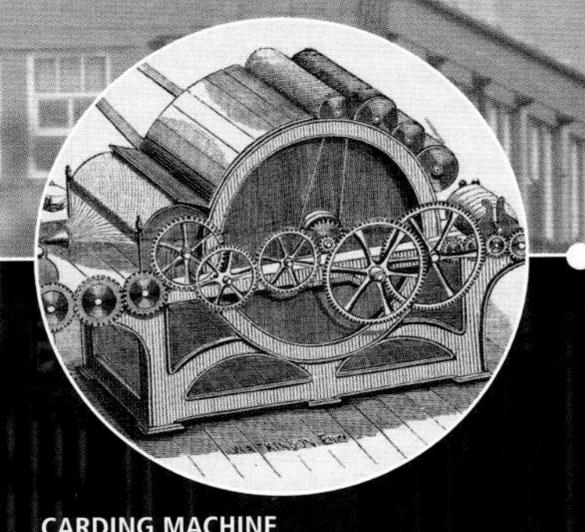

CARDING MACHINE
A machine for carding (the disentangling of raw fibres), which was traditionally carried out by children by hand with a carding brush, was invented by Lewis Paul in 1751.

Strangely enough, the revolution in machines that spread from Britain to Europe began in Italy, in Padua. It was there that a silkmaker called Vittorio Zonca (1568–1603) developed a machine for weaving silk back in the early 1600s. The silkmakers of Padua kept Zonca's machine a secret for a century, until English woolmaker John Lombe (1693–1722) managed to purloin the plans, and used them to develop his own machines. In 1719, Lombe and his brother set up Britain's first textile mill, the aptly named "Italian Works", on the River Derwent in Derby, where the fast flowing waters drove the wheel and powered the machines.

Silk was a luxury product for the majority, and it was never going to become the mass-market product that would flourish with

WOOD AND METAL LATHES
The switch from wooden lathes (left) to all-metal lathes (right), made possible by new iron casting techniques in the late 1700s, allowed machinists to achieve the accuracy essential for making complex machines industrially.

MAUDSLAY LATHE
By making his lathe entirely out of metal, Maudslay was able to machine-make tools and components with unprecedented accuracy – vital for the new machines.

JOSEPH WHITWORTH
Whitworth was the British engineer who, more than anyone, established the standards of precision vital for making efficient machines.

SETTING THE SCENE

- A cotton cloth, **calico**, originating in Calicut (now Kerala), is produced in India. **It is imported into England in vast quantities by the British East India Company** until imports are banned in 1700 to protect the British textile industry.
- **Before the machine revolution**, English cotton and wool are spun and woven into cloth by tens of thousands of **spinners and weavers** working independently in their own cottages.
- The **advent of cotton-making machines** runs in parallel with the **emergence of the factory system**.
- For the first time, **measurements can be made in not just tenths of an inch, but thousandths**, as the new machines need to be made according to previously unimaginably fine tolerances.
- Although certain names are associated with the invention of the machines of the **early Industrial Age, they are often developed and improved by countless unsung mechanics and engineers** in factories and workshops across the world.

mechanization. But cotton could be. The import of Indian printed cottons was already banned under the Calico Acts because they had proved to be so popular that they threatened the British textile industry.

The age of factories began with machines designed for spinning cotton, operated initially by water wheels and later by steam. Soon, making and assembling machines became easier with the accurate standardization of parts.

FIRST SPINNING MACHINES

The only answer was for Britain to make its own cotton prints – and this was where John Kay (1704–64) made his mark. In the 1730s, Kay produced the first of the many cotton-making machines that transformed the British textile industry. Kay's machine was a shuttle, which mechanized the weaving process in a loom that enabled weavers to make an important new cloth, called broadloom. Soon after, James Hargreaves (1720–78) introduced his "jenny" to spin cotton, instantly multiplying the amount of cotton a spinner could make to eight-fold. However, it was confined to producing cotton weft, and was unable to produce yarn of sufficient quality for the warp.

Kay's shuttle and Hargreaves's spinning jenny both demanded considerable skill to operate. They were also powered entirely by hand. But they were the first step in the process, which saw machines developed to take cotton from spinning to weaving to finished cloth automatically, driven first by water wheels and then by steam – all using relatively unskilled labour, including children. In the US, Eli Whitney (see pp.148–49) even developed a machine, the gin, for stripping the cotton fibres from the plants.

The development of cotton-making machines, including Arkwright's water frame for spinning with water-powered machines and Cartwright's loom (see pp.146–47), went hand-in-hand with the creation of the factory system. This transformed textile-making from a cottage craft to a massive industry in which vast mills packed with whirring machines churned out textiles at a previously unimaginable rate.

PRECISION TOOLS

The mechanization of production did not stop with the cotton industry. Once the profits to be made became clear, engineers such as Oliver Evans (see pp.124–25) and Marc Brunel (1769–1849) introduced machines to industries as diverse as milling corn and making blocks for sailing tackle.

At first, most of these machines were made from wood with metal fittings, but in time they became so complicated that using only iron made sense. However, building such machines was way beyond the skills of the traditional blacksmith.

A crucial part of the development of machines was the parallel development of the techniques used to build them. Henry Maudslay (see pp.150–51), James Nasmyth (see pp.152–53), and Joseph Whitworth (see pp.152–53) led a sea change in the introduction of precision engineering techniques to make machines with the robustness to cope with the stresses and the fine tolerances needed for efficiency.

COTTON MAKER
John Kay is pictured here with his flying shuttle, a machine for weaving cotton. In the 1730s it was the first of the new cotton-making machines that launched the British textile industry.

RICHARD ARKWRIGHT

FIRST FACTORY OWNER AND INDUSTRIALIST

ENGLAND 1732–1892

ENGLISH ENTREPRENEUR AND INVENTOR Richard Arkwright launched the Industrial Revolution. He turned clothmaking from a homespun craft into the first mass industry by developing machines for spinning yarn, called water frames, and placing row upon row of them in factories, which massively boosted production and profits.

Richard Arkwright was born in Preston in Lancashire, England, in 1732. His father was a tailor, and the family was never prosperous enough to afford a proper education for Richard, so he was taught to read and write by his cousin Ellen. At the age of 12, Richard was sent off to nearby Kirkham to train as a barber, but eventually realized that there was more money to be made as a wigmaker. For 15 years, wigmaking was his business.

In 1767, Arkwright met clockmaker John Kay, who, during a conversation in a pub, boasted about an idea for a machine that could spin cotton. The industrialist Thomas Highs (1718–1803) had instructed Kay to make the machine, but it had never been built. Ever alert to a business opportunity, Arkwright quizzed Kay, and then set about building a machine.

There had been earlier spinning machines, notably the famous "spinning jenny" of James Hargreaves (1720–78), which could spin a dozen threads at a time, but it needed a skilled person to drive it. Arkwright's machine, the water frame, could spin 96 threads and could be run by a teenager. The machine had three sets of rollers, allowing it to spin strong yarn from frail English cotton. It could therefore be used to make cotton cloth as cheaply as the highly prized Indian calico cloth. Calico had been banned in England because of its threat to the wool trade. Spun by hand, English cotton yarn was too weak to weave alone, and had to be mixed with linen to make the rough cloth "fustian". Arkwright's machine spun pure cotton cloth that was almost as smooth as silk yet cheap to make. It also got around the calico ban.

A LIFE'S WORK

- Arkwright, **one of Britain's first self-made millionaires**, starts his life as a **hairdresser**, and then becomes a **wigmaker**
- Develops the **water frame for spinning cotton**, based on an idea by Thomas Highs
- **Creates the world's first large factories** to maximize the potential of his machine, **increasing cotton production** dramatically
- In about 1775, he **opens the first large steam-powered factory** in Manchester
- In 1785, his **rivals lodge a case of fraud against him** in court, which he loses
- He is **knighted in 1785** for his work

BURGEONING FACTORIES

Arkwright's cotton was available in larger quantities because of his greatest innovation – the factory. Until then, yarns were spun mostly by men and women working alone at home with spinning wheels, or even hand spindles. Even the spinning jenny was, like sewing machines later, intended for the home. Arkwright realized that the business would be transformed if he got scores of his water frame machines to work together, night and day, under a single roof. These would be powered by a water wheel, which is why he called the machines water frames. In 1771, less than two years after Kay built the first prototype, Arkwright's mill opened at Cromford on the River Derwent in Derbyshire. It was called a mill because, like a watermill for milling flour, it got its power from a water wheel.

COTTON MANUFACTURING From modest beginnings at Arkwright's Cromford Mill, cotton-making grew into a huge industry using machines developed from his water frame. This illustration shows a cotton factory in 1830, using the mule developed in 1825.

The factory had scores of workers turning out cotton at an unprecedented rate, and at a novel price that millions of people could afford.

Just a few years later, another English inventor, Samuel Crompton (1753–1827) devised a different spinning system, in which half the machine moved forwards and then back again; the cotton "roving" was teased and twisted into thread on the backward stroke, and then wound on to a bobbin on the forward stroke. This came to be called a "mule", and would eventually be able to spin a thousand threads simultaneously. After paying a small fee to see this machine in action, Arkwright immediately bought and installed several mules in a giant factory at Masson Mills.

In 1775, he opened an even bigger factory at Shudehill in Manchester, incorporating another innovation, the returning engine – a steam engine that pumped water non-stop back into a reservoir

EDMUND CARTWRIGHT

ENGLAND 1743–1823

Edmund Cartwright was the son of a landowner from Marnham in Nottingham and became a rector at a small parish in Leicestershire. But, when he was 40, he visited one of Arkwright's factories where machines called mules were in operation. It inspired Cartwright to create a loom – a weaving machine – that would improve the speed and quality of weaving.

Cartwright believed that his loom could, unlike the mule, work entirely mechanically without requiring the continual attention of an operator to keep the tension of the yarn even. This took two years of intense development, and the aid of a blacksmith and a carpenter, due to the complicated nature of the mechanism. By 1787, Cartwright had installed the world's first power looms, driven by a Boulton and Watt steam engine, in a factory in Doncaster. These looms ran almost completely by themselves. In 1799, a factory in Manchester that had installed 400 looms was burned to the ground, probably by workers who feared for their jobs. By the time Cartwright died in 1823, modified versions of his looms were in widespread use.

to drive the mill's water wheel. With a returning engine, factories were no longer restricted to fast-flowing rivers, nor affected by seasonal vagaries of a river's flow. Later, rotary engines would make factories entirely steam-powered.

RIVALRIES AND KNIGHTHOOD

By now, Arkwright's business was booming. More than 5,000 people were employed in half a dozen factories. But rivals were keen for a share of his success, and, in 1785, Arkwright launched a bruising court case to stop them. One aspect of the dispute was related to whether Arkwright actually invented the water frame, or merely stole the idea from Highs. Arkwright found that Kay, Highs, and even Hargreaves's widow had taken the witness stand against him, arguing that the water frame was not his idea.

Arkwright lost the case, as the court found the description of the machine in his patent too vague. But this defeat earned him an unexpected ally in Scottish inventor James Watt (see pp.118–21), who was worried about the court case's implications for inventors' rights. Together, the two compiled a document making the case for tightening patent laws, but their efforts went unheeded. By that time they were both wealthy men, heading two of the world's biggest businesses. In 1786, Arkwright, one of Britain's first self-made millionaires, was knighted.

Half a century after Arkwright's death, Scottish historian Thomas Carlyle (1795–1881) identified him as the man who transformed the world with the power of cotton. Yet even Carlyle described Arkwright as "a plain, almost gross ... much inventing man and barber", while others insisted he was a thief of other people's ideas. Arkwright made many enemies in his lifetime, and even after his death in 1792, his reputation see-sawed as many tried to knock him down as an exploiter, while others built him up as a hero for his vision.

ORIGINAL SPIN
The first prototype model built to show the principle of Arkwright's system looks very simple, but it developed into a machine that made him rich and Britain "Great".

ELI WHITNEY

INVENTOR OF THE COTTON GIN

UNITED STATES 1765–1825

A LIFE'S WORK

- Whitney, a young entrepreneur, starts his own **nail-making and hatpin business** while still a teenager
- After graduating from Yale, he arrives in the South and **begins to develop the cotton gin** with fellow graduate Phineas Miller
- His cotton gin, used for mechanically removing seeds from cotton, revolutionizes the American cotton industry – **cotton exports expand 200-fold** following the gin's introduction
- **Pirated copies** of the gin's basic design appear in Georgia and the Carolinas; Whitney is **unable to enforce the patent** he had applied for, and is left penniless
- He develops the idea of **interchangeable, machine-made parts** to make high-quality items **in bulk with unskilled labour**

A FARMER'S SON from New England, Eli Whitney went south and devised the cotton gin, the little machine for stripping cotton seeds from their fibres, which turned the Deep South into the world's leading cotton growing region in just a few decades. It was one of the key inventions of the Industrial Revolution and shaped the economy of the South before the Civil War.

Whitney is one of the great heroes of America's early years, and there is a legend that when he arrived in the South, all the cotton was deseeded laboriously by hand. The story goes that Whitney immediately devised the machine "gin", which gave rise to "King Cotton" – a slogan used by southerners to support secession from the US by arguing that cotton exports would support the Confederacy. Soon, everyone wanted to grow cotton, and suddenly there was a great requirement for slaves again – this at a time when slavery had been dwindling because nobody had much need for slaves. So, it seems, the clever Whitney not only made the South rich, but also entrenched slavery in the area, making the Civil War inevitable.

Like many legends, this account is only partly true. The acreage under cotton did expand dramatically at about this time, and the spread of cotton plantations did lead to a rise in the demand for slaves in the South. But gins were not invented by Whitney. Simple gins such as the Indian "darka" had been in use for thousands of years, and cotton growers in the American South had long been using foot-operated gins that squeezed the seeds through two rollers like a mangle. Nor did Whitney come up with his new gin instantly – it took two years of painstaking development. And, of course, had the demand for cotton in the new cotton mills of England not been so insatiable, his gin would not have been able to transform the industry as it did.

COTTON GIN
Whitney's gin may look like a simple machine, but it took a long time to perfect – and its impact was huge.

All the same, Whitney was an ingenious inventor, and his gin had a massive impact. He also came up with the idea of manufacturing muskets using interchangeable parts made by unskilled labour – the forerunner of modern mass-production techniques – even if here, too, his contribution is disputed.

FARMER'S BOY

Whitney was born on a farm in Westborough, Massachusetts, on 8 December 1765. As a boy, he had little interest in school work, and spent his time making and mending tools for his father to use on the farm. He was a young entrepreneur, too, setting up his own nail-making and hatpin businesses while still a teenager. It was only when he was 23 that he decided to go to Yale University, paying his way by working as a schoolmaster.

When he graduated in 1792, another Yale graduate, Phineas Miller, sent him an invitation to come to Georgia, where there was work on offer as a tutor. Whitney was nervous about heading to the very different culture of the South with its slaves and its hot climate. Things did not look any better when his ship ran aground outside New York, where he was to meet Miller, and he was forced to wade ashore. He was even less assured when he shook hands with an old friend – only to find that the friend was infected with smallpox.

As he and Miller headed south on the boat to Savannah, however, they developed a strong friendship. By the time he arrived at Mulberry Grove, the plantation where Miller was acting as general help to the owner's widow Caty Greene,

the pair were beginning to hatch plans. They abandoned the idea of tutoring, and the two of them began to work secretly to develop a machine that would make the local cotton so cheap and easy to produce (with slave labour) that American cotton could easily compete with the cheap cotton from Egypt and India, where growing conditions were better.

GIN GENIUS

Whitney realized he needed tools that were only available in the North, so he returned to Yale to work on the gin, leaving Miller to drum up interest. Eventually, after several mishaps, the gin was ready, and Whitney returned to the South in 1795 to market it with Miller, hoping that they could make money by renting it out to farmers who would use their own men to operate it.

But as soon as their basic design was revealed, pirate copies appeared all over Georgia and the Carolinas, and nobody wanted to hire the original. The gin was a huge success – able to generate 23kg (50lb) of cleaned cotton a day – and the American cotton crop multiplied tenfold in just 12 years after its introduction. But Whitney and Miller made virtually no money, and found themselves vilified as they tried to enforce the patent they had applied for.

MUSKET PRODUCTION

It was perhaps in desperation, then, that Whitney turned a few years later to manufacturing muskets. War between France and the US was brewing and it seemed an opportune moment to be making guns. Whitney's clever idea was to assemble muskets from interchangeable parts that could be made on machines by completely unskilled labour – making the whole process fast and cheap. He intended, he said, "to substitute correct and effective operations of machinery for that skill of the artist, which is acquired only by long practice and experience".

To sell his idea, he took to Washington ten pieces for each part of the musket, then grouped them into piles in front of the Secretary of War. To the Secretary's amazement, Whitney then pulled pieces from each pile, apparently at random, and assembled a working musket. In fact, he was faking it, having carefully fabricated the pieces by hand before to ensure they fitted. The charade was useful in order to gain more time and resources for the project, although he still had to create interchangeable parts. But, the ruse got Whitney his government contract, even though it took him eight years to deliver – and Jefferson Davis, Colonel of the Mississippi Rifles, declared that the guns were "the best rifles which had ever been issued to any regiment in the world". Sixty years later, the concept of manufacturing guns using interchangeable parts was put into practice and proved a success during the Civil War. Whitney died of prostate cancer on 8 January 1825.

SAMUEL SLATER

ENGLAND 1768–1835

Born in Derbyshire, England, Samuel Slater worked as a boy at the local cotton mill, where Arkwright's water frame (see pp.146–47) was in use.

After learning about the machines for 11 years, he headed for the US in 1789, where he was able to replicate the water frame, promising an investor, "If I do not make as good yarn, as they do in England, I will have nothing for my services, but will throw the whole of what I have attempted over the bridge". He was as good as his word. He had thoroughly absorbed not only all the technical details of the British cotton machines, but the whole factory system that made them work. When he set up a cotton-spinning factory on Rhode Island, it marked the beginning of the American textile industry and earned Slater the epithet, "Father of the American Industrial Revolution".

COTTON HARVESTING
This illustration from 1884 shows all the classic elements of the American South's cotton plantations of the 1800s – the black plantation labourers (no longer slaves but little better off), the steam-powered processing mill, and the Mississippi river boat in the background to transport the harvested cotton.

BRAMAH AND MAUDSLAY

THE MEN WHO BUILT THE UNPICKABLE LOCK

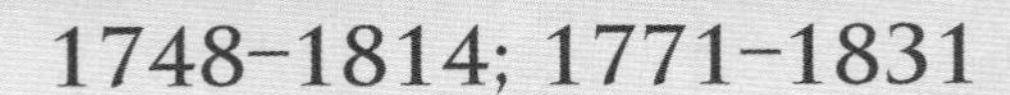

ENGLAND 1748–1814; 1771–1831

JOSEPH BRAMAH

HENRY MAUDSLAY

British engineers Joseph Bramah and Henry Maudslay drove the precision in machine-making that helped trigger the Industrial Revolution – enabling parts for complex mechanisms to be created in bulk by machines rather than individually by hand.

A LIFE'S WORK

- Joseph Bramah grows up in Yorkshire and trains as a **carpenter and cabinet-maker**
- He moves to London and develops the **flush water closet**, a mechanism still used today
- In 1785, he creates his **"unpickable" Challenge Lock**
- Maudslay **trains as a metalworker** at the beginning of his career
- In 1789, **Bramah collaborates with Maudslay** to achieve high precision in manufacturing
- Bramah invents the **fire engine, beer tap, fountain pen, and hydraulic press**
- After his collaboration with Bramah, Maudslay makes the finest measure of the age, **the Lord Chancellor micrometer**
- Maudslay develops **precision bulk manufacture** at the Portsmouth block yard and builds **engines for steamships** at his famous Lambeth Works

In 1790, anyone walking past Joseph Bramah's shop in Piccadilly, London, would have seen in the window a large padlock surrounded by an inscription in gold lettering that boldly claimed, "The artist who can make an instrument that will pick or open this lock, shall receive 200 guineas the moment it is produced". Remarkably, Bramah's famous "Challenge" lock (see box, below) remained unpicked until the Great Exhibition of 1851.

Born in Wentworth, Yorkshire, on 13 April 1748, Bramah had trained as a cabinet-maker before moving to London. His work was to install the new water closets that were then appearing in every well-to-do home. Even before inventing his lock, Bramah had developed a new mechanism for the water closet that alone would have ensured his lasting fame – a hinged-valve flush system that was used for nearly 100 years.

THE UNPICKABLE LOCK

In 1785, with the rise in private property ownership in the growing cities, security was becoming a real issue, and Bramah's lock was very much an invention of its time. The reason it could not be easily picked was its incredibly complex mechanism – made from more than 100 metal pieces to very fine tolerances. At the time, such accuracy could be achieved only by the most skilful craftsmen. Yet the potential demand for Bramah's locks was too great for each to be handmade and so Bramah looked for help. In 1789, he had a momentous meeting with 18-year-old metalworker Henry Maudslay from the Royal Arsenal in Woolwich. Maudslay was already something of a legend for his prodigious skill and mania for precision, and when Bramah offered him the task of creating tools in order to make his complex locks mechanically, Maudslay was keen to take up the challenge.

Maudslay's insight was to recognize the value of iron for construction. The wooden-mounted lathes of the time were unable to make cuts with a margin of error of less than 1.6mm (1/16in) – which was too great for Bramah's lock. Maudslay realized that using iron instead of wood could provide a completely rigid and stable platform. By building lathes – and later drills, planing machines, and other tools – entirely of iron, Maudslay achieved a huge improvement in accuracy, enabling relatively unskilled workers to create, quickly and repeatedly, precision parts that previously only the finest craftsmen could have made.

THE BRAMAH CHALLENGE

In 1784, Bramah started the Bramah Locks Company and designed a lock, receiving a patent for it (and another for it in 1798). The locks produced by his company were noted for their resistance to lock-picking and tampering. The company famously had a "Challenge Lock" displayed in the window of their London shop from 1790. The challenge stood for over 67 years until, at the Great Exhibition of 1851, the American locksmith Alfred Charles Hobbs was able to open the lock and, following some argument about the circumstances under which he had opened it, was awarded the prize. Hobbs's attempt required some 51 hours, spread over 16 days – not much use for a smash-and-grab burglar.

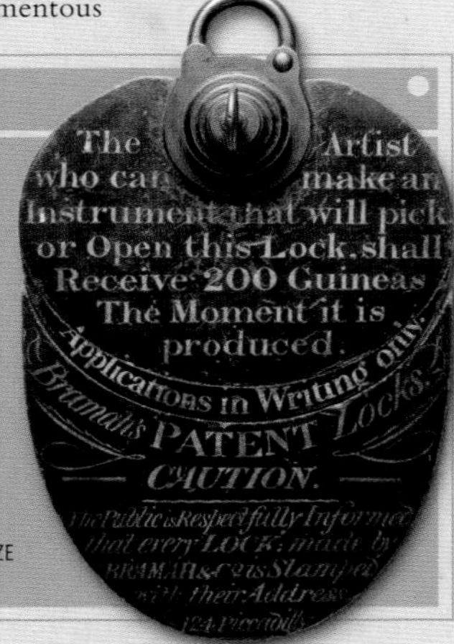

CONVINCED OF THE INVINCIBILITY OF HIS LOCK, BRAMAH OFFERED A 200 GUINEA PRIZE TO ANYONE WHO COULD PICK IT. NO ONE SUCCEEDED FOR HALF A CENTURY.

INDIVIDUAL GLORY

Maudslay worked for Bramah for a decade, and the machine tools he developed to make Bramah's locks in the 1790s were, in their own way, as crucial to the progress of the Industrial Revolution as James Watt's famous steam engine (see pp.118–21). Yet Bramah never quite appreciated Maudslay's contribution, paying him only a meagre wage, and, in 1799, Maudslay left to set up his own business.

The end of the partnership, however, did not bring an end to Bramah's incredible inventiveness. While Maudslay was developing the machine tools to make the lock, Bramah continued to create a string of other inventions still in use today, including the fire engine, the beer tap, the fountain pen, and, most importantly, the hydraulic press. One of Bramah's last inventions was a hydrostatic press that could uproot trees. It was while working with this in the Holt Forest in Hampshire that Bramah caught pneumonia, and he died on 9 December 1814.

PRECISION AND ACCURACY

Meanwhile, after ending the partnership with Bramah, Maudslay's work had become even more influential. He was not only taking the accuracy of the lathe to new levels, but in order to measure its accuracy, he developed a micrometer that was years ahead of that created by James Watt. Maudslay's micrometer was nicknamed the Lord Chancellor because it was the final judge on precision and measurement accuracy. According to some reports, the Lord Chancellor could give its verdict with an accuracy of less than 0.0025mm (1/10,000in). The invention's importance was that by measuring with such exactitude, it allowed small incremental improvements to be recorded and applied persistently, helping manufacturing standards to move forward.

Maudslay's Lambeth Works in Westminster Road, London, became a mecca for young engineers wishing to learn about precision at the hands of the master. Maudslay trained some key men of the next generation, such as Richard Roberts (1789–1864), David Napier (1788–1873), Joseph Clement (1779–1844), Joseph Whitworth (see pp.152–53), James Nasmyth (see pp.152–53), and Joshua Field (1786–1863).

The Lambeth Works specialized in the production of steam engines for ships, and after Maudslay's death, it produced the engines for I K Brunel's SS *Great Western*, the first transatlantic steamship. It also built the innovative tunnelling shield designed by Brunel's father, Marc Isambard, which was crucial in the digging of the Thames Tunnel at Rotherhithe, completed in 1842.

Few engineers have been so universally loved and respected by those they worked with as Maudslay. His famous protégé Nasmyth later wrote of him, "The defatigable care which he took in inculcating and diffusing among his workmen, and mechanical men generally, sound ideas of practical knowledge, and refined views of construction, has rendered, and will continue to render, his name identified with all that is noble in the ambition of a lover of mechanical perfection". Maudslay died on 15 February 1831 from a chill caught while crossing the English Channel, and was buried in the churchyard of St Mary Magdalen, Woolwich, London.

MAUDSLAY MARINE ENGINE
Maudslay's Lambeth Works became famous for its superb steam engines for ships. They specialized in "side-lever" engines (shown here) in which the beam is mounted beside the engine to save space in the confines of the engine room. By 1850, the Works had made engines for more than 200 steamships.

NASMYTH AND WHITWORTH

THE MASTERS OF PRECISION ENGINEERING

SCOTLAND 1808–90
ENGLAND 1803–87

JAMES NASMYTH

JOSEPH WHITWORTH

TWO OF HENRY MAUDSLAY'S students, James Nasmyth and Joseph Whitworth, became great engineers in their own right. Nasmyth gained lasting fame for his invention of the steam hammer, and Whitworth for his establishment of new standards of precision engineering.

In 1829, a 20-year-old James Nasmyth arrived in London after a rough four-day journey by boat from his Edinburgh home. It was a Saturday afternoon, but Henry Maudslay (see pp.150–51) was busy as usual when Nasmyth arrived at the famous Lambeth Works, clutching a letter, asking Maudslay to teach him how to be an engineer. Maudslay was no longer taking on trainees but, impressed by Nasmyth's eagerness, he agreed at least to take a look at Nasmyth's engineering drawings and the beautiful working model of a steam engine he had brought. After careful study, Maudslay told Nasmyth, "I wish you to work beside me, as my assistant workman. From what I have seen there is no need of an apprenticeship in your case".

What both Nasmyth and Joseph Whitworth, also working in the Lambeth Works, learned under Maudslay's exacting eye was the value of precision in engineering. Invention and design were all very well, Maudslay made clear, but they were worthless without execution to the highest standards. Precision engineering was vital for everything from stopping leaks in high-pressure steam engines to making guns reliable. Both of the young engineers learned the lesson well, and were at the forefront in setting benchmarks in British engineering that made Britain the envy of the world over the next half century.

A LIFE'S WORK

- **Whitworth and Nasmyth** learn **precision engineering** at Maudslay's Lambeth Works
- Nasmyth builds a **full-size working steam carriage** while training as an art student in Edinburgh
- In 1845, he **invents the steam hammer and a steam piledriver**
- After retiring, he builds a **reflecting telescope**, and **maps the Moon**
- Whitworth establishes a **standard set of dimensions for nuts and bolts**, thereby greatly improving the precision of assembly
- He invents a **micrometer** capable of measuring to **one-millionth of an inch**
- He **makes high-precision rifles**, which are used in the **American Civil War**

SETTING STANDARDS

Born in Stockport, Cheshire, on 21 December 1803, Whitworth was five years older than Nasmyth. By the time Whitworth arrived at the Lambeth Works in the mid 1820s, he had already served an apprenticeship as a cotton spinner and worked as a mechanic in a Manchester factory. As he worked alongside Maudslay, he developed a method of milling surfaces perfectly flat, and created machine tools for turning, shaping, milling, slotting, gear cutting, and drilling – all to a high degree of accuracy.

WHITWORTH ON DISPLAY
The sheer quality of Whitworth's work shone through at London's Great Exhibition in the Crystal Palace in 1851. The high precision of his tools for planing, slotting, drilling, and boring was clear. His micrometer, capable of measuring with an accuracy of one-millionth of an inch, caused astonishment.

Whitworth left Maudslay's and worked briefly with Charles Babbage (1791–1871) on his "Difference Engine" – a mechanical calculating machine. The cogs and wheels of Babbage's machine demanded engineering of an unprecedented accuracy, and Whitworth was one of the few people who could achieve it. But Babbage's project was never properly completed, and Whitworth went on to establish his own factory at Openshaw in Manchester to make high-quality lathes and machine tools.

Whitworth realized that one of the problems with making and assembling machines precisely was the nuts and bolts. Each nut and each bolt was made by hand, with the result that they rarely fitted together tightly, and were never interchangeable. In 1841, Whitworth set a standard for all nuts and bolts, with a fixed angle of 55 degrees for the screw threads, and a standard pitch for each size of bolt. When all the railway companies decided to adopt this standard, everybody else followed suit, and Whitworth's standard, the British Standard Whitworth (BSW), became the norm.

To achieve this kind of precision, Whitworth needed to make exact measurements, and so he developed the most accurate micrometer yet

built. As he demonstrated to great acclaim at the Great Exhibition of 1851, this could measure up to one millionth of an inch.

In the 1850s, Whitworth was commissioned by the army to make rifles. Whitworth's new rifle, which had a hexagonal bore, proved to be incredibly accurate and had a high velocity – but the British Army rejected it because it was too expensive, and was prone to fouling. The French Army took it up, however, and some rifles were sent to America, where Confederate snipers used "Whitworth's sharpshooters" to great effect in the Civil War. At the Battle of Spotsylvania in 1864, Unionist Major General John Sedgwick's famous last words, as he rode in front of his troops, were, "They couldn't kill an elephant at this distance". Moments later he was fatally shot in the head by a Confederate Whitworth.

THE "RIGHT" HAMMER

Meanwhile, Nasmyth was building his own business near Manchester at a new site he found at Patricroft, next to the Bridgewater Canal. The foundry became famous for its steam engines, built both for British companies and for export all around the world. It also made machines such as balers for compressing cotton into bales at high speed. Nasmyth was instrumental both in designing these machines and ensuring they were built to the highest precision.

His greatest triumph was his steam hammer (see pp.154–55), which made him world-famous. He appeared in *Punch* magazine in 1883 in a cartoon captioned, "The Man Who Knows How To Knock Metal On The Head With The Right Hammer". The main purpose of the steam hammer was shaping metal, but Nasmyth realized that the principle could be used in other ways as well, notably in pile driving. The standard pile for building docks was a wood block 21m (69ft) long and 45cm (18in) square – it took 12 hours to hammer with a traditional piledriver. When Nasmyth unveiled his new steam piledriver in 1845 at Plymouth dockyard, it hammered in the wood in just over 4 minutes. Soon, orders came in from all around the world.

Not long after, Nasmyth, still in his early forties, decided to retire, saying, "I have now enough of this world's goods: let younger men have their chance". He took up painting and astronomy, both of which he excelled at, and also had a clandestine affair with a woman named Virtue Squibb. Whitworth died in January 1887; Nasmyth died three years later in May 1890. Both had seen British engineering reach its zenith.

> I HOPE I HAVE BEEN ABLE TO LEAVE ... USEFUL CONTRIVANCES WHICH ARE IN MANY WAYS HELPING GREAT WORKS OF INDUSTRY
>
> **JAMES NASMYTH**

NASMYTH WITH TELESCOPE
An engraving from Nasmyth's biography shows him star-gazing through his precision-built 51-cm (20-in) reflecting telescope in the grounds of his home in Patricroft, Manchester. He used it to calculate the height of lunar features by the length of their shadows.

BREAKTHROUGH INVENTION

NASMYTH'S STEAM HAMMER

In 1838, engineer Francis Humphries was wrestling with a problem. His task was to build the engines and propulsion for I K Brunel's ambitious steamship, SS *Great Britain*, which would be the biggest ship of any kind yet built. But the huge size of the paddle shaft was causing problems. "I find," he wrote to James Nasmyth in despair, "there is not a forge-hammer in England or Scotland powerful enough to forge the paddle-shaft! What am I to do?"

The problem was the tilt-hammer that foundries used to forge iron. Resembling a giant hand-held hammer, it was lifted by water or steam and allowed to swing down repeatedly under its own weight and beat the iron into shape. Its swing was so small, however, that there wasn't room to forge huge pieces, nor was there any way of adjusting the force of the blow.

Nasmyth's solution was simple but brilliant. Rather than swinging down, the hammer could drop vertically and so be lifted to any height and adjusted for forging even the largest pieces. What's more, as Nasmyth later realized, the hammer need not rely on its own weight for power – it would be like a piston, accelerated downwards by expanding steam, yet adjusted to be brought up short at any point. In a famous demonstration, Nasmyth used the steam hammer to crack an egg in a wineglass, without damaging the glass – then replaced the glass with a white-hot iron block, which the hammer pounded with such force that the building shook.

In the event, propellers, not paddles, were adopted for the SS *Great Britain*, and Nasmyth's idea was not taken up straightaway. However, in 1842, Nasmyth visited the Schneider ironworks at Le Creusot in France and saw a vertical hammer in use that, he believed, was based on his own sketches. He returned to England at once, took out a patent, and began producing steam hammers at his Patricroft foundry near Manchester.

The steam hammer's ability to shape metal with huge power and precision had a revolutionary impact on the manufacture of machines and components. Soon steam hammers were pounding out everything – from railway wheels to the first steel hulls for ships – on a previously unimaginable scale.

ONE OF THE MOST PERFECT OF ARTIFICIAL MACHINES ... THAT MODERN ENGLISH ENGINEERS HAVE YET DEVELOPED

CHARLES TOMLINSON, *CYCLOPEDIA OF USEFUL ARTS*, 1852

DESIGNED WITH AN ARTIST'S EYE
This is Nasmyth's own painting of the steam hammer made for his autobiography. By 1856, Nasmyth's company had produced 490 such hammers that were sold worldwide.

BUILDING BRIDGES

THE INDUSTRIAL AGE SAW MANY WIDE RIVERS SPANNED FOR THE FIRST TIME WITH GIGANTIC BRIDGES. BUT IN BUILDING THESE BRIDGES THE ENGINEERS' TECHNICAL INGENUITY WAS STRETCHED TO ITS LIMITS AND THE WORK INVOLVED IN CONSTRUCTION WAS OFTEN HEROIC.

FROM BUDA TO PEST
William Tierney Clarke's chain suspension bridge linked the cities of Buda and Pest across the River Danube for the first time. Here, the foundation stone is being laid in 1840.

The great bridges of the Industrial Age are that era's most awe-inspiring legacy. Even by today's standards, bridges such as the Forth Rail Bridge in Scotland by John Fowler (1818–98) and Benjamin Baker (1840–1907), the Garabit Viaduct in France's Auvergne Mountains by Gustave Eiffel (see p.301), and the Brooklyn Bridge by the Roeblings (see pp.166–67) in the US are breathtakingly gigantic.

A RISKY BUSINESS

In the 19th century, however, these mighty bridges were not simply the largest engineering structures that had ever been built, but the most difficult to build as well. Some of them crossed over deep water with strong currents. So an engineer's ability was tested not only on the strength of the finished bridge, but on his understanding of building in such dangerous conditions. The several bridge-failures of the time, such as those of the Basse-Chaîne Bridge in France and the Tay Bridge in Scotland, bear witness to the fact that bridge engineers of the time were stretching their ability to the limit.

The fact that the engineers were willing to face these challenges indicates the strong demand for transport during the emerging Industrial Age. Slow river crossings that had seemed minor inconveniences in an earlier, slower age suddenly became intolerable obstacles, and engineers wanted to bridge the widest gaps to speed up transport by road or rail.

> **A demand for speed and infrastructure during the Industrial Age resulted in the construction of grand bridges. The availability of iron made it possible to design new types of bridges over previously unbridgable obstacles.**

COMMITTED SUPPORTER
Although the Brooklyn Bridge was engineered by her father-in-law John Roebling and her husband Washington, Emily Roebling's determination saw the project through to completion.

IRON BRIDGES

Masonry materials and wood had been the main bridge-building materials of earlier eras, and Robert Mylne's Blackfriars Bridge of the 1760s (see pp.158–59) proved that innovations with these materials were still possible.

But the greatest development of the age was the use of iron, which was suddenly available in sufficient quantity to be used as a structural material, thanks to the new coke-fired blast furnaces. Iron offered strength, rigidity, and lightness that was impossible to achieve with masonry. It allowed engineers to build unsupported spans of the length of a bridge – a feat that could not have been performed with masonry. Also, iron bridges could be constructed across rivers where it would have been extremely difficult, if not impossible, to build multiple masonry piers.

The first use of iron was in cast-iron arch bridges, beginning with the famous Iron Bridge built in Coalbrookdale, England, in 1779 – designed by Thomas Pritchard (1723–77) – and it reached its zenith with the beautiful iron bridges built by Thomas Telford in Scotland, such as the Craigellachie Bridge (see pp.160–63).

Telford was also at the forefront of the age's second great innovation in bridge building – the suspension bridge – and his Menai Bridge of 1826 is an engineering landmark. With their light, hanging decks, suspension bridges could be built not just over wide rivers, but also over deep chasms, where no bridge could possibly have been built before, such as Bellin and Berthier's Charles Albert Bridge of 1839, soaring 147m (482ft) above the Usses River in the Alps.

BRIDGING VAST SPANS

In the early suspension bridges, the decks were hung from chains of wrought iron, as in the Menai Bridge. Meanwhile, Marc Séguin developed the idea of wire-cable bridges, which were lighter and stronger, and could be used to

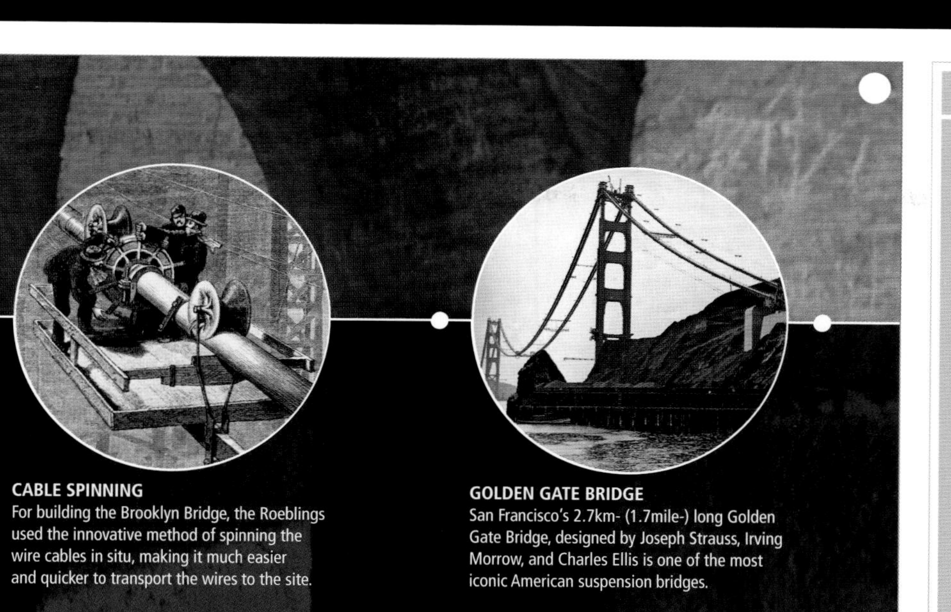

CABLE SPINNING
For building the Brooklyn Bridge, the Roeblings used the innovative method of spinning the wire cables in situ, making it much easier and quicker to transport the wires to the site.

GOLDEN GATE BRIDGE
San Francisco's 2.7km- (1.7mile-) long Golden Gate Bridge, designed by Joseph Strauss, Irving Morrow, and Charles Ellis is one of the most iconic American suspension bridges.

SETTING THE SCENE

- Croatian polymath Faust Vrancic (1551–1617) conceives the idea of **metal cable-stayed suspension bridges as early as the 1580s**.
- **Masonry materials and wood** are the main bridge-building materials of earlier eras, but with the **availability of iron**, bridges spanning wide rivers, including ones with strong currents, become possible.
- The **first known metal suspension bridge** is built in 1801 by American inventor James Finlay (1762–1828) over **Jacob's Creek**, Pennsylvania.
- One of the **first suspension bridges in Europe** is the **Union Bridge** built by Samuel Brown (1776–1852) over the River Tweed in Scotland in 1826.
- Increasingly, many of the solid workaday bridges of the era are robust **"truss" bridges, built with a lattice of steel girders**, such as the Forth Rail Bridge in Scotland.

bridge even more extreme spans (see pp.164–65). French engineers Joseph Chaley (1795–1861) and Louis Vicat (1786–1861) took the technology a step further by spinning the cables in situ, which dramatically simplified the building process. This was later crucial in the success of the Brooklyn Bridge in New York.

But the human cost involved in building these grand bridges was high. Twenty-seven men died during the construction of the Brooklyn Bridge, and every other bridge took its toll in terms of severe injuries, if not fatalities. Yet the countless bridges – both large and small – that survive to this day are ample testaments to the skill and courage of the engineers who constructed these timeless pieces.

HIGH AND MIGHTY
Gustave Eiffel and Maurice Koechlin's iron-truss Garabit Viaduct in the middle of France's Auvergne Mountains, completed in 1885, was for a long time the world's highest bridge – towering 124m (406ft) above the Truyère River.

ROBERT MYLNE

A MULTI-FACETED ENGINEER

SCOTLAND 1733–1811

Architect and civil engineer Robert Mylne's Blackfriars Bridge, over the River Thames in London, with its innovative flat elliptical arches, was a triumph of 18th-century bridge-building. Less well known, but equally important, were his key contributions to Britain's waterborn infrastructure – its bridges and rivers, its canals and harbours, and its water supply.

A LIFE'S WORK

- Mylne becomes the first Briton to win the **triennial architecture competition** at the **Accademia di San Luca** in 1758
- Wins the competition for designing the new **Blackfriars Bridge** in London
- Gains respect as the builder or renovator of **St Cecilia's Hall in Edinburgh, Greenwich hospital**, and the **Gloucester and Sharpness Canal**
- Becomes an engineering consultant on everything from **harbour works** to **canal construction**
- Designs the monument to **Admiral Nelson** in **St Paul's Cathedral**
- Joins John Smeaton, one of the founding members of the **Society of Civil Engineers**, the first engineering society in the world, established in 1771

Robert Mylne cannot have been an easy man to deal with, even for his own family. After he finished Blackfriars Bridge, his family clearly found him, to say the least, high and mighty.

Born in Edinburgh on 4 January 1733, Robert Mylne became an apprentice mason at 12, and was soon a skilled carpenter and woodcarver. Often considered proud and overbearing – his brother William referred to him as "his Honour of London", while his sister Anne described him as "the Bashaw" (the Pasha). He was frequently involved in bitter lawsuits and contractual scraps. His diaries convey the impression of someone who was often angry.

After hiding behind the anonymity of the name Publicus while campaigning for his design for Blackfriars, he became steadfast in maintaining his integrity and plain-spokenness. Fellow professionals, including John Smeaton (see pp.100–01) and James Watt (see pp.118–21), spoke highly of him and vouched for the quality of his work.

Structural engineering and bridge-building – or masonry as it was then called – flowed deep in the Mylne family blood. Robert's ancestor John Mylne I was Master Mason on the celebrated 11-arch Bridge of Tay at Perth and his son was responsible for the upkeep of all royal castles in Scotland, while his great nephew Robert developed Edinburgh's water supply and built a revolutionary bridge over the River Clyde in 1682. Robert also developed the Leith docks. Mylne's father, Thomas, was the project's surveyor and also a magistrate in Edinburgh.

ITALIAN LESSONS

When Mylne was 21, his father sent him to join his younger brother William, who was studying architecture in Europe. The brothers travelled to Rome together, where Robert began to learn architectural drawing under the guidance of the Italian master Giovanni Piranesi (1720–78). The future architect Robert Adam (1728–92) was studying there at the same time and, with an allowance 20 times the size of the Mylne brothers, was somewhat scathing about their frugal ways – although he did admit that Mylne "begins to draw extremely well".

Whatever Adam felt, Mylne established a strong relationship with Piranesi and learned a great deal studying Roman aqueducts. In 1758, Mylne won the prestigious Concorso Clementino architecture prize, the first Briton to do so, and was presented with the award in front of 20 cardinals and James Stuart, the "Old Pretender", still known in Rome as King James III of England.

After a leisurely architectural tour through Europe, Mylne arrived in London in July 1759. He heard about a competition to design a new bridge over the River Thames at Blackfriars. Mylne had just a few months to complete his design, and was competing with luminaries such as Smeaton. His revolutionary design for a graceful bridge with flatter, elliptical arches excited the judges, despite public attacks from critics who argued that the flat arches would collapse.

MR MYLNE WAS A RARE GENTLEMAN ... AS HOT AS PEPPER AND AS PROUD AS LUCIFER

IRISH CONSTRUCTION WORKER DESCRIBING HIS BOSS ROBERT MYLNE

BUILDING BLACKFRIARS
Mylne's mentor Giovanni Piranesi followed his progress on Blackfriars Bridge closely. In this illustration, he captures the moment when the first arch of the bridge nears completion.

Mylne was awarded the job and work began in October 1760. He bought a new house on the river and employed a full-time gondolier to take him on site. He introduced a number of technical innovations in bridge-building, such as removable wedges in the framework that supported the arches while they were being built. Mylne also developed the idea of the "caisson" – a heavy watertight box high enough to reach from the surface to the river bed and filled with air, so that builders could construct the bridge foundations right in the middle of the river (see p.167). Caissons have become an integral part of bridge-building.

Blackfriars Bridge was completed in 1764 to widespread acclaim, and repaid its building costs through tolls within a few years. All the same, Mylne had to fight hard to get his fee paid in full.

After the success of Blackfriars, commissions for new bridges flowed in, including Middle Bridge in Romsey, Hampshire, and an interesting, though never completed, design for an all-iron bridge for Inverary Castle in Scotland, five years before the world's first iron bridge was built in Shropshire.

Although he was the architect of a number of fine country houses, Mylne's work, like Robert Adam's house, was focused much more on civil engineering.

WATER ENGINEER

Before long, Mylne was the consultant of choice on major engineering projects involving water, from water supplies and harbour works to river maintenance and canals. Taking on board his lessons from Rome, for instance, Mylne became involved in developing London's water supply and was appointed for life as surveyor of the New River, which carried water into the city from deep in the countryside north of London.

Some of his most spectacular work was carried out with canals, such as the Dearne and Dove and the Gloucester and Sharpness. Cutting off a treacherous and winding section of the River Severn, the Gloucester and Sharpness was once the broadest and deepest canal in the world.

CIVIL ENGINEERS

When Mylne was growing up, most buildings were designed by architects and engineering works were mostly carried out by the military. But the coming of the industrial age created a need for civilian experts who could build factories and transport links, water supplies and harbours, and everything else required to service the growing towns. The need for a knowledge of design, the technicalities of mechanics, hydraulics, and structural analysis, as well as mathematics, surveying, geology, and materials and costing, created a new kind of profession – the civil engineer. Mylne's friend John Smeaton was the first to call himself a "civil" engineer in a 1763 directory. In 1771, with Mylne and a few others, Smeaton formed the world's first civil engineering society, the Society of Civil Engineers, later the Institute of Civil Engineers, as a fortnightly dining club.

In an age when corruption was rife and standards in engineering were far from high, it was Mylne's belligerence and self-assertion that helped him achieve success in life. He died at New River Head on 5 May 1811, and was interred in Westminster Abbey near the tomb of Sir Christopher Wren (1632–1723), the architect of St Paul's Cathedral.

DESIGN FOR A BRIDGE
This illustration of Mylne's proposal for Blackfriars Bridge, made shortly after the design was approved, shows just how elegant the bridge was, with Mylne's ingenious elliptical arches lending it a graceful quality.

THOMAS TELFORD

GREAT ENGINEER OF THE EARLY INDUSTRIAL PERIOD

SCOTLAND 1757–1834

THE SON OF A POOR SHEPHERD from the Scottish border hills, Thomas Telford grew up to become the greatest civil engineer of the early industrial age. His fantastic bridges still inspire awe for combining structural mastery and innovation with an aesthetic grace that ensures they adorn the landscape as impressively as the cathedrals of the Middle Ages.

The first piece of masonry little "Tammie Telfer" ever did was carve the inscription on his father's gravestone in the churchyard at Westerkirk in Eskdale. Shepherd John had died eight years earlier, leaving his young widow and her three-month-old baby in poverty. Although the mother and son were very poor, Tammie's charm won him many friends, who found him an apprenticeship with a local mason making roads and bridges for the Duke of Buccleuch's estate.

In 1781, Telford got the chance to deliver a horse to London, where he found work on the building of Somerset House. His skills as a mason resulted in him being recommended to William Pulteney, a rich landowner known to him from Eskdale. Through Pulteney's patronage, Telford was engaged to work first at Portsmouth dockyards, then in the renovation of Shrewsbury Castle in Shropshire.

"THE STREAM IN THE SKY"

As Telford's reputation grew, he was invited to engineer the new Ellesmere Canal to carry iron from the Welsh border hills. No one had built a canal through such terrain before, but the novice Telford proved himself a master. Although the network was never finished, what Telford achieved was breathtaking. The Pontcysyllte Aqueduct that carries the canal across the Dee Valley is impressive even today. The 19 soaring brick arches, each hollow yet tied together with iron, were revolutionary in construction, and the entire canal was carried on top in an iron trough, cast by William Hazledine (1763–1840), whom Telford fondly called "Merlin". The iron trough was a masterstroke. Only with this cast channel could the canal be carried so high on such slender masonry.

Although a mason, Telford quickly grasped the structural potential of iron, using it to build bridges and canal installations with the aid of Merlin and the crack team of craftsmen he built up around him. His design for a new London Bridge in a single 200-m (656-ft) iron span, might have been the most graceful iron bridge ever built if it had gone ahead. Many other Telford iron bridges do exist, however, just as solid and as elegant as on the day they were built.

TELFORD IT WAS, BY WHOSE PRESIDING MIND, THE WHOLE GREAT WORK WAS PLANN'D AND PERFECTED ...

ROBERT SOUTHEY, "AT BANAVIE"

A LIFE'S WORK

- Telford travels to London and **establishes himself as a master mason**
- **Builds the Ellesmere Canal**, including the ground-breaking **Pontcysyllte Aqueduct**
- **Builds the Scottish Highlands' entire infrastructure of roads, bridges, harbours** – and even churches
- **Cuts the Caledonian Canal** right across the Scottish Highlands
- Presents a **plan to improve London**
- **Builds the Holyhead Road** through the mountains of north Wales, including the **Conway and Menai suspension bridges**
- Constructs the **Birmingham and Liverpool Junction Canal**

MENAI SUSPENSION BRIDGE
Telford's Menai Bridge, completed in 1826, was for many years the world's longest bridge. To hang the deck, Telford used 16 chains, each made of 935 bars of iron soaked in linseed oil then painted.

The Pontcysyllte Aqueduct made Telford's name. Telford's biographer, L T C Rolt, paints a picture of the opening ceremony on 26 November 1805, as six boats set out across the bridge at sunset to the sounds of cannon firing and a band playing, "To those standing by the river gazing skyward at the boats as they moved slowly along their iron channel, dwarfed by the tall perspective of Telford's towering piers, the wonder of it was almost past belief. This was Pont Cysyllte, the magical stream in the sky". Yet Telford's greatest pride was that there had been just a single casualty in the whole project – unique in those times.

HIGHLAND ROADS

Telford was soon commissioned for many new projects – including water supplies for Liverpool and Glasgow. But the greatest challenges lay back home in Scotland. Far away from the big cities, the country did not receive the attention it deserved. And, for Telford, it was a mission to give his blighted country a future.

WILLIAM TIERNEY CLARKE

ENGLAND 1783–1852

Born in Bristol, William Tierney Clarke was the pioneer of chain-suspension bridge design, as opposed to the iron links and pins favoured by Telford.

His first chain bridge was London's Hammersmith Bridge, the first suspension bridge over the Thames, opening in 1827. But his most famous is the Szechenyi Bridge in Budapest, built to make it possible to cross the Danube between Buda and Pest all year round. It opened in 1849. Tierney Clarke is also known for his schemes for pumping water from the great London reservoirs.

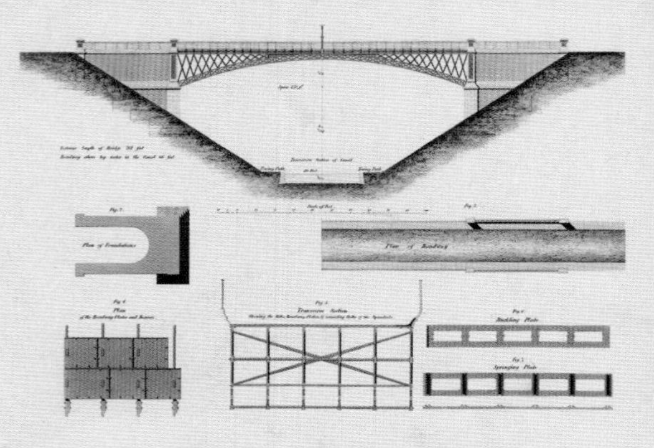

THE GALTON BRIDGE
This is Telford's design for the cast-iron Galton Bridge over his new canal at Smethwick in England's West Midlands. At the time it was built it was the world's longest arch bridge (46m/151ft) and the highest bridge overall.

Back in 1784, Scottish MPs argued that the only way to stem emigration from the Highlands was to develop the Highland fishing industry and give the local communities an economic future. But roads were needed to get the fish to market.

Telford was commissioned to survey Highland communications in 1803, and returned with a report so convincing that the parliament granted permission for a gigantic engineering project. Telford was to build a road system for an entire country – in some of the wildest terrains in the world. Not only that, he was to build a canal cutting right across the Highlands.

Most of the road system in the Highlands today is a direct result of the achievements of Telford and his crew in the early 1800s. It was an extraordinary feat of organization, technical ingenuity in driving through mountain passes and over wild rivers, and sheer human effort and endurance – creating, in all, nearly 1,500km (930 miles) of new roads. Telford and his men built more than 1,200 bridges to carry his roads across the Highlands' plentiful waters, from simple stream crossings to mighty spans across estuaries. Some are masterpieces of masonry. Others were ground-breaking structures of cast iron, such as the gracious Craigellachie Bridge.

But if the roads and bridges were a monumental achievement, then carving the Caledonian Canal through 161km (100 miles) of mountainous country, in often bitter weather, far from centres of supply, was perhaps the greatest engineering feat of the canal age – even though circumstances ultimately rendered it something of a white elephant. The construction of the lock sequence known as Neptune's Staircase at Banavie and the Laggan summit cutting rival any engineering feats of the modern age.

THE HOLYHEAD ROAD

Most of Telford's projects were in such wild settings that one wonders if he was drawn to these places, or if he was the only engineer capable of overcoming the challenges found there. From the Highlands of Scotland, Telford moved to the mountains of Wales where he constructed the last and most arduous section of the road from London to Holyhead and the ferry port to Ireland.

TELFORD AND SOUTHEY

As a young shepherd boy, "Tammie Telford" came across Milton's *Paradise Lost* in the parlour of his friend, Miss Pasley. Poetry became very important to him. He was a poet himself, and a sincere admirer of his fellow lowlands poet Robert Burns. Later in life, he became firm friends with the poet laureate Robert Southey. They went on a walking tour of Scotland together and Southey witnessed Telford at work on the Holyhead Road, calling him punningly the Colossus of Roads and *Pontifex Maximus* ("the giant bridge-builder"). When Telford died, he left Southey a large, and much needed legacy of £840, and Southey repaid him by writing the first biography of him.

ENGINEERING TIMELINE

1757–81 Orphaned as a baby, shepherd boy Thomas Telford obtains an apprenticeship with a local mason building small roads and bridges

1782–87 Telford moves to London, where he shines as a mason on the new Somerset House and is offered building projects in Portsmouth Dockyards

1788–93 Telford is appointed surveyor for the county of Shropshire, where his tasks involve renovating Shrewsbury Castle and building 40 new bridges

1794–99 He becomes the chief engineer on the building of the Ellesmere Canal, where his work includes the spectacular and innovative Pontcysyllte Aqueduct

1800 While completing the Ellesmere Canal, Telford works on water supplies for Liverpool and London

Telford's roads, however, never achieved the fame of those of his compatriot John McAdam (1756–1836). With their carefully constructed stone foundations, they seem over-engineered for the densely populated flatter regions. In these areas, McAdam's quick levelling of the ground and covering with stamped-down aggregate was cheap and effective.

Nevertheless, the Holyhead Road is still in use, providing gentle gradients, for example, for the main road through the high Nant Ffrancon pass. It also incorporates two of Telford's best-known engineering triumphs – the suspension bridges over the Menai Straits to the island of Anglesey and over the River Conwy next to Conwy castle. Telford's Menai Bridge, which many said could never be built, inspired awe when it was finally opened in 1826, carrying a single 176-m (577-ft) span more than 30m (98ft) above the Straits, and allowing tall ships to glide clean underneath in full sail.

There were still more projects to come, including St Katharine's Dock in London, the Birmingham and Liverpool Junction Canal, and the 3,000m- (9,840ft-) long second Harecastle canal tunnel, an engineering feat that would have got more attention if the canals had not been eclipsed soon after by the railways.

By the mid-1820s, Telford was nearly 70 and decided to settle in London, although he continued to work on projects such as Galton Bridge, the highest and longest arch bridge in the world at the time. He died on 2 September 1834 and was buried in Westminster Abbey.

PONTCYSYLLTE AQUEDUCT
At over 300m (1,000ft) long, the awe-inspiring Pontcysyllte Aqueduct in North Wales is still the longest cast-iron aqueduct in the world. Its 19 arches tower almost 40m (130ft) above the River Dee flowing below.

1801–06 Begins a massive programme to build roads and canals in the Scottish Highlands, including the great Caledonian Canal

1807–10 Telford works on the Göta Canal linking Gothenburg and Stockholm in Sweden

1811–22 He rebuilds the London to Holyhead Road through North Wales; the project includes building the Menai Suspension Bridge

1823–28 Telford works on numerous projects including London's St Katharine's docks and the Birmingham and Liverpool Junction Canal

1829–34 He finally buys a home in Westminster, London, but continues to work, until his death in 1834

MARC SÉGUIN THE ELDER

PIONEER OF FRENCH TRANSPORT

FRANCE 1786–1875

A BRILLIANT FRENCH ENGINEER who pioneered wire-supported suspension bridges, Marc Séguin gave his country its first significant railway and its first locomotive with an innovative multi-tube boiler that eventually made the vehicle faster than a horse. This achievement inspired English engineers George and Robert Stephenson to invent the famous locomotive, *Rocket*.

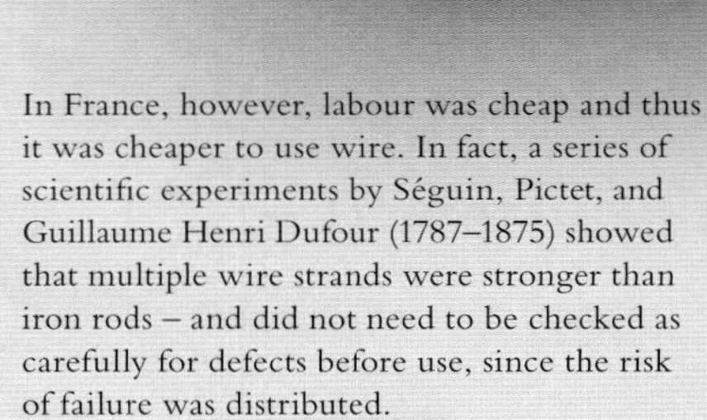

When the teenaged Marc Séguin arrived in Paris in 1799, his great-uncles Joseph and Jacques Montgolfier (see pp.172–73) were still the talk of the town for their hot-air balloon. Links between the Séguin and Montgolfier families were strong, and Séguin enrolled at the l'Institut et Conservateur des Arts et Métiers, where he was taught science by his great-uncle Joseph. A striking aspect of Séguin's engineering marvels is that they were all achieved in less than 15 years. By 1835, he had finished everything that he wanted from a practical view point and retired to concentrate on theory – making a major contribution to the concept of the conservation of energy.

Although the details are unclear, it appears that Séguin continued his studies in Paris until he was 35, returning to his home town of Annonay in Savoy in 1820. There, Marc the Elder, as he was known, began to work alongside his younger brothers, Camille, Jules, Paul, and Charles. Marc was the leader, but all five brothers often worked so closely that it is hard to establish their individual contributions. Their father Marc-François was a textile manufacturer, and his factory in Annonay gave the Séguins the financial support to develop their ideas. Marc the Elder was meant to assume control of the family firm on his return, but engineering projects soon took up his time and energy.

A LIFE'S WORK

- Séguin builds the **world's first major wire-suspension bridge** with his brothers
- He is responsible for the **first successful steamship in France**
- Engineers **France's first railway**, from St Étienne to Lyons
- Develops **France's first steam locomotive**, pioneering an **innovative multi-tube boiler**
- Contributes towards **conservation of energy**

INNOVATIVE BRIDGE

In the autumn of 1822, distinguished Swiss scientist Marc-Auguste Pictet (1752–1825) travelled from Geneva to Annonay to see something that the Séguin brothers had set up. It was a narrow, wire-test bridge suspended between two rocks over the River Cance. Pictet was fascinated because he had himself experimented with iron in suspension bridges, and was at that time writing about Thomas Telford's new Menai Straits Bridge (see pp.160–63). The Séguin brothers' test bridge was little more than a catwalk, but it was revolutionary because it used iron wires, not chains, to suspend the deck of the bridge.

Telford had thought about using wires earlier, but abandoned the idea because the labour needed to spin wires was prohibitively expensive in Britain, while iron was cheap. Therefore, it made more sense for British engineers to use robust chains of iron rods. In France, however, labour was cheap and thus it was cheaper to use wire. In fact, a series of scientific experiments by Séguin, Pictet, and Guillaume Henri Dufour (1787–1875) showed that multiple wire strands were stronger than iron rods – and did not need to be checked as carefully for defects before use, since the risk of failure was distributed.

Following Pictet's recommendations, the Séguins worked with Dufour to build the world's first full-scale wire-suspension bridge across fortifications in Geneva in 1822. But the Séguins already had plans for a bigger bridge between Tournon and Tain across the River Rhône. Tournon–Tain was a large bridge, with two 89m (292ft) spans supported by 12 cables, each consisting of 112 wires. When completed in 1825, it was a triumph for the cause of the wire-suspension bridge. The Séguin brothers went on to build a number of wire bridges – perhaps more than 180 – and by 1831, there were eight such bridges on the River Rhône alone. The Andance bridge in the Ardeches, built by the Séguins in 1827, is the oldest suspension bridge still in use in continental Europe.

STEAMSHIPS AND RAILWAYS

Meanwhile, Marc Séguin's interests were already developing in other areas. In 1824, he built one of France's first steamships, *Le Voltigeur*, which chugged along the River Rhône, powered by three innovative boilers. The point of distinction of *Le Voltigeur*'s boilers was that water was heated by directing the heat of the fire backwards and

forwards in tubes through the water. At the same time, the Séguin brothers started developing plans for a railway, the first in France, running 56km (35 miles) from St Étienne to Lyons.

The railway, completed in 1832, was a massive engineering undertaking, and involved digging the longest tunnel in Europe, 1.5km (1 mile) long, under Terrenoire. Originally, trains were intended to glide down the line to Lyons, and then be hauled back up by horses. But before work started in 1826, Séguin travelled to England to see George Stephenson's steam engine, the *Locomotion*, (see pp.190–91) on the Stockton and Darlington Railway. He was so impressed that he began to develop his own locomotives at once.

In building France's first steam locomotives, adapted from two supplied by Stephenson, Séguin made several key innovations, including a multi-tube boiler developed from his steamship, which fed hot gases to and fro through the tubes to heat the water. The locomotives also had a unique pair of rotary fans on either side of the tender to boost the supply of air into the furnace. The improvements in performance were so dramatic that they boosted the vehicle's speed from a little more than walking pace to over 40kph (25mph). Stephenson was so impressed that he incorporated Séguin's multi-tube boiler into *Rocket*, helping it to victory in the Rainhill trials, which finally assured the future of steam locomotion. Séguin's investors, however, took a little longer to be persuaded, and it was not until 1833 that locomotives were working on the St Étienne–Lyons line.

Not long afterwards, Séguin retired to his desk to concentrate on theoretical studies. He was still only 49, but perhaps it was no surprise that he decided to retire. He already had 13 children by his first wife, and, only four years later, he married his 20-year-old second wife, Augustine de Montgolfier, by whom he had another six children. Séguin was still busy writing and inventing when he died in 1875, aged 89.

IT IS ESSENTIAL FOR A RAILWAY TO HAVE AN EASY GRADIENT

MARC SÉGUIN ON THE ST ÉTIENNE–LYONS RAILWAY

WIRE CROSSING
Completed in 1825, the Séguin brothers' bridge linking the ancient towns of Tain L'Hermitage and Tournon across the River Rhône was the first major suspension bridge to have its deck suspended from wire cables.

FRANCE'S FIRST RAILWAY

Building the first railway in France, running from St Étienne down to Lyons, was a huge undertaking. Séguin's main concern here was to get a shallow gradient for the entire 56-km (35-mile) distance, starting at 1.2–1.4 per cent, and then levelling out towards Lyon. When the line opened in 1832, trains ran down the slope to Lyons at a speed of up to 28kph (17mph), and then horses hauled them back up to St-Étienne at 3–4kph (2–2.5mph). By 1844, steam locomotives hauled the trains both ways. Originally, the line was intended for freight only, but it was soon taking passengers.

THIS IS A MODEL OF SÉGUIN'S FIRST STEAM LOCOMOTIVE, BUILT FOR USE ON THE ST ÉTIENNE–LYONS LINE.

THE BROOKLYN BRIDGE REQUIRED OF ITS BUILDERS FAITH IN THEIR ABILITY TO CONTROL TECHNOLOGY

JOHN PERRY BARLOW, AMERICAN POET AND ESSAYIST

THE BROOKLYN BRIDGE RISES
In 1876, a catwalk was strung precariously between the two towers of the Brooklyn Bridge while the suspension wiring was being put in place. For a while, New Yorkers paid to cross the river this way, until it was deemed too dangerous and the public were banned. In 1883, the bridge was opened for vehicular and pedestrian passage.

THE ROEBLINGS

VISIONARIES OF THE BROOKLYN BRIDGE

UNITED STATES 1806–1926

JOHN A ROEBLING

THE CREATION OF THE BROOKLYN BRIDGE was an engineering feat of heroic proportions. It was not just the scale of the bridge, by far the largest yet built, but the sheer human endurance that went into building it that makes it such a monument. At its heart were John and Washington Roebling, father and son, who sacrificed life and health to see their design come to fruition.

A LIFE'S WORK

- John Roebling **emigrates to the US**, intending to establish a **German farming settlement** in Pennsylvania
- **Introduces wire suspension bridges** to the US, and, in 1851, builds the **first rail and road suspension bridge over the Niagara Falls**
- Constructs the **world's longest suspension bridge** at the time over Ohio River at Cincinnati, **now called the John A Roebling Bridge**
- In 1867, he submits his plan for the **Brooklyn Bridge, the largest bridge ever built**
- Roebling **dies in 1869 of tetanus** and his son **Washington steps in** to complete the bridge
- Washington succumbs to **"caisson disease"**, but carries on with the help of his **wife, Emily**
- In 1883, Brooklyn Bridge is **finally completed**

When German-born John Roebling suggested spanning the East River to link New York and Brooklyn with a vast suspension bridge, everyone knew it was needed, but few believed it could be done. Suspension bridges still had a reputation of being fragile, and the waters of the East River, with their racing currents, whirlpools, and treacherous ice in winter, were among the most dangerous in the world. But Roebling was not just an extraordinary engineer: he was also a man of steely determination.

He had pioneered the use of wire cables to support suspension bridges – and the idea of winding the cables in situ – and had proved the workability of his technique repeatedly since arriving in the US from Germany in 1831. By the time he proposed the construction of the Brooklyn Bridge in 1867, he had already built the first road and railway suspension bridge over the Niagara Falls, and the world's longest bridge over the Ohio River at Cincinnati. But the Brooklyn Bridge would be on a scale that would push even Roebling's skill to its limits.

THE FIRST VICTIM

Such was the force of Roebling's will and the thoroughness of his proposal that backers were won over, and in 1869 he got the approval for his visionary design. The main span would be more than 488m (1600ft) long, supported between two huge towers, 84m (276ft) high. The towers were built on massive foundations that were sunk to the bedrock at 12m (40ft) below the water on the Brooklyn side, and at 21m (70ft) on the New York side. Two giant anchorages would hold the cables on either side, and for the first time ever, the cables would be of steel, not iron.

But just as the work was beginning, a ferry coming into land crushed Roebling's foot. Unwilling to let his mind be clouded by opiate even for a moment, he had his mangled toes amputated without anaesthetic. Tetanus set in, and he died in agony three weeks later. With the project in danger of collapsing, Roebling's 35-year-old son, Washington Roebling (1837–1926) stepped into the breach to complete the bridge, determined to fulfil a promise made to his dying father.

OPEN CAISSON

Caissons are watertight structures that can be sunk to the river bed to allow bridge foundations to be dug midstream. In shallow water, they can project above the surface and be open at the top. In deeper water, "pneumatic" caissons, which were used on the Brooklyn Bridge, are submerged like an upturned box, and are filled with compressed air to balance the pressure of water outside. This made them highly dangerous.

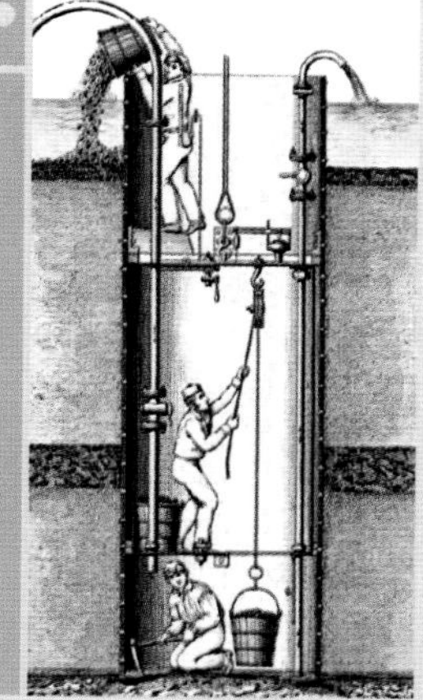
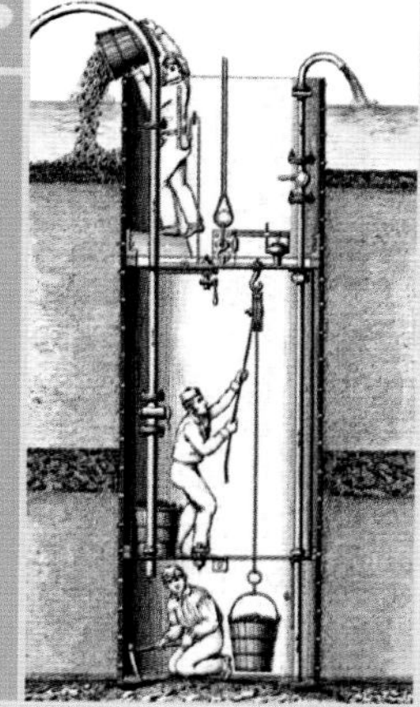

HERE, BRIDGE WORKERS DIG FOUNDATIONS IN THE RIVERBED INSIDE A SIMPLE OPEN CAISSON.

THE DEADLY BOX

Washington knew his father's work well, and had the same iron spirit. In August 1869, he took over as chief engineer. The first task was to dig the foundations inside submerged "caissons", inverted wooden boxes like giant diving bells, dropped onto the river bed (see box, below). Apart from the ever-present dangers of explosions, blow-outs, and flooding, the compressed air in the caisson also posed a terrifying threat of "caisson disease" – now known as the "bends" suffered by divers when they surface too fast. Two men had already died from the disease. In 1872, Washington succumbed, too. In excruciating agony and virtually paralyzed, it seemed likely he would follow his father to the grave. Out of sheer doggedness, he pulled through, but was left barely able to move or speak.

SUCCESS AT LAST

It was at this crucial moment that a third Roebling, Washington's young wife, Emily Roebling (1843–1903), took up the mantle with remarkable skill and authority. With Emily as his eyes and ears, the bed-ridden Washington drove the project on, despite being in almost continual pain.

The foundations were completed and the great towers soared above them, and, in August 1876, the first wires were strung between the towers, allowing chief mechanic Edmond Farringdon to tread across like a tightrope walker. Then one by one the wires were spun, and the bridge carriageway was suspended beneath them. Finally, the mighty Brooklyn Bridge was completed, and, on 24 May 1883, it was declared open amid huge crowds and spectacular fireworks. The bridge had cost the lives of 20 men, including John Roebling, and almost killed his son, but it was not just an engineering triumph – it was the moment the US realized that it was at the forefront of progress and growth.

SUSPENDED OVER THE EAST RIVER

The Manhattan Bridge, which crosses the East River in New York City, is seen here during construction in 1909. Its main span is 448m (1,470ft) long and it has two levels, the lower level carrying trains as well as traffic. The design engineer, Leon Moisseiff (1872–1943), became known for his work on "deflection theory", which held that the longer a bridge, the more flexible it could be. The idea was disastrously put to test with his Tacoma Narrows Bridge in Washington. A few months after opening in 1940, the bridge began wobbling violently during a storm, and collapsed.

THE CARSTEN
MOQUIN, OFFERMAN
WELLS COAL COMPANY.
1896

BALLOONS AND AIRSHIPS

THE INVENTION OF THE BALLOON IN FRANCE IN 1783 CAUSED SUCH EXCITEMENT THAT FOR DECADES PARIS WAS GRIPPED BY WHAT MANY CALLED BALLOON MANIA. FOR THE FIRST TIME HUMAN BEINGS HAD BROKEN FREE FROM THE EARTH AND HAD SOARED INTO THE HEAVENS.

BLANCHARD'S CHANNEL FLIGHT
Frenchman Jean Blanchard and his companions made the first crossing of the English Channel in a gas balloon in 1785. They only made it after jettisoning all their ballast.

It is hard to imagine just what a stir the balloon caused in 18th-century Paris. On 19 September 1783, the scientist Pilâtre de Rozier (see p.172) launched the first hot-air balloon called "Aerostat Reveillon". The passengers were a sheep, a duck, and a rooster. The balloon stayed in the air for a grand total of 15 minutes before crashing back to the ground. The first manned balloon ascent was made in the Montgolfier brothers' hot-air balloon, in November 1783 (see pp.172–73). This was so amazing that a quarter of a million people turned out to witness the historic event. Barely two weeks later, when Jacques Charles (1746–1823) and Noël Robert (1760–1820) made the second manned ascent in a hydrogen gas balloon, 400,000 people watched – perhaps half the entire population of Paris. The streets were crammed as spectators scrambled up trees and lamp-posts, climbed out of windows, and shinned over roofs to get a glimpse of the wonderful phenomenon.

Paris went balloon-mad, as everything from fans and hats to tables and chairs were decorated with balloons. The American statesman Benjamin Franklin (see pp.136–37), who was in Paris at the time, wrote that "all our circle of friends, at all our meals, in the antechambers of our lovely women, as in the academic schools, all one hears is talk of experiments, atmospheric air, inflammable gas, flying cars, and journeys in the sky". It was truly the birth of hot-air ballooning.

Early balloonists burnt materials onboard the balloon to generate heat for propulsion. Later, gas and helium designs were introduced, which were considered safer and more reliable than flying with an open flame.

RIP AND DRAG

Technically, balloons advanced little after that first launch in 1783, and the craze died down. For his first balloon, Jacques Charles came up with the idea that hydrogen would be a suitable lifting agent for flying. The Robert brothers (see pp.174–75) had introduced rubberized silk for making the bag, while Charles had included the valve to let gas out for descent, and sand ballast for controlled ascent – all the key elements of a balloon. Otherwise, the only significant additions came half a century later with the drag rope introduced by Briton Charles Green (1785–1870) and the rip panel by American John Wise (1808–79). The drag rope is a rope that can be trailed on the ground to control the height of the balloon with the rope's weight, saving on gas and ballast. The rip panel is the section of the balloon that can be torn open quickly with a tug on a cord to let gas rush out and settle the balloon when it lands.

AIRSHIP GLOBETROTTER
A matchbox commemorates the 1929 circumnavigation of the globe by the *Graf Zeppelin*, the first journey of its kind in an aircraft, commanded by Dr Hugo Eckener.

DIRIGIBLE BALLOONS

At first, balloonists had hoped to gain a measure of flight control, and tried paddles, oars, and various other contraptions to steer and propel, or as the French said, make it "dirigible". But they quickly realized that none of these had any effect, and balloons had to drift at the mercy of the winds.

Finally, in 1852, Frenchman Henri Giffard (1825–82) created the first dirigible balloon, adding a steam engine and a giant propeller to a cigar-shaped balloon. It was the world's first passenger-carrying airship. Giffard's dirigible could chug along at barely 10kph (6mph), and having a steam engine puffing away beneath a giant bag of highly inflammable hydrogen was risky. Dirigibles only became a reality right at the end of the 19th century when the light and powerful internal combustion engine was added by both Alberto Santos-Dumont (see pp.176–77) and Hungarian David Schwarz (1852–97). Schwarz's dirigible was the first rigid airship, in which the gas was contained not in a flexible bag but inside an envelope created by laying the gasproof fabric over a rigid metal framework. Count von Zeppelin (see pp.178–79) went on to make enormous, rigid airships – by far the biggest aircraft that have ever flown.

MILITARY RECONNAISSANCE
Thaddeus Lowe was a ballooning pioneer in the US. His courageous exploits with balloons in the Civil War showed how valuable aerial reconnaissance could be in wartime.

ZEPPELIN CONTROL CAR
By 1914, Zeppelins were 150m (500ft) long and could carry loads of 9,000kg (20,000lb); they could even carry passengers across oceans at up to 80kph (50mph).

SETTING THE SCENE

- English scientists **Henry Cavendish and Joseph Priestley discover the gas hydrogen, and call it "inflammable air"** because it catches fire so easily. They are the first to **distinguish this "inflammable air" from ordinary air** and investigate its specific properties. They realize it is much lighter than air, and Cavendish astonishes dinner guests by sending bags full of hydrogen floating up to the ceiling.
- French **scientist Antoine Lavoisier confirms how buoyant inflammable air** is, and renames it, more tamely, as hydrogen.
- **Joseph Montgolfier** notices his wife's washed underwear billowing upwards as it hangs over the fire to dry, and **thinks of hot-air balloons**.
- Later, English scientist **Michael Faraday** (see pp.140–41) **uses sheets of raw rubber** ("caoutchouc") to make containers for hydrogen, and so **invents the toy balloon**.

ZEPPELIN UNDER CONSTRUCTION
The airships of the 1920s and 1930s were so big that gigantic hangars had to be built to construct them. Here, the aluminium framework of the USS *Shenandoah* nears completion in 1923. It crashed two years later.

MONTGOLFIER BROTHERS

PIONEERS OF THE FIRST MANNED AIR ASCENT

FRANCE 1740–1810; 1745–99

JOSEPH-MICHEL

JACQUES-ÉTIENNE

WITH THEIR FANTASTIC hot-air balloon, the brothers Joseph-Michel and Jacques-Étienne Montgolfier allowed humankind to defy gravity and take to the air for the first time. Their achievement created a sensation in pre-revolutionary Paris and paved the way for other aeronauts.

JEAN-FRANÇOIS PILÂTRE DE ROZIER

FRANCE 1754–85

After his historic first flight, Pilâtre de Rozier came to be known as "the intrepid Pilâtre, who never loses his head".

On 15 June 1785, however, he decided to cross the English Channel from France in a hot-air balloon with a second, hydrogen-filled envelope. With the winds blowing inland, de Rozier had to wait. Meanwhile, his rival Jean-Pierre Blanchard (1753–1809) used the winds to cross from the English side in a hydrogen balloon. Frustrated, de Rozier set off despite the unfavourable weather. But the combination of brazier and inflammable hydrogen gas proved fatal. The balloon went up in flames, and as it plummeted, he was killed.

The Montgolfiers were a long-established family of paper manufacturers from Annonay in the Ardèche mountains of southern France. They were a big family of 16 siblings – Joseph Montgolfier was the twelfth and Étienne Montgolfier the youngest. Étienne trained in Paris as an architect, but after the death of most of his elder brothers, he came back to Annonay to run the family paper business with Joseph. The business did so well, with Étienne managing and the wayward but inventive Joseph introducing the latest technologies, that the brothers were awarded a grant by the French government to act as a business model for all French paper-making industries. This knowledge of paper-making became an asset for the Montgolfier brothers in their experiments with hot-air balloons.

A LIFE'S WORK

- Born into a family of paper manufacturers, the Montgolfier brothers become **the foremost paper-makers** in France
- They conduct experiments in which **bags filled with hot air are lifted into the air**
- The brothers amaze the people of their home town of Annonay with the **flight of a giant balloon filled with hot air**
- In a demonstration of their hot-air balloon in front of the French king, they **send a sheep, a duck, and a rooster into the air in the balloon**
- In 1783, they create the **hot-air balloon** in which two men make the **first successful human ascent** into air

BALLOONING IDEAS

The idea of balloons started when in 1766 English chemists Henry Cavendish (1731–1810) and Joseph Priestley (1733–1804) discovered the existence of "inflammable air" – the gas now known as hydrogen. One of the things that intrigued people most about hydrogen was the realization that it is much lighter than air. Scottish scientist Joseph Black (1728–99) amazed his dinner guests by letting small bags filled with hydrogen float to the ceiling. People in England and in France – where Antoine Lavoisier (1743–94) gave hydrogen its name in the mid-1770s – were soon wondering if they could create larger, lighter-than-air balloons.

The problem with hydrogen, however, was that it was quite expensive to make in large amounts. It also tended to leak from any container not completely gas-proof, and, as Cavendish realized, it was highly flammable and very dangerous. Hot air, on the other hand, was easy to generate in large quantities with just a fire, was much safer than hydrogen, and, best of all, could be collected in a simple paper bag – which is just what the Montgolfiers were good at making.

TRAPPING AIR IN PAPER

Joseph said he got the idea for hot-air balloons in 1777 when watching his wife's underwear billow upwards near the fire where she hung it to dry. The brothers' experiments with balloons began indoors, with a very light box made from a thin wooden frame and taffeta walls – like a Chinese lantern. Joseph filled this box with hot air by burning some crumpled paper. The box shot up so quickly that the brothers believed they had discovered a new, extremely light gas, similar to hydrogen, which they named, with their eye for publicity, Montgolfier gas. They joked about "putting a cloud in a paper bag" and Joseph imagined the day when giant paper bags would float the entire French army into Gibraltar and seize it from the British.

FLYING ANIMALS

By 1782, the brothers were ready to try a bigger version. The results were so spectacular – the taffeta box flew nearly 2km (1.2 miles) – that they decided to make a public demonstration on 4 June 1783. They made a pear-shaped balloon from four sections of paper-lined taffeta buttoned together. The balloon was huge, more than 9m (30ft) high, and filled with more than 566 cu m (20,000 cu ft) of hot air from burning braziers. A large crowd gathered to watch the balloon climb 2,000m (6,562ft) into the air and fly 2km (1.2 miles) in a flight lasting 10 minutes.

News of the Montgolfier brothers' invention of flight travelled fast, and they were invited to demonstrate their balloon before the king at Versailles. They spared no expense and got royal wallpaper-manufacturer Jean-Baptiste Reveillon to make a stunning blue and gold balloon of taffeta, varnished with alum to make it fireproof.

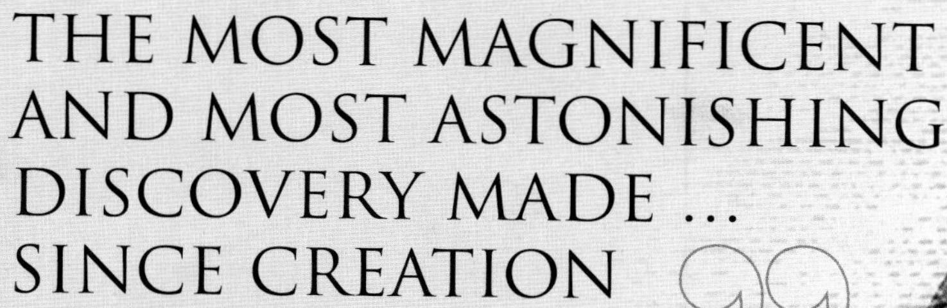

THE MOST MAGNIFICENT AND MOST ASTONISHING DISCOVERY MADE … SINCE CREATION

THE GENTLEMAN'S MAGAZINE ON BALLOON FLIGHTS, 1807

Everyone wanted to know if it could carry people aloft, but for this first trial, the Montgolfiers played safe and sent up a sheep, a duck, and a rooster. On 19 September 1783, King Louis XVI and his queen, Marie Antoinette, witnessed the giant balloon and its menagerie ascend gracefully into the air, climb 450m (1,476ft) and come down safely 3km (2 miles) away after eight minutes.

FIRST HUMAN FLIGHT

Soon there was a clamour to be the first human "aerostatists", and in the end the flamboyant scientist Pilâtre de Rozier (see box, opposite) and officer Marquis d'Arlandes were chosen. An even bigger balloon was made, this time with a ring-shaped gallery for the two pilots. After a couple of brief tethered test flights, first with Étienne, and then with de Rozier, on 21 November 1783 the balloon was finally set free to float aloft with de Rozier and d'Arlandes aboard.

D'Arlandes later published an account of that historic first flight, which began at the Chateau de la Muette (near the present location of the Eiffel Tower) and ended 27 minutes later outside the city at the Butte aux Cailles. D'Arlandes describes how they soared 300m (984ft) high, drifted across the Seine, swooped low over the houses of St Germain, and flew over the Luxembourg park. D'Arlandes paints a vivid picture of the cool aplomb of de Rozier, who was standing on the other side of the gallery. D'Arlandes admits being scared stiff, shouting frequently, "We must land now!" as de Rozier calmly fed the brazier. At one point, when the balloon was shaken by a gust of wind, he cried out to de Rozier, "Stop dancing!" while de Rozier calmly told him to enjoy the view. When the balloon finally landed, d'Arlandes scrambled out in utter relief, while de Rozier folded his coat and stepped out sedately to meet the cheering crowds.

FIRST FLIGHT
The Montgolfier balloon floats high above Paris and over the Seine on 21 November 1783, carrying humans aloft for the first time. The Marquis d'Arlandes and Pilâtre de Rozier wave from the balloon's gallery.

ROBERT BROTHERS

PIONEERS OF HYDROGEN-BALLOON FLIGHT

FRANCE 1758–1820

Just ten days after the Montgolfier brothers stunned Paris with their balloon, a second manned balloon went up, made by the brothers Anne-Jean and Nicolas-Louis Robert for scientist Jacques Charles, and filled not with hot air but with hydrogen gas. This balloon also incorporated all the main features of modern helium balloons – a basket suspended underneath, bags of ballast to jettison for ascent, and a valve to release gas for descent.

The Robert brothers built the the world's first hydrogen balloon for professor Jacques Alexandre Charles (1746–1823). Born in Beaugency sur Loire on 12 November 1746, Charles came to Paris to work in the Bureau of Finances. He probably lost his job in 1779, at about the time scientist Benjamin Franklin (see pp.136–37) visited Paris. Charles was so fascinated by Franklin's discoveries that he resolved to become a scientist himself, and was soon giving scientific demonstrations.

Charles heard about the work on gases of English scientists Robert Boyle (1627–91) and Henry Cavendish (1731–1810), and the Italian scientist Tiberius Cavallo (1749–1809), and realized that hydrogen, because it is much less dense than air, would be capable of lifting a balloon off the ground. The main problem was creating a leak-proof skin, since hydrogen escapes rapidly through the smallest holes. What is more, to make enough hydrogen gas to fill a balloon took more than a day of pouring sulphuric acid on to a mass of scrap iron.

Charles began to work with the Robert brothers, who ran a workshop near St Victoire in Paris. Their solution was to dissolve rubber in turpentine, then brush the solution onto a silk bag to make a gas-tight envelope. The Montgolfiers could rely on natural cooling of the air to bring their hot-air balloon down, but Charles had to incorporate a valve to release gas slowly. He also realized that with hydrogen he could not simply stoke a brazier to gain height; instead he had to include sand-filled ballast bags that he could jettison one by one. Finally, with no need to lift fuel into the brazier, the pilots could travel in a basket suspended beneath the Montgolfier balloon, rather than in an awkward ring-shaped gallery around the balloon's neck. So, between them, Charles and the Robert brothers came up with all the key elements of a modern balloon's design – the gas valve, the ballast system, and the suspended basket.

A LIFE'S WORK

- The **Robert brothers are employed by Jacques Charles** to make a hydrogen balloon
- Charles and the Robert brothers **devise all the features of modern balloon design**: the gas valve, the ballast system, and the supended basket
- The Robert brothers and Charles **beat the Montgolfier brothers in launching the first balloon**
- Nicolas-Louis Robert joins Jacques Charles on the **first hydrogen-balloon flight**
- **Charles discovers** that **gas volume** is always **proportional to temperature**
- Charles is the **first solo balloonist**
- The Robert brothers make the **first balloon flight of 100km (60 miles)**

UP AND AWAY

The trio actually stole a march on the Montgolfiers on 27 August 1783, by launching a small trial balloon from Paris's Champ de Mars – a fortnight before the Montgolfiers' demonstration. Even though their balloon was just 2m (7ft) in diameter, such was the lifting power of hydrogen that it travelled 21km (13 miles), far outside the city to the village of Gonesse, where local villagers attacked it with pitchforks, leading the government to issue a proclamation that balloons were perfectly safe.

The Montgolfiers beat the trio with the first manned flight. All the same, ten days later, on 1 December, Charles and Nicolas-Louis took off from the Tuilieries gardens, watched by a gigantic crowd of 400,000 – half the population of Paris. The pair's jubilation lasted two hours, and they landed in a field more than 40km (25 miles) from Paris. As they landed, Robert leaped out, and, relieved of his weight, the balloon shot up again, sending Charles 3,000m (10,000ft) in less than ten minutes. By default, Charles became

DESTRUCTION OF A BALLOON
Jacques Charles and the Robert brothers launched the world's first hydrogen balloon on 27 August 1783. The balloon flew north for 45 minutes, and landed 21km (13 miles) away at the village of Gonesse, where terrified peasants destroyed it.

the first solo balloonist. The control and duration of their flight was in such contrast to the Montgolfiers' wayward progress that it soon became clear that hydrogen was the preferred option for balloonists, and, as Paris became gripped with balloon-mania, it was hydrogen balloons that lifted Parisians into the skies.

Charles, however, never flew again, and turned back to science. The Robert brothers, meanwhile, conducted further flights in a balloon they called the *Charliere*, in honour of their former colleague. Their last memorable experiment was in developing a cigar-shaped balloon that would be dirigible (see box, right) – that is, it could be both propelled and steered, rather than simply drifting on the winds.

ROBERTS'S DIRIGIBLE

Drawing on an idea by French engineer Jean-Baptiste Meusnier (1754–93), the Robert brothers tried to develop a dirigible balloon. It incorporated internal gas cells or ballonets in the main bag, a rudder, and oars to "row" through the air. The oars and the rudder were useless, and the lack of a gas valve meant that a flight on 15 July 1784 had to be curtailed by slashing the ballonets with a knife, thousands of metres above land. Two months later, however, the brothers made the first balloon flight of more than 100km (60 miles) in a more conventional balloon.

THIS AIRSHIP DESIGNED BY MEUSNIER CARRIES A GONDOLA.

ALBERTO SANTOS-DUMONT

PIONEER OF THE FIRST WINGED AIRCRAFT FLIGHT

BRAZIL **1873–1932**

For a while, Alberto Santos-Dumont, the flamboyant heir of a wealthy coffee-growing family from Brazil, was the most famous aeronaut in the world. He was celebrated both for the dirigible (powered and steerable) balloons he built and flew around Paris with such panache, and for making the first flight in Europe of a winged aircraft with his *14-bis* in 1906.

Alberto Santos-Dumont was born on 20 July 1873 in Cabangu near São Paolo in Brazil. He was the youngest of 11 children of a wealthy French-born coffee magnate, known as the Coffee King of Brazil. Monsieur Dumont was keen on technology, and Alberto grew up surrounded by the latest farm machines.

When Alberto was 17, a fall from a horse left his father paralysed. Monsieur Dumont decided to move to France, taking Alberto with him. But this tragedy opened up the playground of Paris to the wealthy teenager Alberto. At once, he bought himself a car, one of the first on the road. Then he took lessons in ballooning, which was still very much in vogue. He rapidly became an expert pilot and soon knew enough about balloons to design his own, named the *Brésil*, in 1898, when he was 24.

A LIFE'S WORK

- Santos-Dumont **designs personalized dirigibles**, and becomes a sensation for his **flights through the Paris boulevards**
- He builds **Europe's first ever winged aircraft, the *14-bis***
- **Makes several flights in his *14-bis***, the third and longest flight being 60m (197ft) in 1906
- In 1908, he creates the ***Demoiselle***, the **world's first personal, cheap sports plane**
- His ***14-bis*** aircrafts are used extensively during **World War I**
- Is **depressed by the way his "babies", the *14-bis* aircraft, are used in World War I** – not to bring a moral rebirth as he had hoped, but for death and destruction

BALLOONS FOR EVERYONE

Santos-Dumont believed in the democratization of flight. He dreamed of a world where everyone had their own balloon, which they could tether to lamp-posts as they tethered horses. So his balloon was made of a special, light silk, and could be folded so small that it would fit into a handbag.

He took the same approach when he started making dirigible balloons. Unlike the cumbersome airships of the day, Santos-Dumont's dirigibles were compact, personal transport. Just how personal, he demonstrated in his own style. On fine mornings, he would step into the one-man basket with a picnic box, a flask of Brazilian coffee, and a decanter of Chartreuse, and glide above the boulevards of Paris, or out into the countryside to drop in on some chateau. Occasionally, he would simply land on the sidewalk and take lunch at a café before setting off again. Once he even dropped in on his apartment at No 9, Rue Washington.

Santos-Dumont was determined to win the 100,000-franc prize set up by the financier Henri Deutsch for the first man to fly from the

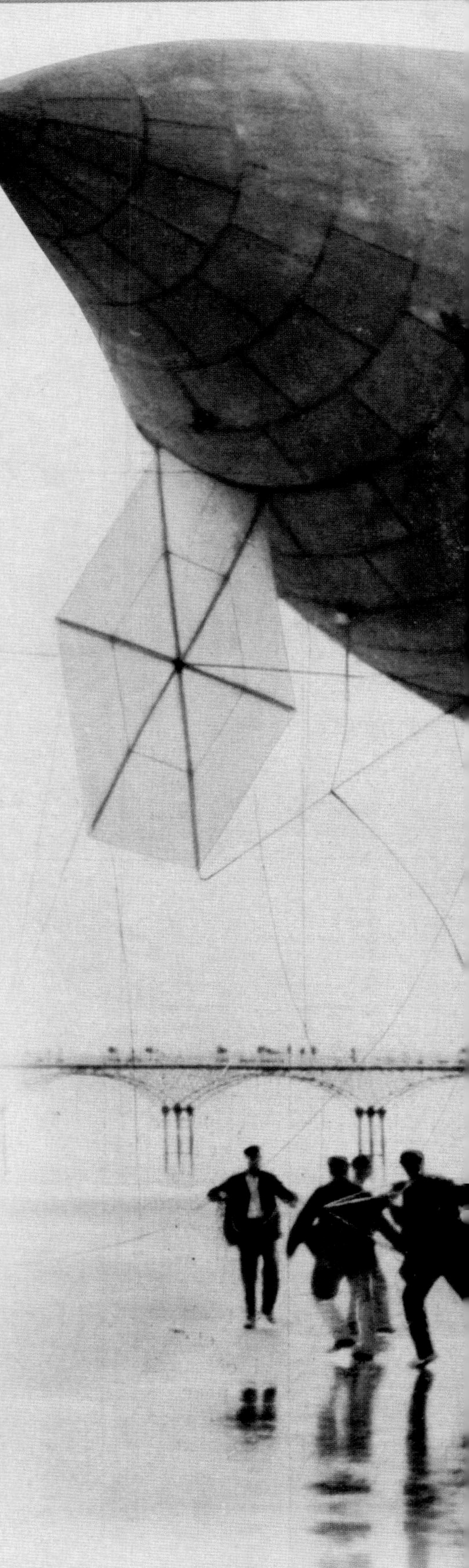

COME IN, NO 14!
The last of Santos-Dumont's airships, No 14, flies above the sands at Trouville, France, on 1 September 1905. No 14 was the dirigible that was meant to lift the *14-bis* aircraft into the air.

THE HOT-AIR BALLOON SEEMS TO STAND STILL IN THE AIR WHILE THE EARTH FLIES PAST UNDERNEATH

ALBERTO SANTOS-DUMONT

Aéro Club de Paris's headquarters in St Cloud to the Eiffel Tower and back in under 30 minutes. After two failed attempts, Santos-Dumont succeeded – but almost lost the prize because his theatrical gesture of circling the Eiffel Tower meant that he ran a few seconds late. After much consternation, the judges awarded him the prize, which he gave to charity and to the workers who had helped him build his airships.

GIVING WINGS TO FLY

By now, Santos-Dumont was a celebrity. In 1901, *The Times* of London wrote that when the names "of those who have occupied outstanding positions in the world have been forgotten, there will be a name that will remain in our memory, that of Santos-Dumont". They were wrong, but in 1901 Santos-Dumont was a superstar. Although just 1.5m (5ft) tall, he cut a distinctive figure with his trademark floppy panama hat, high-collar shirts, and dark, pinstriped suits. His image appeared on everything from matchboxes to dinner plates. He popularized the wristwatch, after Cartier designed one for him personally when he complained that he could not check the time without taking his hands off the controls.

Soon, Santos-Dumont moved on from lighter-than-air balloons to the problems of winged flight, and by 1905 he had built his first aircraft. The following year, he made several flights in his *14-bis*, the third and longest flight being 60m (197ft) on 23 October 1906. *Bis* means "encore" in French, and it was originally meant to be lifted into the air by his No 14 dirigible. But Santos-Dumont soon realized that his aircraft could take off on wheels under its own power. This led to some dispute over the claim to having undertaken the first flight of this kind.

The Wright brothers (see pp.324–27) made the first sustained, powered controlled flight with their *Flyer* at Kitty Hawk in the US in 1903. But many claim that Santos's third flight was the world's first proper flight because the *14-bis* took off on wheels under its own power, while the *Flyer* was catapulted into the air. Moreover, Santos's flight took place in front of a crowd with officials present, whereas the *Flyer*'s took place in a remote place in secrecy. But Santos's claims were completely eclipsed when the Wright brothers came to Europe.

Undaunted, in 1908, Santos-Dumont launched a pioneering little aircraft called the *Demoiselle*. Nicknamed the grasshopper, it was the world's first personal sports plane. It was cheap, simple, and ultralight and Santos-Dumont gave the design freely to anyone who wanted to have a go at making it, as thousands did.

By 1910, Santos-Dumont, still in his mid 30s, had developed multiple sclerosis, and he spiralled into depression. As his health declined, he returned to Brazil, where he died in 1932, largely forgotten by everyone outside his home country.

ROBUR'S ALBATROSS

Santos-Dumont's dreams of flying were inspired by reading Jules Verne's novel *Robur the Conqueror*. In it, the Weldon Institute is convinced that only lighter-than-air dirigibles are suitable for flight. But then Robur reveals his Albatross, a flying clipper ship, with rotors instead of masts to provide both lift and forward propulsion. After a thrilling tale, Robur's helicopter-like craft proves far superior to the dirigible balloon.

THIS PICTURE DEPICTS THE FANTASTIC FLYING CLIPPER OF THE CLOUDS FROM JULES VERNE'S 1886 NOVEL.

FERDINAND VON ZEPPELIN

DESIGNER OF THE FIRST SUCCESSFUL AIRSHIP

GERMAN CONFEDERATION 1838–1917

GERMAN NOBLEMAN FERDINAND VON ZEPPELIN was a cavalry officer for more than 30 years, but when he retired, he realized a long-held ambition to build balloons that could be steered (dirigibles). The airships he created were so successful that they were to open up long-distance air travel to fare-paying passengers for the first time, beginning a global revolution in communication.

A LIFE'S WORK

- In the 1860s, Zeppelin **joins the Union army as a volunteer during the Civil War**, and while there, meets Thauddeus Lowe and witnesses the military use of balloons for the first time
- Meets Lowe again in 1869 and **learns all he can about ballooning** and of Lowe's ambitions for transatlantic travel
- **Retires early in 1891**, and begins to work first with engineer Theodor Gross, then with Theodor Kober, **to develop large airships for the military**
- On 2 July 1900, **the LZ1, his first successful rigid airship**, has its first flight
- His Zeppelins are soon in civilian as well as commercial use, **taking paying passengers on long airflights** – eventually across the Atlantic – for the first time
- Introduces the world to **long-distance air travel** with the first large airships

Born on 8 July 1838, on an island in Bodensee (Lake Constance) in what is now Baden-Württemberg, Ferdinand Graf von Zeppelin could trace his noble heritage back to the 14th century. His father was a minister and Hofmarschall for the court of Württemberg, and when he grew up, von Zeppelin became a cavalry officer just as so many of his aristocratic contemporaries did.

FATEFUL MEETING

In 1863, the young cavalry officer volunteered to fight for the Unionists in the American Civil War, which proved to be a life-changing experience. He was there for three years, and when he returned to Germany, he took up his military career where he had left off, but while in the US, he had had a crucial encounter with the remarkable Thaddeus Sobieski Coulincourt Lowe (1832–1913).

Thaddeus Lowe was a self-educated scientist and inventor of extraordinary energy, who began his career by giving spectacular chemistry displays before turning to ballooning. His great ambition was to fly across the Atlantic, and for this purpose he built a gigantic hydrogen balloon, the *City of New York* (later named the *Great Western*). He made several attempts in 1860, before the American Civil War brought a halt to his plans. In the war, Lowe set up the world's first aerial reconnaissance corps, and personally made many balloon flights over dangerous areas to provide vital observations for the Unionists. Zeppelin met Lowe during the Peninsular Campaign, and Lowe sent the young German officer to St Paul, Minnesota, where fellow German John Steiner gave him his first ride in a balloon. The effect was intoxicating and in 1869, when the war was over, Zeppelin returned to the US to meet Lowe again and learn all he could about ballooning.

RISE OF THE AIRSHIPS

For the next 22 years, Zeppelin served his time as a cavalry officer, rising to the rank of general, but the idea of large, dirigible airships – lighter-than-air crafts that could be steered – continued to occupy his thoughts, and when he finally retired in 1891, he set about turning his ideas into reality. He did not want to build a small dirigible balloon with a light wooden framework; he wanted to build something huge, with a strong framework of aluminium, made up of dozens of individual gas cells. That way his envelope could contain enough hydrogen to give it immense lifting power – and thus carry powerful engines and the fuel for long-distance flights. It was Lowe's invention of the water gas process in 1873 that made hydrogen available cheaply in large quantities.

After seven years of preparation, Zeppelin came up with the design for the LZ1, which was the world's first airship. It was gigantic, 130m (427ft) long and 12m (38.5ft) in diameter, and able to hold almost 11,000 cu m (400,000 cu ft) of hydrogen. The problem was where to build such a huge craft. In the end, Zeppelin constructed a giant wooden

AIRSHIPS AT WAR

Alberto Santos-Dumont (see pp.176–77) was distraught when he learned that airships were to be used for military purposes. In World War I, the Germans began to use Zeppelins to combat British seapower, scouting for and attacking British ships. Zeppelins were used for the first aerial bombing raids on British cities. But after the invention of incendiary bullets, their inflammable hydrogen gas put them at risk of being shot down in flames. Hence they became ineffective. The British, meanwhile, built hundreds of semi-rigid airships or "blimps". In World War II, the Americans used airships successfully to protect convoys from submarine attacks.

ZEPPELINS ON BOMBING RAIDS COULD EASILY BE BROUGHT DOWN IN FLAMES BY AIRCRAFT FIRING INCENDIARY BULLETS.

hangar on Bodensee, near his birthplace. The movable, floating hangar could be swung around so that the ship could enter or leave into the wind. Construction began in June 1898 and was finished in the winter of the following year, but Zeppelin decided to wait until the summer before attempting the first flight. In June 1900, the gigantic shape was inflated with hydrogen gas and on 2 July, the LZ1 emerged to make its maiden flight over the lake.

There were many problems with the design, and over the next eight years Zeppelin went on to build the LZ2, 3, and 4 with the help of brilliant engineer Ludwig Dürr (see box, right), who replaced Zeppelin's tubular aluminium framework with a dramatically stronger network of triangular girders. The LZ4 was spectacularly destroyed by a fire during a storm at Echterdingen on 5 August 1908. By that time, however, the German people were so excited by Zeppelin's vast ships of the air that the crowd witnessing the disaster rallied round to donate 6 million deutsch marks in what became known as the "Miracle at Echterdingen", enabling Zeppelin to set up the Zeppelin Construction Company for building airships.

Just a year later, DELAG, the world's first airline, was set up and began flying passengers. By World War I, they had seven airships in operation. The military, meanwhile, took 14 for reconnaissance, and later bombing. They were typically 160m (525ft) long and flew at up to 80kph (50mph). Zeppelin died on 8 March 1917, while the war was still going on, but by then the future of airships seemed, for the moment, assured. Dr Hugo Eckener took over, and, in 1928, made sure that Lowe and Zeppelin's dream of flying across the Atlantic became a reality.

LUDWIG DÜRR

GERMANY 1878–1956

None of Zeppelin's airships would ever have been quite so successful without the help of Ludwig Dürr. He was the chief designer of all the Zeppelin Company's airships except the very first – the LZ1 – and made key innovations from the moment he became designer on the LZ2.

The LZ1 had a framework of tubular girders that were not rigid enough to prevent its hull from bending and twisting in flight – Dürr replaced them with stiffer triangular girders. He also became one of the chief Zeppelin test pilots. The most successful of his designs was the *Graf Zeppelin* – it made its first trip across the Atlantic in 1928, flew around the world in 1929, and, in 1931, flew over the North Pole. Dürr remained chief designer until the Zeppelin Construction Company closed in 1945.

ZEPPELIN OVER LAKE CONSTANCE
Spectators observe Count von Zeppelin's second airship, the LZ2, hovering over Lake Constance, 17 January 1906. It was destroyed on its first flight following an engine failure. The "wings" at the front and rear were designed to provide extra lift, but these became redundant as engines became more powerful.

THE LARGEST FLYING OBJECT EVER BUILT
The passenger Zeppelin *Hindenburg*, seen here in its gigantic hangar at Frankfurt, Germany, measured 245m (804ft) in length, only 24m (79ft) shorter than that other doomed ship, *Titanic*. At its widest point in the middle, it had the height of a 13-storey building. The *Hindenburg* caught fire as it came into land at Lakehurst, New Jersey, probably when a static electric spark ignited leaking hydrogen. At the time, it was the fastest vessel over the Atlantic, travelling from Frankfurt to New York in just 59 hours and to Rio de Janeiro in 104 hours.

THE MACHINE AGE

THE MACHINE AGE

1780

1787
US inventor **John Fitch's** (1743–98) steam canoe, which is propelled by 12 oars and driven by a steam engine, is demonstrated on the Hudson River.

1788
Scottish engineer **William Symington** (1764–1831) demonstrates his first steamboat on Dalswinton Loch in Scotland.

1808
US entrepreneur **Robert Fulton** introduces the world's first steamship for passengers on the Hudson River (see pp.188–89).

▲ 1811
The **Fulton Company** builds the first Mississippi steamboat.

◀ 1816
Scottish engineer **Robert Stirling** creates his heat engine, now known as the Stirling engine (see pp.214–15).

1820

1825
George Stephenson's Stockton and Darlington line, the world's first steam railway, opens, carrying trucks and passengers in cars hauled by steam locomotives (see pp.190–91).

▲ 1829
Robert Stephenson's ***Rocket*** wins the Rainhill locomotive trials (see pp.190–91).

▼ 1830
The world's first passenger railway, the **Liverpool and Manchester**, opens amid much acclaim, despite the accidental death of MP William Huskisson, knocked down by the *Rocket*.

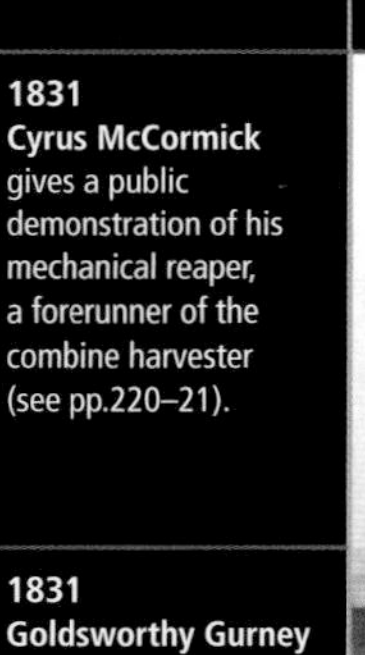

1831
Cyrus McCormick gives a public demonstration of his mechanical reaper, a forerunner of the combine harvester (see pp.220–21).

1831
Goldsworthy Gurney establishes a regular steam-carriage road service to carry passengers between Cheltenham and Gloucester in England (see pp.216–17).

1836
US industrialist **Samuel Colt** introduces the six-shooter revolver pistol, developing an idea that was first thought of by Elisha Collier in 1818 (see pp.224–25).

1840

► 1849
French horticulturalist **Joseph Monier** and others in France, such as **François Coignet** (1814–88), develop reinforced concrete, initially for making stronger flower pots (see pp.244–45).

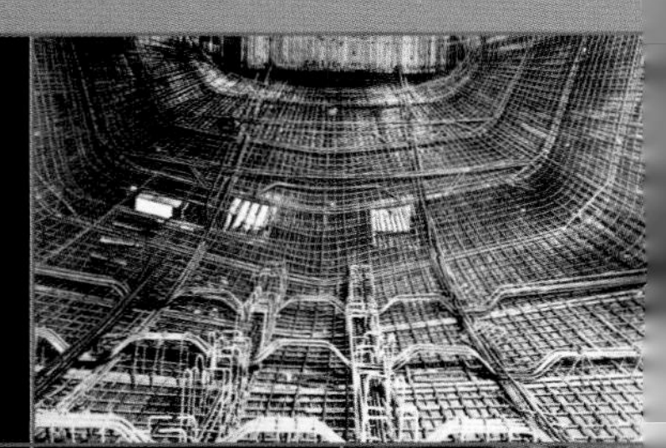

► 1855
Henry Bessemer invents the Bessemer converter, making it possible to create large quantities of steel cheaply by blowing hot air through molten pig iron (see pp.228–29).

1859
English engineer **I K Brunel**'s wrought-iron Royal Albert Bridge over the River Tamar at Saltash, in southwest England, is opened (see pp.194–97).

► 1859
Joseph Bazalgette commences work on the remarkable sewage system that solves London's sewage problem and sets the trend for sewers around the world (see pp.252–53).

The Industrial Revolution provided the launchpad for the emergence of a huge range of new and often astounding machines during the 19th and early 20th century. Many of these were connected with transport and communication. Steam power was developed first to power boats and then to drive locomotives, bringing the railways that changed the world forever. But if the speed of the railways seemed fantastic to an earlier age, it was nothing compared to the miracle of instant communication unveiled first by the electric telegram, then the telephone, and finally the radio, which saw telegraphs beamed across the Atlantic for the first time in 1902. Meanwhile, the newly discovered power of electricity spawned other key developments, from electric railways to the electric bulb.

1860

► 1860
German schoolteacher **John Philipp Reis** (1834–74) invents the "*telefon*" for transmitting speech electrically through wires.

1865
French engineer **Eugène Belgrand** completes the first of the aqueducts that eventually deliver clean running water to tens of thousands of Parisians for the first time (see pp.250–51).

▲ 1866
Scottish physicist **William Thomson** oversees the completion of a cable beneath the Atlantic, establishing a continual transatlantic telegraph service (see pp.232–33).

1875
Joseph Monier builds the world's first concrete bridge (see pp.244–45).

► 1876
Scottish inventor **Alexander Graham Bell** makes the world's first phone call (see pp.258–59).

1879
German industrialist **Werner von Siemens** demonstrates the world's first electric railway at the Berlin Trade Fair (see pp.234–35).

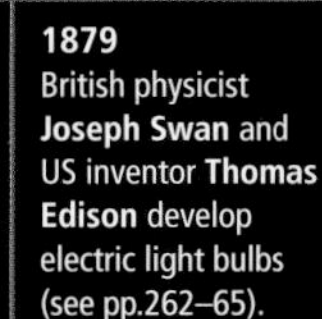

1879
British physicist **Joseph Swan** and US inventor **Thomas Edison** develop electric light bulbs (see pp.262–65).

1880

► 1882
Croatian-US scientist **Nikola Tesla** develops his alternating current induction motor (see pp.266–67). This AC motor eventually wins over direct current in electricity transmission.

◄ 1887
Belgian engineer **Charles Van Depoele** creates the first urban electric tramway system in Montgomery, Alabama (see pp.236–37).

1890
John Fowler and Benjamin Baker design the all-steel **Forth Rail Bridge** in Scotland, erected by William Arrol's company (see pp.246–47).

1894
Italian-American engineer **Guglielmo Marconi** (see pp.268–69) develops wireless telegraphy with the help of radio waves.

1893
Thomas Edison introduces the **kinetoscope**, one of the world's first successful moving picture systems (see pp.262–65).

1895
Nikola Tesla builds the world's first hydro-electric power station at Niagara Falls (see pp.266–67).

1900

1901–02
Guglielmo Marconi makes the first **transatlantic radio transmissions**, first from Cornwall to Newfoundland, then from Nova Scotia to Cornwall (see pp.268–69).

1902
Spanish army officer **Julio Cervera** (see pp.270–71) establishes regular radiotelegraphic services between the Spanish islands.

1914
The Panama Canal, designed by US engineer **John Frank Stevens** (see pp.202–03), opens a sealink between the Atlantic and Pacific Oceans.

▼ 1915
During World War I, English tractor-maker **William Tritton** (see pp.206–07) develops caterpillar-tracked gun carriages and then the first tanks.

1950
US roboticist **Joseph Engelberger** (see pp.238–39) develops the *Unimate*, the first industrial robot.

A TRANSPORT REVOLUTION

THE CENTURY FOLLOWING THE INDUSTRIAL REVOLUTION IN BRITAIN AND ELSEWHERE BROUGHT DRAMATIC CHANGES IN TRANSPORT – NEW ROADS AND CANALS CUT ACROSS THE LANDSCAPE AND STEAM-LOCOMOTION MADE RAILWAYS A PRACTICAL POSSIBILITY.

ENTERPRISING RAIL WIDOW
When her husband Solomon died within two weeks of arriving in India in 1855 to construct the railway line from Bombay to Poona, Alice Tredwell took over the project and completed it.

In the 1750s, a coach journey from London to Edinburgh took 12 exhausting days, and that was considered fast. In winter, however, the roads would often become impassable, and a traveller might have to take shelter in a remote inn for days until the weather improved. Even when conditions were good, coaches would rarely travel faster than walking pace. Roads were essentially little more than cart-tracks.

THE ROAD TO BETTER TRAVEL

As Britain's trade grew in the 18th century, the need for better roads led to the establishment of turnpike trusts that collected road tolls to maintain roads. There was a marked improvement, and engineers such as Blind Jack Metcalf (1717–1810) built Britain's first properly surfaced roads since Roman times. In France, Pierre-Marie-Jerome Tresaguet (1716–96) made similar improvements with roads built on solid stone foundations with proper drainage. Scottish engineer Thomas Telford (see pp.160–63) took Tresaguet's methods further and built excellent new roads through remote areas of Scotland and Wales.

Scottish road-builder John Loudon McAdam (1756–1836) further improved the roads by making a tough, almost waterproof, cambered surface of small compacted stones. By the 1830s, Britain had thousands of miles of "macadamized" roads, and the journey from London to Edinburgh by coach now took less than two days.

SHIP BUILDER
Robert Napier's magnificent marine steam engines and innovative ironclad ships put Scotland's Clydebank at the forefront of shipbuilding in the mid-19th century.

DEVELOPMENT OF CANALS

The last quarter of the 18th century, meanwhile, had seen a huge expansion of the canal network, because horse-and-cart transport was difficult on the dreadful roads, and canal barges were able to carry much greater quantities of raw materials and products of the Industrial Revolution. Transport by water was important in both Britain and France, but even by canals, the speed achieved was that of a steady trudge of a horse. Sailing ships on the sea were faster, but sea voyages were long and at the mercy of bad weather.

> Owing to the poor conditions, transport by road was difficult. But with the development of steampower, steam locomotion became possible, and the world was convinced of the viability of the railways.

POWER OF STEAM

The impact of steampower on this slow-paced world of muscle and sail was dramatic. Steam was first used for motive power on boats, where the heavy weight and low power of early steam engines did not matter much. In the American Midwest, where there were no roads or canals, steamboats turned the land's great rivers into major transport links that opened up the interior of the country almost overnight.

It was Cornish inventor Richard Trevithick (see pp.128–29) who introduced the steam locomotive to the world in 1804. But developing high-power compact steam engines that were practical – and could outrun a horse – took a while longer. Even in 1830, in Baltimore, Maryland, Peter Cooper (1791–1883) was compelled to race his *Tom Thumb* locomotive against a horse to prove its worth.

The famous Rainhill Trials to assess pioneering steam locomotives in 1829, was the turning point. The success of the Stephensons' *Rocket* (see pp.190–91) convinced the world of the viability of steam railways – and the world responded.

GROWTH OF RAILWAYS

Soon Britain, and then many other parts of the world became enthralled by railway mania, as engineers from Germany to

TARMACADAM
In about 1820, John Loudon McAdam developed a way of making good roads so quickly and cheaply that "macadamizing" became the standard method for decades.

HIGH-SPEED CLIPPER
With their high fuel costs and limited speeds, early steamboats failed to dominate the seas; in the 1860s, sailing clippers, such as the *Taeping*, were still the leading ship designs.

SETTING THE SCENE

- In 1750, people and goods travel at a **speed little faster than walking pace**, and travel is often **impossible during winter**.
- Scores of railways built by mining companies **use horses to pull trainloads** of minerals over short distances. **Steam locomotives** are first thought of simply as a **replacement for horses** on these short tracks.
- With the advent of the Industrial Revolution, it becomes **imperative to transport goods in greater quantities**, and more smoothly over the dreadful rutted roads, and hence vast networks of canals are built.
- Engineering techniques developed in building canals lay the foundation for the **rapid extension of the rail network** from the 1830s.
- The turning point in railway developments comes with the extraordinary **success of the Stephensons' *Rocket***.
- The **first dedicated passenger railway opens** in Britain in 1830; the US, 1830; France, 1832; Russia, 1833; Belgium and Germany, 1835; Italy, 1839; India, 1851; and China, 1876.

China set to work building vast rail networks, and investors fell over themselves to pour money into railway schemes.

In Britain, by 1838, just nine years after the Rainhill Trials, there were already 800km (500 miles) of railway tracks. Ten years later there was ten times as much, and, by 1860, there were a staggering 16,000km (10,000 miles) of railway tracks in Britain. The expansion of railways in the US, though it began later, was even more rapid. In India, under British engineer Robert Brereton (1934–1911), 14,500km (9,000 miles) of railways were built between 1857 and 1880.

The impact of these developments on movement and transport was dramatic. By the 1880s, the distance from London to Edinburgh could be covered by rail, in comfort, in just over eight hours.

TOM THUMB
This is a replica of Peter Cooper's *Tom Thumb*, America's first steam locomotive, built in 1830. Pitted against a horse-drawn carriage, it broke down and lost the race, but still managed to prove to investors that steam railways were the future.

ROBERT FULTON

PIONEER OF THE FIRST PASSENGER STEAMSHIP

UNITED STATES 1765–1817

A FLAMBOYANT artist-turned-engineer, Robert Fulton built a submarine, and torpedoes for it, and tried selling them to warring parties Napoleon Bonaparte and Lord Nelson. But he is best remembered as the man who gave the United States its steamships, setting up the new nation's first steamboat service, *North River*, on the Hudson River in 1807.

A LIFE'S WORK

- Fulton starts out as a **painter of miniatures and sets up his own studio**
- He **develops spectacular panoramic paintings** in Paris in 1800
- **Devises weapons for underwater warfare** – the submarine and torpedo
- He **tries to sell his devices** for underwater warfare first **to Napoleon, then to Nelson**
- **Launches a paddle-steamboat on the River Seine** in Paris in 1803
- He **introduces the world's first steam passenger service** on the Hudson River in 1808
- His company **builds the first Mississippi steamboat** in 1811

Unlike contemporaries such as Jacob Perkins (see pp.126–27), Robert Fulton was not a brilliant engineer. However, he had an acute eye for an opportunity. He is sometimes seen as the inventor of both the submarine and the steamship, not because he thought of them first but because he seized on what seemed to be a good idea and made sure that the right people knew about it.

The son of a tailor, Fulton was born on 14 November 1765 in Little Britain, Pennsylvania, and was brought up in Lancaster. As is the case with many famous engineers, there are tales of Fulton's inventiveness as a child, such as his making a rocket to light up Lancaster on Independence Day when there was a candle shortage. He later went on to train as a painter of miniatures. In fact, as a teenager, he must have been a talented miniaturist because he moved to Philadelphia to set up a studio and was soon able to support his entire family financially. However, he was not interested in simply making a living; he wanted to be rich, and decided to seek his fortune in Europe.

EUROPEAN ENCOUNTERS

Fulton arrived in London in 1787, where fellow American painter Benjamin West helped him set up as an artist. Fulton painted very little, but met and charmed many influential people. One of them was the openly homosexual Earl of Devon, William Courtenay, who created a huge scandal with his relationships and had to flee the country. Fulton spent 18 months with Courtenay at Powderham Castle in Devon, and when the disgraced Courtenay ended up in New York, Fulton was one of his few supporters. It was while at Powderham that Fulton turned to invention. He came up with an award-winning device for cutting marble, as well as the idea of a cargo haulage slope (similar to a funicular railway), to save building lock staircases on canals.

By 1794, Fulton was trying to get people interested in his canal schemes despite his lack of experience in the field. Nothing came of his ideas, however, and, by 1797, he was in Paris, apparently in a *ménage à trois* with fellow Americans Joel Barlow and Ruth Baldwin, who nicknamed him "Toot". There, Fulton turned his inventive mind to entertainment, wowing Parisians with giant panoramic paintings that featured changing lighting effects inside two drum-shaped buildings.

UNDERWATER WARFARE

With this success under his belt, Fulton tried to persuade Napoleon Bonaparte that he could defeat the

> ... HOW TO RAISE A SUM IN THE DIFFERENT STATES HAS BEEN MY GREATEST DIFFICULTY
>
> **FULTON** TO GEORGE WASHINGTON

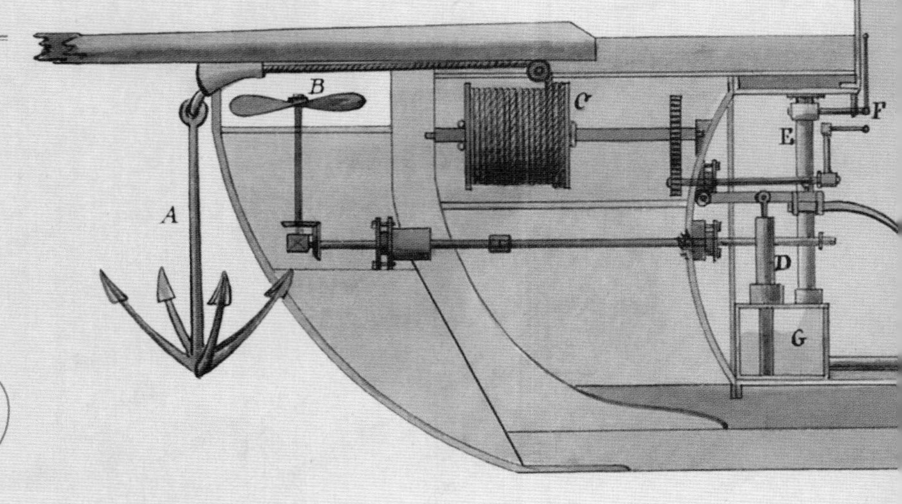

FIRST STEAMBOAT SERVICE
A contemporary illustration of Robert Fulton's *Clermont*, the steamship in which, from 1808 onwards, he began the first regular steam passenger service in the world on the Hudson River.

British navy with the help of his ideas on underwater warfare. Fulton was not the first to come up with the idea of a submarine – fellow American David Bushnell (1742–1824) had created a similar craft 25 years earlier – but Fulton's *Nautilus*, with its cigar-shaped metal hull, conning tower, propeller (driven by a hand-crank), and compressed air supply is recognizably the forerunner of modern submarines. He invented the torpedo too – a submerged, tube-shaped craft, packed with explosives and designed to hit a ship below the waterline and detonate with a clockwork timer.

After several abortive attempts to demonstrate the effectiveness of his torpedo with respect to attacks on British warships, Fulton was later approached by the British, who allowed him to experiment in the naval dockyard. There he gave spectacular demonstrations of blowing up ships, reported excitedly in *The Times*. The British probably had no intention of developing the torpedo, however, especially after their naval victory over the French at Trafalgar in 1805 – neither did they want the French to take it. With interest in his ideas for underwater warfare dwindling, Fulton turned his attention to an idea suggested to him in 1801 by the American ambassador to Paris, Robert Livingston – the steamboat. Livingston had already enquired about an engine from Boulton and Watt (see pp.118–21), but it was easier in Paris to get one from a local manufacturer named Monsieur Périer. Together, they had built a 20-m (66-ft) paddle-steamer, which, in August 1803, made a few successful runs on the River Seine in front of various dignitaries, before sinking.

FULL STEAM AHEAD

With the failure of his British torpedo venture, Fulton returned to New York at Livingston's suggestion and set about building a much bigger steamboat, powered this time by a Boulton and Watt engine. Although far too cumbersome for steam carriages, low-pressure engines, such as Boulton and Watt's, worked well on a boat, and only moderate power was required to propel the boat through the water. Named first *North River*, then *Clermont*, Fulton's new steamboat made its first voyage on the Hudson River in 1807.

By 1808, it was running services 500km (300 miles) upriver, from New York to Albany, in 32 hours. It was the first regular steam passenger service in the world, and Fulton celebrated by marrying Livingston's niece, Harriet. Three years later, Livingston and Fulton's company launched *New Orleans*. It was the first of the legendary Mississippi paddle-steamers that would become a symbol of the opening up of the American heartland. Fulton died in 1815, aged just 50, but by that time steamboats had become an established feature of American waterways.

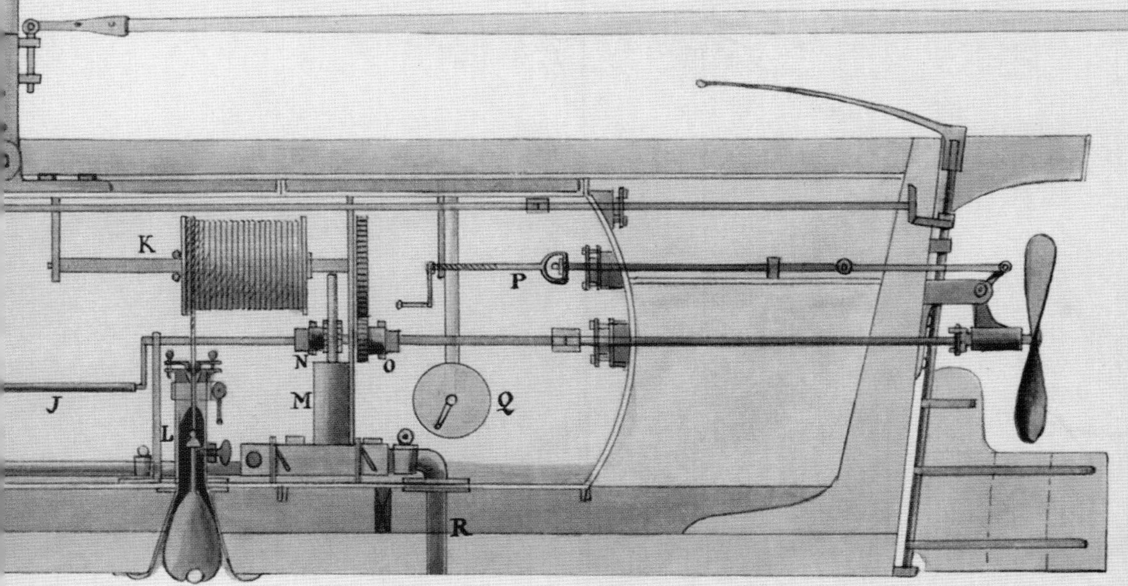

FULTON'S SUBMARINE
This illustration by Fulton shows his second design for a submarine, which he hoped to sell to the British. This was much bigger than his earlier *Nautilus* and looked more like a conventional boat, but it still had the hand-cranked propellers and conning tower for operation underwater. It was never built, since the British lost interest after their great naval victory at Trafalgar in 1805.

THE STEPHENSONS

FATHER AND SON WHO REVOLUTIONIZED RAIL

ENGLAND 1781–1848; 1803–59

GEORGE STEPHENSON ROBERT STEPHENSON

George Stephenson, the father of modern railways, conceived a vision of a network of railways connecting the main centres of population, carrying trains driven by steam locomotives. Together with his son Robert, he translated this vision into reality.

A LIFE'S WORK

- George is engineer for the Stockton and Darlington Railway, the **world's first public railway**, opened in 1825
- George establishes **Robert Stephenson & Co** engine builders in 1823
- Robert's **revolutionary *Rocket*** locomotive **wins the Rainhill trials** in 1829
- With George as engineer and with Robert's locomotives, the **Liverpool and Manchester Railway opens in 1830** to public acclaim
- The **London and Birmingham Railway** in 1837 **establishes Robert's reputation** as one of the great railway engineers of the age
- Robert's **Britannia Bridge** in North Wales **breaks boundaries** on large bridge design, with its novel, wrought-iron box section

George and Robert Stephenson were close companions and collaborators throughout their careers, and had great mutual respect for one another. When Robert was born near Newcastle upon Tyne, his father, George, was employed in operating steam-winding engines at local collieries. A devoted father, George took on extra work in shoemaking and clock-repairing to provide income for his son's good schooling.

With no formal education, George learnt his trade as an engine wright at Killingworth Colliery, Northumberland. Horse-drawn railways were well established at that time, while engines were being introduced. In 1821, George surveyed the 43km- (27-mile) long Stockton and Darlington Railway, which first established his name. The line was planned for passengers as well as goods. He persuaded the railway directors to adopt locomotive haulage for most of the route, and introduced wrought-iron rails, which were less brittle than cast-iron ones.

ENTREPRENEURIAL MOTIVES

The line's opening in 1825 created interest among financiers for more ambitious projects, and George surveyed several trunk routes. Meanwhile, in 1823, he established Robert Stephenson & Co as an engine-manufacturing company in Newcastle upon Tyne, with his son, Robert, as managing partner.

In 1826, George was appointed engineer for the 50-km (31-mile) Liverpool and Manchester Railway. It took more than four years to build and was a major civil engineering project for the time: the route included two large masonry arch bridges, the nine-span Sankey viaduct, 21m (70ft) high, the Olive Mount rock-cutting, 1.6km (1 mile) long and 21m (70ft) deep, a rock tunnel near Liverpool 2.1km (1.3 miles) long, and the crossing of the treacherous Chat Moss peat bog, 6km (4 miles) wide where the track was supported on a bed of wooden frames, brushwood, shingle, and cinders, with parallel drainage ditches. He adopted the standard gauge of 1.5m (4ft 8in) from the horse-drawn railways of Northeast England. The highly engineered route, with shallow gradients and open curves, set the standard for a generation to come.

NOVEL BRIDGE
Robert Stephenson's wrought-iron Britannia Tubular Bridge (in the foreground) across the Menai Strait was completed in 1850. At the time, it far surpassed the length of other cast-beam or plate-girder bridges.

IN HIS FATHER'S FOOTSTEPS

George campaigned tenaciously for locomotive power on the Liverpool and Manchester route, though some doubted its feasibility. A series of trials was arranged at Rainhill, with engineers invited to submit their locomotive designs. Robert's *Rocket* was the triumphant winner, and he developed this into his Planet class locomotives, powering the line's first trains.

The Liverpool and Manchester was the world's first dedicated passenger railway – the dawn of a new era in transport – and led to a flurry of railway proposals, in Britain and across the world.

These included the first of the trunk routes outside London, the London and Birmingham Railway, with Robert as engineer-in-chief from 1833. Robert made detailed plans and working drawings for the route, a practice that civil engineer I K Brunel (see pp.194–97) and others followed. He then divided the 179-km (112-mile) route into five districts – one managed by himself, and the others delegated to trusted engineering assistants. Despite difficulties with the notorious 2.2km- (1.36-mile) long Kilsby Tunnel, the completion of the line sealed his reputation and led to further commissions in Britain and elsewhere in Europe.

NEW BRIDGES FOR RAIL

During the 1840s, Robert's reputation reached a new peak with his ground-breaking railway bridges that used cast iron to withstand compression, and wrought iron to carry tension. This was followed by Robert's wrought-iron, box-girder bridges on the Chester and Holyhead line: the Conway Bridge and the larger and more ambitious Britannia Bridge over the Menai Strait, connecting the island of Anglesey with mainland North Wales.

Working with engineer and shipbuilder William Fairbairn (1789–1874), Robert went on to create large, rectangular box beams, the largest measuring 137m (450ft) in span, through which trains would pass. These massive beams, each weighing about 1,525 tonnes (1,500 tons) were fabricated on the shore, floated into position, and jacked up the masonry piers into position.

The Britannia Bridge was opened in 1850, and Robert later designed the 3.2km-(2-mile) long Victoria Bridge across the St Lawrence River at Montreal for the Canadian Grand Trunk Railway, using the same principle.

Towards the end of his career, Robert noted that the first 25 years of railway construction, until the end of 1854, had seen 12,886km (8,054 miles) of route constructed in Britain – about the diameter of the globe. He added that the amount of earth moved in the works was the equivalent of a mountain 2.4km (1.5 mile) high with a base measuring 0.80km (0.5 mile) in diameter – shifted almost entirely by men and horses without mechanical equipment.

George died in 1848 aged 67 while Robert died on 12 October 1859 at his London home aged 55. Robert was buried in Westminster Abbey and in his eulogy he was called "the greatest engineer of the present century".

STEAM HISTORY
This late-19th century photograph shows engineers driving George Stephenson's history-making "Locomotion No. 1", first demonstrated at the opening of the Stockton and Darlington Railway in 1825. It used high-pressure steam to drive two vertical cylinders. In 1828, the boiler exploded, killing the driver.

STEPHEN HARRIMAN LONG

UNITED STATES 1784–1864

An extraordinarily multi-talented explorer and engineer, Colonel Stephen Harriman Long was a pioneer of the American West.

His feats as an explorer included his 1817 expedition to the Upper Mississippi region and his 1820 search for the source of the Platte River in the Rocky Mountains.

In 1826, he was appointed as consulting engineer for the Baltimore and Ohio Railroad, which ran for 320km (200 miles) through mountain wilderness. This railroad was the exemplar of US railways, the first leg in the national rail system, and the key to opening the West to development. Here he designed timber bridges and formulated his *Rail Road Manual*, the bible of railroad design. He also built forts and dams, improved steamboat engine design, and advanced river navigation.

IN THEIR OWN WORDS

THE STEPHENSONS

As Robert Stephenson described it to the biographer Samuel Smiles, finalizing the boiler for the locomotive *Rocket* was not easy: "...water squirted out at every joint, and the factory floor was soon flooded. I went home in despair and, in the first moment of grief, wrote to my father that the whole thing was a failure. By return of post came a letter, telling me that despair was not to be thought of, and that I must try again". Robert did, of course, and the *Rocket* was triumphant. It competed successfully in the Rainhill Trials where none of its competitors were able even to finish the competition.

A TRIALS NOTEBOOK
These are pages from the notebook of observations on the Rainhill Trials kept by one of the judges, John Rastrick. It shows details of the *Rocket* and its boiler design.

B DESIGN FOR THE *ROCKET*
A technical drawing of the design for the *Rocket* was prepared by the Stephenson Company for the Rainhill Trials in 1829.

C BRASS RULER
Draughtsmen at the Stephenson Company used simple brass rulers. The drawing office was disciplined and hierarchical, with employees working 12-hour days.

D RAINHILL TRIALS
A contemporary illustration of the Rainhill Trials in October 1829 shows the *Rocket* out in front. It eclipsed the *Novelty*, the runner up.

E 1818 DESIGNS FOR LAMPS
While working in the Killingworth Colliery, George Stephenson invented a safety lamp (first left) that would burn without an explosion. His first locomotive in 1814 was built at the pit's workshop.

F FIRST INTERCITY RAIL
A map of 1830 shows the new railway that ran for 56km (36 miles) between Liverpool and Manchester, the world's first intercity rail map.

G KILLINGWORTH LOCOMOTIVE
George Stephenson made this drawing of the Killingworth Locomotive in about 1815. At the time he was an engineer at the Killingworth Colliery, near Newcastle upon Tyne.

H THE *ROCKET'S* FIREBOX
Another sketch from the notebook of John Rastrick shows the dimensions of the *Rocket*'s firebox.

ISAMBARD KINGDOM BRUNEL

TITAN OF THE ENGINEERING WORLD

ENGLAND 1806–59

THE MOST INNOVATIVE of all the great engineers of the Railway Age, Isambard Kingdom Brunel captured the public imagination in his day, and has continued to be a source of inspiration ever since. In everything he did – his railways, his bridges, and his ships – he challenged conventional thinking, which resulted in triumphant successes as well as some spectacular failures.

The only son of the engineer Sir Marc Isambard Brunel, Isambard was educated in England and France before starting work in his father's office in 1822. He was soon involved with the Thames Tunnel in Wapping, East London, and, at the age of 19, was appointed engineer on site. His youthful energy and enthusiasm kept the project going in appalling conditions – he was seriously injured and nearly killed in a great inundation in 1828. Brunel was later honoured for risking his life to save others in the tunnel.

INGENIOUS DESIGNS

While recovering his health, Brunel visited Bristol and submitted several designs for a new Clifton Bridge in the city. His suspension bridge, with a span of 214m (702ft), was the boldest design, and, despite opposition from Thomas Telford (see pp.160–63) – the greatest engineer of the day, who had considered the design impractical – Brunel was appointed chief engineer for its construction. The abutments were completed and the chains made before the project was abandoned due to lack of funds in 1843. It was completed only after Brunel's death.

By the early 1830s Brunel was making a name for himself in Bristol, and, in 1833, he was appointed as engineer for the Great Western Railway with the task of creating the railway from London to Bristol. Not content with following the precedents set by George Stephenson (see pp.190–91), he innovated in all aspects of the project. He pressed for a high-speed railway with very flat gradients and a broad gauge of 2.1m (7ft). His novel track construction involved rails laid on longitudinal timbers and braced by cross-timbers, with the whole framework supported on piles. Later, the piles were abandoned in favour of more conventional ballast support. His specifications for locomotives challenged many of the basic principles of the Stephenson Railway, but early prototypes did not perform well and had to be modified.

Structures on the railway included the eight-arch Wharncliffe Viaduct in brick, impressive stone viaducts through Chippenham and Bath, and a two-arch skew bridge across the Avon in Bristol in laminated timber. His Maidenhead Viaduct across the River Thames was the longest span in brick in Britain. It was also extraordinarily flat – Brunel used his understanding of arch behaviour to create twin semi-elliptical arches for the main spans. The challenging 2.8km (1.75 mile) Box Tunnel through difficult ground required a workforce of up to 4,000 navvies. The line was completed in 1841.

HIS THREE GREAT SHIPS

Ambitious Bristol merchants were keen to capitalize on the benefits of the new railway by challenging Liverpool for transatlantic trade, and it was probably Brunel himself who

A LIFE'S WORK

- Aged 19, he **becomes chief engineer** for the construction of his father's Thames Tunnel
- He is appointed engineer for the Great Western Railway with the task of **surveying and building the line from London to Bristol**, in 1833
- His innovative bridge designs culminate in the **wrought iron Royal Albert Bridge** over the River Tamar in Cornwall, opened in 1859
- The *Great Eastern*, **the largest and most innovative ship ever built**, is launched in London in 1858

THE *GREAT EASTERN*
Photographed here at Millwall Docks in London as she neared competition in 1858, the *Great Eastern* was the largest ship of Victorian times. She served as a passenger liner, cable-layer, as a concert hall, and as an advertising hoarding until her break up in 1889.

IN ALL THAT CONSTITUTES AN ENGINEER IN THE HIGHEST, FULLEST, AND BEST SENSE, BRUNEL HAD NO CONTEMPORARY, NO PREDECESSOR

THE ENGINEER, 1910

suggested extending the railway from Bristol to New York by the use of steamships. This led to the design of his three great ships for the Great Western Steamship Company.

The first ship designed by Brunel was the *Great Western*, which had a conventional timber hull, but trussed with iron and wooden diagonals to resist racking forces, and with rows of long longitudinal iron bolts to strengthen the ship across its length. With twin paddle wheels and auxiliary sails, it was the largest steamship of the day, with a length of 72m (235ft) and weighing 1,800 tonnes (2,000 tons). However, the ship caught fire shortly before her first voyage. In the confusion, Brunel fell from a ladder and nearly drowned in the flooded engine room.

PADDINGTON STATION

The London terminus of the Great Western Railway, Paddington Station, designed by Brunel with Sir Matthew Digby Wright (1820–77), and opened in 1854, was intended to create a sensation of grandeur and space. The original building comprised a shed made from three wrought-iron barrel arches, which supported a glazed roof. Although they have been refurbished, the roof still retains its original shape. Examples of Brunel's designs for smaller stations on the Great Western Railway, which survive in good condition, include Mortimer, Charlbury, and Bridgend.

THIS ENGRAVING DEPICTS PADDINGTON STATION – BRUNEL'S GREAT RAILWAY TERMINUS IN LONDON.

Nevertheless, the *Great Western* immediately set a new speed record for crossing the Atlantic, and had many years of successful service.

Next was the *Great Britain*, the first big ship both to be built in iron and to be fitted with a screw propeller. His engine included four inclined cylinders driving upwards to a common crankshaft – developed from his father's idea – and was the largest marine propulsion unit of its day. At 98m (322ft) long, the *Great Britain* was larger than any other ship afloat.

But Brunel's boldest creation was his *Great Eastern*. Nicknamed the Leviathan, it was 162m (692ft) long, and with a displacement eight times greater than the *Great Britain*, was by far the biggest ship ever built, and aimed at the growing market for emigrants travelling from Europe to North America. It was built with a double skin hull and was propelled both by paddle wheels and by screw, each with independent sets of steam engines. The project faced numerous technical and financial problems. In particular, the difficult sideways launch from its slipway at Millwall into the River Thames took three months to achieve, and a massive explosion on her maiden voyage killed five stokers. The ship finally achieved success laying submarine telegraph cables, starting with the 1865 transatlantic cable.

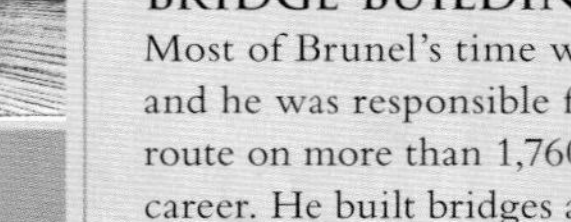

A MAN OF FIRSTS

During his short career, Brunel achieved many engineering "firsts", including assisting in the building of the first tunnel under a navigable river and the development of the first propeller-driven ocean-going iron ship.

BRIDGE-BUILDING

Most of Brunel's time was spent on railways, and he was responsible for laying out the rail route on more than 1,760km (1,100 miles) in his career. He built bridges and viaducts in brick and stone, and generally avoided cast-iron girders. However, his most innovative bridges were in wrought iron. His Hungerford Footbridge across the River Thames in London used suspension chains and anchorages based on his Clifton Bridge design, and his railway crossing of the River Thames at Windsor in 1849 was a bowstring skew bridge. The same year, he experimented with a wrought-iron truss for a swing bridge in Bristol.

Brunel built his much larger bridge across the River Tamar in Cornwall, opened in 1859. The two 139-m (455-ft) spans of the Royal Albert Bridge were formed of large tubular arches carrying the compression forces with chains beneath carrying the tension, held together by cross bracing and supporting the bridge deck via hangers. Just as innovative was the foundation for the central masonry pier. Here Brunel used a giant wrought-iron cylinder, lowered vertically to the river bed. An annular space around the bottom allowed water to be expelled by air pressure, so allowing work on the foundations to progress.

Brunel again challenged convention by championing the atmospheric system of propulsion for railways. There was no locomotive. The train was pulled by an iron rod attached to a piston, which ran in a cast-iron pipe laid between the rails. Steam engines beside the track pumped out the air in front of the piston, and atmospheric pressure pushed it forward. Stephenson dismissed the idea as "the atmospheric caper", but in 1847 it ran for 35km (22 miles) from Exeter to Newton Abbot. Persistent problems, particularly with

ENGINEERING TIMELINE

1825–34

In 1833, Brunel is made chief engineer of the Great Western Railway and begins his survey of the proposed route across southern England

1835–41

Box Tunnel on the Great Western Railway is completed in 1841; at the time of opening, it was the longest railway tunnel in the world the valve along the top of the pipe, obliged Brunel to abandon the system. Later too, his broad gauge was abandoned in favour of Stephenson's standard 1.435m (4.7ft) gauge.

Brunel was strongly self-motivated, and rarely delegated responsibility to his assistants. His staff considered him to be distant and autocratic, and he made little effort to train them. Most of his schemes ran well over cost and schedule, and he had a difficult relationship with his contractors, some of whom refused to work for him, as his treatment of them could be acrimonious, bordering on unethical. Worn out by hard work and stress, Brunel suffered a heart attack on board the *Great Eastern* during the ship's proving trials. He died two weeks later, aged 53.

ROYAL ALBERT BRIDGE
Made of wrought iron, the Royal Albert Bridge, spanning the River Tamar between Devon and Cornwall in southwest England, was built in 1859 to continue the route of Brunel's Great Western Railway. Twenty thousand people attended the official opening ceremony, which was performed by Prince Albert.

1842–55
Work on the Thames Tunnel resumes; it is fitted out with lighting, roadways, and spiral staircases, and is finally opened to the public on 25 March 1843

1856–59
The SS *Great Eastern* is launched; Brunel had hoped to conduct the launch with a minimum of publicity, but many thousands of spectators witness the event

1860–64
Clifton Bridge in Bristol is completed in 1864, after Brunel's death – it is opened in his honour

LAUNCH OF A LEVIATHAN
In 1858, the SS *Great Eastern* was by far the most modern ship ever built, with sail, paddle, and screw propulsion. Her designer, Isambard Kingdom Brunel, seen here in the centre of the picture, referred to her as "Great Babe". The birth was not without problems. The ship's launch into the Thames failed on three attempts, with the steam winches unable to move her immense tonnage down the slipways. The fourth attempt, on 31 January 1858, succeeded, but on her maiden voyage, a massive boiler explosion killed six crew. Only the strong bulkheads stopped the ship from being blown apart.

ENGINEERING INNOVATIONS

THE RAILWAYS

The invention of the steam locomotive and the iron railway led to a transport revolution that quickly spread from Britain to the rest of the world. In the century following the building of the first revolutionary railway systems, trains became the primary mode of land transport, moving goods and passengers across distances and at speeds that were previously unimaginable. Even after the advent of the motorcar and the aeroplane, railways have continued to be an essential component in our modern infrastructure.

MODEST BEGINNINGS

The first horse-drawn railways evolved from the wooden wagonways that had been used in mines to carry coal since the 16th century. Initially the wooden rails were simply reinforced by nailing iron plates onto them, but during the Industrial Revolution in Britain they were replaced with iron tracks. Tramways to carry horse-drawn wagons for both goods and passengers appeared at the beginning of the 19th century. These were the ancestors of the modern railway systems.

At the same time as the development of iron rail tracks, engineers were improving the design of steam engines, and the idea of using them to replace draught animals became a viable proposition. The first successful steam locomotive, built by Richard Trevithick (see pp.128–29), was tested in 1804 and marked the beginning of the railway era. Improvements were made to the design of the locomotives by George and Robert Stephenson (see pp.190–91), and different track designs were tried and tested. (These included a rack and pinion system, now

1600s – MODEL TRACKS
To carry coal from mines across rough and muddy ground, horse-drawn wagons in Germany are pulled along tracks consisting of wooden rails laid on sleepers.

1770s – IRON RAILS
Wooden wagonways in British mines are reinforced with cast-iron plates, and soon flat iron rails replace the wooden rails.

1790s – EDGE RAIL TRACKS
Horse-drawn carriages on edge rail tracks are used to transport goods for longer distances, connecting mines and mills with the canal network.

1801–03 – FIRST RAILWAY
A 29-km (18-mile) tramway, from Wandsworth to Croydon near London, is opened as the Surrey Iron Railway.

1812 – STEAM POWER
The *Salamanca*, a two-cylinder steam locomotive, goes into service, running on a rack and pinion railway.

1813 – HORSE SUBSTITUTE
William Hedley (1779–1843) builds *Puffing Billy*, a smooth-wheeled locomotive, to replace horses on the wagonway from Wylam Colliery to the docks at Lemington-on-Tyne in Northumberland, England.

1825 – LOCOMOTION NO 1
The world's first public railway, the Stockton and Darlington Railway, opens in England, carrying both goods and passengers on trains drawn by George and Robert Stephenson's locomotive *Locomotion No 1*.

1826 – AMERICAN STEAM
John Stevens builds a steam locomotive and demonstrates it on a track at his estate in Hoboken, New Jersey (see pp.202–03).

1828 – RAILROAD CONSTRUCTION
Construction of the Baltimore and Ohio Railroad begins, and the first section opens in 1830.

1830 – FIRST AMERICAN LOCOMOTIVE
Tom Thumb, built by industrialist Peter Cooper (1791–1883) for the Baltimore and Ohio Railroad, is the first American-built locomotive to go into service on a public railway.

1851 – RAILWAY IN INDIA
A railway opens in Roorkee, in northern India, and is followed two years later by India's first passenger line.

1600 1700 1800 1810 1820 1830 1840 1850

Richard Trevithick builds a locomotive with a high-pressure steam engine, which is demonstrated the following year on a cast-iron tramway at the Pen-y-darren Ironworks near Merthyr Tydfil in Wales (see pp.128–29). The unnamed locomotive pulls an iron load of 9 tonnes (10 tons), 70 men, and five wagons for 16km (10 miles), winning a 500-guinea bet for its sponsor, Samuel Homfray.

1803 LOCOMOTIVE SUCCESS

George and Robert Stephenson build the *Rocket*, with which they win the Rainhill Trials, organized by the Liverpool and Manchester Railway to find the best design of locomotive (see pp.190–91). The *Rocket's* innovations become standard in steam locomotives, including a multi-tube boiler and near-horizontal pistons. In 1830, it is used for opening the Liverpool and Manchester Railway, and remains in service for another 10 years.

1826–29 THE *ROCKET*

used for mountain railways, as it was thought the smooth wheels and track would not give enough grip.) By the 1820s, the commercial possibilities of railways had become obvious, and companies were set up for building the new rail networks.

AGE OF THE RAILWAY

Within a short time, there was a "railway mania" that exceeded the "canal mania" of the previous century, with a network of railways across Europe that transformed industry and enabled more people to travel than ever before.

News of the steam railways had already spread to the US, and charters to build the railroads were granted as early as 1815, with construction taking off in the 1830s. By mid-century, trains were appearing elsewhere, too, notably in India, where British engineers began the most ambitious programme of railway building to date, covering the whole of the subcontinent with a network of lines within 50 years. Throughout the 19th century, the design of locomotives slowly evolved, from the Stephensons' *Rocket*, but steam power was eventually superseded by the diesel engine and electric motor. In the 20th century, engineers have continued to incorporate advances in technology into train design, improving the performance and speed, and introducing innovations such as the linear induction motor, which allows magnetic levitation. Although many other forms of mechanized transport have appeared since the appearance of the first steam trains, the age of the railway seems far from over.

TRACK DEVELOPMENT

Tracks were crucial to the progress of railways. The original flanged cast-iron plates could not carry heavy steam locomotives and were superseded by edge rails, laid on sleepers on a base of ballast. These wooden sleepers supported the rails and also maintained the correct distance between them, the gauge. Many different gauges were used initially, but eventually standard gauges were established – in Britain it was settled at 1,435mm (4ft 8½in).

RAILS ARE SUBJECT TO VERY HIGH STRESSES AND HAVE TO BE MADE OF A HIGH-QUALITY STEEL ALLOY.

1858 – INJECTOR
French engineer Henri Giffard invents the injector to improve steam-engine efficiency.

1863 – UNDERGROUND TRAIN
The Metropolitan Railway, an underground railway between Paddington and Farringdon Street in London, opens to the public after three years of construction.

1879 – ELECTRIC TRAIN
In Berlin, Werner von Siemens (see pp.234–35) demonstrates a passenger train driven by electricity from a third rail. It is put into public service in 1881.

1891–1916 – LONGEST RAILWAY
The main route of the Trans-Siberian Railway is built from Moscow to Vladivostok on the Sea of Japan, making it the longest railway in the world.

1901 – SUSPENSION SUCCESS
The first section of an electric suspension railway, the *Schwebebahn* (Floating Tram), designed by Eugen Langen, opens in Wuppertal, Germany.

1930s – DIESEL LOCOMOTIVES
The Burlington Railroad and Union Pacific Railroad introduce high-speed diesel-electric locomotives, known as "streamliners", on their passenger lines.

1964 – SUPERSONIC TRAINS
In Japan, the so-called "bullet train" makes its first journey on the Tokaido Shinkansen, part of a network of high-speed railway lines designed to carry trains at up to 300kph (186mph).

1979 – SPEEDING UP EUROPE
The first European high-speed trains (*Trains à Grande Vitesse* or TGV) with electric motors are built in France.

The famous "Last Spike" is driven in the rails of the American transcontinental railroad where the final sections meet at Promontory Summit, Utah. The Pacific Railroad connects Omaha, Nebraska and Sacramento, California, with an extension to the coast at Oakland, California. It replaces the slower and more dangerous stagecoach routes, and reduces the journey time from coast to coast to one week.

1869 PACIFIC RAILROAD

A Maglev (magnetic levitation) railway in Shanghai becomes the fastest passenger-carrying train service, travelling at speeds of up to 400kph (250mph). Its system uses a linear induction motor to propel and suspend trains above a guideway, and is virtually friction free. Maglev railways were developed in the US and Germany in the 1970s. The first commercial lines were opened in the 1980s.

2004 MAGLEV RAILWAYS

JOHN FRANK STEVENS

ENGINEER WHO SAVED THE PANAMA CANAL

UNITED STATES 1853–1943

RAISED IN RURAL MAINE, and largely self-taught as an engineer, John F "Big Smoke" Stevens was the foremost railroad engineer of his day – two US presidents asked him to lead major construction projects. He was tough, decisive, direct, and down-to-earth. But he was also extremely modest, and quick to credit others, which is perhaps why history has largely forgotten him.

A LIFE'S WORK

- Stevens builds **1,600km (1,000 miles) of track for the Great Northern Railway**, the finest track in the country, across vast plains and mountain ranges **under the employment of "empire builder" James Hill**
- Brings order and motivation to the chaos and despondency in the **Panama Canal Zone** by **building rather than excavating** any further, and by **improving the living standards** of the workers
- **Persuades the US government to accept his plans for a lock canal** instead of the "big ditch", which would have been too difficult owing to the treacherous Chagres River
- **Resigns from the Panama Canal project**, once the success of the project is assured
- **Despite war and famine**, keeps the Trans-Siberian Railway running during the Russian Revolution, and **the lines of communication open within Russia**

Stevens' involvement in the expanding railway industry began in 1875. He worked for several companies, gradually earning a reputation as a creative and intrepid location and construction engineer. In 1889, he joined the Great Northern Railway owned by the celebrated "empire builder" James Hill. Snaking from Minnesota to Montana, the Great Northern was headed westwards across the Great Divide to the Pacific. Before the end of the year, Stevens had found the elusive Marias Pass across the Rocky Mountains, removing the need for a tunnel or long diversion. The following year he discovered another key route, named "Stevens Pass" against his wishes, through Washington's Cascade Mountains.

When he left Hill's employment 16 years later, Stevens was general manager and had built 1,600km (1,000 miles) of the Great Northern Railway, then considered the finest track in the country. Hill referred to Stevens as "the best engineer in the world", and it was on Hill's recommendation that Theodore Roosevelt chose Stevens as the man to rescue the Panama Canal.

AVERTING DISASTER IN PANAMA

In 1905, the US's Panamanian canal project – the most expensive government undertaking outside the US – was heading for disaster. The chief engineer had quit unexpectedly and even Roosevelt admitted that his pet project was in "a devil of a mess". Initially reluctant, Stevens agreed to be chief engineer on the condition that he would have a free reign, and would stay until he could determine the project's success or failure.

On his arrival, Stevens found the Canal Zone even worse than he had expected. There were no proper plans or equipment, and morale had hit rock bottom. Most of the 17,000 workers lived in appalling conditions and at risk of contracting yellow fever, which had already killed 34 men. Despite Roosevelt's wishes to "let the dirt fly", Stevens promptly stopped excavation and instead started building. He built houses, schools, hospitals, and water and

UP AND OVER
Although Stevens left Panama before most of the canal was built, his lobbying in Washington for a high-level canal with locks and gates was crucial to the project's success.

sanitation works. He even built clubhouses and refrigeration plants so the workers could socialize and eat decent food. Despite much doubt about the link between mosquitoes and yellow fever, he also gave his full support to a massive – and successful – mosquito eradication programme.

He realized that the main task was earth removal rather than digging. So as well as procuring excavation equipment, Stevens amassed railway carriages and locomotives and overhauled the Panama Railroad, devising an ingenious flexible track system to haul away rocks and dirt at great speed. He also realized that the expected sea-level canal was unrealistic, mainly because of the strong Chagres River. So he travelled to Washington to lobby for a high-level canal, with locks and a huge artificial lake to buffer the river's flow. The design was less popular, but Congress agreed.

SHOCK DEPARTURE

In January 1907, just three months after Roosevelt made a high profile visit to the Canal Zone, Stevens resigned. He pointed out that many others were now able – and willing – to complete the canal work, so he returned to the American railroad. Citing personal reasons, Stevens never revealed exactly why he quit. George Goethals (1858–1928), an army engineer, finished the canal ahead of schedule, but it appears Roosevelt never forgave Stevens for leaving, and perhaps also for the typically irreverent tone of his resignation letter. When describing the canal in his autobiography, Roosevelt failed even to mention Stevens.

A SIBERIAN ADVENTURE

After Imperial Russia fell in 1917, the provisional government asked the US for help in maintaining the Trans-Siberian Railway. After years of war, this was barely functioning and massive stockpiles of munitions were needed from Vladivostok in the Far East. President Woodrow Wilson sent Stevens, now 64, to Siberia as chairman of a railway advisory commission. Stevens stayed to carry out its recommendations, with the help of a quasi-military unit of American railway workers. He was soon caught up in the revolution, with the Bolsheviks seizing power days before the men arrived.

When Russia withdrew from World War I, the US's role in the region became ambiguous. Stevens headed a new Inter-Allied Technical Board, set up to protect Allied interests along the railroad. Despite famine, disease, and anarchy, he got the railways working and kept them running. Trains brought in food and other supplies to Siberia, saving thousands of lives. Later, they also enabled the withdrawal of Allied troops.

Stevens returned home in 1923, and, in 1927, became president of the American Society of Civil Engineers. He died at the age of 90, having received many honours, including the John Fritz and Hoover Medals, the US Distinguished Service Medal, the Franklin Institute's Gold Medal, and a host of honorary memberships and doctorates.

> THERE IS NOT A MAN WHO EVER WORKED FOR ME WHOM I CANNOT NOW CALL MY FRIEND ... MY GREATEST TRIUMPH
>
> **JOHN FRANK STEVENS**

TREKKING SOLO

When John Frank Stevens found the Marias Pass in 1889, the only suitable gap through the Rockies for a railroad, the area was shrouded in fear and Indian legend. Abandoned by his guide, Stevens trekked on alone. He spent the night on the summit, walking back and forth in the bitter cold. Even on the plains, the temperature that night dropped to almost −40°C (−104°F) – and his constant pacing probably saved his life. His only comment on his achievement was that it took a strong man to carry it out.

THE MARIAS PASS IS THE SHORTEST ROUTE FOR THE GREAT NORTHERN RAILWAY ACROSS THE CONTINENTAL DIVIDE.

A MEETING OF THE RAILS
Railway workers from the Central Pacific Railroad (from the west) and the Union Pacific Railroad (from the east) gather at Promontory Point, Utah, on 10 May 1869 to celebrate the "driving of the golden spike", which marked the completion of the US Transcontinental Railroad. The transcontinental journey time from coast to coast was reduced from six months to just one week.

WILLIAM TRITTON

MASTERMIND BEHIND THE FIRST TANKS

ENGLAND 1875–1946

In 1915, the British navy contracted a small engineering firm to construct a machine that could break the deadlock of trench warfare on the Western Front. The idea behind the tank was not new, but it was William Tritton who made it a working reality. Protecting crews from machine-gun fire while moving across trenches, these tanks changed the face of modern warfare.

A LIFE'S WORK

- Developing business contacts all over the world, Tritton **reverses the fortunes** of struggling **agricultural machinery firm William Fosters & Company**
- He becomes the man **Winston Churchill depends on** to turn dreams of **land battleships and armoured tractors into reality** during World War I
- When ready-made chain-link tracks prove too weak, Tritton takes just a few days to **revolutionize the caterpillar track design**
- Designs and develops the world's first successful tracked armoured vehicle, the **"Little Willie"**
- He **builds almost all the tanks** Britain deploys in World War I, including the famous **Mark I and Mark IV tanks**
- Tritton is **knighted** in 1917 for his tanks

The son of a London stockbroker, Tritton had a varied early career that helped him learn about business while honing his engineering skills. After an apprenticeship in hydraulic engineering, he inspected steel rails, built pumps in torpedo boats, and worked shifts at the Metropolitan Electric Supply Company. In 1899, he joined the new UK arm of American printer-manufacturer, Linotype. By 1905, he was general manager of Fosters & Co, a Lincolnshire firm that produced threshing machines and tractors.

The appointment proved a great success for the struggling company. Tritton was an energetic and able manager – even-headed and meticulous. He turned the company's fortunes around, and by 1911, had risen to the post of managing director.

Soon after the outbreak of World War I, the navy approached Tritton for ideas of how to transport the heavy artillery howitzers. Tritton built large tractors that carried the massive guns working in tandem. Winston Churchill, then First Lord of the Admiralty, saw the potential of these tractors, and asked if one could be built that would cross trenches. Tritton designed one where the tractor used its own portable bridge that it laid out to cross the trenches and then retrieved the bridge afterwards. Unfortunately it was heavy, and performed poorly on rough ground.

Churchill was not alone in considering armoured tractors. Shocked by conditions in the trenches, war correspondent Earnest Swinton contacted the War Office with the idea of modifying the American Holt tractor. This ran on caterpillar tracks and was proving effective for carrying artillery. The War Office tested the Holt but it got stuck in a ditch, so the idea was dropped.

LITTLE WILLIE AND BIG WILLIE

In February 1915, Churchill formed the Landship Committee. It commissioned Fosters to build prototypes of two "land battleships", which would hold 50 men. One ran on tracks, the other on three huge wheels. Neither was a success. With track technology in its infancy, the available tracks were far too weak. The wheeled design was practicable, but its height would have made it an easy target for enemy fire.

By the end of July, Churchill's committee had teamed up with the War Office and was reconsidering armoured tractors. They gave Tritton a list of criteria, including a top speed of 6.4kph (4mph), and the ability to cross a 2.4m (8ft) gap and climb a 1.5m (5ft) parapet. The vehicle would run on ready-made tracks imported from the US.

Tritton set to work with his team, which included chief draughtsman William Rigby, who was adept at producing detailed drawings quickly, and Walter Wilson (see box, right), a talented naval engineer assigned to the project by the Landship Committee. The team took just 37 days to produce the first prototype – the Number One Lincoln Machine was a steel box with a turret, powered by the Daimler engine and transmission system Fosters had developed

TANK PROTOTYPE
The world's first fully tracked armoured vehicle was a simple box on tracks. It was soon superseded by the more stable rhomboid design, which could cross trenches.

to haul howitzers. But the tracks sagged over ditches and often jammed or fell off. They were also insufficiently robust for the job. A few days later Tritton had designed the tracks now considered the key factor in British tank development. He and Wilson were also working on a different design. With tracks running around side-mounted guns, the Lincoln Machine had a low centre of gravity and high manoeuvrability.

After testing Tritton's tracks on the Lincoln Machine, now renamed "Little Willie", they built a new prototype with a rhomboid shape. In January 1916, "Big Willie", or "Mother", was shown to senior politicians and military men and Swinton, who described the "gigantic cubist steel slug" rearing over the parapet. The gathering was impressed, and shortly afterwards the government ordered 100 units.

DEPLOYMENT AND REFINEMENT

For security reasons, the story goes, the huge steel boxes were called "water tanks for Mesopotamia", which was shortened to "water tanks" and then just "tanks". They were designed to cross trenches, resist small-arms fire, travel over difficult terrain, carry supplies, and to capture enemy positions. The tank was certainly a surprise to the German army when first deployed on the Somme that September. However, Tritton had little time to iron out its design flaws – so it was unreliable, underpowered, hard to drive, and very uncomfortable. He went on to make many refinements, and in swift succession produced Mark II and III, then Mark IV, which participated in the first really successful tank offensive in Cambrai in 1917, and finally the Mark V. He also developed the lighter, faster "Whippet tank".

Tritton was knighted for his work on tanks. There were numerous claims to the invention, but a Royal Commission on Inventors set up to assess these conferred on Tritton and Wilson by far the largest award. Tritton gave his share to his employees. He helped with the City of Lincoln's hospital extension schemes and was a Justice of the Peace from 1934.

BREAKING THE DEADLOCK
Tanks were first used on the Somme. Beset by mechanical problems and inexperienced crews, only a few broke ahead of the infantry. But those that did made a profound impact.

WALTER WILSON

IRELAND 1874–1957

With a strong character and a creative brain, Walter Wilson was a great mechanical innovator. He developed sophisticated engine and gearing systems for cars, planes, and trucks, and played a vital role in building the first tanks.

In 1897, Wilson co-founded a company to build the first internal combustion aircraft engine. But when his partner died in a gliding accident, he shifted his attention to cars. He pioneered many advanced features, including electric ignition and an epicyclic gearbox he later modified for the Mark V tank. When seconded to Fosters & Co, Wilson was already an expert in armoured cars. But it was his creativity, coupled with Tritton's tenacity, that enabled them to realize their designs so quickly. The rhomboid shape was Wilson's idea, which is why "Mother" was also dubbed the "Wilson".

A FASTER, MORE MANOEUVRABLE TANK

MARK V TANK

INTRODUCED IN 1918, the Mark V Heavy Tank was the last rhomboid tank deployed in large numbers in World War I. Four hundred were produced – 200 of the "female" type with machine guns, and 200 "males", which also boasted six-pounder (57mm) guns. Faster, more reliable, and much easier to manoeuvre than earlier models, Mark V gave the British Tank Corps a significant advantage in battle. Among its new features was a raised cabin behind the engine for a machine-gunner and the tank commander. The machine guns could now swivel on ball mounts, rather than through loopholes, giving better protection for the gunner and a wider field of fire.

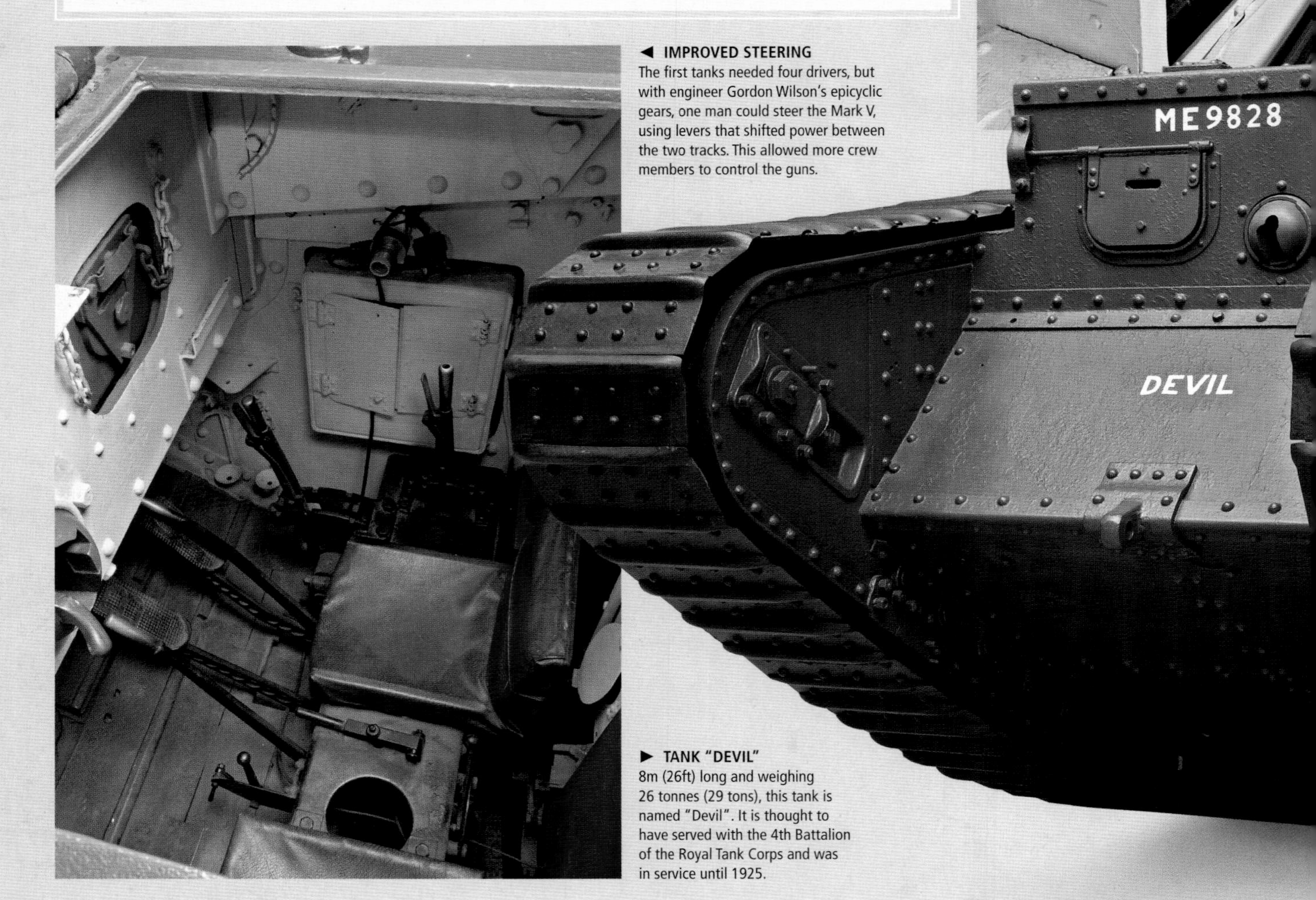

◄ **IMPROVED STEERING**
The first tanks needed four drivers, but with engineer Gordon Wilson's epicyclic gears, one man could steer the Mark V, using levers that shifted power between the two tracks. This allowed more crew members to control the guns.

► **TANK "DEVIL"**
8m (26ft) long and weighing 26 tonnes (29 tons), this tank is named "Devil". It is thought to have served with the 4th Battalion of the Royal Tank Corps and was in service until 1925.

◄ **RICARDO ENGINE**
The Mark V Tank's six-cylinder engine, purpose-built by engine expert Harry Ricardo (1885–1974), was more powerful and reliable than the modified tractor engines that it replaced. The tank's engine was exposed, allowing for easy maintenance.

▼ **SPONSON GUNS**
Guns were mounted on housings called sponsons on the tank's sides. Four of the eight-man crew managed the two large Hotchkiss guns in this male tank, which was also equipped with four machine guns on the rear side.

◄ **INSIDE THE TANK**
The new steering system was more compact, making the Mark V Tank less cramped. However, with the exposed engine in the centre and extremely poor ventilation, the interior was hot and filled with fumes.

► **TRITTON'S TRACK**
The track consisted of tough metal plates, with ridges to improve traction. The plates slotted onto the track frame, which had tension adjusters to keep the track tight.

PIERRE MICHAUX

DESIGNER OF THE FIRST PEDAL BICYCLE

FRANCE 1813–83

THE MODERN PEDAL BICYCLE rode into public consciousness in the 1860s, ushering in an age of individual mobility, as well as a new sport. But the bike was not born overnight. In 1864, Pierre Michaux was approached by the Olivier brothers to make parts for a new version of the hobbyhorse, or *draisienne* – the forerunner of the modern bicycle. Within a couple of years, Michaux wheeled the first pedal bicycle out to the Parisian public and a new craze was born. However, ousted by his partners, Michaux's fame faded into oblivion, although his invention is one of the most famous in the world.

A LIFE'S WORK

- Michaux starts his career as a **blacksmith in Paris**, making ironwork for carriages
- He is approached by the **Olivier brothers** to **make parts for a pedal-powered device** that would become the bicycle
- Helps **work out the design for the modern bicycle** and begins to mass produce it
- Becomes **a partner** with the Olivier brothers in the **bicycle firm, Michaux et Compagnie**
- After disagreements over the bicycle's design, René Olivier **issues a contract that gives the Olivier brothers**, including a third brother Marius, **a 70-per-cent share in the company**
- After being **given money to exit** the company, Michaux **launches his own company**, for which the **Olivier brothers sue him**; he dies a **bankrupt and forgotten man**

Michaux was a successful blacksmith who forged iron parts for carriages from his workshop on the Champs-Elysées in Paris. In 1864, he was approached by three young men who wanted him to make parts for what appeared to be a new type of *draisienne*, otherwise known as the "hobby-horse". This device was a crucial intermediary step on the way to the pedal-powered *vélocipède* ("fast foot"), later called the bicycle. The *draisienne* was the work of German inventor Karl von Drais (see box, opposite), who had unveiled his *Laufmaschine* ("running machine") in 1817. He connected two small carriage wheels to a wooden frame that had a cushioned seat. Sitting astride the device, the rider powered it by using one foot at a time, like walking, and it could be steered, thanks to a pivoting pole. He was granted a patent for the device in 1818, and there was a short-lived craze in Europe for the *draisienne*, or *vélocipède*.

British inventors were also intrigued, and London coach-maker Denis Johnson (1760–1833) began to produce his own, which he marketed as the "pedestrian curricle", more commonly known as "hobby-horses". There was a brief fashion for the device in Britain as well, although only the wealthy could afford it. After Johnson, other inventors continued to experiment with the design, although the device's popularity had faded. It took the bold and necessary addition of pedals to resurrect it.

MICHAUX VÉLOCIPÈDE
This 1869 wood-and-iron *vélocipède*, or "boneshaker", manufactured by Michaux & Co in Paris, was the first two-wheeled device to be powered by pedals.

MURKY COMPETITION

Although Michaux is often credited as being the first person to use pedals, at the time there were competing claims. A man named Karl Kech claimed to have attached pedals to the front of a *draisienne* in 1862. But it was Pierre Lallement (1843–91) who actually held the only patent for the bicycle, which he obtained in the US in 1866.

In 1862, Lallement was working for a firm that made prams and children's tricycles when, he claimed, he designed a pedal-powered device. He moved to Paris and completed a prototype, working at two workshops – one at the Faubourg Saint-Martin and the other at the École Central. The latter location is significant because sons of a wealthy industrialist, Aimé and René Olivier, were studying there. Although there is no direct evidence of a collaboration, the two boys must have known of Lallement's public exhibition of his device in 1864.

Lallement decided to leave for the US and, a short time later, Aimé, René, and a classmate, Georges de la Bouglise, decided to build their own prototype. But first they needed someone who knew how to make the parts, which is how René ended up in Michaux's shop in 1864.

BICYCLE INVENTOR UNSEATED

Soon the prototype, with pedals attached to the front hub, was finished. Aimé added brakes and strengthened the frame. Then, in 1865, René, Aimé, and Georges rode the test models from Paris to Marseille. They worked out the design problems when the bikes came apart and, by 1866, they and Michaux were ready for production. In May 1867, *Le Moniteur Universel du Soir* began carrying advertisements for "pedal *vélocipèdes*" produced by Michaux. There was a soaring demand despite the 250-franc price, and Michaux could scarcely keep up. In November 1868, *The Times* of London observed that "Anybody who has visited Paris within the last few months cannot have failed to notice the large number of *vélocipèdes* ... a recent police edict compels the riders to affix a lamp to them in consequence of the accidents that have happened from their use".

However, things were not faring so well behind the scenes. René and Michaux fought over the design – early models had malleable iron serpentine frames and René wanted to switch to a diagonal frame of wrought iron, but Michaux refused. René issued a contract in 1868 that gave the Olivier brothers a 70-per-cent share in their company and kept Michaux out of technical matters. They still called themselves Michaux et Compagnie, but the tensions continued.

Soon after, the brothers accused Michaux of embezzlement. They gave him 150,000 francs to leave, and renamed the firm Compagnie Parisienne. Michaux took the money and launched his own company, forcing the brothers to sue him in 1869 for breach of contract. They won, and Michaux faded into obscurity, though his contribution to the development of the pedal-powered bicycle has not been forgotten.

KARL VON DRAIS

GERMANY 1785–1851

Although inventor Karl von Drais, whose regular job was as a forest inspector, is best known for his *draisienne* – a version of the modern bicycle, he initially designed a four-wheeled machine.

In 1813, Drais built a vehicle that was powered by one of the two to four passengers powering it with cranked axle and another person using a tiller to steer. However, his patent was rejected. He revisited the invention a few years later – the attempt, it has been suggested, triggered by the Mount Tambora volcano eruption in Indonesia in 1815, which caused snow in Europe and subsequent crop failures. Drais observed that humans had to feed their horses scarce and valuable grain (though some people even ate their animals in desperation), and so he wanted to develop a horse-less method of transportation.

> GET A BICYCLE. YOU WILL NOT REGRET IT IF YOU LIVE
>
> **MARK TWAIN**

RIDING GEAR
This bicycle-riding uniform of the 1880s was advertised by the sporting and working equipment manufacturer John Wilkinson in its US catalogue.

INDUSTRIAL INNOVATIONS

THE PERIOD FROM THE MID-19TH CENTURY TO THE START OF WORLD WAR I IS AT TIMES REFERRED TO AS THE "SECOND INDUSTRIAL REVOLUTION", WHEN A SERIES OF INNOVATIONS ACCELERATED INDUSTRIALIZATION AND USHERED IN A NEW ERA OF STEEL, ELECTRICITY, AND COMMUNICATIONS.

COILING THE CABLE
In 1857, the crew of HMS *Agamemnon* wound 1,100 tonnes (1,250 tons) of telegraph cable around a drum before attempting to lay it across the Atlantic.

In 1851, the high point of Britain's industrial dominance was marked by a great showcase of technological invention – the Great Exhibition in London's Hyde Park. Some 13,000 exhibits, including machines and contraptions of all sorts were displayed – and the greatest wonder of all was the building that hosted it, the Crystal Palace. Designed by Joseph Paxton (see pp.218–19), it was essentially a glasshouse, but recreated on a gargantuan scale, measuring 564m (1,851ft) by 124m (408ft). Built in just nine months using the relatively new concept of sheet glass, it seemed the ultimate symbol of modernity.

ERA OF STEEL

The frame holding all those panes in position was made of cast iron, whose time as a major building material was very nearly over. Cast iron, although strong enough to support heavy loads, was prone to structural deficiencies that often made it brittle and liable to break.

Just four years after the Great Exhibition, Henry Bessemer (see pp.228–29) patented a process for turning iron into steel – a much stronger, more structurally sound material – both cheaply and in bulk. The ready availability of steel allowed for the construction of new types

PEOPLE IN GLASSHOUSES
A technological marvel built to hold technological marvels, the Crystal Palace played host to the 1851 Great Exhibition in Hyde Park before being moved to South London – as shown here – where it burnt down in 1936.

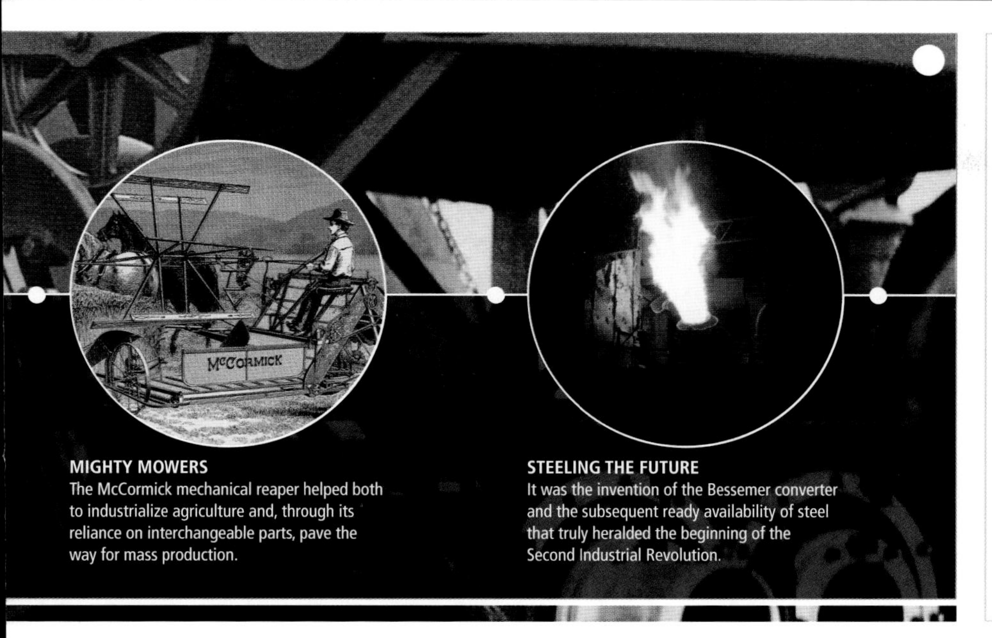

MIGHTY MOWERS
The McCormick mechanical reaper helped both to industrialize agriculture and, through its reliance on interchangeable parts, pave the way for mass production.

STEELING THE FUTURE
It was the invention of the Bessemer converter and the subsequent ready availability of steel that truly heralded the beginning of the Second Industrial Revolution.

SETTING THE SCENE

- Britain's triumphs of the Industrial Revolution are celebrated at the **Great Exhibition of 1851** where the technological wonders of the age are displayed inside one of the most ingenious buildings ever created – **the Crystal Palace**.
- By the mid-19th century, the technical advances of the early Industrial Revolution – centred on steam power, cast iron, and textile production – give way to the new technologies of electrical power, steel, and international communications ushering in the **"Second Industrial Revolution"**.
- The latter half of the 19th century sees the **emergence of new industrial powers**. Germany leads in the fields of telecommunications and electrical engineering, while the US develops the revolutionary process of mass production.
- **Industry and science grow closer**. Many of the advances in this era are a result of a greater understanding of the different fields of science, including electromagnetism and thermodynamics.

of structures and machines – larger ships and bridges, taller buildings, stronger engines, longer railways, and more powerful armaments.

The second half of the 19th century saw innovations that were spurred on by the methodical understanding of the various branches of science, including physics, thermodynamics, chemistry, and electromagnetism. Consequently, industry grew.

IMPACTFUL IDEAS

Among the many American inventors who displayed their products at the Exhibition were Cyrus McCormick (see pp.220–21) and Samuel Colt (see pp.224–35), who demonstrated, respectively, a mechanical reaper and a repeat-action revolver. It was the method of their construction that had lasting impact. Both were created using interchangeable parts, a process known as the "American System", which would eventually evolve into the mass production assembly line techniques of the 20th century.

At about the same time in the 1850s, an American financier, Cyrus Field (1819–92), and a Scottish university professor, William Thomson (see pp.232–33), were working on solving the problem of how to establish a reliable telegraph system linking Britain with the US. About a decade later – and after many mishaps and setbacks – the first transatlantic cable was laid, and a successful intercontinental telegraph service began. Over the ensuing decades, more long-distance telegraph cables were laid and news started travelling faster between countries.

What distinguished industrial progress in the latter half of the 19th century from development in the earlier period was the greater understanding of the science that underpinned many of the new innovations. The trial-and-error mechanistic approach of the early Industrial Revolution had given way to more rigorous methods based on science.

THE SCIENCES

By the last quarter of the century, advances in a number of scientific disciplines had pushed forward industrial technology across a range of spheres. A deeper knowledge of thermodynamic principles helped improve the efficiency and safety of steam engines, and led to the development of the steam turbine in 1884. Invented by Sir Charles Parsons (1854–1931), and improved by Gustav de Laval (1845–1913), the steam turbine brought the era of steam to its pinnacle.

An enhanced appreciation of chemical principles proved vital in the creation of new metal alloys and the development of the petroleum industry, which grew rapidly following the opening of the first oil well in Pennsylvania in the US in 1859. In the beginning, the industry was primarily concerned with the manufacture of kerosene for lamps and heaters. It was the invention of the internal combustion engine in the late 19th century that gave petrol a central role in the motorcar industry.

Another important scientific breakthrough in this period was the formulation of electromagnetic theory, which led to the invention of the electric dynamo by Werner von Siemens in 1866 (see pp.234–35). This device finally made it possible to use electricity as a source of industrial power, thus paving the way for the creation of electric lighting, electrically powered transport – as pioneered by Siemens in Europe and Charles Van Depoele (see pp.236–37) in the US – and the full-scale adoption of mass production techniques using an assembly line.

As the 19th century drew to a close, the applications of electromagnetic theory would be extended further to include the sending and receiving of radio waves, heralding the next great leap forward – the electronics revolution of the 20th century.

ROCKET MAN
Swedish engineer Gustav de Laval invented a revolutionary type of nozzle for use in steam turbines that increased the steam jet to supersonic speeds. These nozzles are still used in rocket engines.

ROBERT STIRLING

INVENTOR OF THE HOT-AIR ENGINE

SCOTLAND **1790–1878**

IN 1816, THE CLERGYMAN Robert Stirling invented the "hot-air engine" – today almost universally referred to as the "Stirling engine". Steam engines drove the Industrial Revolution, but they were not without their problems. Boilers regularly exploded, often with fatal consequences. Stirling's hot-air engine provided a significantly safer alternative.

A LIFE'S WORK

- Stirling shows a **keen interest in mechanical devices** from an early age, helping maintain his family's various agricultural contraptions
- He enrols as a **student of divinity** at Glasgow University, and, in 1816, is **ordained as a minister** in a church in Kilmarnock
- In 1816, he **invents the hot-air engine**, which is meant to provide **a safer alternative** to the explosion-prone steam engines
- Stirling's engine is **installed in an iron foundry in Dundee**, but this engine **breaks down frequently**
- With the revival of steam engines, his engines are **used for small tasks** such as driving church organs or drawing water from wells
- Philips Electronics **give hot-air engines the name "Stirling engines"**

Robert Stirling's engine worked by using an external power source to alternately heat and cool a fixed quantity of air within the engine. This caused the air to expand and contract, driving a piston that could then perform mechanical work. Much safer and quieter than their steam equivalents, Stirling's hot-air engines looked set to become the dominant industrial force of the mid-19th century. Unfortunately for Stirling, after some early signs of promise, his invention failed to make a mark, and steam engines continued to be favoured in factories, owing to the improvements they were undergoing.

DEVOUT AMBITIONS

Stirling was born in a small village just outside Perth to an ingenious family. His grandfather, Michael Stirling, had invented a threshing machine in 1758, and Robert showed a keen interest in mechanical devices from an early age, helping his father maintain the family's various agricultural contraptions. He attended Edinburgh University from 1805 to 1808, where he studied Latin, Greek, logic, law, and, perhaps of most use to his later engineering pursuits, mathematics. However, like many other great engineers of his day, Stirling's forays into inventing were of a purely amateur nature. He had a much loftier ambition for his professional life. In 1809, he enrolled as a student of divinity at Glasgow University, and, in 1816, he was ordained as minister of Laigh Kirk, a church in Kilmarnock.

During the seven years of religious studies, he pursued his interest in engineering, trying to solve the problem of how to create a safer alternative to the steam engine. Shortly after becoming a clergyman, he filed his first patent application for his hot-air engine – the device that, he believed, would provide the answer.

CHANGING FORTUNES

With the patent granted in 1816, Stirling constructed many small working models of his engine, followed by a successful full-size version, which was used to pump water from a quarry. In 1819, he got married to Jean Rankin, the daughter of a local wine merchant. During that

THE FREE-PISTON STIRLING ENGINE IS CHANGING HOW THE WORLD USES ENERGY

INFINIA CORP

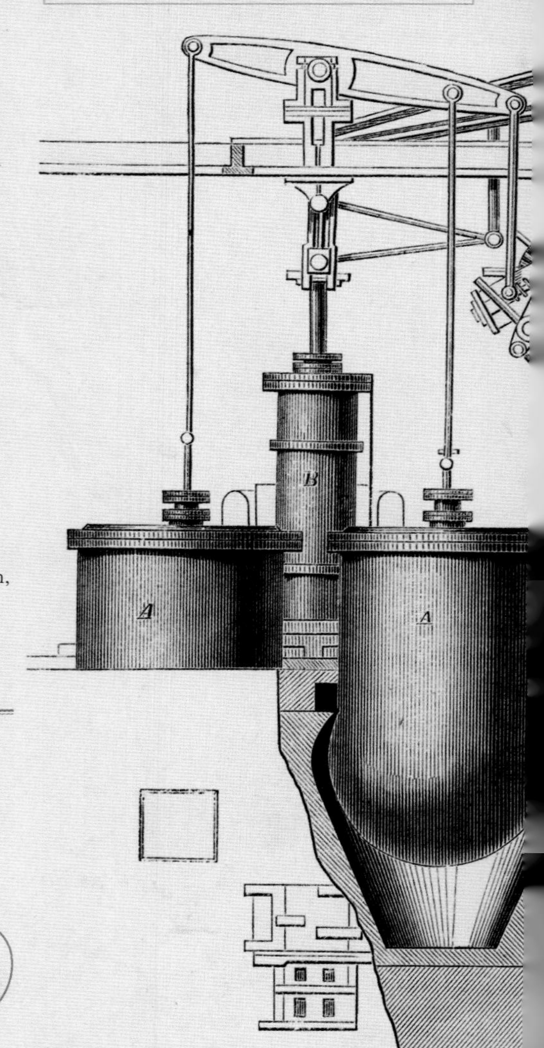

time, things were looking positive for Stirling. However, his invention did not turn out to be quite the runaway triumph he had anticipated.

Although it had been proved that Stirling's engine could work, the hot-air engine did not produce much power, and certainly not enough to compete with the best steam engines of the day. Improvements were obviously required, and to make these changes, Stirling engaged the services of his younger brother James – the family prodigy who had been accepted to Edinburgh University at the early age of 14.

James thought the engine's design might be improved by increasing the pressure of the air contained within it. The pair developed the design over the next few years, filing patents for further adaptations in 1827 and 1840.

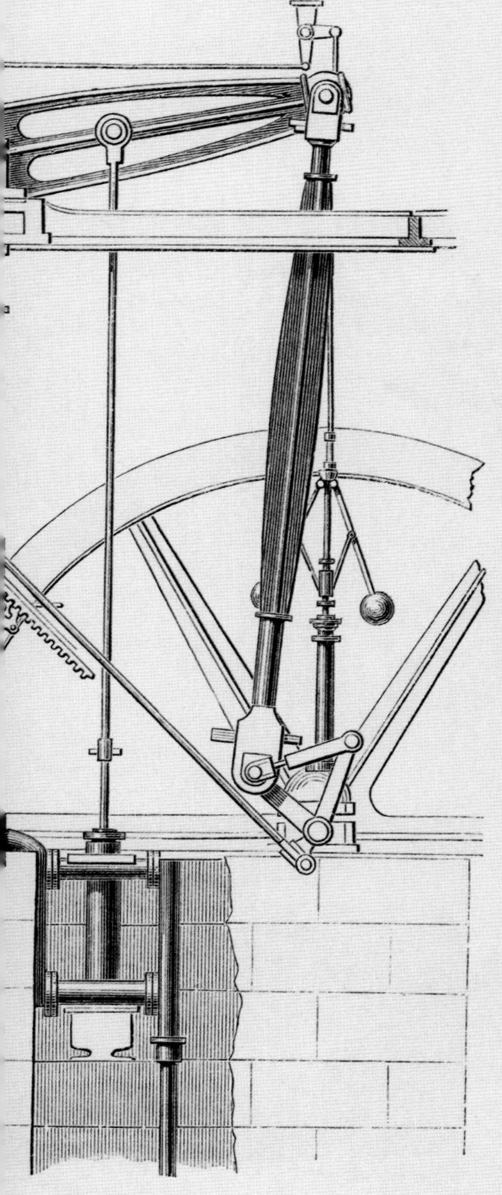

With its power significantly increased, the engine was given its most important commission in 1843, when it was installed at a Dundee iron foundry to drive the machinery. It was not a success. Indeed, the Dundee experience demonstrated exactly why the engine was not more widely adopted. Despite being quieter, safer, and more efficient than the steam engine, it was also more prone to breakages. While these breakages were usually not as dangerous as steam-boiler explosions, they seemed to happen more regularly.

To operate efficiently and maximize the work it could do, the hot-air engine had to be run very hot indeed, which often caused the cast iron to fail. The Dundee engine broke down repeatedly, and, after four years, it was replaced by a steam engine – the rival it was meant to vanquish. Indeed, by this time steam engines had become much safer, with boiler explosions considerably less common.

HOW STIRLING ENGINES WORK

Stirling engines comprise the engine itself, an external power source, and a regenerator – Stirling called it an "economizer". These engines work using an external power source to alternately heat and cool a fixed quantity of gas (or, in Stirling's day, air) enclosed within the engine. This causes the gas to contract and expand, thereby driving a piston (or pistons) and performing the task. The "economizer" stores and recycles the heat that would otherwise be lost as the gas passes between the hot and cold spaces, thereby increasing the engine's efficiency.

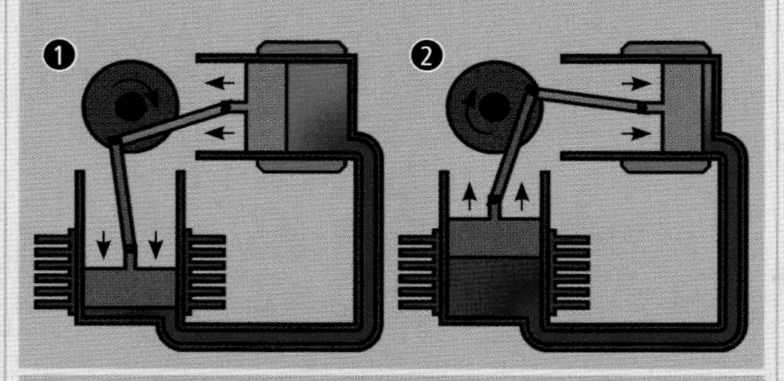

THE EXPANSION CYLINDER (RED) IS MAINTAINED AT HIGH TEMPERATURES WHILE THE COMPRESSION CYLINDER (BLUE) IS COOLED.

DIMINISHED USAGE

Dundee seems to have been the end of the road for the Stirlings, the moment they had to give up on their dream of making commercially successful hot-air engines. Although hot-air engines never became the great industrial behemoth that Stirling had hoped for, they were not completely abandoned either.

One notable characteristic of Stirling engines is that they are remarkably quiet. They therefore found widespread use for driving church organs in the 19th century. Thundering steam engines would drown the organ music, but Stirling engines were ideal. In modern times, they have been used by Swedish submariners. For charging at high speed, submarines use big diesel engines, but enemy submarines are usually detected by the noise of these engines. So, if they want to approach enemy territory without being detected, they switch off the diesel engines and switch on the Stirling engines, which are almost silent.

The engines, however, did have one notable supporter – William Thomson, later known as Lord Kelvin (see pp.232–33) – who, in 1847, found one of Stirling's original model prototypes lying forgotten in a storeroom at Glasgow University, where he was Professor of Natural Philosophy. Fascinated by its thermodynamic possibilities, Thomson began firing up the contraption in his lectures, and it is believed to have influenced his work – indeed it prompted him to investigate the science of thermodynamics.

Stirling always hoped that his engines would one day become important, and in a surprising way they have. An intriguing aspect of Stirling engines is that they can be run backwards – turn the wheel, so that the gas inside is compressed and expanded, and one end gets hot while the other end gets cold. This can be a most convenient way of achieving low temperatures. Today there are satellites in orbit that take photographs of the Earth's atmosphere – to function efficiently they need to operate at a low temperature, and this is achieved by the use of Stirling engines running in reverse.

STIRLING AIR ENGINE
By the time this US drawing was made in 1893, the Stirling engine was already an industrial has-been. It has enjoyed a resurgence in recent years, however. Clean and quiet, because it has no exhaust fumes or internal combustion, the engine is now used to power everything from submarines to computer fans.

A FINAL TRIBUTE

Stirling, who died in Galston, East Ayrshire, in 1878, did achieve one final, posthumous victory. During his lifetime, his contraptions were usually referred to as hot-air engines. However, in the 20th century, as other gases such as helium began to be used, the engines needed a new name. The Dutch electronics giant, Philips, which funded significant research into the engines in the mid-20th century, began calling them "Stirling engines" after the great man, and the name stuck.

GOLDSWORTHY GURNEY

PIONEER OF STEAM-POWERED ROAD TRANSPORT

ENGLAND **1793–1875**

A MAN FOR WHOM SUCCESS always seemed tantalizingly out of reach, Goldsworthy Gurney was an early, if ultimately unsuccessful, pioneer of steam-powered road travel. His most popular invention – a very bright oil lamp – was widely credited to somebody else. Disputes over the ownership of certain inventions would become a running theme in Gurney's life.

A LIFE'S WORK

- Gurney **becomes a doctor**, practising first in the West Country and then in London
- In the 1820s, he tries to **perfect a steam-powered road vehicle** with the intention of setting up a **transport business**
- His company produces **steam-carriages** resembling horse-pulled stage coaches; the coaches are **not popular with passengers** fearful of riding on top of a steam boiler as they tend to explode, leading to fatalities
- The **proprietors of horse coaches** see Gurney's machines as a **threat to their livelihoods** and petition Parliament; a **high toll of £2** is enforced on steam carriages that makes Gurney's business financially unsupportable and **he goes bankrupt**
- Creates intensely **bright limelight**, used in theatres, and **Bude light**, used for lighting London streets and the House of Commons
- Invents the **Gurney Stove** that is used to heat large buildings, including several churches

Gurney was born in Padstow, Cornwall, to a very well-to-do family whose noble lineage could be traced back to the Norman conquest, and who clearly thought nothing of giving their family member one of the more unusual first names of the age. Goldsworthy was the surname of Gurney's grandmother, who had been maid of honour to Queen Charlotte, wife of George III.

Goldsworthy showed a keen interest in science as a child, and at the age of eight was taken to Camborne in Cornwall to see a demonstration of Richard Trevithick's Puffing Devil (see pp.128–29), a steam-powered road locomotive. It was an event that clearly had a profound influence on him. Gurney grew up to become a doctor, practising first in the West Country and then, from 1820, in London. Apart from his medical studies, however, he spent much of his time pursuing scientific interests.

He was particularly intrigued with the mechanics of steam engines, and, in the 1820s, tried to perfect a steam-powered road vehicle with the intention of setting up a transport business. It certainly seemed like the right time to attempt such a venture as steam travel was in vogue. In 1825, George Stephenson opened the Stockton and Darlington Railway, the world's first steam train to carry passengers (see pp.190–91), and it seemed obvious that somebody would try to do the same for road transport. Unfortunately, Gurney would not be that man, although not for any want of effort.

STEAM-CARRIAGE TRIALS

From its base at Regent's Park in London, the Gurney Steam Carriage Company churned out a succession of contraptions. The early steam-carriages resembled horse-pulled stage coaches, and could reach a top speed of 32kph (20mph). One coach even travelled all the way from London to Bath on a test run, although the journey was not an unqualified success – the steam carriage was attacked by locals just outside Bath, presumably as a sort of Luddite-style protest against technological advancement.

The coaches did not prove to be popular with passengers either – they were fearful of riding on top of a steam boiler, which in the early 19th century tended to explode fairly regularly, often leading to fatal consequences.

So Gurney devised a new vehicle, known as the *Gurney Steam Drag*, which people rode in a carriage towed behind the engine. Two of these vehicles were built and shipped to Glasgow, but before they were up and running, one of the boilers exploded, injuring several people, and the interest in this carriage service understandably waned.

INTENSE LIGHT
Bude lights, such as this one in Trafalgar Square, London, were easier to manage, since oil could be pumped in continuously, while the solid lime gradually wore away, and had to be mechanically replaced.

POLITICIZED TUSSLE

Even if Gurney could have perfected his designs – or at least managed to get his vehicles to explode less often – he would have failed to make profits, since powerful forces were lined up against him. Among them were the proprietors of the nation's horse coaches, who saw Gurney's machines as a threat to their livelihoods. In 1831, when a regular steam carriage road service was finally established between Cheltenham and Gloucester, running four times daily, the stage-coach proprietors petitioned Parliament. Their friends in high places ensured the passing of the Turnpike Acts, enforcing a £2 toll per steam carriage journey – horse-drawn carriages typically paid a twentieth of this or less – which instantly made the service financially unsupportable. Gurney was left bankrupt with debts amounting to £232,000.

However, not every politician was against Gurney. Indeed, many found the Turnpike Acts grossly unfair. A parliamentary select committee, convened to look into the case, concluded in its final report that Gurney's "steam carriages were safe, quick, cheap, and less damaging to roads than horse carriages [and] that they would be a benefit to the public and the prohibitive tolls should be removed". The House of Commons agreed, and passed a bill abolishing the tolls. The House of Lords, however, opposed it. And that was as far as Gurney's involvement in steam-powered road transport went – although Parliament did finally agree to pay him some compensation for his lost earnings.

INTENSELY BRIGHT LIGHT

Unhappy with this defeat, Gurney moved back west to Bude in the 1830s where he built himself a castle on the sand dunes; it still stands and is a heritage centre today. There, he invented a new type of lamp that employed an oxy-hydrogen blowpipe – a device invented by American chemist Robert Hare (1781–1858) – to heat a lump of quicklime (calcium oxide) to a very high temperature, which created an intensely bright flame. Indeed, it was so bright that it was used in theatres to pick out performers on stage – which is where the phrase "in the limelight" comes from – and in lighthouses, as it could be seen by ships many miles away. The device was popularized by the Scottish civil engineer Thomas Drummond (1797–1840), who devised an adaptation for his trigonometrical survey of the country. Although Drummond never claimed to have invented the lamp, nobody seemed to believe him, and so it came to be known among the general public as the "Drummond light".

Gurney made more lamps by blowing oxygen into a flame of burning oil. This made the oil burn at a much higher temperature than when it was just in air, and so produced an intensely bright flame. Gurney called these "Bude lights", after his home town. They were used for street lighting, and also, for 60 years, to light the House of Commons, where three Bude lights replaced 280 candles.

HEATING UP ENGLAND

Gurney had married in 1814, but when his wife died in 1837 he moved out of "the Castle" and built himself a new house, Wodleigh Cottage, just outside Bude. There, with the devoted support of his daughter Anna Jane, who was a great defender of his work (see box, right), he continued developing his inventions. The most successful of these was the Gurney Stove, which had an innovative ribbed outer casing to increase its surface area – and by extension, the amount of heat it emitted – and proved a popular means of heating large buildings, including several churches. Gurney stoves can still be found in the cathedrals of Chester, Ely, and Durham, among others. Gurney also devised new heating and ventilation systems for the Houses of Parliament. He was knighted in 1863. However, Gurney became much less active over the last decade of his life after suffering a stroke.

GURNEY'S CHAMPION

Never has an inventor had such a dedicated admirer as Gurney did in his daughter, Anna Jane Gurney. She was a fierce defender of his right to be credited as the inventor of the blastpipe, or steamjet, which other authorities credit to either Stephenson or Trevithick. The pipe increased the amount of air flowing through a steam engine's chimney, thus improving its power and efficiency. This was what allowed steam locomotives to travel quickly, as Anna pointed out in Gurney's gravestone inscription: "To his inventive genius the world is indebted for the high speed of the locomotive, without which railways could not have succeeded and would never have been made".

GURNEY'S STEAM CARRIAGES
Although Parliament decided that Gurney's carriages (shown here in this 1827 engraving) were "safe … less damaging to roads than horse carriages", the operators of those horse carriages conspired to drive the inventor out of business.

JOSEPH PAXTON

DESIGNER OF THE SPECTACULAR "CRYSTAL PALACE"

ENGLAND 1803–65

RISING FROM HUMBLE BEGINNINGS, Joseph Paxton used his considerable imagination and intelligence to develop a career encompassing horticulture, landscape design, architecture, engineering, publishing, and railway promotion. His revolutionary design for the 1851 Great Exhibition demonstrated for the first time the role engineers could play in modern building.

A LIFE'S WORK

- Paxton gets his first opportunity when he is **employed as a garden labourer** for the Horticultural Society in London
- He is appointed **head gardener for Chatsworth House** in 1826, starting off his **fascination with glasshouses**
- His **Great Conservatory at Chatsworth, the largest greenhouse ever built**, wins critical acclaim
- He gets the contract for constructing the building for the **Great Exhibition of 1851** – his design is approved within a month of submission
- The triumphant success of the 1851 Great Exhibition, housed in his innovative **"Crystal Palace", secures Paxton a knighthood** and heralds the engineer's involvement in the design of modern buildings

The eighth child of a Bedfordshire agricultural labourer, Joseph Paxton was only seven when his father died. Paxton spent his childhood close to poverty. After a number of gardening jobs, he was initially taken on as a labourer at the Horticultural Society's gardens at Chiswick, on land rented from the Duke of Devonshire, who owned nearby Chiswick House. The Duke spotted the young man's talents, and, in 1826, appointed him as head gardener at his great house at Chatsworth, Derbyshire.

Arriving at Chatsworth two days later at 4:30 in the morning, Paxton explored the gardens, set the staff to work, ate breakfast with the housekeeper, and met her niece Sarah Brown, his future wife – completing his first morning's work before 9 o'clock.

BUILDING GLASSHOUSES

With the Duke's support, Paxton built increasingly sophisticated glasshouses at Chatsworth, producing a ridge and furrow roof design to minimize loss of light when the Sun's rays struck obliquely. The concept was not original, but Paxton developed the shape into a practical structural form with timber gutter beams and slender glazing bars – each made from solid timbers. He also developed ingenious steam-driven machinery to produce these complex sections. His thinking contained all the elements of future standardization and mass production.

These glasshouses culminated in his Great Conservatory, the "Great Stove of Chatsworth", measuring 69m (227ft) by

CRYSTAL PALACE
This lithograph shows the Crystal Palace during the "Great Exhibition of the Works of Industry of all Nations" in 1851. It was the world's first prefabricated ferrovitreous (iron and glass) structure, and Paxton's crowning achievement. It was dismantled easily after the event.

29m (123ft) in plan – by far the largest greenhouse ever built. This was followed by the Victoria Regia Lily House, which for the first time used his ridge and furrow glazing on a flat roof.

His water works at Chatsworth included the Emperor Fountain, with a jet about 51m (270ft) high – fed through cast iron pipes from a large reservoir Paxton had created on the hill overlooking the house. With the Duke's approval, Paxton also edited and contributed to several magazines including *Paxton's Magazine of Botany*, set up an architectural practice, and became adept at landscape planning. He laid out the picturesque village of Edensor near Chatsworth. Meanwhile he became interested in railways, got to know both George and Robert Stephenson (see pp.190–91), became a director of the Midland Railway, and invested extensively in railway development.

GIANT CRYSTAL DESIGN

Paxton's masterpiece, however, was the building for the Great Exhibition of 1851. At the height of Britain's industrial and engineering achievement in the 1840s, it was proposed that a "great exhibition of the works of industry of all nations" be held in London. A Royal Commission was appointed; the site, Hyde Park, was chosen, and a design competition held. However, all 245 entries were rejected, and the Commission's own alternative design received a lukewarm response when published in June 1850.

Paxton had a better solution based on his experience with his glasshouses at Chatsworth, and he worked fast. He made his first sketch on 11 June, and more developed sketches of his "Crystal Palace" – his best known design – were published in the *Illustrated London News* on 6 July. His plans to use glass and iron extensively were exciting, modern, appropriate, and captured the public imagination. By using cast iron and mass production techniques, the huge building – measuring 563m (1,848ft) long, 139m (456ft) wide, and 41m (135ft) high – could be created quickly. This was an important consideration, since at this stage, there were only 10 months until the opening – an incredibly short time, even by today's standards.

GREAT CONSERVATORY
Paxton worked on the Great Conservatory at Chatsworth House from 1836 to 1841. At the time, it was the largest glasshouse in the world – costing a staggering amount. It was demolished in 1920.

With the help of Charles Fox (1810–74) from iron founders and contractors Fox Henderson & Co, a minimal range of standard components was designed, which could be mass produced and speedily assembled. The contract was given in July, work started in August, and the building was completed by March. The Great Exhibition opened on 1 May 1851 to international acclaim, and 6 million people attended the exhibition between May and October. Paxton and Fox were both knighted for their contributions. The surplus money from the venture was used to found London's Victoria & Albert Museum, Science Museum, and Natural History Museum.

MASTER OF MANSIONS

The building was intended to be temporary and, because of its method of construction, it was easy to dismantle. But Paxton and many others argued that it should be retained as a permanent winter garden in Hyde Park. This was not to be, and from 1852 to 1854 the building was reconstructed – larger, and with many changes – at Sydenham in South London. Paxton was involved in designing the layout of the gardens. But it was never a success with the public and the building was destroyed by fire in November 1936, although Crystal Palace park still remains.

The success of his building for the Great Exhibition brought Paxton worldwide fame and, with it, an even more hectic lifestyle. While continuing to develop the Duke's estates at Chatsworth and at Lismore Castle in Ireland, he designed two mansions for different branches of the Rothschild family. In 1854, he was elected a Member of Parliament for Coventry, a position he held for more than a decade. Subjects on which he campaigned included improved land drainage, the use of sewage as a fertilizer, and improved public health. Paxton also fought for improvements to the River Thames, which was a stinking sewer at that time.

Paxton's interest in railways developed into concern about transport in towns – road, rail, and pedestrians. His plan for the Great Victorian Way – a 16-km (10-mile) long elevated walkway with incorporated rail – was debated seriously. Glazed and ventilated to keep it dry and protected from the surrounding smoke and smell, the walkway would have connected London's main rail stations and other key points. Paxton retired from Chatsworth when the Duke died in 1858, but carried on working on various projects, such as the Thames Graving Dock. He died in 1865.

A TRULY AMAZING AMOUNT OF COST AND SKILL IN DESIGN, AND TASTE IN EXECUTION ... RENDERS THE ESTABLISHMENT WITHOUT A PARALLEL

THE LEISURE HOUR REFERRING TO THE CRYSTAL PALACE

CYRUS McCORMICK

THE MAN WHO REVOLUTIONIZED REAPING

UNITED STATES 1809–84

THE "PERFECTER", if not the originator, of the mechanical reaper, Cyrus McCormick enabled American farmers to harvest grain on an unprecedented scale and turned the prairies of the Midwest into the "breadbasket of the world". His invention was also seen as one of the prime means through which the United States would achieve its "Manifest Destiny".

A LIFE'S WORK

- McCormick **files for a patent** for his version of the mechanical reaper in 1834, which is **granted the same year**
- With his reaper farmers can cut about **5 hectares (12 acres) of wheat per day**
- **Expands his company** from a small farm workshop to a **large factory in Chicago**
- His reaper wins the **gold medal at London's Great Exhibition of 1851**, the greatest industrial show of the time
- He has a **keen sense of marketing** and introduces several innovative campaigns and schemes to attract buyers
- Helps accelerate the **settlement of the American West**
- The McCormicks become **one of the richest families in the US** by the 1870s
- His company continues to survive today as a **global transportation firm** called **Navistar**

McCormick's father had been working on the design of a horse-drawn mechanical reaper for about a decade before Cyrus was born. However, it would take the younger McCormick's grit and commercial acumen – as much as his engineering ability – to make the project a success.

McCormick was born on 15 February 1809 on the family farm, Walnut Grove, in Virginia's Shenandoah Valley. Receiving little formal education, he spent most of his youth helping out on the farm. As a young man, he began to take an interest in the mechanical reapers being developed in the family workshop and soon realized their potential. In 1831, while still in his early 20s, he performed a trial of his new and improved version of the machine at a neighbour's oat field to drum up interest among the local farmers. However, the machine kept breaking down, and when it did "work", it either sank into the mud or ripped up the oat stalks completely rather than neatly shearing off their tops, as was intended.

A DETERMINED EFFORT

McCormick was not disheartened, though, and continued refining his machine, filing for – and being granted – a patent in 1834. But despite numerous public demonstrations, he could not sell his first reaper until 1840. This did not exactly herald a boom, as he sold none the year after.

Things did pick up after that, with seven machines being sold in 1842, 29 in 1843, and 50 in 1844, all of them assembled by Cyrus and his brothers at the workshop on the family farm. Once people realized the potential of the reaper, it began to be employed in a big way. Before the reaper's invention, harvesting was a slow, labour-intensive process. Farmers could typically cut, gather, and bind just 0.2 hectare (0.5 acre) of wheat per day. Using McCormick's invention, this increased to about 5 hectares (12 acres) of wheat per day, allowing for the cultivation of much larger areas than had previously been thought possible.

However, just when it looked as if McCormick was finally making progress, his fledgling business was almost derailed by competition. When he tried to re-file his reaper patent in 1848, he was informed by the US Patent Office that it now believed that a rival design by Obed Hussey (1792–1860), patented in 1833, should be given precedence as the "first" reaper. McCormick's application was, therefore, turned down. However, McCormick was able to use his public demonstration of 1831 as proof that his reaper had been the earlier invention. After considering the merits of both the cases, the US Patent Office reversed its decision and re-granted McCormick his patent.

In the late 1840s, with a new patent in hand, the family reaping business, now named Cyrus H McCormick and Brothers, moved its headquarters from the small farm workshop in Virginia to a vast purpose-built factory in Chicago. The move was market-led, as most orders for reapers were from the country's Midwest, where the great size of the farms made any labour-saving device particularly appealing. Chicago was chosen for its transport links, as

REAPING THE WEST

McCormick's reaper formed the basis of all subsequent reapers and harvesters. As the reaper moved through the field, a rotating wheel pulled the crop stalks against a vibrating cutting bar. The cut ears dropped onto a platform, from where labourers would rake them off onto the ground to be gathered and bound. Threshing took place later. The late 19th and early 20th centuries saw the introduction of reapers that automatically bound the wheat, resulting in the "combine harvester", which could cut, bind, and thresh the wheat on its own.

AMERICAN FARMS HAD PLENTY OF LAND BUT NOT ENOUGH LABOUR, ENABLING THE DEVELOPMENT OF MACHINES.

raw materials could be transported from New York in the east along the recently built Erie Canal and the Great Lakes, while the finished reapers could then be transported westwards on the railroads.

CONFRONTING COMPETITION

The 1850s were a time of increasing success for McCormick. He displayed his invention to gold-medal-winning acclaim at the greatest industrial show of the age, the Great Exhibition of 1851 at London's Crystal Palace, which led to the tremendous growth of his business.

This growth was helped in no small part by McCormick's appreciation of salesmanship. He was an enthusiastic marketer, creating eye-catching advertising campaigns and employing an army of salesmen to ride the railroads and demonstrate his equipment first-hand to sceptical farmers. He also introduced many forward-thinking business practices, such as setting a fixed price of $120 for his reapers ("One Price to All and Satisfaction Guaranteed" ran the advertising blurb), developing interchangeable replacement parts, and allowing farmers to buy on credit and pay over time. As a result, by 1860, McCormick was selling about 4,000 reapers a year.

It has been claimed that McCormick's mechanical reaper contributed greatly to the westerly migration of the US population during the latter half of the 19th century, as it enabled settlers rapidly to adapt large tracts of land to agricultural use. In fact, the governor of New York in the 1860s, William H Seward, believed that the invention moved the "line of civilization … westward thirty miles each year".

A FAMILY AFFAIR

In 1858, aged 48, McCormick married for the first time and spent much of the next two decades travelling abroad, expanding business, and setting up a distribution system in Europe, while leaving the day-to-day running of the factory to his brothers James and William. The company, although headed by Cyrus, had always been a family affair. By the 1870s, the McCormicks had become one of the richest families in the US, which enabled them to rebuild the business when the Great Chicago Fire of 1871 destroyed the main factory.

However, the other members of the family felt that they did not receive the credit that they deserved, and, in 1879, Cyrus's younger brother Leander got the company's name changed to the McCormick Harvesting Machine Company to put the family's contribution on a more equal footing. He was still dissatisfied, however, and eventually left the company, embittered.

The firm, however, continued to thrive long after the original McCormicks had died, and has carried on in one guise or another to the present day. It is now known as Navistar.

McCORMICK REAPER
This is a hand-coloured engraving of Cyrus McCormick's mechanical reaper in action. The first successful reaping machine, it won a gold medal at the Great Exhibition in London in 1851 and made McCormick one of the richest men in America.

WILLIAM ARMSTRONG

PIONEER OF THE HYDRAULIC CRANE AND ARMAMENTS

ENGLAND 1810–1900

Baron William Armstrong of Cragside began his career as a lawyer before becoming an innovative civil, mechanical, and electrical engineer. He invented the hydraulic crane and pioneered hydro-electric power. He founded one of Britain's greatest Victorian industrial enterprises in Newcastle upon Tyne, manufacturing armaments and ships, and employing up to 25,000 men.

A LIFE'S WORK

- Armstrong's innovative **hydraulic crane is installed on Newcastle's quayside**, and proves an instant success
- Is elected a **Fellow of the Royal Society in 1846, an unusual distinction** at the time for someone whose profession was the law
- **Forms Armstrong & Co in 1847** to design and manufacture hydraulic machinery
- Is **appointed as the British Government's Engineer for Rifled Ordnance** and starts manufacturing armaments in 1859
- Begins **warship construction in 1868**
- His house at Cragside, Northumberland, becomes the **first house in Britain with electric light**, powered by a **hydroelectric generator**
- In 1897, his **company is merged** with the interests of Sir Joseph Whitworth to form Sir W G Armstrong Whitworth & Co Ltd

The son of a prosperous corn merchant and local politician in Newcastle upon Tyne, William Armstrong qualified as a solicitor, and, in 1833, became a partner in a local firm. He seemed destined for a successful career in law.

While practising in the late 1830s, he started experimenting with machines that used water to generate motive power, and later with hydro-electricity. In 1846, he was elected a Fellow of the Royal Society, sponsored by Michael Faraday (see pp.140–41), who argued that membership of the society should be open to non-scientists – "for men such as Armstrong".

HYDRAULIC CRANE

In November 1845, Armstrong installed a hydraulic crane on Newcastle's quayside "with the view of increasing the rapidity and lessening the expense of the operation of delivering ships". It was a success, and three more were installed in 1847. In the same year, Armstrong & Company was formed to design and manufacture hydraulic machinery, with a site chosen at Elswick.

The principle of his crane was straightforward. In his own words: "In order to lift the weight, the water is admitted into a cylinder by means of a slide valve, and it exerts its force upon a piston. The rod, to which the piston is attached, is connected to the hoisting chain, so that as the piston recedes from the pressure the weight is lifted. When the weight is to be lowered, the water is allowed to escape from the cylinder by a different movement of the same valve". Hydraulic power had existed before, used mainly for the drainage of mines. Armstrong's contribution was to develop practical applications, especially through the use of the "jigger", converting linear motion to rotary, and later the "accumulator", a device that enabled higher hydraulic pressures to be used without the need for elevated reservoirs. This revolutionized dock operation, but also found uses in industrial engines, locomotives, the operation of dock gates and swing bridges, plants for mine drainage and ore crushing, and for railway cranes and turntables. Orders poured in, and by 1852 Armstrong was employing some 350 men.

ARMSTRONG 100-TON GUN

In the late 1800s, large muzzle-loading cannons were built by Armstrong's Elswick Ordnance Company for the British government. They were built on the Armstrong system of a primary steel tube, with successive, shorter, wrought-iron tubes, heated and shrunk on the main tube. They had a 45-cm (18-in) bore, a little over 9m (30ft) long. The guns were first fired in 1884, but the weapons were not fully operational until 1889 owing to hydraulic system problems. The four original British guns were divided between Gibraltar and Malta and none of them were ever fired in anger.

THIS GIANT ARMSTRONG MUZZLE-LOADING GUN IS FIRED DURING A HISTORICAL RE-ENACTMENT IN MALTA IN 2006.

STEADY PROGRESS

During the Crimean War (1853–56), Armstrong turned his hand to armaments, particularly the development of an effective field gun. His rifled 3-pounder coil breech-loader, closely followed by an 18-pounder, were adopted by the army. Armstrong was appointed as engineer for Rifled Ordnance and superintendent of the Royal Gun Factory at Woolwich in London, and his company was given a monopoly of government orders. Soon, he established the Elswick Ordnance Company to manufacture guns for the British government. However, in the early 1860s, the arms orders dried up, and in 1863 Armstrong resigned. This freed him to build a new international armaments trade, supplying many countries with heavy guns.

In 1868, Armstrong's company started building ships. At first this was achieved in partnership with others. Most successful were his warships, helped by Armstrong's connections with British Admiralty officials and by the fact that, by then, all warships relied on Armstrong's hydraulic systems of gun handling. He built warships for the navies of many countries, including Japan and Italy. In 1885, Amstrong opened a factory for the manufacture of guns at Pozzuoli, near Naples. In 1897, after his retirement from active work, his company was merged with the interests of Sir Joseph Whitworth (see pp.152–53) to form Sir W G Armstrong Whitworth & Co Ltd.

Meanwhile, Armstrong was developing his estate at Cragside, Northumberland. The hunting lodge was successively improved and enlarged under architect Richard Norman Shaw, but Cragside's greatest distinction was the engineering works that Armstrong introduced. He first built a reservoir with an 11m- (35ft-) high embankment dam to supply water to power a hydraulic engine that in turn supplied spring water to the house. The water from two further reservoirs was used to power a turbine and generate electricity – one of the first ever hydro-electric systems. The electricity lit the house, originally with arc lamps, and later with some of the earliest incandescent bulbs, installed by his friend Joseph Wilson Swan, making Cragside the first house in Britain with electric light. He also installed a telephone system, although he could call no one, since he was the first one to possess the instrument.

TYRANT OR BENEFACTOR?

At one level, William Armstrong – or Baron Cragside, as he became from 1887 – fitted the image of the tyrannical, insensitive, and single-minded Victorian industrialist, and he amassed plenty of criticism during his lifetime. The water company of which he was chairman was seen to be providing water of poor quality. In 1871, when his workers at Elswick came out on strike (along with others on Tyneside), demanding a reduction in employment to a nine-hour day, Armstrong organized strong and inflexible opposition by local employers (although eventually concessions were made). He did not build housing for his employees, as some of his contemporaries did. He came under attack for the morality of arms manufacture, a charge to which he was sometimes sensitive, although he argued his case forcefully.

On the other hand, besides being a gifted civil, mechanical, and electrical engineer, Armstrong also showed enlightenment and generosity. After his knighthood, he donated his patents to the nation – three for hydraulic machinery and 11 for ordnance and projectiles. He supported the education movement that produced schools and free libraries, and established a Mechanics Institute at Elswick to educate the children of his workforce. He promoted apprenticeships and those at Elswick became much sought after: one of the most eminent beneficiaries was Charles Algernon Parsons (1854–1931), who later invented the steam turbine.

Armstrong was also Newcastle's greatest benefactor. He purchased and gave to the city an 11-hectare (28-acre) park, which was named after him. He transformed the land adjacent to his Newcastle house into a 38-hectare (93-acre) park, and presented this to the city, along with an ingenious wrought-iron bridge manufactured at Elswick and designed to articulate so as to avoid any future problems due to colliery subsidence beneath. He established and funded a technical college, now part of the University of Newcastle upon Tyne. The later years of Armstrong's life were spent in his magnificent parkland mansion of Cragside near Rothbury in Northumberland.

BATTLESHIP PRODUCTION LINE
Here, gun turrets are assembled in Armstrong's works in Newcastle upon Tyne. From 1864 he began producing naval guns and warships. The first vessels produced were the torpedo cruisers *Panther* and *Leopard* for the Austro-Hungarian navy. Japan was another major customer – armed with Armstrong's warships, it defeated the Russian Empire at the Battle of Tsushima in 1905.

SAMUEL COLT

THE MAN WHO ARMED THE WEST

UNITED STATES 1814–62

AN INVENTOR AND INDUSTRIALIST, Samuel Colt was a man who, depending on your point of view, either invented or improved the revolver. What is agreed upon is that, after numerous false starts, Colt's creation eventually made him the world's leading private arms manufacturer. His innovative contributions to the arms industry shaped the destiny of American firearms.

There is a fine line between invention and innovation – between creating something wholly new, and creating something that is just an improvement on an existing idea. It is a distinction that is neatly illustrated by the life of Samuel Colt.

One of nine children, Colt was born in Hartford, Connecticut, the son of a textile manufacturer. Colt's mother died before he was seven years old. Put to work at an early age, both as an apprentice in his father's dyeing mills and at a nearby farm, he soon became adept at fixing objects – including farm machinery and his father's firearms – taking them apart to see how they worked and then reassembling them. He also apparently spent his free time reading a popular encyclopedia, *The Compendium of Knowledge*, which provided him with useful scientific insights. This comprehensive encyclopedia contained articles on Robert Fulton (see pp.188–89) and gunpowder, both of which provided inspiration and ideas that would influence Colt throughout his life. *The Compendium* revealed that Robert Fulton and several other inventors had accomplished things deemed "impossible – until they were done". Later, after hearing soldiers talk about the success of the double-barrelled rifle and the impossibility of a gun that could shoot five or six times, Colt decided to invent and create the "impossible" gun.

INSPIRATION AT SEA

Colt's father, however, was keen that he should become a sailor, and so, in 1830, at the age of 15, Samuel was enrolled to study navigation at Amherst Academy in Massachusetts, but he was soon expelled for some unknown misdemeanour. His father decided the boy might as well take a more practical approach to learning his trade, and so signed him up for the crew of the *Corvo*, a ship making a year-long voyage to India and back.

The legend goes that it was while sailing the high seas that Colt had his "*Eureka!*" moment, coming up with his idea for a gun with a revolving action after observing the ship's wheel, and the way it could either spin freely or be locked into position by use of a clutch. As Colt himself described it, "regardless of which way the wheel was spun, each spoke always came in direct line with a clutch that could be set to hold it ... the revolver was conceived". He fashioned his first wooden prototype while still at sea. This was a model gun with a revolving cylinder containing six bullet chambers. Many historians, however, believe that Colt is more likely to have found inspiration on dry land. It is an indisputable fact that the first revolving pistol was invented by the Bostonian Elisha Collier (1788–1856), and patented in 1818. The gun was principally used by the British army in India, which is probably where Colt saw it, and where he began solving the problems of its design – particularly its unreliable flintlock firing mechanism.

ARMING THE UNION
This daguerreotype photograph shows a Union soldier armed with a Colt Dragoon revolver, which was widely used during the American Civil War of 1861.

Back in the US, Colt's father financed the manufacture of some metal prototypes, and the young inventor filed for patents in Britain in 1835, and in the US the following year. He used the family's money, in 1836, to establish Colt's Patent Arms Manufacturing Company, where he began producing five types of "revolving" weapons – three pistols and two rifles. These weapons could fire six shots without reloading, at a time when it typically took around 20 seconds to reload a standard weapon.

Colt thought his invention would appeal to the US government, as it would give its military a significant advantage over a less well-equipped enemy. The trouble was, he could not convince them, at least not in the beginning. Although President Andrew Jackson was impressed by a demonstration of the weapons given by Colt

A LIFE'S WORK

- In 1830, Samuel Colt **enrols at the Amherst Academy in Massachusetts**, to study navigation
- He **improves and solves the problems of the design** of the first revolving pistol invented by Elisha Collier
- Establishes **Colt's Patent Arms Manufacturing Company in 1836**, where he produces three pistols and two rifles
- His company **receives an order of 1,000 guns from the Texas Rangers** for use in the Mexican-American War, the proceeds from which are used to establish **Colt's Patent Fire-Arms Manufacturing Company**
- Recruits Eli Whitney Blake to **help him with the firearms business**

THE ROOT REVOLVER
Designed by Colt, the Root pocket revolver was named after his factory foreman, Elisha K Root. A .28 calibre weapon, it came with a choice of ivory or walnut handle (as here). This model was made in the two calibres of .28 and .31. About 40,000 of them were produced between 1855 and 1861.

PEOPLE OF THIS WORLD ARE NEVER SATISFIED WITH EACH OTHER AND MY ARMS ARE THE BEST PEACEMAKERS

SAMUEL COLT, LETTER TO CHARLES MANBY (18 MAY 1852)

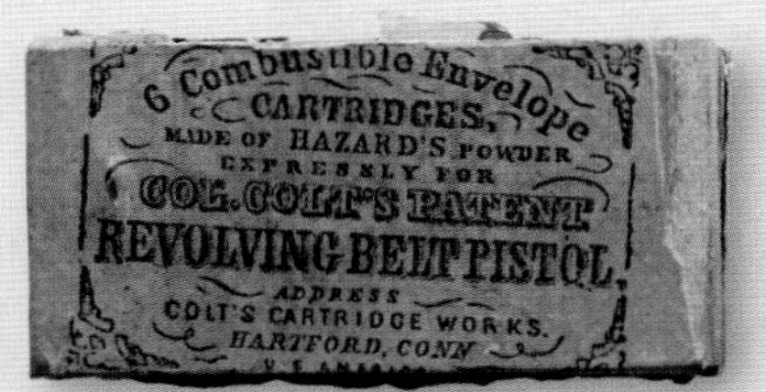

in Washington DC, those in charge of military contracts were less easily swayed. Believing the new weapons to be so new as to be potentially unreliable, the army's top brass declined to place any orders, putting a great financial strain on the fledgling business.

To raise money, Colt toured the country giving demonstrations of the effects of nitrous oxide (laughing gas). He did manage to sell some guns along the way too, particularly during the Florida Seminole War (1835–42), but not enough to keep the business afloat. In 1842, the factory closed and its assets were sold off.

Colt then turned his attention to designing an underwater mine, but this proved no more successful than his guns. Despite impressing yet another president – this time John Tyler – the government again declined to invest.

AN ORDER FOR WAR

Colt's luck finally turned in 1847 when the Texas Rangers placed an order for 1,000 guns for use in the Mexican-American War (1846–48). The large order allowed Colt to re-establish his firearms business. However, as he no longer had a factory, Colt hired another inventor, Eli Whitney Blake (1795–1886), nephew of the inventor of the cotton gin (see pp.148–49), to make them for him, as Blake was already established in the arms business. The company then received an order for 1,000 more guns; Colt took a share of the profits at $10 per pistol for both orders. He then used the proceeds to establish Colt's Patent Firearms Manufacturing Company, which grew from strength to strength as the orders, which had barely been a trickle over the previous decade, flooded in. Colt's guns became hugely popular, particularly during the Californian Gold Rush, allowing him to expand his business into Europe. In 1855, a vast new armaments factory – the largest in the world in private hands – was built in Hartford, overlooking the Connecticut River.

Colt's guns were a major success not only because they were the right product at the right time, but because he had set up a precision production system, so that all the parts were interchangeable. By the time the Civil War began in 1861, Colt's factory employed 1,000 people, and he was one of the richest men in America. He ceased supplying armaments to the Confederacy soon after dealing mainly with the Union. However, he did not live to see the conflict's conclusion. Aged just 47, Colt succumbed to complications arising from his chronic rheumatism in 1862, leaving behind a fortune of about $15 million. His company, however, continued to prosper. Still trading today, Colt's Manufacturing Company has to date produced more than 30 million firearms.

HELPING MAKE AMERICA

Colt contributed greatly to the subsequent economic success of the US. The processes he adopted – using interchangeable parts, precision machinery, and an organized assembly line – paved the way for the advent of mass production in the 20th century. He was also instrumental both in the rise of gun ownership in the US, and in forging the mythology of the righteousness of gun ownership. He commissioned paintings by the "Old West" artist George Catlin that showed Americans creating their own destiny, gun in hand, and of settlers taming the wilds armed only with their own ingenuity, and of course, a Colt revolver.

FEMALE WORKERS ON A 1940s ASSEMBLY LINE IN A COLT PATENT FIREARMS MANUFACTURING COMPANY FACTORY INSPECT THE COMPONENTS OF COLT .45 AUTOMATIC PISTOLS.

MEN OF PROGRESS
Gathered before a portrait of Benjamin Franklin (see pp.136–37) are some of the men who enriched the United States with their technological and industrial innovations in the mid-19th century. Included are Samuel Colt (see pp.224–25), the gun manufacturer (third from left); Cyrus McCormick (see pp.220–21), standing before a model of his threshing machine (fourth from left); Joseph Henry (see pp.142–43), inventor of the electric motor (standing, centre left); and Samuel Morse (1791–1872), inventor of the electric telegraph (seated at the front, centre right).

HENRY BESSEMER

INVENTOR OF THE BESSEMER CONVERTER

ENGLAND 1813–98

An inveterate inventor, Henry Bessemer held 129 patents during his lifetime. The most significant was the one granted in 1855 for the Bessemer converter, the device that finally made it possible to create large quantities of steel cheaply, thereby ushering in the so-called "Second Industrial Revolution" – and making Bessemer an extraordinary fortune.

A LIFE'S WORK

- Bessemer gains experience as a **caster of decorative metal medallions**
- Creates a new type of un-forgeable stamp for official deeds – for which he receives no money
- Astonished at the high price of gold paint bought for his sister, he makes his own **cheaper version – the "bronze powder"**
- Keeps the **formula a secret** for 35 years
- Makes a fortune with his invention of his eponymous **converter for turning iron into steel**, this time by licensing the patent
- Makes other inventions such as a hydraulic **device for extracting more sugar** from sugarcane, and **a ship**, which proves to be highly unstable, making only a single voyage
- In 1879, he is **knighted** for his contributions

Bessemer grew up in the small country village of Charlton in Hertfordshire, where his father owned a type foundry. Young Henry became fascinated by the cutting of dies and making of moulds that took place in the foundry, and grew up to become a skilled caster of decorative metal medallions. He utilized these skills during his first major project when he created a die for a new type of stamp for official deeds. The stamps cost £5, but to avoid paying this charge, people often peeled the stamps from old deeds and placed them on new ones – a fraud that was estimated to cost the exchequer some £100,000 a year. Bessemer's die, which punched 400 holes into the stamp in an uncopiable pattern, made them impossible to reuse.

A suitably grateful government offered him the post of "Superintendent of Stamps" at a grand salary of £700 a year. Bessemer then had a better idea, albeit one he later wished he had kept to himself. He argued that it would be even easier to stop forgeries if the date was printed on all the stamps. The government agreed, adopted his idea, and then told him they no longer needed him as their superintendent. He ultimately received nothing for either innovation. From that point on, Bessemer did his best to make sure his inventions were always properly rewarded.

BRIGHT IDEA

Bessemer built his fortune not from dies and moulds, but by making gold paint, otherwise known as "bronze powder", which was used to give objects a gilt-like finish. Decorative powders of this sort were available at the time, but were painstakingly created by hand and fiendishly expensive. In an early example of "reverse engineering", Bessemer worked out how the product was made, and then reconstituted his own version using finely powdered brass. He realized that if the process could be mechanized, vast profits awaited.

A proponent of the "pile 'em high, sell 'em cheap" approach to business, Bessemer developed steam-powered machinery to churn out the powder on an industrial scale, which he then sold for 1/40th of the price of the equivalent handmade alternative. The Victorians loved to give and receive gifts with gilt finishes – they bought the product in large quantities and Bessemer's coffers swelled.

GLITTERING SECRET

Bessemer kept his "bronze powder" recipe a secret by regulating its manufacturing process. His machines were made in separate sections in separate factories and then assembled by him. His employees were told only their own role and knew nothing of what anyone else did, and security at the factory was tight – only five people were ever allowed in the building.

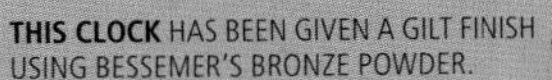

THIS CLOCK HAS BEEN GIVEN A GILT FINISH USING BESSEMER'S BRONZE POWDER.

When Bessemer created his bronze powder he did not take out a patent – which would have revealed his formula to potential copyists – but chose instead to control the manufacturing process, so as to keep the recipe secret (see box, below). So tight were his measures that he was able to hold on to the secret recipe for 35 years – much longer than the terms of a standard patent. Similar practices have been used since by other "secret formula" firms, such as Coca-Cola.

THE BESSEMER CONVERTER

Bessemer used the money he earned from his popular "bronze powder" to fund the invention for which he is now best known, the Bessemer converter – used for creating steel (see pp.230–31). He had become interested in steel production through his attempts to create a sturdy gun capable of firing a new type of heavy shell that he had devised for the British forces in the Crimean War in 1853.

The brittle cast iron available at the time simply was not up to the job. Hard-wearing, structurally strong steel would have been ideal, but steel was incredibly expensive to make because of the long hours and high temperatures involved, and only 25kg (55lb) could be made in one batch. At the time, steel was used only for small goods, such as knives and razors.

Having no great knowledge of steel-making, but all the zeal of an enthusiastic amateur with a large fortune, Bessemer threw himself into solving the problem. Steel was traditionally made via a long process that involved heating a crucible of molten iron in a furnace for three hours to burn away its impurities. But Bessemer came up with a device that greatly speeded up the process. His converter simply blew air directly into the molten pig iron, producing a high-intensity flame, which quickly burnt out most of the carbon and turned the brittle iron into tough steel. Suddenly, it was possible to produce large volumes of steel easily and inexpensively.

Bessemer patented his converter in 1855, and soon afterwards opened a steelworks in Sheffield to demonstrate its effectiveness. Able to under-sell all his competitors, Bessemer began receiving applications for licences from other manufacturers hoping to produce their own versions, with the result that the price of steel fell from £40 a ton to just £6 a ton.

Steel gave industry a huge boost, ushering in the "Second Industrial Revolution". The best converters could produce 27 tonnes (30 tons) of high-grade steel in half an hour. Once it became widely available, steel became a major material of infrastructure and was used to make everything from guns, railway tracks, and ships to bridges, engines, and buildings.

The Bessemer converter was particularly popular in the US where more than half a dozen towns were named after its inventor. Indeed, it had a huge influence on the look of US cities. Enabling the construction of strong, light, load-bearing steel frames for the first time, it was the device that made the skyscraper possible.

QUIRKY INVENTIONS

It may have been his most successful invention, but the converter was by no means the last of Bessemer's contributions to industry. The incorrigible tinkerer also created a hydraulic device for extracting more sugar from sugarcane, a steam-driven fan for ventilating mines, a furnace for making sheet glass, a process for compressing waste graphite powder into solid graphite so as to form lead pencils, and, in an attempt to cure his own sea-sickness, a ship – the SS *Bessemer,* which used hydraulics to keep the decks level no matter how choppy the seas were. Unfortunately, it proved highly unstable, demolishing a pier at Calais on its maiden (and only) voyage.

There were many other inventions, most of them unsuccessful, but the failures were more than compensated by the profits accrued from his bronze powder and steel maker. W M Lord said with regard to this success that "Sir Henry Bessemer was somewhat exceptional. He had developed his process from an idea to a practical reality in his own lifetime and he was sufficiently of a businessman to have profited by it. In so many cases, inventions were not developed quickly and the plums went to other persons than the inventors". In 1879, Bessemer was knighted for his contribution to science, and was made a fellow of the Royal Society. In 1895, he was elected a Foreign Honorary Member of the American Academy of Arts and Sciences. Bessemer died in 1898 in Denmark Hill in London.

STEADY AS SHE GOES
This 1875 diagram shows a cross-section of the inventor's ill-fated ship, the SS *Bessemer*, illustrating how hydraulics would keep its lavish saloon level, even in rough seas.

BREAKTHROUGH INVENTION

BESSEMER CONVERTER

It was the crowning glory of Henry Bessemer's inventing career: a large, egg-shaped furnace in which molten iron is turned into steel, known as the Bessemer Converter. Hot air is blown upwards through the metal, burning off impurities, including carbon, before other materials are added – including, intriguingly, carbon, though in much smaller proportions. The effect is to improve the alloy's strength and malleability.

By the mid-19th century, the deficiencies of cast iron, the main material of industrial progress at that time, were becoming all too apparent. Although strong and capable of bearing heavy loads, cast iron is brittle, particularly when under tension. Structural failures were commonplace: railway boilers exploded, train tracks cracked, support pillars gave way, and bridges occasionally collapsed – as happened with tragic consequences on the Ashtabula River in Ohio, US, in 1876 and the Tay Bridge in Scotland in 1879.

These problems were solved by the introduction of steel, cast iron's stronger and more malleable cousin. Both are alloys of iron and carbon; the key difference is the amount of carbon in each. If the carbon content is more than 2.1 per cent by weight, brittle cast iron is produced; if it is between 0.1 per cent and 1.5 per cent, the result is sturdy steel. Bessemer's invention enabled metal-workers to control carbon input, allowing the bulk production of cheap steel for the first time.

The converter fundamentally redrew the technological (and physical) landscape, but its discovery owed something to good fortune. No other mid-19th century metallurgist would have attempted what Bessemer did, since blowing air into molten pig iron might have resulted in a colossal explosion. The only reason Bessemer tried it was because he was not fully aware of what he was doing, being one of the Victorian gentlemen amateurs who dabbled in engineering. The iron, of course, did not explode – but Bessemer's bank balance did.

I HAD AN ADVANTAGE ... AS I HAD NO FIXED IDEAS ... AND DID NOT SUFFER FROM THE GENERAL BELIEF THAT WHATEVER IS, IS RIGHT

HENRY BESSEMER

STEEL FOUNDRY
Here, molten steel from a Bessemer converter is poured into moulds. The invention reduced the long process of making steel to a mere half hour, using far less coke (the fuel used to melt pig iron). Bessemer looked for an efficient method to produce steel after being told by an army officer that it was unsafe to fire a 14kg (30lb) shot from a 12-pounder cast-iron gun.

WILLIAM THOMSON

HERO OF ABSOLUTE ZERO

SCOTLAND 1824–1907

Principally remembered as a physicist – and the man who came up with the concept of absolute zero – William Thomson (later Lord Kelvin) was also a celebrated engineer. He achieved worldwide fame in the mid-19th century for his contributions to the laying of the transatlantic telegraph cable, a project that fostered in him a love for the sea that lasted a lifetime, and inspired a host of inventions.

A LIFE'S WORK

- Aged 22, William Thomson is made **Professor of Natural Philosophy** at Glasgow University
- Develops the **"mirror galvanometer"**, a highly sensitive detector of electrical current
- He is **noted for his work on the mariner's compass**, and the binnacle, for better navigation on iron ships
- **Publishes more than 600 scientific papers and applies for 70 patents** (not all are issued)
- **Sails onboard the cable-laying ship HMS *Agamemnon*** in August 1857, but the voyage ends after 610km (380 miles) when the cable breaks
- **Invents a super-sensitive mirror galvanometer** to detect signals, and a siphon recorder to record them, and **succeeds in laying a long-lived transatlantic cable** from the SS *Great Eastern*
- **Is knighted** on 10 November 1866

Born in Belfast, Thomson moved to Glasgow at the age of eight. He was enrolled in the university's elementary school and began publishing articles on applied mathematics while still a teenager. By 1846 he was made Professor of Natural Philosophy at Glasgow University, a position he would hold for more than half a century.

Thomson then dedicated himself to the study of physics, elaborating on the Second Law of Thermodynamics – which states that in any closed system, everything will eventually reach the same temperature – and proposing a temperature scale, which was named after him. The Kelvin scale uses the same units as the Celcius scale, but it starts at absolute zero, the temperature at which everything, including air, freezes solid.

MESSAGES OVER THE ATLANTIC

Thomson remained a brilliant, but fairly obscure, scientist until the mid-1850s. What brought him to international attention was his involvement in the laying of the first transatlantic cables. His scientific expertise and great problem-solving ability were considered just what this hugely ambitious and technically demanding project required, and, in 1856, he was asked to join the board of directors of the Atlantic Telegraph Company to lay the first permanent communications link between the US and Britain.

The logistical problems were immense. Connecting the two continents would require 3,500km (2,200 miles) of cable weighing 4,500 tonnes (5,000 tons). Since no ship at the time was big enough to carry such a cargo, the cable had to be split between two ships and then spliced together mid-ocean. Furthermore, no one had yet worked out how to send and receive a clear signal over such a long distance. The longer the cable, the weaker the signal would become, making it very difficult to detect. Both Edward "Wildman" Whitehouse, the chief electrician of the Atlantic Telegraph Company, and Thomson came up with possible solutions. The trouble was, they proposed exactly opposite ideas, and one of them was dangerous. Unfortunately, the company initially opted for the wrong solution – Whitehouse's – almost scuppering the project.

THE TIDE PREDICTOR
Thomson invented the first tide-predicting machine in 1872. A sailor would turn the handle of the contraption, which was loaded with oceanographic and astronomical data, to get tide predictions for up to a year ahead in any particular port.

Whitehouse's idea was to boost the voltage to overcome the cable's resistance, effectively creating a bigger, more easily detectable signal. Unfortunately, the extra current might well damage the cable. Thomson, meanwhile, approached the problem from the opposite end. Rather than boosting the signal, he thought the solution was to improve the sensitivity of the device detecting the signals. And this is exactly what he did, developing an instrument called the mirror galvanometer, which could pick up even the faintest of signals.

Inspiration for the device allegedly came to Thomson as he stood idly by his desk, twirling his monocle. He noted that the glass was reflecting a spot of sunlight around the room. He thought something similar could be made to work for the telegraph, and so created a highly sensitive contraption that used a mirror glued to a magnet suspended on a fine wire inside a coil of wire. The slightest current in the wire caused the magnet to twitch. A beam of light reflected from the mirror would move right across a screen for each twitch. The mirror galvanometer was able to detect signals a thousand times weaker than previous instruments.

A BIGGER SHIP

Thomson made two journeys aboard the British cable-laying ship HMS *Agamemnon*. The first had to be aborted when the cable broke, but the second, in 1858, was more successful. The triumph, however, proved to be short-lived. Whitehouse's insistence on using a

CABLE-LAYER
This shows the cable-laying machinery installed on the SS *Great Eastern* in the 1860s. The biggest ship in the world, it could carry the entire transatlantic cable (unlike the smaller ships used in previous expeditions), which made the laying process much more straightforward.

high current to transmit the signals soon caused the cable to fail, as Thomson had predicted it would. The whole thing would have to be done again. This time, however, the company would take Thomson's advice. Thomson sailed on the two expeditions of the SS *Great Eastern* – then the world's biggest ship (see pp.194–97) – which finally resulted in the establishment of a successful, continuous, transatlantic telegraph service in 1866.

In addition to the mirror galvanometer, Thomson made several other telling contributions to the project, including the invention of a siphon recorder (a device for recording the receipt of a telegraph message) and insisting that the cable-laying company use the highest-quality copper possible, so as to increase the strength of the signal.

The success of the project brought great public acclaim to those involved, and Thomson was knighted in 1866 for his contribution. His involvement in long submarine telegraph systems was far from over, however. He took part in cable-laying enterprises for various countries over the next couple of decades and even met his second wife during a stop-off on Madeira.

These experiences fostered in Thomson a great love for the sea, and many of his later inventions were of a nautical nature. He devised a new method for deep sounding – finding out how deep the water was below a ship – by using a piano wire, which allowed the measurements to be taken while the ship was going at full speed (known as flying sounding). He also created a tide-predicting machine and a new type of mariner's compass adapted to overcome inaccuracies created by iron ships' own magnetism.

He developed much of the theoretical work that would lead to the invention of refrigeration, and also contributed towards an understanding of the age of the Earth. His studies on thermodynamics had led him to surmise correctly that the Earth must have once been much hotter than it currently was and that it must have taken several million years to have cooled down. He eventually came up with an age for the Earth to be between 100 million and 400 million years, but later revised this down to just 20–40 million years.

Despite his many great breakthroughs, Thomson became less willing to embrace new scientific developments as he grew older. He disagreed with Darwin's theory of evolution, not because of his religious convictions, but because he believed that the Earth simply was not old enough for it to be possible. He also scorned the possibility of powered flight, famously declaring, "Heavier-than-air flying machines are impossible".

President of the Royal Society from 1890 to 1895, Thomson was made Baron Kelvin in 1892, taking the name from the river running through the grounds of his beloved Glasgow University. His title died with him in 1907, whereupon he was buried in Westminster Abbey next to that other great epoch-defining scientist, Sir Isaac Newton (1642–1727).

WERNER VON SIEMENS

THE MASTER OF "ELEKTROTECHNIK"

GERMAN CONFEDERATION 1816–92

Industrialist and electrical engineer Werner von Siemens rose to prominence in a period that saw major changes in Germany. Between the 1790s and the 1850s, feudalism was abolished, commercial laws were improved, new technologies were introduced, and railways were developed. These changes enabled German states to shake off their social conservatism, unify, and industrialize.

The process of German industrialization began when Siemens was a young man, looking to make his way in the world. In fact, Siemens's rise mirrored – and in part contributed to – the rise of his country. By the time of his death, Germany had emerged as one of the world's leading industrial nations, while Siemens had built his company into a global engineering behemoth.

One of 14 siblings born into a farming family in Lenthe near Hanover, Siemens received only basic schooling. Most of his education was provided by the army, which he joined aged 18, and where he developed an aptitude for engineering, studying at the artillery and engineering school of the Military Academy in Berlin. His skills helped him become a supervisor of the artillery workshops where his inventing career began. While still in the army, he used a cigar box and some copper wire to create his first great invention, the pointer telegraph. Designed to help transmit messages more reliably over long distances, it used a needle to point out letters rather than Morse code. It proved popular, and in 1847 Siemens, along with his partner, mechanic Johann Georg Halske (1814–90), formed a telegraph-manufacturing company, Telegraphen-Bauanstalt von Siemens & Halske.

With his reputation growing, Siemens was given the task of laying Germany's – and in fact Europe's – first major telegraph line, covering the 500-km (300-mile) distance between Berlin and Frankfurt am Main in 1848.

A LIFE'S WORK

- Siemens **invents the pointer telegraph** in the 1840s and **founds the company** that still bears his name
- His brother **Carl opens the company's St Petersburg office**, while his brother **Wilhelm opens an office in London** giving the company a **pan-European reach**
- He pioneers electrical engineering in the 1870s and 1880s, creating the **first electric railway**, the **first electric elevator**, and the **first electric street car**
- Introduces an **employee pension scheme** and a profit-sharing model for his employees
- Siemens continues to be **one of the world's largest engineering giants** with **revenues of more than $70 billion** and operations in nearly 90 countries

A FAMILY CONCERN

Siemens left the army in 1849 to concentrate full-time on the business, and soon received an order from Russia to build a telegraph linking up the Baltic Sea and the Black Sea. Werner's brother Carl (1829–1906) was sent to oversee the project, while another brother, Wilhelm (1823–83), travelled to England to open an office in London and increase business, which he duly did, successfully pitching for contracts to lay submarine telegraph cables. Within a

TAKING THE LIFT
At the 1880 Mannheim Trade Exhibition, Siemens demonstrated the world's first electric elevator (shown here), which utilized his electric dynamo, taking visitors up to the top of a temporary viewing tower.

TECHNOLOGY NOW HAS THE MEANS TO GENERATE ELECTRICAL CURRENT OF UNLIMITED STRENGTH

WERNER VON SIEMENS

few years, the company had expanded from a small-time local operation to a major international concern, which was rare in the mid-19th century. However, Siemens viewed this development not so much as extending the influence of his company as that of his family. He regarded the Fuggers, the German family of financiers who took over from the Medici as Europe's most influential bankers in the 16th century, as the template for his business, appointing his relatives to many senior positions. Even as his company grew into a huge industrial giant, he continued to see it primarily as a family enterprise, almost like an heirloom to be passed down the generations.

DYNAMO-ELECTRIC PRINCIPLE

Siemens's interests were not limited to the field of telegraphy, and extended, in particular, to electrical engineering or *Elektrotechnik* (the word was coined by Siemens himself). In 1866, building on the work of physicist Michael Faraday (see pp.140–41), he discovered the dynamo-electric principle, which suggested the possibilities of using electricity as a source of industrial power for the first time. This discovery was also made independently by scientists Samuel Alfred Varley (1832–1921) and Charles Wheatstone (1802–75) at about the same time. Siemens's and Wheatstone's breakthroughs were made public on the same day, 17 January 1867. At the meeting of the Berlin Academy, Siemens displayed his latest invention, a motor known as a "dynamo-electric machine", which used electromagnets to vastly increase power output. The motor paved the way for the development of power engineering and the vast extension of the applications of electrical power.

Siemens's most prolific period came at the turn of the 1880s, when, in swift succession, he created the first electric streetlights for the Kaisergalerie in Berlin; the first electric railway at the Berlin Trade Fair; the first electric elevator in Mannheim; and the world's first electric street car, the Elektromote, in Berlin. But apart from being one of Germany's leading industrialists, Siemens was an enlightened employer for his time, believing that treating his employees well was the best way to ensure their loyalty. He introduced a kind of profit-sharing model and set up an employee pension scheme more than a decade before it became a statutory requirement.

Siemens devoted the latter part of his life to encouraging the teaching of electrical engineering in Germany. He helped establish Germany's Physical and Technical Institute in Berlin in 1887, contributing both land and money for it. He was ennobled in 1888, becoming Werner von Siemens, and retired from business two years later, passing it on to his relatives, in accordance with his family-oriented plan. However, he continued to take a keen interest in the firm until his death in 1892. Today, Siemens AG is still the largest engineering conglomerate in Europe.

ALFRED NOBEL

SWEDEN 1833–96

The son of the inventor of plywood, Alfred Nobel achieved worldwide renown for inventing the powerful explosive dynamite.

Invented in 1867 as a way of making the compound of nitroglycerine safe to use, dynamite was a huge commercial success. Nobel followed it up in 1876 with an even more powerful explosive, gelignite. Together, the inventions made him extremely rich. In 1889, however, he got a glimpse of how he would be remembered after his death when a newspaper inadvertently published his obituary instead of that of his brother, calling him a "merchant of death". So, in order to be remembered in a positive way, he left most of his wealth to establish a number of prizes, for Physics, Chemistry, Medicine, Literature, and the most prestigious of all, Peace, awarded to whoever prevented the most explosions that year.

CHARLES VAN DEPOELE

THE ELECTRIC-TRAM MAN

BELGIUM 1846–92

ALTHOUGH NOT THE ORIGINATOR of electrically powered transport, Charles Van Depoele was one of its most important pioneers, particularly in North America where his trolley car systems got the public moving around cities and towns across the continent. Unfortunately, he died before seeing the age of electric-tram transport, which his inventions had helped bring about.

A LIFE'S WORK

- Van Depoele creates a **Bunsen battery with 40 cells** at the age of 15
- Emigrates to the US, where, after an early career as a furniture-maker, he becomes a **full-time electrical engineer**
- Forms his own business, **Van Depoele Electric Manufacturing Company**
- Creates an **electric railway for Chicago** in 1883
- He creates a system in 1887 that makes **Montgomery**, Alabama, the first city in the world to have a **citywide electric transport system**
- **Sells his company to Thomson-Houston**, which later merges with Edison General Electric to form the **General Electric Company**, the world's leading electric tram company
- **Files for more than 200 patents** related to lighting and transport, as well as a coal-mining machine

Van Depoele's father was the master mechanic of the telegraph system of the East Flanders railway near the small town of Litchervelde, northwest of Brussels. As a child, Van Depoele was deeply interested in his father's work, tinkering with his tools and instruments, and at the age of 15, he even made a Bunsen battery, a battery using a carbon electrode, which consisted of 40 cells. His father, however, discouraged such activities, seeing no future for the boy in railways, and instead had his son apprenticed as a furniture-maker. Adept with his hands and possessing a sharp, practical mind, Van Depoele excelled at the profession.

In 1864, the family moved to Lille, France. There, the 18-year-old Van Depoele was employed by the renowned local sculptor Charles Rigot-Busine to create altar pieces and religious statuary for local churches, although he also attended lectures at the Imperial Lyceum. A few years later, deciding that his ambitions would be best served in another country, Van Depoele moved to the US, settling in Detroit in 1869. He started his own church furniture business, which was a success and funded his further forays into electrical engineering. He began experimenting with arc lamps, and, following some advertising of his products around the city, received his first professional commission in 1878. He was asked to install his new lights for a visiting circus. The illuminations showcased his talents, and, within a short period of time, he was employed to provide lights for the Detroit Opera House and excursion boats on the Detroit River.

ELECTRIC AVENUES

Now that his business was established, Van Depoele became interested in electrically powered transport. He conducted extensive research into the field in 1880, just a year after Werner von Siemens had demonstrated the first electric railway at the Berlin Trade Fair (see pp.234–35). Convinced that it was possible to build citywide transport systems powered by electricity, he moved to Chicago, forming the Van Depoele Electric Manufacturing Company. He had trouble finding financial backers, but persevered, demonstrating an electric streetcar of his own design at the Chicago Inter-State Fair of 1883. New, improved models were also shown at the next two Toronto Annual Expositions, by which time trolleys could travel at up to 50kph (30mph).

His first commission to construct a commercial line came from South Bend, Indiana, in 1885. It used adapted motorized horse cars, and was soon followed by similar small systems in Scranton in Pennsylvania, Appleton in Wisconsin, and St Catherine's in Ontario, Canada. The following year saw Van Depoele tackle his most ambitious project yet, with the Capital City Street Railway Company of Montgomery, Alabama, employing him to electrify its entire horse line, making Montgomery the first city in the world to have a citywide electric transport system. A glut of orders followed, and, within a few years,

> HE IS ENTITLED TO MORE CREDIT THAN ANY OTHER ONE MAN FOR THE EXPLOITATION OF ELECTRICITY AS A MOTIVE POWER
>
> **GEORGE HERBERT STOCKBRIDGE**, POET

Van Depoele's electric-trolley cars were trundling backwards and forwards (and up and down – they could cope with gradients of up to 9 degrees) across the streets of US and Canadian towns, including Dayton, Detroit, and Washington, DC.

Van Depoele designed the entire systems – from the cars and motors to the rails and overhead electricity lines – which delivered power via spring-loaded devices that pushed up from the car against the wire and were known as "Van Depoele Poles".

SELLING UP

Although delighted to have found a market for his work, Van Depoele lacked the funding to fulfil the orders that were now coming his way. So, in 1887, he sold his company to Thomson-Houston Electric Company of Lynn, Massachusetts. Now holding Van Depoele's patents and with the capital to invest in further expansion, the company became one of the dominant forces in the electric tramway industry, installing systems throughout the US and beyond. These systems were particularly popular in France, Van Depoele's former home. The following year, Van Depoele moved to Lynn, where he was employed by Thomas-Houston as the company's inventor-in-chief. In 1892, Thomson-Houston merged with Edison General Electric to form the General Electric Company. Edison held the patents to a rival electric tram system developed by Frank J Sprague (1857–1934), whose trolley cars served more than 100 cities around the globe. The company remained the electric tramway industry's dominant player for decades to come.

In his lifetime, Van Depoele filed more than 200 patents. Many were, of course, related to lighting and transport (in addition to trolley cars, Van Depoele was also involved in the development of aerial tramways known as telpher systems, and underground electric railways), but he also patented designs for hammers and drills, as well as a coal-mining machine.

Unfortunately, Van Depoele did not live to see the great era of electric tram travel that he had helped to introduce, dying of heart failure in 1892 at the age of just 46.

THE ELECTRIC TRAM

Various inventors contributed to the development of the electric tram. Werner von Siemens is widely regarded as having developed it first, exhibiting an electric passenger train at the Berlin Trade Fair of 1879. In the US, its development is credited to both Charles Van Depoele and Frank J Sprague. British engineer Magnus Volk's efforts (1851–1937) are also noteworthy: the line he laid out at Brighton in 1883 is the world's oldest operating electric tramway.

THIS LATE-19TH CENTURY PICTURE SHOWS A CAR ON VOLK'S RAILWAY TRUNDLING OVER BRIGHTON'S ROUGH SEAS.

CROSS-TOWN TRAFFIC
In this picture from 1911, electric trolley cars travel alongside horse-drawn carriages in downtown Chicago. Charles Van Depoele installed his first electric railway in the city in 1883.

JOSEPH ENGELBERGER

THE FATHER OF ROBOTICS

UNITED STATES B.1925

A PHYSICIST, ENGINEER, and entrepreneur, Joseph Engelberger held a lifelong dream of using robots to help humankind. With his flair for business, he pioneered a company whose innovations inspired the growth of industrial robots and reshaped production lines around the world, earning him the title of "Father of Robotics".

A LIFE'S WORK

- Engelberger meets George Devol in 1956 and forms a lifelong partnership; **they found Unimation Inc. and work on their first robot, the Unimate**
- In 1961, **an industrial robot is installed for the first time on an assembly line**; it lifts and stacks hot castings from moulds at General Motors
- A **robot appears for the first time on public TV**, giving a star turn on *The Tonight Show* starring Jonny Carson, **demonstrating golfing and music skills**
- The **Robotics industry honours Engelberger by creating the Joseph Engelberger Award** for outstanding achievement in robotics in 1977; he writes a key book on robots applauded by roboticists, and devotes his life to the service industry
- He **forms a new company in 1984, to build HelpMate robots**, which navigate hospital corridors while loaded with supplies

Born in Brooklyn, New York, Engelberger's family moved when he was a young boy to Connecticut, where they endured the Great Depression of the early 1930s. As a teenager, Engelberger escaped by immersing himself in science fiction. His favourite author was Isaac Asimov, who wrote stories about robots, but Engelberger did not simply want to read about robots, he wanted to build them.

AN ENTERPRISING IDEA

When the draft age for American men was lowered to 18 in 1942 in response to the deepening crisis of World War II, the US Navy anticipated a shortage of college-educated officers for its operations, and the government stepped in to create the V-12 Navy College Training Programme. Engelberger was enrolled in the accelerated programme at Columbia University in 1945, attending the first course ever given on servo theory, whereby the function of systems is corrected through error-sensing. He was following in the footsteps of his hero, Asimov, who had also studied there. After graduating in 1946, he served briefly in the navy.

The US Navy was at the cutting edge of technology. Engelberger worked on aerospace and nuclear power projects, including early nuclear tests held at Bikini Atoll in the Pacific, where the handling of radioactive material spurred the development of automated systems.

After completing his military service, Engelberger returned to Columbia to complete a Master's degree in physics and electrical engineering. He found that physicists were in demand and fell into a job with Manning, Maxwell, and Moore (MM&M) in Connecticut, developing controls for jet engines and nuclear power plants. He soon became chief of engineering. In 1956, during his time at MM&M, a fortuitous meeting took place at a cocktail party. George Devol (1912–2011), a talented inventor, had been working on introducing automation in industry, and was looking for backers. He had applied for a patent for his controlled machine-tool and explained the concept to Engelberger. They discovered a shared vision: bringing robots to life.

THE UNIMATE

George Devol developed a robot called Unimate alongside Engelberger. The machine undertook the job of transporting die castings from an assembly line and welding these parts onto auto bodies, a dangerous task for workers, who could be poisoned by exhaust gas or lose a limb if they were not careful. The original Unimate consisted of a large computer-like box joined to another box, and was connected to an arm, with systematic tasks stored in a drum memory.

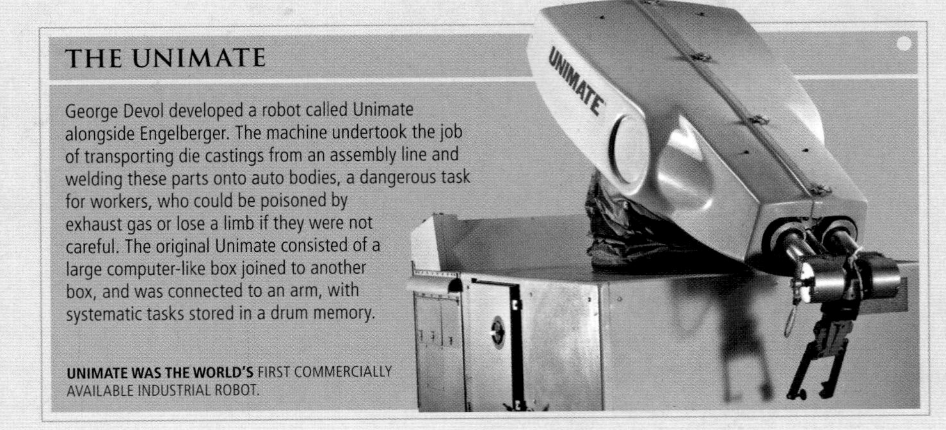

UNIMATE WAS THE WORLD'S FIRST COMMERCIALLY AVAILABLE INDUSTRIAL ROBOT.

"ROBOTIC" SUCCESS

It was the beginning of a lifelong partnership. Engelberger and Devol founded "Unimation Inc", which stood for "universal automation", in Danbury, Connecticut. The pair visited 15 local companies to establish whether their idea could work, and came up with their first design – the Unimate. The invention was a type of extending mechanical arm that could be instructed to carry out a series of repetitive commands stored on a magnetic drum.

In 1961, General Motors put the first Unimate on an assembly line at the company's plant in New Jersey. The robotic arm was to remove hot castings from moulds. Most welcomed the new robot as it tackled jobs that were dirty, dull, and dangerous. Chrysler and Ford followed, in the face of resistance from labour unions worried about men losing their jobs. Business trickled in, and by 1966 other Unimates were in production, which could spray-paint and spot-weld.

The general public paid scant attention to the innovation. Engelberger also found it difficult to convince US industrialists that Unimates were a good investment. By the mid 1960s, Unimation had developed a more human-looking robot called PUMA. Engelberger, desperate for publicity, agreed to appear on the popular *The Tonight Show* starring Johnny Carson. The robot, weighing 1,200kg (2,700lb), was rehearsed for weeks, and performed three acts – putting a golf ball, taking part in a beer commercial, and conducting the band. The audience loved it.

Despite this coup, American industry took a long time to catch on. Engelberger worried that his product seemed like a novelty toy, such was its demand at fairs. General Motors ordered an additional 36 robots, which generated wide publicity, but it was not what Engelberger had hoped. However, the publicity caught the attention of Japanese industrialists, and as a result, the government of Japan invited Engelberger to give a lecture on his work. At the time, Japan enjoyed full employment, but was reluctant to bring in foreign workers to ease the labour shortage. They saw in Unimate the answer to their problems. Engelberger was startled to find an eager audience of more than 700 delegates, and was overwhelmed by five hours of questions from the floor.

Japanese industry embraced robotics while the rest of the world watched. Unimation finally made a profit in 1975 after 19 years, but a sudden surge in competition meant the field became crowded. Engelberger and Devol had built the world's largest robotics factory, but after 30 years decided that their enterprise was over. They sold the company in 1983 for $107 million.

However, Engelberger's passion for robots did not flag. He wrote *Robotics in Service*, describing the opportunities for robotics to enter the service industries. In 1984, he formed a new company to build HelpMate robots that would navigate hospital corridors while loaded with supplies. It was the first successful service robot in healthcare. Lauded as the "Father of Robotics", Engelberger won many plaudits. In 1977, the Robotics Industry Association established the Joseph Engelberger Award for outstanding achievement. He continued to push the concept of robots in space, deep sea exploration, and healthcare. The world's first industrial robot, Unimate 001, lasted 100,000 hours before retiring to the Smithsonian Institution in Washington, DC.

YOU END UP WITH A TREMENDOUS RESPECT FOR A HUMAN BEING, IF YOU ARE A ROBOTICIST

JOSEPH ENGELBERGER, ***ROBOTICS AGE***

HOSPITAL HELP
Engelberger's HelpMate was designed to be a robotic courier, used in hospitals to carry medical records, lab samples, medicine, equipment, and even meals, avoiding fixed or moving obstacles as it travelled.

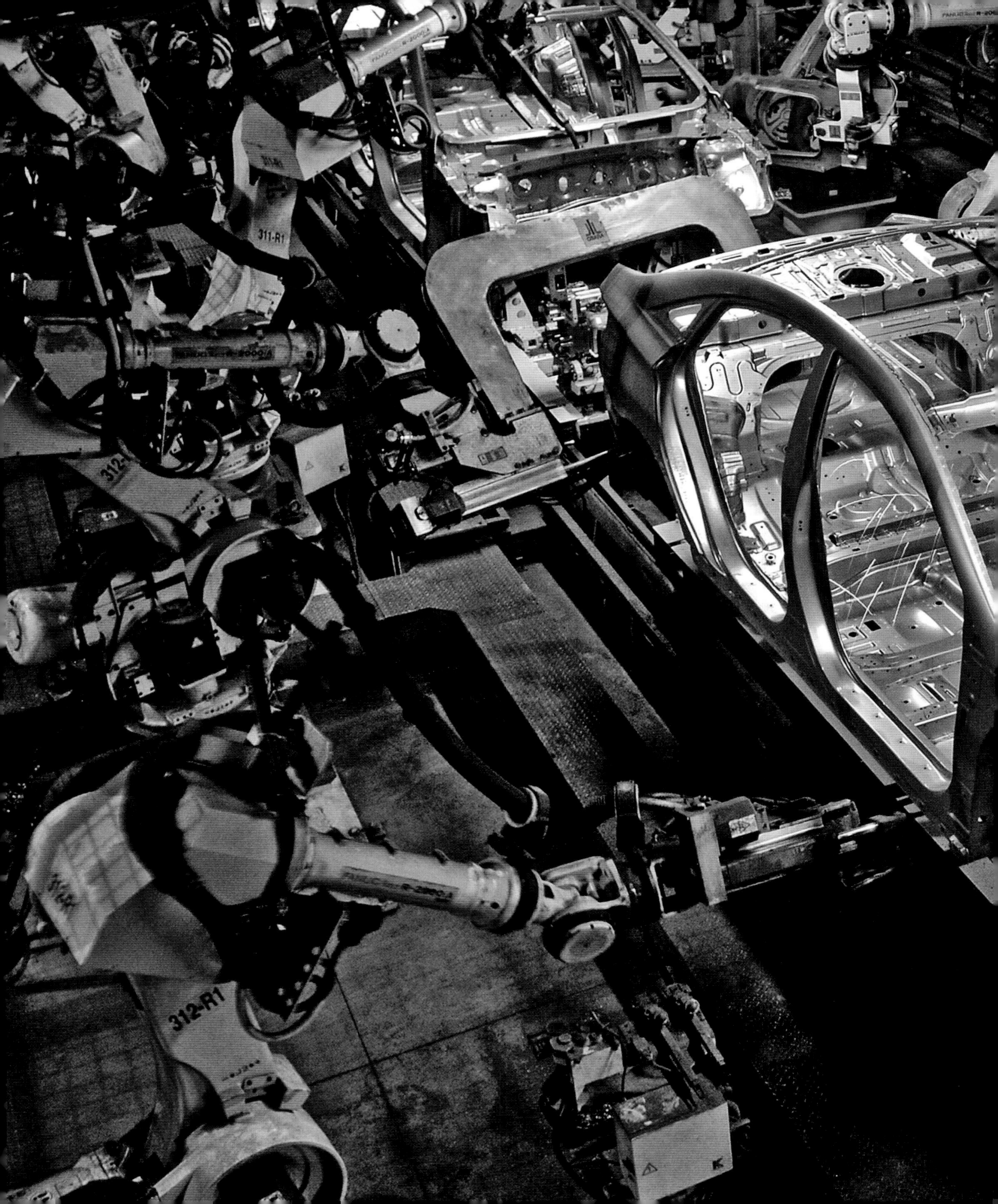
311-R1
312-R1

ROBOT WORKFORCE
Here, rows of articulated welding robots are busy on the assembly line of the Hyundai car plant in Alabama. They comprise jointed arms that can be programmed to perform dangerous, repetitive functions with accuracy. Other types of industrial robot are more sophisticated and can identify objects and tasks, with computers acting as their "eyes". Such basic artificial intelligence is becoming increasingly important in manufacturing.

GREAT CONSTRUCTIONS

THE LATE 19TH CENTURY WAS A TIME OF IRON AND STEEL, AS THESE WERE THE MATERIALS USED TO MAKE BRIDGES, BUILDINGS, AND RAILWAYS. BUT THERE WAS ANOTHER DEVELOPMENT UNDERWAY THAT REVOLUTIONIZED THE BUILDING INDUSTRY, AND IT STARTED WITH A CONCRETE FLOWER POT.

TOWER BRIDGE
Reinforced concrete was essential in building this London landmark. The building material can bear more weight, which allows structures such as these to be built taller than before.

Concrete is a composite material made up of several substances bound together by cement. When concrete hardens into stone, it can bear weight and be used as a building material. The ancient Egyptians are credited with creating a substance that resembles modern concrete. They used lime (known as calcium carbonate and usually derived from limestone) and gypsum as cement. The ancient Romans and Greeks built concrete structures as well, although by the Middle Ages most buildings were constructed from stone and wood. By the modern era, some man-made materials, such as bricks, were being mass produced, but concrete was seldom used.

There were very few advances in concrete technology until the 18th century. In 1755, John Smeaton required a cement that could bond under water as he needed to secure the foundation of the Eddystone Lighthouse (see pp.100–01) that sat on a rock outcrop 22km (14 miles) off the coast of Plymouth, England. He used Aberthaw

CONCRETE SHIP
Shipbuilders work on a steel mesh in preparation for laying reinforced concrete. The idea was to make ships from a material more durable than wood, but this was solved by the emergence of ironclad ships.

TAY BRIDGE
The strength of William Arrol's Tay Bridge also came from the use of reinforced concrete. Once engineers realized what the material was capable of, its use in public works proliferated.

FRANÇOIS HENNEBIQUE
This Frenchman developed construction methods based on the use of reinforced concrete that revolutionized architecture and public works projects.

SETTING THE SCENE

- In the beginning of the 19th century, developments are being made in the **materials used in cement**. The invention of Portland cement provided builders with stronger mortar and tougher **concrete**.
- Joseph Monier makes **strong flower pots with concrete and iron mesh**. This idea is adapted to larger slabs of concrete, and soon the technique is applied to large projects.
- This concrete excites people in civil engineering because of its ability to **bear more weight** and is soon put into use in buildings and bridges.
- François Hennebique **franchises** his technique of **building with reinforced concrete**. By 1911, about 24,000 new concrete buildings in the world used his methods and designs.
- Civil engineers everywhere are soon making **bridges and skyscrapers** with innovative new designs, thanks to the strong reinforced concrete.

limestone from Wales, which had a high clay content and so could be used as a waterproof cement, and he completed his lighthouse to much acclaim. The next significant development was the creation of "Portland cement" in 1824. Joseph Aspdin (1778–1855) experimented with cement-making by heating a mixture of chalk and clay to a high temperature, producing a much stronger material.

But the real leap forward for concrete did not come until it was merged with iron. Like many other advances in construction, the discovery of this new composite material came about through practical problem solving – in this case, the need to make sturdier flower pots.

INGENIOUS FLOWER POTS

French gardener Joseph Monier (see pp.244–45) needed stronger tubs to hold his plants as the clay and wooden containers he used would break or rot. In 1849, he decided to experiment with putting iron rods and mesh into a cement mixture. The results were promising – this new material seemed capable of bearing more weight. Monier patented the tubs in 1867. He was not, however, the first person to strike upon this idea. Joseph Louis Lambot (1814–87) had built a *ferrocement* boat, which was made out of cement and two layers of iron bars and a mesh of wire. This dinghy was displayed at the International Exhibition in Paris in 1855. A few years earlier, industrialist François Coignet (1814–88) had built a four-storey house of reinforced concrete in Paris. Although the *béton armé* house was intended to be a publicity stunt to promote Coignet's cement business, he takes the credit for building the world's first structure with reinforced concrete.

> **The 18th century saw innovations for strengthening concrete. Initiated by the need to strengthen flower pots, reinforced concrete was developed and soon used to build skyscrapers, bridges, and stadia.**

SOARING SKYLINES

What made this composite material different from regular concrete was that the use of rods and mesh allowed it to withstand wind, vibrations in the earth, and, most importantly, bear more weight. Civil engineers were quick to see the benefits, and it paved the way for skyscrapers.

Although much of the initial development happened in France, the use of reinforced concrete soon became widespread thanks to the efforts of François Hennebique (1842–1921). He began to experiment with reinforced concrete and realized that rods near the supports on concrete floors needed to be bent upward in order to absorb more stress. He created his own building system based on these T-beams, and, by 1892, he had closed his existing business, become a consultant, and developed an innovative marketing strategy. He would sell a licence for about 10 per cent of the project budget, and for that money, he would give contractors a set of drawings. These contractors were part of his network of agents, and he ran special training courses for them and his engineers. He also published a magazine called *Le Béton Armé*. The results were impressive – by 1909, Hennebique had 62 offices worldwide, and, by 1911, about 24,000 buildings in the world had used his system.

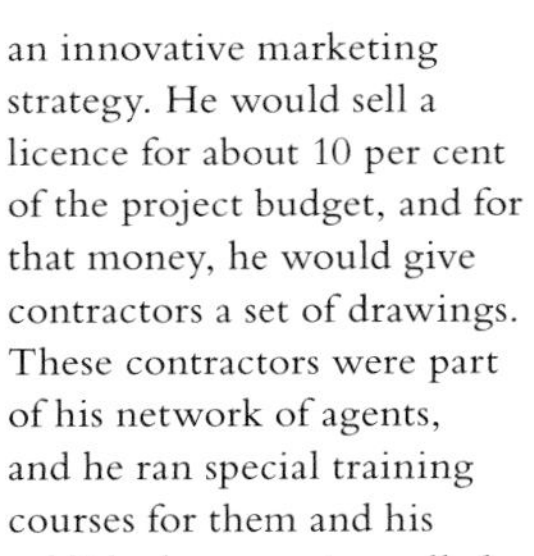

REINFORCED CONCRETE
This cross-section of reinforced concrete shows the metal mesh that is the key to the material's strength and durability.

In 1909, Hennebique's agent Louis Mouchel used reinforced concrete to build the first skyscraper in Europe – the Royal Liver Building in Liverpool. It reached the then-dizzying heights of 95m (311ft). In the US, Ernest Ransome (1852–1917) made improvements by using twisted rods for reinforcement. He is best-known for the 54-m (177-ft) tall, 16-storey Ingalls Building in Cincinnati, Ohio.

During the 20th century, multi-storey, high-rise buildings became a part of cityscapes all over the world, with few people, if any, walking in the shadows of skyscrapers realizing that the genesis of these buildings was a concrete boat and some flower pots.

MONIER AND FREYTAG

PIONEERS OF REINFORCED CONCRETE

FRANCE 1823–1906
GERMANY 1846–1921

JOSEPH MONIER

CONRAD FREYTAG

ALTHOUGH JOSEPH MONIER, a Parisian gardener, is credited with being the first to patent reinforced concrete, it took the prowess of Conrad Freytag, a German concrete supplier, to put this material to practical and commercial use, thereby transforming European civil engineering.

Monier began with a modest ambition – to make sturdier flower pots for his Parisian landscaping business. He wanted to put his plants in large tubs, but he found the existing materials of clay or wood to be too weak. Clay was easily broken and wood weathered badly, and the tubs could be broken by plant roots. So he began to experiment with cement, trying a new technique – using iron bars and a thin iron mesh to make the structure stronger.

Monier was not the first person to explore the possibilities of metal and concrete (see pp.242–43), but his results were impressive. He saw great potential in the technique, and promoted it extensively. The important point of Monier's idea was that concrete is strong in compression, but weak in tension; the iron provided the composite material with tensile strength. He started producing concrete garden tubs, acquiring a patent for his "system for portable iron and cement tubs for use in horticulture".

THE BRILLIANCE OF CONCRETE

What made this building material so revolutionary was its strength. The addition of iron allowed the tubs to hold much more weight and still maintain structural unity. Monier soon began to apply this technique to the construction of water towers (see box, right). Indeed, earlier developments in the composition of concrete and cement (which is used to bind the particles that make up concrete) meant that he could use the material to build structures capable of storing liquids. By 1873, Monier had obtained further patents for the use of concrete in arched bridges and beams, after realizing that the material could withstand heavy pressure. He put his new ideas into practice in 1875, building the world's first bridge from reinforced concrete. Constructed on the estate of the Marquis de Tilière in Chazelet, central France, the bridge spanned a castle moat, measuring 16.5m (54ft). He used concrete beams that contained thin iron rods. Although Monier had called himself a *rocailleur en ciment* ("sculptor in cement") it was clear that there were more practical implications for his work. Companies in Belgium, Holland, Austria, and Germany soon heard about his work and began to enquire about obtaining licences to his patents.

FREYTAG AND WAYSS
This is the headquarters of the Freytag and Wayss Company in Frankfurt. By the 1890s, Conrad Freytag had built more than 300 bridges using the Monier system.

MEETING OF MINDS

The involvement of German engineers became crucial in the application of Monier's ideas, but the person who did more than any other did not even have a scientific education. Conrad Freytag was the owner of a building materials business, Freytag & Heidschuch, which specialized in concrete for all kinds of constructional projects. In 1884, on a trip to Trier, in southwest Germany, he noticed a new water tower under construction on the Bahnhofstrasse. Struck by the design and materials, Freytag was told that it was being built according to Monier's technical specifications. Freytag immediately saw the potential of this new building system, and was soon on the train to Paris. In September 1884, he visited Monier and obtained rights to the patent that would allow him to build around most of southern Germany with the option for the rights for the rest of the country. Soon his firm was involved in the construction of water tanks in the area.

A LIFE'S WORK

- Monier needs stronger material to hold large tubs of flowers and other plants, so he **experiments with mixing iron and concrete**
- He begins to patent more materials and **builds the world's first concrete bridge** in 1875
- On a business trip in Germany, Freytag **sees a water tank being built and is impressed by its construction**; he is told it is Monier's design and immediately goes to Paris to see the patent; once back in Germany, he builds a dog kennel to test the design himself
- Along with his partner Wayss, Freytag uses Monier's reinforced concrete to **build more than 300 bridges by 1899**; they also later publish a manual on how to use the Monier technique, ***Das System Monier***
- Freytag and Wayss are forced to **compete against Hennebique's growing network of agents** to win contracts in Germany
- Monier is **unable to achieve the success he hoped for** and dies in obscurity and poverty on 13 March 1906

Another German engineer, Gustav Adolf Wayss (1851–1917), who later claimed he "discovered" what was called the Monier system, met Monier at an 1885 industrial exhibition in Antwerp. He conducted further research in the use of reinforced concrete as a building material, and wasted no time in obtaining the necessary licences to use Monier's techniques, approaching Freytag for the rights to north Germany. Freytag had so much work in southern Germany that he gave Wayss the rights for free. Soon Wayss and Freytag's firms began to design load-bearing and fireproofing tests on the materials in order to persuade the authorities that reinforced concrete was safe to use for buildings. Also around this time, architect Matthias Koenen (1849–1924) became involved with Wayss and Freytag. After they compiled the results from the testing of the materials, Koenen wrote *Das System Monier*, which explained how to build with reinforced concrete. Wayss, who was eager for the building material to gain approval for use in civic projects, had this *Monier-Broschüre* published in 1887 and the 10,000 copies were circulated around Europe. The publication was not limited to discussing buildings, but also listed the many applications of this technique to other industries, such as shipbuilding and horticulture.

Eventually, Freytag and Wayss merged their businesses in 1893, a year after French stonemason-turned-engineer François Hennebique began to promote his own building system (see pp.242–43). Freytag and Wayss had no intention of paying Hennebique for his patented building method. However, they were soon forced to compete against Hennebique's growing network of agents to win contracts in Germany. But their earlier work stood in their favour – they had built more than 300 bridges using the Monier system – and the firm continued to thrive.

Monier, however, was unable to achieve similar success – his patents did not give him as much money as he had hoped. In retirement, he was harassed by bailiffs and the tax office, which reasoned that he was receiving large commissions from his many foreign patents. As the reinforced concrete industry grew, it left him behind and he died in relative obscurity and poverty. But his legacy lives on in the skyscrapers that sprung from his humble concrete tubs.

HOLDING WATER

Monier had no technical training, so when he decided to build a reservoir or tank to collect rainwater for use in gardening, he was not sure what to expect in the absence of theoretical calculations. What he instinctively understood, however, was that this type of concrete had more tensile strength than previous forms, which meant that it could withstand much more pressure. This had been clear with the flower pots, but a water tank was a much larger undertaking. However, the principle held. His first tank, completed in 1872, was more than 15m (50ft) in diameter.

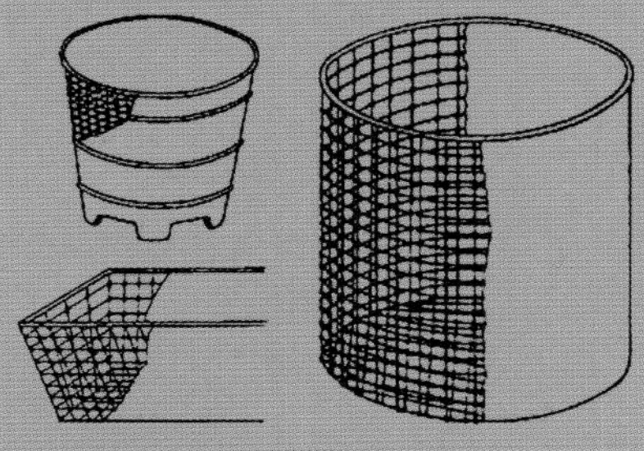

MONIER'S DESIGNS FOR HIS PATENT APPLICATION SHOWED HOW IRON MESH COULD BE INCORPORATED INTO CONCRETE.

REINFORCED CONCRETE BRIDGE
Within only a few years of Monier's first concrete bridge of 1875, bridges constructed with his reinforcing technique had proliferated across Europe. This one is in Earl's Court, London.

WILLIAM ARROL

MASTER OF BRIDGE ENGINEERING

SCOTLAND 1839–1913

From his humble roots in Renfrewshire as the son of a cotton spinner, William Arrol rose to become one of Scotland's greatest engineers, working on some the most famous bridges in Britain: the Tay, the Forth, and London Bridge. He had a reputation for being ingenious and energetic, and was one of the most successful railway contractors of his time.

A LIFE'S WORK

- Arrol starts out as a **blacksmith's apprentice when he is only 14**, and receives no further formal education
- Becomes a foreman at a boiler-works; a few years later, **he sets up his own boiler-making business in Glasgow**, and is soon winning contracts to undertake engineering jobs
- His firm's **first major contract is for the Bothwell Viaduct in 1875**, whose successful completion earns him more offers
- In 1879, the **Tay Bridge over the Forth collapses**, killing 75 people; Arrol is called in to rebuild it, which he does in 1887; **he also completes a bridge across the Firth of Forth in 1890**, and these two bridges remain his most famous works

Like many young men of a similar position, Arrol did not have access to the same kind of education as his middle-class counterparts. Arrol attended primary school until the age of 11 while working in a local cotton mill. When he was 14, he began an apprenticeship with a local blacksmith before earning his living as a journeyman blacksmith, travelling all over the country before returning to Paisley, Scotland, in the 1860s to look for work in the shipyards.

Young, intelligent, and enterprising, Arrol managed to find a job at the engineering firm Blackmore and Gordon. By 1863, he was working for Glasgow engineers R Laidlaw, where he was appointed boiler-shop foreman. He was beginning to understand the engineering business and gradually learned enough to set up on his own.

Five years later, Arrol left Laidlaw and began his own boiler-making business in Glasgow – William Arrol & Co. He used his life savings – £85 – to set up the enterprise. His first engine and boiler cost him £43, just over half his capital. From this precarious beginning grew an enterprise that would one day be the largest steel works in Britain. He soon began to win contracts for numerous engineering projects, including the suburban railway in Edinburgh.

The Bothwell Viaduct is considered to be Arrol's first major project, which he undertook in 1875. The North British Railway company needed it for one of its branch lines that was to run over the River Clyde. He developed a technique that involved building the bridge on land, and then used rollers to move it out on the water. The finished viaduct was 221.5m (727ft) long, and the project was a success. This set the pattern for future work – not only did he complete the job, but he also developed new ways of working on labour-saving devices as part of the project.

Indeed, for his next job, a bridge over the Clyde at Broomiclaw, he invented a new type of mechanical drill. Soon the railway companies were hiring him to build more viaducts and bridges. While he was working on another

THE MOST IMMEDIATELY AND INTERNATIONALLY RECOGNIZED SCOTTISH LANDMARK

COLLINS ENCYCLOPEDIA OF SCOTLAND ON THE FORTH BRIDGE

viaduct for the Caledonian Railway in 1876, he won the commission to build a railway bridge across the Firth of Forth.

DANGEROUS DESIGNS

A bridge across the Forth had already been designed by Thomas Bouch (1822–80), who had earlier been behind the design of the celebrated Tay Bridge. Bouch's bridge across the Tay, which opened in May 1878, was almost 3.2km (2 miles) long, and had 85 spans, making it the longest bridge in the world at the time. Queen Victoria made a journey across it in 1879, and gave him a knighthood for his efforts. With that project completed, he turned to the question of the Firth of Forth. But then tragedy struck.

On 28 December 1879, a violent storm hit the area and caused some of the spans on the Tay Bridge to collapse while a train was passing over it. Seventy-five people were killed in the disaster. An inquiry found that some of the bracing of the ironwork on the bridge was to blame, as well as some problems with the casting of the columns. Bouch was taken off the Forth job, and forced into retirement. He died the following year.

This turn of events, however, benefitted Arrol, and, in 1882, he was awarded contracts for the Tay Bridge. The contract for the Forth Bridge was a staggering £3,000,000. However, it was not a straightforward job. The designers, John Fowler (1817–98) and Benjamin Baker (1840–1907), wanted a cantilever system. This bridge was also going to be made of steel, rather than wrought iron, though in order to build the supports, Arrol needed to design a specific drilling machine, as well as a machine for riveting. The Forth Bridge opened in 1890, and in the meantime Arrol managed to complete the rebuilding of the Tay Bridge, which reopened in 1887. In 1890, he was knighted for his contributions.

Arrol began to win numerous contracts outside Scotland, most notably the rebuilding of Tower Bridge in London from 1886–89, and his firm was recognized as Britain's best bridge builder. But he did not stop there – he built more viaducts, workshops, steel roofs, and hydraulic riveting machines, among other structures and inventions.

In his later years, with his profitable business running smoothly, he decided to enter politics. He was opposed to Irish Home Rule. This issue concerned the repeal or reforms of the Act of Union of 1800, freeing Ireland from British control, and this heated issue dominated political life in the second half of the 19th century. During this period, Arrol became a member of the Unionist party, which was against any breakup of the United Kingdom. He stood in South Ayrshire as a Unionist candidate in 1892, but won a seat only in 1895. He did not remain an MP for long, standing down in 1906, and retiring to Ayrshire, where he lived until his death in 1913.

FORTH BRIDGE CONSTRUCTION
Here, Arrol's Forth Bridge is near completion in the late 1880s. It was one of the first major structures to be built entirely of steel. Its contemporary, the Eiffel Tower, was made of wrought iron.

DISASTER ON THE TAY

After the Tay Bridge collapsed in 1879, it was found that Bouch's design was top-heavy and vulnerable to high winds. In fact, he had made no allowance for wind at all. Neither did he regularly visit the on-site foundry. The cast-iron columns supporting the 13 longest spans of the bridge were of poor quality. Bouch had cast many of them horizontally with the result that the walls were of uneven thickness.

AN ENGRAVING FROM DECEMBER 1879 SHOWS A STEAM BARGE AND BOATS SEARCHING FOR SURVIVORS.

PUBLIC HEALTH

HAUSSMANN'S RECONSTRUCTION
French engineer Georges-Eugène Haussmann, right, is seen here with his patron Napoleon III. Their collaboration led to the rebuilding of Paris, from the sewers to the streets.

ONE OF THE GREATEST ENGINEERING MARVELS OF THE 19TH CENTURY WAS LOCATED WHERE FEW COULD SEE IT. UNLIKE SKYSCRAPERS OR BRIDGES, SEWERS WERE HIDDEN UNDER CITY STREETS. THESE NETWORKS OF TUNNELS AND PIPES TRANSFORMED URBAN LIFE AND PUBLIC HEALTH.

Paris and London, like many growing cities, had become increasingly crowded and polluted over the course of the 19th century. Throughout this period, thousands of people died from a range of diseases, particularly those caused by contaminated drinking water, such as cholera. Doctors did not realize that a disease could be passed through water. The prevailing scientific belief of the time was that disease was transmitted in the air, through what were called "miasmas", which is why London relied on "nightsoil" men, who would cart away sewage from cesspools near homes. However, poor people could not afford to pay for this service, while the growing middle classes began installing water closets, which added strain on the already overburdened sewers. So, in 1815, the law was changed to allow the discharge of effluent through sewers that ran into the River Thames. Soon waste material began filling up the river.

The first cholera epidemic hit London in 1831, killing thousands of citizens. But there was not only cholera in the water – people suffered from other diseases, such as diarrhoea, typhoid, and dysentery. A study was undertaken and the findings were issued by social reformer Edwin Chadwick (1800–90) in his 1842 report "Sanitary Conditions of the Labouring Population and on the Means of Improvement". Chadwick outlined the many problems of public health facing the city, and one of the responses was the establishment of a Metropolitan Sewer Commission. The measures were not enough to stop another cholera outbreak in 1848, which killed more than 14,000 people.

INVITATION TO THE SEWER
This is an invitation to the 1865 opening of the Southern Pumping Station at Crossness, southeast London. The rebuilding and extension of London's sewers won much public acclaim.

The situation in London reached a crisis in 1858. There was a prolonged, hot summer, and the rotting levels of sewage in the River Thames were so high that this period was known as the "Great Stink". Parliament, sickened by the smell, demanded remedial measures.

The improvements were put under the supervision of civil engineer Joseph Bazalgette (see pp.252–53), who envisioned building a network of sewers that would move the pollution away from the city. To do this, he transformed the Thames by building embankments – Victoria, Albert, and Chelsea – on reclaimed riverbed. His work began in 1859, and, by 1866, the numbers of cholera cases had dropped to about 5,500. Further, the outbreaks were only seen in areas that were not connected to Bazalgette's network.

GLOBAL SANITATION

Chadwick, meanwhile, crossed the English Channel to see what had been accomplished in Paris. Like London, this city had been struggling with poor health and sanitation problems. Victor Hugo (1802–85), in his novel *Les Misérables*, describes the Paris sewers as "... an ancient and formidable thing. It has been a sepulchre, it has served as an asylum ... The sewer is the conscience of the city. Everything there converges and confronts everything else".

Under the direction of Georges-Eugène Haussmann (see pp.251), who was rebuilding Paris on the orders of Napoléon III, sewer engineer Eugène Belgrand (see pp.250–51) expanded the sewage system extensively in the 1850s and 1860s, finding new sources of fresh water and building a system of aqueducts to supply it to the city. The zeal with which

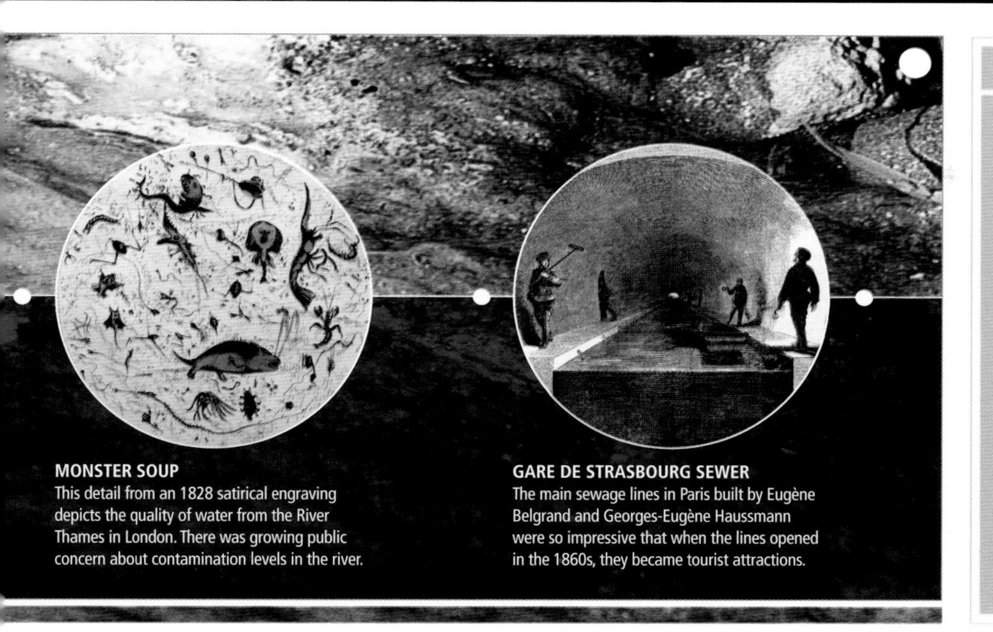

MONSTER SOUP
This detail from an 1828 satirical engraving depicts the quality of water from the River Thames in London. There was growing public concern about contamination levels in the river.

GARE DE STRASBOURG SEWER
The main sewage lines in Paris built by Eugène Belgrand and Georges-Eugène Haussmann were so impressive that when the lines opened in the 1860s, they became tourist attractions.

SETTING THE SCENE

- In **London**, restrictions on the discharge of effluent into sewers are lifted in 1815, which leads to the **contamination of the River Thames**.
- In 1858, the **level of pollution in London is so great that it fills the city with a stench known as the Great Stink**. In response, **Joseph Bazalgette designs a system** in 1859 that drains sewage away through a system of **embankments** and allows it to flow into the River Thames much further down stream.
- Georges-Eugène Haussmann **extends the sewage system in Paris** in the 1850s, with the goal of having fresh water piped into every home.
- In **Bombay**, plans are made to build a water system to service the disease-ridden British areas. Modelled on the Glasgow Loch Katrine reservoir, **the Vihar water scheme brings fresh water to parts of the city in 1859**.
- **Tokyo** starts **constructing sewers** in 1884.

sanitary engineering was being embraced was not limited to Europe.

American cities also experienced similar cholera epidemics from 1832 until the 1870s. City planners in Chicago, New York, and Philadelphia decided to reform the sewage system and looked to European models for inspiration. They built their own networks throughout the latter half of the 19th century, by which time sewers were becoming a global phenomenon. Tokyo, Japan, started work on its own system in 1884.

> **The link between water and disease was finally suspected in the 1850s. Millions of people continue to die each year from preventable water-borne diseases as a result of inadequate sanitation and clean running water.**

British colonial administrators in India and China were concerned about high mortality rates from disease. They drew up plans for water systems in cities such as Bombay and Hong Kong. Even nurse and health reformer Florence Nightingale (1820–1910) took an interest in the problem and corresponded with British colonial officers and Indians about the issue of village sanitation. In Bombay, the work of John Frederick La Trobe Bateman (1810–89), who built the Loch Katrine system in Glasgow, was used as the model. The chief engineer, Henry Conybeare (1823–84), wanted to build a similar project in 1854, known as the Vihar Water Scheme. It ran into controversy with many Indians because it only delivered water to areas with a high population of British residents. Nonetheless, it was opened in 1859.

The Vihar project was not the only one that faced controversy. All over the world, civil engineers faced the prospect of delivering clean water and fighting disease but at the expense of the destruction of neighbourhoods and the displacement of residents. Through these sewer projects, people finally had access to clean, safe drinking water, transforming their lives and health.

BUILDING SEWERS TO LAST
This photograph of 1862 shows the construction of sewage lines running to the Abbey Mills Pumping Station, designed by Joseph Bazalgette, in East London.

EUGÈNE BELGRAND

MODERNIZER OF THE PARISIAN SEWER SYSTEM

FRANCE 1810–78

Plucked from obscurity by the great French city planner Baron Georges-Eugène Haussmann, hydrographer Eugène Belgrand became one of the most celebrated engineers of Paris in the latter half of the 19th century. Through his work he helped expand and transform the city's sewer system and constructed aqueducts to supply clean water.

THE SEWERS OF PARIS
This 1870 illustration shows fashionable members of the public being taken on a tour of Paris's new sewer system. The reconstructed sewer system was considered to be an engineering marvel, and was a popular tourist attraction.

Belgrand first came to the attention of Georges-Eugène Haussmann (see box, opposite) when the latter was on a visit to the Burgundian town of Avallon. Haussmann was surprised to see a fountain in the town providing clean drinking water. He was intrigued – the fountain's builder would have had to know how to source this water, and thus have a sound knowledge of the region's geological formations. The engineer was Belgrand, who had studied at the prestigious École Polytechnique, which emphasized the importance of geography. Haussmann was struck by Belgrand's engineering capabilities, and, later, when he required an engineer to rebuild sewers in Paris, remembered Belgrand and appointed him for the role in 1855. Napoléon III had come to power in 1848 and was rebuilding Paris. The parts of the city that needed immediate attention were the crumbling sewers that emptied into the River Seine. Haussmann was planning to dig up the narrow streets of the French capital, so it was the perfect opportunity to rebuild the sewers.

The reconstruction of the city also allowed Belgrand the opportunity to indulge in another passion: archaeology. In fact, discoveries about the city's history by Belgrand and two other contemporary engineers and archaeologists, Theo Vanquier and Charles Sellier, are exhibited at The Carnavalet Museum in Paris.

A LIFE'S WORK

- Belgrand's **fountain at Avallon**, which draws water from a natural source, attracts the attention of Georges-Eugène Haussmann
- Haussmann appoints him as **head of the sewer project in Paris** in 1855
- Finds a **source of clean water** for Paris in the nearby Marne Valley, **using the streams of Dhuys and Vanne**
- He **builds a large aqueduct** to carry the water, and at the same time **enlarges the sewer system** in Paris, taking waste out of the city and providing homes with **clean running water**
- To commemorate his work, Belgrand's **name is engraved on the Eiffel Tower in Paris** and the main gallery of the Paris Sewer Museum is named after him

PURE WATER FOR PARIS

In the 1840s, drinking water was carted up from the Seine or a nearby canal by men with barrels and then sold to households. In addition, public fountains and wells fed from the river. However, there was not enough water to wash the grime and slurry off the streets, adding to the general filth and contamination.

Although no scientific link had been made between contaminated water and disease, Haussmann realized that the system had to change. He was interested in sewage developments in Britain (see pp.248–49) and dispatched another member of his team to tour the sewage systems of London, Manchester, Liverpool, Edinburgh, and Glasgow.

Haussmann and Belgrand were surprised to hear that the London authorities had rejected the idea of using water from remote springs – which was a crucial part of the French engineers' plan. Yet, like their London counterparts, Haussmann and Belgrand faced difficulties in getting the city council to believe that unclean water could be a health hazard – its members, too, believed in the "miasma theory" (which held that epidemics were caused by a noxious form of "bad air").

Paris had suffered deadly cholera outbreaks in 1832 and 1849, yet no one realized the connection with water. There was also the cost consideration: maintaining the existing system of crumbling sewers, contaminated wells, and water-carriers was far cheaper than Haussmann and Belgrand's proposals. However, the growth of Paris, especially to the hilly suburbs of Montmartre and Belleville, made the councillors change their mind. These suburbs

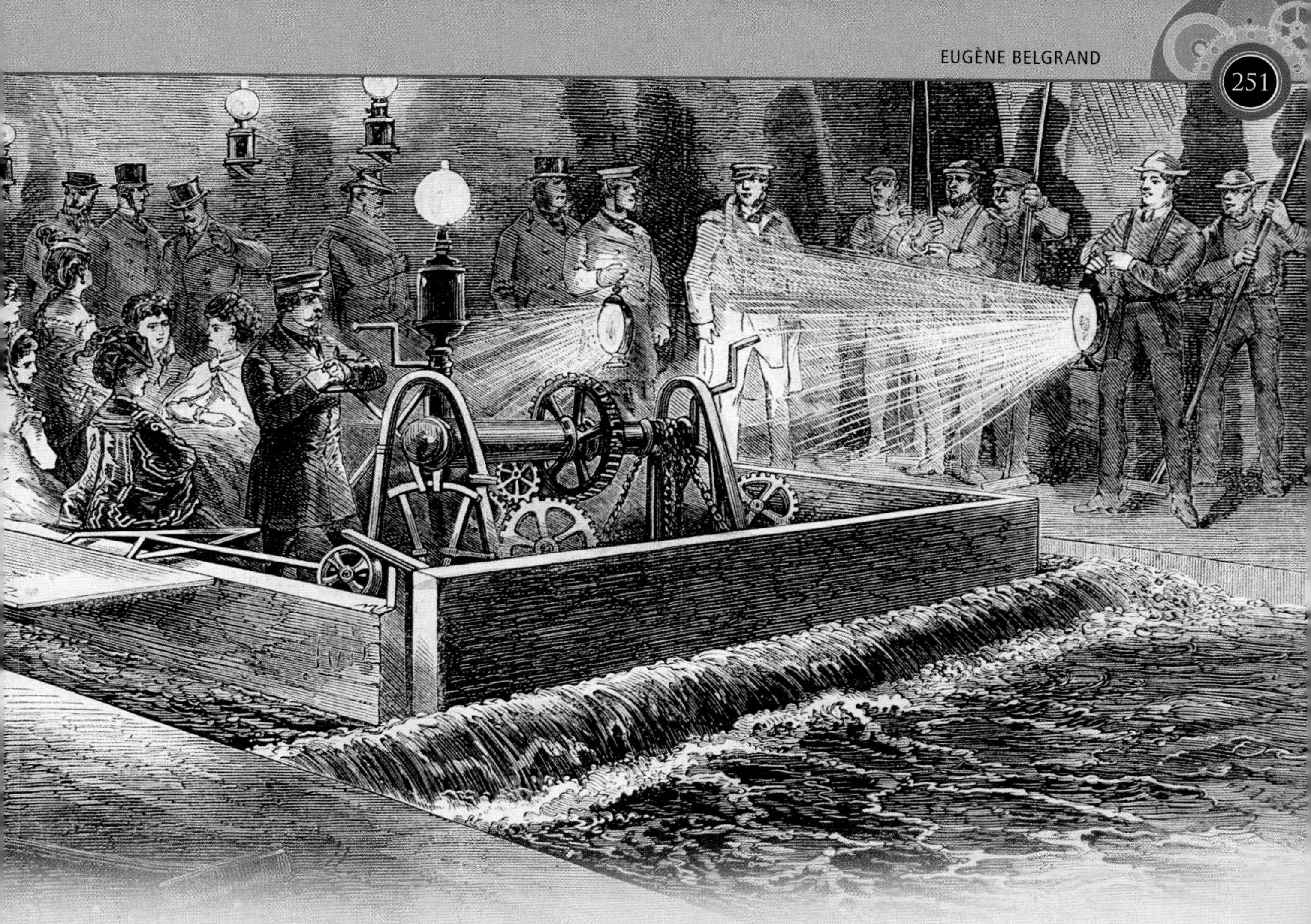

would need a pumping system to link them to the Seine, so a plan to obtain clean water would be better in the long run.

ROMAN METHODS

Belgrand soon found his source of water – the springs near Châlons-sur-Marne, about 160km (100 miles) east of Paris. But Napoléon III needed the support of this province and the residents did not want their springs to be tapped, so Belgrand was forced to continue his search. He hit on a source in the Marne Valley – the streams of Dhuys and Vanne, which were situated at a point that would help take full advantage of the flow of gravity. Belgrand decided to follow the Roman model to harness the water – and build aqueducts.

By 1865, Belgrand had completed an aqueduct 130km (81 miles) long, which had 32 tunnels and more than 20 bridges that connected to a reservoir in Ménilmontant, east of Paris. The water was delivered from a height of 108m (345ft) into the city, and the pressure was high enough for water to reach the upper floors of multi-storey buildings. Within the sewers, the pipes had a slope of about 3 per cent, and the smaller pipes fed into three main collectors. Another aqueduct was built in the Vanne Valley, finished in 1874, which was 173km (107 miles) long, and ran up to Arcueil, to the south of Paris. The number of people with running water in their houses increased from 6,000 to about 34,000, but, even in 1870, half of the city's homes were still not supplied with water through a tap. However, with the introduction of overground aqueducts and reservoirs, as well as underground sewer tunnels and conduits, Belgrand laid the basis for a system that could be extended over the following decades, reaching 1,214km (754 miles) by 1911.

The effect was profound – disease reduced, people could drink clean water, and, for the first time, there was enough surplus to clean the streets. From his provincial start, Belgrand had managed to transform the lives of Parisians with his knowledge and vision. To acknowledge his contribution to the city, his name is among the 72 engraved on the Eiffel Tower in Paris.

BARON HAUSSMANN

FRANCE 1809–91

Civil engineer Georges-Eugène Haussmann was appointed prefect of the Seine department and charged with the task of modernizing Paris.

Napoléon III wanted to make it difficult for people to revolt and barricade the streets. So he charged Haussmann with the task of laying wide boulevards – making it easier to mobilize and deploy troops – in place of narrow medieval streets. Haussmann's "improvements" to the city were applauded by some, and detested by others, who saw these measures as an attack on the poor. That apart, his years working under Napoléon III radically transformed the city.

JOSEPH BAZALGETTE

PIONEER OF THE MODERN SEWAGE NETWORK

ENGLAND 1810–91

HAILED AS THE MAN WHO SOLVED the problem of the polluted and stinking River Thames, Joseph Bazalgette was responsible for creating London's great system of interceptor sewers. An expert in devising efficient sewer systems, he transformed the appearance of London's riverside and saved thousands of lives by preventing cholera and other waterborne diseases.

Born in north London, Bazalgette trained in Ireland before setting up as a consulting engineer in London in 1842. At the height of the "Railway Mania" he was thrust into survey work on railways. Overwork and stress caused a serious breakdown in his health, and he left London to recuperate.

Returning in 1849, he was appointed assistant surveyor to the Metropolitan Commission of Sewers. By 1851, the commissioners had clearly established the broad outlines of the scheme of interceptor sewers to divert sewage eastwards from the River Thames with separate systems north and south of the river and with remote outfalls to the east of London. In 1852, Bazalgette was appointed engineer to the commissioners.

A LIFE'S WORK

- Bazalgette sets up as a **consulting engineer** in London in 1842, but is forced to retire and leave the city because of overwork and stress
- He returns in 1849 as **assistant surveyor to the Metropolitan Commission of Sewers**, and starts work on London's main drainage scheme
- After a **cholera epidemic** in 1853–54 kills more than 10,000 people, the **Metropolitan Board of Works** is established in 1856, with **Bazalgette as its engineer**
- His **system of interceptor sewers** is accepted by Parliament after the **Great Stink of 1858**
- His **Victoria Embankment** contains an **underground railway** as well as a **sewer and a new road** to relieve traffic congestion
- He **receives a knighthood** in 1874

ESCAPING STINKS AND DISEASES

In the mid-19th century, Londoners were suffering from frequent cholera epidemics – in 1853–54, more than 10,000 people were killed by the disease. Sewage in the streets and a stinking river prompted a national outcry. Consequently, the Metropolitan Board of Works (MBW) was established in 1856 with Bazalgette as its engineer. Its first task was to complete the design and implement the plans for the main drainage of London. The Great Stink of 1858 (see box, right) prompted Parliament to pass enabling legislation, and work began in 1859.

The southern system included the Deptford Pumping Station, the Southern Outfall Sewer, and the Crossness Pumping Station. The northern system, which included pumping stations at Abbey Mills in Stratford and at Pimlico, was delayed by the complexities of integrating a low-level sewer, the Victoria Embankment, and the Metropolitan District Railway. The whole system comprised 2,100km (1,300 miles) of sewers, and 131km (82 miles) of large west-east intercepting sewers. The generous size of his sewers and the high standards of construction ensured that the system lasted for many generations.

As part of the main drainage system, Bazalgette fulfilled the long-cherished plan of embanking the River Thames in central London. This had three major parts. The first was the Victoria Embankment – 2km (1.2 miles) long – running between Westminster and Blackfriars bridges. Lengthy negotiations with private property owners, coal wharf operators, the City Gas Company, and Metropolitan Railway directors took place before work began in 1864. Behind the curved, granite river wall was the northern low-level sewer, a service subway, and the underground railway.

The Albert Embankment, on the south bank between Westminster and Vauxhall bridges, was opened in 1868 and Chelsea Embankment was completed in 1874.

Apart from tidying the mud banks on the river edge, the embankments improved road traffic flow and provided additional building land – Bazalgette explained that the embankments extended "35 miles along the river, and they had reclaimed about 52 acres of land". Bazalgette was knighted in 1874.

Between the early 1870s and the early 1880s, London's death rate improved by the equivalent of about 12,000 lives per year, and Bazalgette

ABBEY MILLS PUMPING STATION
Bazalgette built the pumping station at Abbey Mills in the soaring style of London's Gothic Revival, adorned with Moorish chimneys and pointed turrets. Its steam-engine pumps were replaced by electric pumps in 1933.

THE GREAT STINK

The growth of London in the early 19th century, combined with the introduction of flushing toilets and the development of industry, led to more and more sewage and other pollutants contaminating the city and flowing untreated into the River Thames. The scandal came to a head during the unusually hot summer of 1858, when the stench from the river was so overwhelming that it affected the work of the House of Commons, where curtains soaked in chloride of lime were hung over the windows. In the same year, Parliament passed an enabling Act for London's main drainage, approving Bazalgette's interceptor sewer system.

A CARTOON OF 1858 DEPICTS DEATH AS A SILENT HIGHWAYMAN, STALKING THE THAMES.

> ONE OF LONDON'S MOST STARTLING SITES – ABBEY MILLS ... NICKNAMED THE CATHEDRAL OF SEWAGE
>
> **SARA McCONNELL**, JOURNALIST, *THE GUARDIAN*

observed that "it may not be unfair to claim for those works a considerable share in this decrease in deaths".

CONGESTED LONDON

Bazalgette continued to work on improving London's infrastructure. An Act of 1877 enabled the MBW to purchase 12 River Thames bridges and free them from tolls. He decided to replace three of these with new bridges of his own design – the masonry arch bridge at Putney, the steel-link suspension bridge at Hammersmith, and the iron arch structure at Battersea.

In an effort to ease the congestion in London due to horse-drawn traffic, Bazalgette initiated a major programme of design and construction of new thoroughfares, including Southwark Street, Queen Victoria Street, Northumberland Avenue, Shaftesbury Avenue, and Charing Cross Road. His department reported that there were plans for about 3,000 new streets, and that the cost of street improvement works was about £14 million.

An important part of Bazalgette's work for the MBW was to monitor the progress through Parliament of private bills that would impact the public amenities of London. These included railways, tramways, docks, water supply, and energy utilities such as gas, electricity, and hydraulic power. He observed that "private individuals are apt to look after their interests first ... and it is necessary that there should be somebody to watch the public interests".

Bazalgette was also consulted about town drainage in at least 30 British towns between 1858 and 1874. He also designed a new bridge over the River Medway for Maidstone Council, opened in 1879. He retired when the MBW was replaced by the London County Council in 1889.

An obituary for Bazalgette, who died in 1891, said, "Sir Joseph never appeared to be an ambitious man, and yet he became identified with undertakings which the most ambitious might have been proud to accomplish".

HERTHA AYRTON

PIONEERING WOMAN ENGINEER OF THE 19TH CENTURY

ENGLAND 1854–1923

FROM MODEST BEGINNINGS, Hertha Ayrton became a noted engineer, mathematician, and inventor. Her significant contributions to science and technology, including her experiments with electricity and the invention of an anti-gas fan for use in the trenches during World War I, paved the way for more women to participate in the field.

A LIFE'S WORK

- Ayrton **reads mathematics at the new Girton College for women in Cambridge** from 1877 to 1881
- When her husband, William, goes to Chicago to attend the Electrical Congress in 1893, **she continues his experiments on the electric arc**, and publishes the results in "The Electrician" journal in 1895–96
- She is the **first woman ever to read her own paper** before the Institution of Electrical Engineers (IEE), in 1899
- Experiments with sand and waves, and later **gives an influential paper on the subject** to the Royal Society in 1904
- During World War I she **develops a type of fan that can help push gas out of the trenches**, saving thousands of soldiers' lives

Ayrton's scientific interests were wide-ranging, and she established her reputation by working on the electric arc, which is a current passing from one electrical conductor to another through air. It was her work on sand ripples, however, that won her the most scientific acclaim, and her paper to the Royal Society in 1904 was the first ever given by a woman.

UNORTHODOX UPBRINGING

Ayrton was born Phoebe Sarah Marks on 28 April 1854 to a seamstress and a watchmaker. In 1861 her father – who had emigrated from Poland to escape the pogroms – died, leaving her mother to raise eight children. When Ayrton was nine, she was taken in by two aunts who ran a school in London. She was educated at home and one of her tutors was Eliza Orme (1816–92), who taught her mathematics. Although Ayrton began teaching at 16 to help her mother, her friends and family encouraged the bright young woman to apply to the newly established Girton College in Cambridge. Indeed, the novelist George Eliot (1822–80) met Ayrton and took a keen interest in her education. Around this time, one of her friends began to call her Hertha, after the poem by Algernon Charles Swinburne, and the nickname stuck. Her time at Girton, from 1877 to 1881, was a success – she read mathematics and developed her first inventions, a device for recording pulse beats and a tool that could divide lines into equal parts. The latter she patented in 1884. At that time, Cambridge gave only certificates and not degrees to women. She successfully completed an external examination and received a BSc degree from the University of London in 1881.

In 1884, she began to attend lectures at the Finsbury Technical College given by William Edward Ayrton (1847–1908). William was immediately struck by her intelligence and the two married the following year. In 1888, she began to give lectures to women about electricity. Teaching women about science was a novel thing to do at the time, so she had to focus on the domestic uses of this new technology.

When William went to Chicago for the Electrical Congress in 1893, Ayrton continued the experiments on the electric arc that he had started. He wanted to give a paper on his work but needed to finish his experiments. She did those for him, posting the results in bi-weekly letters, relaying the information he needed to complete his presentation. But Ayrton did not stop there – she claimed she "should like to solve the whole mystery of the arc from beginning to end" – and continued researching. She was asked to submit her findings to "The Electrician" journal in 1895–96, and her articles were later turned into a book, *The Electric Arc*, in 1902. The tendency of electric arc lamps to flicker and hiss was a major problem. She examined this issue and explained that this was due to oxygen coming into contact with the lamp's carbon electrodes, showing that the sounds did not occur if the arc did not come into contact with air. The British Admiralty became interested in the implications of her work on searchlights. She was nominated for Royal Society membership, but her candidature was turned down because she was a married woman and as such had no legal standing, and so was not eligible for admission. Although the law – and social attitudes – later changed, no other woman was proposed until 1944. Ayrton, though denied membership, was in 1906 awarded the prestigious Hughes Medal for her work.

AYRTON FAN
This fan was specifically designed to generate air vortices that would push away toxic gas from the trenches during World War I.

William began to suffer from ill health, and the couple moved to the coast – Margate, Kent – to aid his convalescence. On the beach, Ayrton became fascinated by the ripples in the sand left by the outgoing tide. She worked out the mathematics involved in wave motion, and borrowed her landlady's zinc bath to experiment with waves and sand. When they returned to London she continued her work, using glass tubs in her attic, and in 1904 gave her ground-breaking paper on the subject to the Royal Society.

WAR AND SUFFRAGE

As World War I raged, Ayrton began to think about the properties of the gas that soldiers were inhaling in the trenches. After working with sand and water vortices, it occurred to Ayrton

[THE EXPERIMENTS] WERE CONVINCING AND MANY INTELLIGENT OFFICERS WERE IMPRESSED

A P TROTTER, PRESIDENT OF THE INSTITUTION OF ELECTRICAL ENGINEERS

that some of the same principles could be applied to air. She tested these ideas using smoke to imitate gas, and discovered that air vortices could be used to push gas away. Her solution was a portable fan. It was about 930 sq cm (1 sq ft) in size, made of cotton with a wooden T-shaped handle. When used according to her instructions – they were not simply to be waved about, for example – they created a vortex in the air, which drove away the denser poisonous gas.

However, Ayrton had to fight for the device's acceptance, although once she organized the production and distribution of some 100,000 fans, there were many soldiers in the trenches who owed their survival to her work. A letter in *The Times* of London on 4 May 1920 from Major H J Gillespie (who first introduced the fans in France) – recalls that "my battery escaped without a casualty" during a gas bombardment of Armetières in 1917.

Although Ayrton was succeeding in a field that was mostly closed to women, she was not complacent about the progress being made towards equality. A supporter of the suffragists, she often marched alongside them and helped the movement when she could. She became close friends with the scientist Marie Curie (1867–1934) – a fellow advocate for women's rights – whom Ayrton met in 1903 when Marie and Pierre Curie (1859–1906) were in London to speak at the Royal Institution. Ayrton lived to see the vote extended to women over the age of 30 in 1918, but died in 1923 before the arrival of universal suffrage.

GAS ATTACK
This photo from World War I shows Australian troops wearing gas masks in the trenches. Deadly gas attacks were common during the war. Soldiers not killed by the gas would suffer from temporary blindness or respiratory problems.

RADIO AND SOUND

THE KEY DEVELOPMENTS IN SOUND TECHNOLOGY IN THE LATE 19TH CENTURY – THE TELEPHONE, THE PHONOGRAPH, AND THE RADIO – WERE INVENTED FOR BUSINESS, AS A MEANS OF PASSING MESSAGES, RECORDING INFORMATION, AND SHARING DATA. BUT THESE PURPOSES WOULD SOON CHANGE.

TOO LATE
This 19th-century engraving shows Elisha Gray arriving at the US Patent Office with his patent application two hours after Alexander Graham Bell had filed his.

In fact, these devices achieved mass popularity only when people began to use them as ways of passing their time, rather than merely organizing it – sharing gossip on the telephone, listening to music on the phonograph, and tuning in to news on the radio.

RINGING THE CHANGES

The first of these inventions to appear was the telephone of Scottish inventor Alexander Graham Bell (see pp.258–59). Few people, however, realize that several other inventors almost got there first. In fact, on 14 February 1876, the day that Bell filed his patent for an "apparatus for transmitting vocal or other sounds telegraphically", American electrical engineer Elisha Gray (1835–1901) also made his own patent application for a version of the telephone – but crucially a few hours late. And so the spoils went to Bell. Also, Gray's attempts to market his own device, in conjunction with American inventor Thomas Edison (see pp.262–65) and Western Union, were met with implacable resolve by Bell's lawyers.

Edison may have been forced to accept the demise of his telephone, but a year later he unveiled his own contribution to sound technology – the phonograph. "The Wizard of Menlo Park", as the US press dubbed him, was an extraordinary character. Although officially the fourth most prolific inventor in history, he did not always fully appreciate the possibilities of his (or others') discoveries. He believed that the future of concrete lay in creating prefabricated homes, complete with concrete furniture; that the future of cinema lay in small, "what the butler saw" style machines, rather than in large screens; and, surprisingly, he saw no commercial future for aircraft.

In spite of their prowess and many talents, Alexander Graham Bell, Thomas Edison, and Guglielmo Marconi were not true pioneers. Rather, they put together existing technologies in the most commercially appealing forms.

It is therefore perhaps not surprising that he thought the phonograph would principally serve as a dictating machine. In fact, it was the device's potential as a medium of entertainment that really enthused the public. And in the late 19th century, with the transition from Edison's cylinder-based phonograph to the disc-spinning gramophone of German-born American inventor Emil Berliner (see p.259), an international recorded music industry was born.

SPINNING AROUND
Emil Berliner's gramophone used discs to reproduce sound, which could be produced more cheaply and in greater volume than Edison's cylinders.

MAKING WAVES

Edison was by no means alone in his failure to spot an invention's potential. In 1888, when Heinrich Hertz (see p.270) built the first instrument capable of sending and receiving radio waves, thereby confirming Scottish scientist James Clerk Maxwell's (1831–79) classical theory of electromagnetism, Hertz commented: "It's of no use whatever … This is just an experiment that proves Maestro Maxwell was right."

Other inventive minds had slightly more imagination, and, in the 1890s, a number of gifted individuals raced to create machines capable of exploiting radio waves. Ironically, the competition was won not by the most gifted – that was undoubtedly Croatian-born Nikola Tesla (see pp.266–67) – but by the most commercially savvy, Italian engineer-entrepreneur Guglielmo Marconi (see pp.268–69).

By the early 20th century, Marconi was firmly established as the fledgling radio industry's dominant – and richest – player. However, it is worth noting that the one thing neither Tesla's nor Marconi's inventions did is what we usually associate with the word "radio" – make audio broadcasts. Tesla was primarily concerned with the wireless transmission of power, while Marconi's early systems were used solely for the wireless relaying of Morse code signals. Human speech and music would be transmitted by radio only later in the 20th century.

MAKING A NOISE
In this 1878 photograph, Thomas Edison is sitting next to one of his early phonographs, which recorded sound by scratching lines on a tinfoil cylinder.

CALLING LONG DISTANCE
In 1892, Alexander Graham Bell's firm built the first long-distance telephone network. Here, Bell is shown making the first ceremonial call from New York to Chicago.

SETTING THE SCENE

- In 1861, German schoolteacher **Philipp Reis** (1834–74) invents a device he calls the ***telefon***, which can **relay clicks and musical notes**, but only faint, indistinct speech.
- **Alexander Graham Bell** wins the race to create a **functioning telephone** in 1876, patenting his "electrical contrivance for reproducing in different places the tones and articulations of a speaker's voice".
- In 1877, **Thomas Edison** demonstrates his **phonograph**, the first device for **recording and reproducing sound**, getting it to say "How do you like the phonograph?" to startled editors at *Scientific American* magazine.
- **Nikola Tesla** publicly demonstrates the **first radio communication** in 1893, but is soon **bettered by Guglielmo Marconi**, who sends a **wireless telegraph message across the Atlantic** in 1901.
- In 1906, American **Lee De Forest** (1873–1961), invents an **amplifier, the audion**, which boosts the signal that can be picked up by an antenna thus making it possible to **broadcast human speech by radio**.

ANSWERING THE *TELEFON*
Philipp Reis is shown here speaking into his *telefon*, which had a transmitter shaped roughly like the human ear.

ALEXANDER GRAHAM BELL

LEGENDARY CREATOR OF THE TELEPHONE

SCOTLAND 1847–1922

THE NEED TO COMMUNICATE was a driving force in Alexander Graham Bell's life. His speech therapist father and deaf mother had a profound influence on his scientific work, and led to the creation of one of the most important communication devices of modern life: the telephone. Bell has also been described as one of the most influential figures in human history.

A LIFE'S WORK

- While still a child in Edinburgh, Bell **builds a replica head that can "speak" by making human sounds**; he later develops a way for his dog to "talk"
- He moves to Canada and then to the US, where he **sets up a school for deaf people and a training college for the teachers of the deaf** in Boston, in 1872
- He reads widely about physics and is **soon working on what he called a "harmonic telegraph"**, drawing from the new telegraph technology that had appeared in the 1840s
- After many experiments he **manages to transmit sound through wires**
- **Receives his first patent** in 1876
- His device **goes on public display at the Centennial Exposition in Philadelphia**, Pennsylvania, in 1876, to public acclaim
- **The Bell Telephone Company** is established in 1877

Bell was born in Edinburgh, Scotland, during a time of great enterprise. The city had been a hub of the Enlightenment, and innovation continued well into the 19th century, allowing Bell to grow up in a city whose designers, engineers, and inventors were at the cutting edge of their professions at the time.

FAMILY LESSONS

Communication was at the heart of family life. His father, Alexander Melville Bell, gave elocution lessons. He later developed "visible speech", which was an alphabet of all human sounds. His grandfather, Alexander Bell, was also a noted authority in his field, and wrote books about speech. Growing up, when Bell spoke to his mother, he would forgo talking into her ear tube but instead spoke close to her head, using low tones and sound vibrations.

On a visit to London in 1863, he saw a "speaking machine" built by Charles Wheatstone (1802–75), and he and one of his brothers decided to make one themselves. They made a device that had a mouth and a nose, and even a movable tongue, and it could make some human-like sounds, such as "ma-ma". Bell also used the family dog in his experiments – he came up with a way to move the dog's mouth while it was growling, making the animal appear to talk.

In his studies, he became interested in the writings of Hermann von Helmholtz (1821–94), a German physicist who wrote the book *On the Sensations of Tone* in 1863. However, Bell's knowledge of German was basic, and he misunderstood some of von Helmholtz's ideas, thinking that the German had discovered that sounds could be transmitted over a wire. Bell later referred to this as a "very valuable blunder".

After finishing school, the young Bell joined the family business, although he focused on teaching deaf people. By 1870, both his brothers had died of tuberculosis and his family decided to emigrate to Canada. Bell went with them and soon continued his work with deaf people. He began to use his father's visible speech technique to help his students read lips as well as make sounds and speak, which was considered revolutionary at the time. He then moved to Boston, Massachusetts, opening a school for the deaf and a training college for teachers of the deaf in 1872, and later becoming a professor at Boston University.

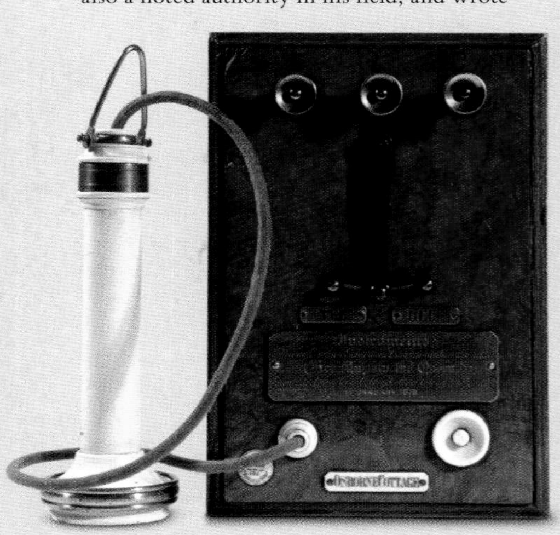

EARLY TELEPHONE
This model was one of the earliest telephones made by the Bell Telephone Company. Bell had received the patent for his invention in 1876.

SOUND EXPERIMENTS

Bell continued to stay abreast of the flurry of experiments and discoveries that were taking place in communications technology. The telegraph had been established in the 1840s, and, by Bell's time, it was part of the growing communication network in the US. Drawing from this, Bell and his assistant, Thomas Watson (1854–1934), wanted to build a "harmonic telegraph". They developed a system of transmitters and receivers and experimented with a variety of metals and liquids.

On 2 June 1875, while trying to unwind wires on a transmitter, Watson produced a sound – a kind of "twang" – that Bell heard in another room, where he was working on a receiver. After further experimentation, Watson told Bell that he could hear vocal tones through the device. Bell worked out that if a vibrating wire was partially covered by a liquid conductor it could transmit sound. He was testing this theory on 10 March 1876, when he was purported to have knocked over some

BELL SHOWS THE WORLD THE TELEPHONE
Alexander Graham Bell astounds an audience at the Lyceum Hall in Salem, Massachusetts, in 1877, during a demonstration of his new device – the telephone.

battery acid. He called out "Mr Watson, come here. I want you", and Watson, who was working in another room, heard him through the wire. Bell had accidentally made the world's first telephone call.

THE TELEPHONE

Soon they were able to demonstrate how speech could be transmitted through wires by using vibration and electric currents, and Bell received patents in 1876 and 1877. He displayed the device at the Centennial Exposition in Philadelphia, Pennsylvania, in 1876, and the public was amazed.

Along with some investors, Bell set up the Bell Patent Association in 1875 to facilitate his research. This turned into the Bell Telephone Company in 1877, and eventually it was called the American Telephone and Telegraph Company, which was officially incorporated in 1885. It did not take long for telegraph companies and other inventors to rush to create a similar technology, though Bell did manage to preserve his patents and defeat hundreds of legal challenges to his claim on the technology.

Meanwhile, the telephone took on a life of its own. Bell had no interest in running a company, and through his patents he could afford to pursue a variety of scientific interests, such as aviation. In 1877, Bell married a former student, Mabel Gardiner Hubbard, who was deaf. His work with deaf people continued throughout his life, and he set up the American Association to Promote the Teaching of Speech to the Deaf in 1890 (later known as the Alexander Graham Bell Association for the Deaf and Hard of Hearing). Bell died on 2 August 1922, and all the buzzing telephone lines in the US were stopped for a minute's silence in tribute.

THE GREAT INVENTION, WITH WHICH HIS [BELL'S] NAME IS IMMORTALLY ASSOCIATED, IS A PART OF HISTORY

MACKENZIE KING, CANADIAN PRIME MINISTER, 1922

EMIL BERLINER

GERMANY 1851–1929

Emil Berliner was a German immigrant who arrived in the US in 1870. He studied physics and was interested in the growing field of sound technology.

Berliner became interested in the new audio technology of the telephone and the phonograph, and invented an improved telephone transmitter. Inspired by Bell's telephone, Berliner began to experiment with methods of sound recording, and, in 1887, he patented a sound recording system using a flat record, or phonographic disc, which would soon replace the rolled cylinders that sound had been recorded on. The discs could be played back on a gramophone, which worked by using a fine needle to read the grooves in the record. The sound was then transmitted through the needle's vibration, and amplified through the speaker.

IN HIS OWN WORDS

ALEXANDER GRAHAM BELL

WITH HIS INVENTION OF THE TELEPHONE in 1876, Alexander Graham Bell was catapulted from an obscure position as a teacher of vocal physiology to one of the most feted men in the world. He left behind a rich legacy of documents and artefacts relating to both his inventions and his work with the deaf. Indeed, his interest in communication technology was largely inspired by his desire to alleviate the deafness of his mother and his wife.

A TEACHING THE DEAF
Bell was deeply interested in silent communication, as this letter on teaching deaf children shows. Rather controversially, he believed that the deaf should lip-read to communicate as he thought that sign language excluded them from mainstream society.

B BUSINESS MODEL
Dating from 1878, these devices are among Bell's earliest telephones. The first models were meant for two-way communication within offices. It was only later that the device came to be used for social interaction.

C THE DRAWING BOARD
This is a sketch made by Bell of his telephone. He writes below: "As far as I can remember these are the first drawings made of my telephone – or 'instrument for the transmission of vocal utterances by telegraph'."

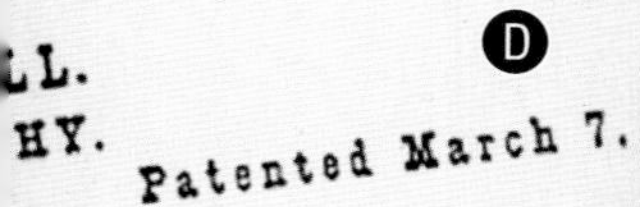

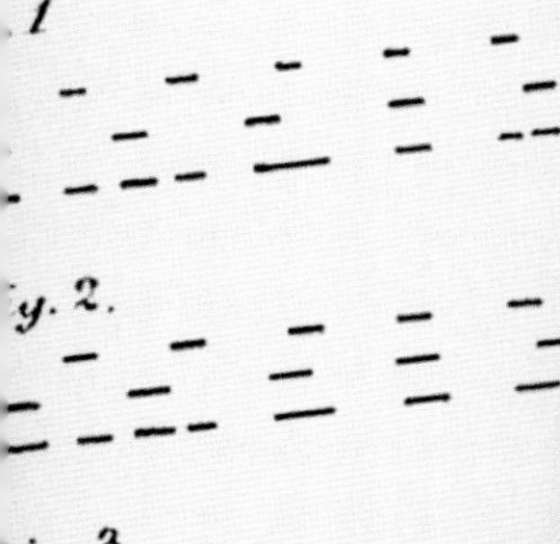

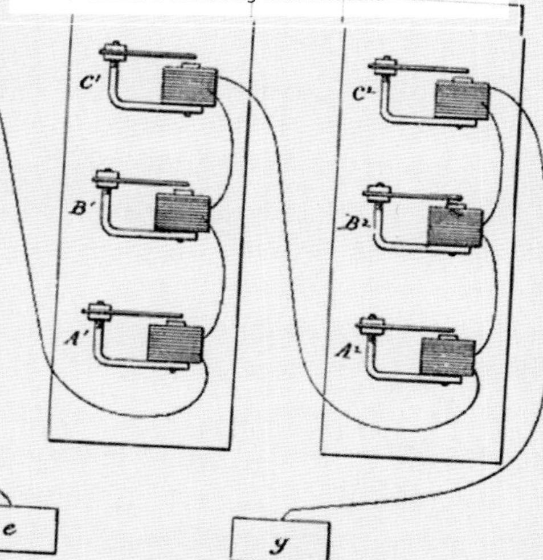

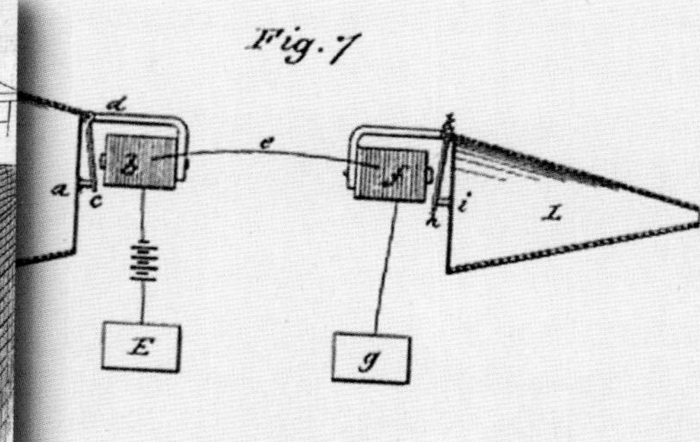

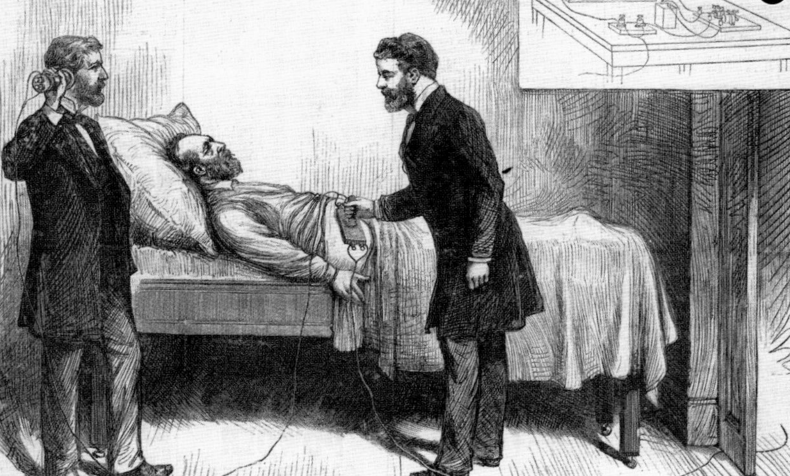

D LICENCE TO PRINT MONEY
These are Bell's drawings for "improvements in telegraphy", otherwise known as the telephone. They formed the basis of the most valuable patent ever granted.

E HELEN KELLER
This 1901 picture shows Bell with the deafblind Helen Keller, who was briefly one of his pupils and went on to become a leading advocate for the rights of disabled people. Bell and Keller continued to remain friends.

F LISTENING IN
In this pair of drawings from 1877, Bell puts his invention to ingenious use, giving a lecture in Salem, Massachusetts, while colleagues in Boston listen via a telephone.

G THE PRESIDENTIAL METAL DETECTOR
In 1881, Bell invented a metal detector to locate the bullet with which President Garfield had recently been shot. However, the device was confused by the metal springs of the president's bed and did not give an accurate reading. This illustration is a rendering of that incident.

THOMAS EDISON

WORLD'S MOST PROLIFIC INVENTOR

UNITED STATES 1847–1931

A PROLIFIC INVENTOR, Thomas Edison deeply influenced modern life. His laboratories delivered 1,093 patents – for electric power and light, sound recording and telecommunications, but also in other, diverse areas such as medicine, defence, transport, and housing. Colourful and unconventional, Edison was internationally revered, and for some time was the most famous American alive.

Thomas Alva Edison was born into a middle class family in Ohio. When he was seven, the family moved to Port Huron, Michigan, and he joined the local school. His teachers soon claimed his brain was "addled". Although very inquisitive, he was easily bored by the school's rote learning methods, and had hearing problems, which undoubtedly affected his behaviour. So his mother taught him at home. He became an avid reader, developing a love for literature, history, and science. By the age of ten, he had a chemical laboratory in the cellar. His lack of formal education and passion for discovery strongly influenced his work as an inventor. Mistrustful of scientific theories, his method was to try everything until he succeeded.

A BUDDING ENTREPRENEUR

At the age of 12, Edison started selling newspapers and sweets on trains. He used an empty baggage carriage as a laboratory and to produce a local newspaper, *The Weekly Herald*, which he sold to customers. He also set up stalls and hired boys to run them. In 1862, he rescued a station official's child from an oncoming train and was rewarded with training in telegraphy, a well-paid and prestigious vocation. He was soon working as a full-time telegraph operator, later moving from city to city taking available telegraphy jobs. All the while he was learning and experimenting. When sounding keys became common, which enabled operators to translate Morse code as clicks rather than printed dots and dashes, he found himself at a disadvantage. So he invented devices to make other tasks easier.

In 1869, having made enough progress on his inventions to become a full-time inventor, Edison settled in New York. His first patented device was an electric vote recorder. But politicians were not interested in speeding up the voting process and he failed to find a buyer. Vowing only to invent things the public would want, he soon came up with his first commercial success: a telegraphic stock ticker to print stock

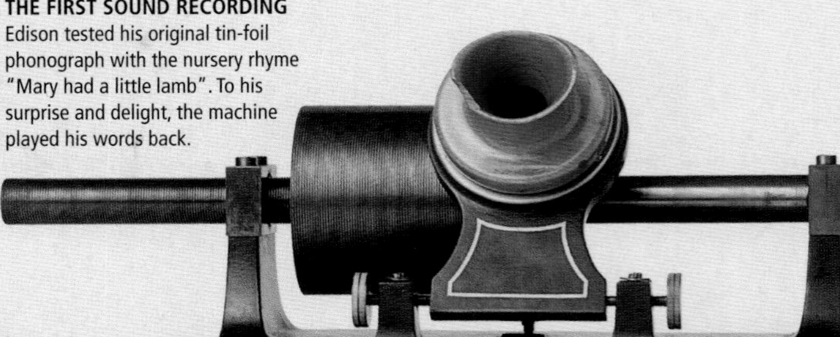

THE FIRST SOUND RECORDING
Edison tested his original tin-foil phonograph with the nursery rhyme "Mary had a little lamb". To his surprise and delight, the machine played his words back.

A LIFE'S WORK

- Becomes a **full-time inventor** in 1869
- **Pioneers modern research and development**, with fully equipped and fully staffed facilities at Menlo Park and West Orange
- **Makes the first recognizable recording of a human voice** with the words "Mary had a little lamb"
- Lights New York with electricity by **making a cheap, long-lasting light bulb** and just about everything else needed to produce a working domestic lighting system
- **Kickstarts the movie industry** by inventing an effective movie camera to shoot celluloid film
- **Holds more than a thousand patents**, the most anyone had ever held

TRY, TRY AGAIN
Edison spent many long hours working in his laboratory, often well into the night. He believed that every failed experiment or test was a step towards eventual success.

GENIUS IS ONE PER CENT INSPIRATION, NINETY-NINE PER CENT PERSPIRATION

THOMAS EDISON

prices for Wall Street brokers. This earned him enough to set up a laboratory in Newark, New Jersey. Soon afterwards he married one of his employees, the 16-year-old Mary Sitwell.

For the next few years he worked as an independent entrepreneur in the telegraph industry. Competition was fierce, and he became adept at pitting companies against each other to bid for his products. He patented many telegraph improvements and other items, such as an electric pen for use in duplicating machines. But he did not just invent; he also set up machine shops and manufactured his inventions. Later, Edison became a highly successful businessman, setting up several factories and offices to make his products and sell them to the public.

THE WIZARD OF MENLO PARK

In 1876, Edison moved his family and assistants to the village of Menlo Park (now Edison, New Jersey), where land was cheap and there was easy rail access to potential investors in New York. There, he built a fully equipped research and development facility, the first of its kind, with a laboratory, machine shop, office, and library.

Menlo Park became known as the invention factory. Edison often worked on several projects simultaneously, with many specialist staff – his "muckers" – helping to put his ideas into practice. He was informal and often entertaining, but a hard taskmaster. He worked very long hours, expecting his muckers to do the same, and was quick to fire them for minor misdemeanours, such as deviating from his instructions. He was less disciplined with himself; if he noticed something interesting, he would not hesitate to explore it further.

His first major success at Menlo Park was the carbon transmitter, a microphone that made Alexander Bell's telephone practical, and was used in telephones until the 1980s. Further experiments with this led Edison to his most original invention, the phonograph – the first machine that could record and play back sound. It caused a sensation. Edison became world famous and was dubbed the Wizard of Menlo Park, although it was ten years before he developed the phonograph commercially. Being very hard of hearing, Edison depended on others to serve as his ears for these projects, but famously learned to "listen" to phonograph recordings through his teeth. He also claimed that his deafness was an asset, helping him sleep easily and work with fewer distractions.

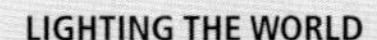

LIGHTING THE WORLD
A practical, long-lasting light bulb was just one of Edison's 389 patents in electric power and lighting.

In 1882, Edison decided to create a commercially viable electric lamp and founded the Edison Electric Illuminating Company with backing from J P Morgan and others. He bought the patent for an existing bulb and proceeded to test thousands of potential filaments. After spending 14 months and US $40,000, he discovered that a carbonized cotton filament burned for 40 hours. Edison declared he would now make one that would last for hundreds of hours, finally settling on carbonized Japanese bamboo, which lasted for 1,200 hours. Edison did not invent the first carbon filament incandescent bulb, but he did initiate domestic electric lighting and is largely responsible for popularizing it. As well as improving the efficiency and longevity of bulbs, he developed numerous other essential devices, such as safety fuses, light sockets, and voltage regulators, and also the first electrical distribution system for lighting. He set up a model plant in New York's financial district in 1882, where he developed the largest ever dynamo to generate electrical power. By 1887, there were already 12 Edison power stations across America.

WAR AGAINST AC TECHNOLOGY

Edison was a ruthless businessman. In 1887, when the Westinghouse Corporation had developed a power service using Alternating Current (AC) electricity – which had significant advantages over Edison's own Direct Current (DC) service – Edison started a smear campaign, claiming that AC technology was unsafe. At a public demonstration, he attached an AC generator to a metal platform and electrocuted various animals. The next year, when New York decided to adopt electrocution for state executions, his employees carried out public experiments with AC and DC to show that DC only tortured animals, while AC killed them instantly – a proof that DC was indeed a safer option. Edison eventually lost his battle against the superior technology.

DESPITE OPPOSING CAPITAL PUNISHMENT, EDISON FUNDED AN ELECTRIC CHAIR SAMPLE AND SOLD PRISONS AC GENERATORS FOR ELECTROCUTIONS.

WEST ORANGE

Soon after the death of his wife Mary in 1884, Edison met Mina Miller, the 19-year-old daughter of another inventor. In 1886, he married her and moved to Glenmont, a large mansion in West Orange, New Jersey. Nearby he built an even bigger laboratory complex. Employing about 5,000 people at its peak, this became the template for modern research and development facilities; some even view this as his greatest invention.

One West Orange project with a huge impact on modern life began in 1888, when Edison applied for a caveat at the patent office to protect

ENGINEERING TIMELINE

1847–70

Born in Milan, Ohio, he is the youngest of seven children, although only the fourth surviving and is himself a sickly child

In 1863, he starts working as an apprentice telegrapher; this involves looking after telegraph equipment, enabling him to learn about electricity

Applies for a patent for the electric vote recorder in 1868, and decides to concentrate on inventions people would buy

1871–76

Patents the Universal Stock Ticker, his first commercially successful invention in 1871

In 1876, opens his "invention factory" in Menlo Park, New Jersey, the first facility of its kind future work on a device that would "do for the eye what the phonograph does for the ear". He tasked William Dickson (1860–1935), an assistant keen on photography, with development, and by 1891 was able to unveil the kinetograph, a motion picture camera, and the kinetoscope, a viewing instrument with peepholes through which a viewer could watch a film (see pp.290–91). In the "Black Maria", a studio that rotated to capture the best available sunlight, Dickson shot films of dancers, boxers, and vaudeville performers, including one of Annie Oakley firing her gun at glass balls.

Edison spent much of the 1900s developing an alkaline storage battery for electric automobiles. Although petrol-powered cars soon dominated the market, he found many other uses for the batteries. But not all of his ideas were successful. He spent a great deal of time and money developing a process to mine iron ore, a venture he finally abandoned. And when Portland cement caught his imagination, he built concrete houses and phonograph cabinets, and even the original Yankee stadium. But cement was expensive and the enterprise eventually failed.

During the 1920s, Edison's health began to deteriorate, although he was convinced the right diet would make him better and largely avoided food for several years in favour of milk. After years of neglecting his family he started spending more time at home. But he never gave up working. His last major project, bankrolled by his high-profile friends Henry Ford (see pp.282–85) and Harvey Firestone (1868–1938), was to find an alternative to rubber that could be produced in the US. He did not live to complete the project. In October 1931, America mourned the death of its favourite inventor. Thousands flocked to West Orange to pay their respects to Edison's body – laid out in an open casket in his laboratory.

PERSONAL FAVOURITE
The phonograph was Edison's favourite invention. Despite his deafness, he loved opera. His phonograph company recorded opera singers from around the world, played on his later models.

In 1877, develops the carbon transmitter for the telephone and the tin foil phonograph, and is soon dubbed the Wizard of Menlo Park

1877–85

Patents the first electrical power system for lighting and founds the Edison Electric Illuminating Company in 1882

Relocates to West Orange, where he earns more than half of his 1,093 patents

1886–95

In 1891, patents the kinetograph and the kinetoscope, which initiates the movie industry

Sets up the Edison Storage Battery Company in 1901, and works to perfect the alkaline battery – his most profitable invention

1896–1915

NIKOLA TESLA

THE BRILLIANT LIGHTNING SCIENTIST

AUSTRIA-HUNGARY 1856–1943

THE MOST GIFTED ENGINEER of the late 19th and early 20th centuries, Nikola Tesla was the inventor of the AC motor and remote control, as well as a major pioneer of radio technology. Today, however, he is perhaps better known as the archetypal "mad scientist", operating in the outer limits of scientific respectability and given to some deeply bizarre behaviour.

A LIFE'S WORK

- Tesla **devises the method for transmitting alternating current** on which modern electrical power systems are based
- Invents the **Tesla coil** in 1891, which he uses to produce, among other things, some of the **first neon fluorescent illuminations, X-ray photographs, and wireless energy transfers**
- He **builds a hydroelectric plant** in the Niagara Falls
- **Showcases a radio-controlled boat in 1898**, which he terms a "teleautomaton"
- Makes important contributions to the fields of **radar and Vertical Take-off and Landing (VTOL)** in later life
- He is credited with about **300 patents** around the world

Tesla's early career in the last two decades of the 19th century is a tale of almost unrelenting success. With his powers at their peak, he seemed to be turning out paradigm-shifting inventions almost by the month. This was followed by a long decline, which, although punctuated by odd moments of brilliance, was generally characterized by feuds, failures, and increasing eccentricity.

Tesla grew up in Austria-Hungary (now Croatia). A supremely gifted young man, he studied at both the Austrian Polytechnic in Graz and the Charles Ferdinand University in Prague. In 1880, he moved to Budapest to become the chief electrician of the National Telephone Company, where he invented his induction motor, a type of alternating current (AC) motor that would revolutionize the industrial world. However, he could not find anyone in Europe willing to invest in it.

By 1882, he was working in Paris at the European outpost of Thomas Edison (see pp.262–65), the one man Tesla was sure would "get" his ideas, if only he could find a way to show them to him. Tesla travelled to the US in 1884 to meet Edison, armed with a letter of recommendation from Charles Batchelor, Tesla's boss at the Continental Edison Company, which reputedly stated, "I know two great men and you are one of them; the other is this young man".

Unfortunately, Tesla's faith in Edison's insight proved somewhat misplaced. Tesla used his brilliant intellect and academic rigour to conceptualize a solution, whereas Edison tended to rely on simple trial and error. Unable to conceive the theoretical possibilities of Tesla's AC motor, Edison was quite unimpressed with the machine. He was, however, impressed with Tesla, offering him a job and a reward of a staggering US $50,000 (more than US $1 million in today's money) if he could improve the efficiency of his direct current (DC) generation plants, then struggling to maintain the supply for his burgeoning electric-light industry. When Tesla did as he was asked, Edison supposedly refused to pay, whereupon Tesla resigned.

THE WAR OF THE CURRENTS

Tesla spent the next few years of his career perfecting his system for AC power transmission. His system provided a direct challenge to the DC system developed and patented by Edison, which had become the standard for electrical distribution in the US. The subsequent so-called "War of the Currents" in the 1880s and 1890s grew to become a worldwide battle for electrical supremacy. Although Edison's economic might and reputation warded off competition for a while, AC eventually won, principally because it was the better system – it could be transmitted over longer distances at higher voltages than DC, used a lower current, and had greater transmission efficiency.

ARTIFICIAL LIGHTNING
These coils, or windings, in a Tesla coil transform low-voltage electricity into high-voltage electricity, which can then be discharged into the air.

While he was busy convincing the world of the benefits of AC, Tesla was also growing increasingly interested in the wireless transmission of energy. His first major breakthrough was the development of the Tesla coil in 1891, a type of transformer that could produce pulses of high-frequency, high-voltage alternating current, which Tesla used to explore a variety of scientific avenues, including fluorescent lighting and X-ray photography. However, Tesla believed his device's greatest application would be in the wireless transmission of electrical energy, both through the ground and through the air. He had a vision of a network of devices spanning the globe to form a giant communication system, with messages being sent and received in an instant.

In 1902, Tesla entered into a public feud with Guglielmo Marconi (see pp.268–69), who would prove to be an even more fearsome adversary than Edison. By the mid-1890s, Tesla had become firmly established as the leading figure in radio-wave technology, having filed his first patent for radio in 1897. However, a fire destroyed much of his research and his place

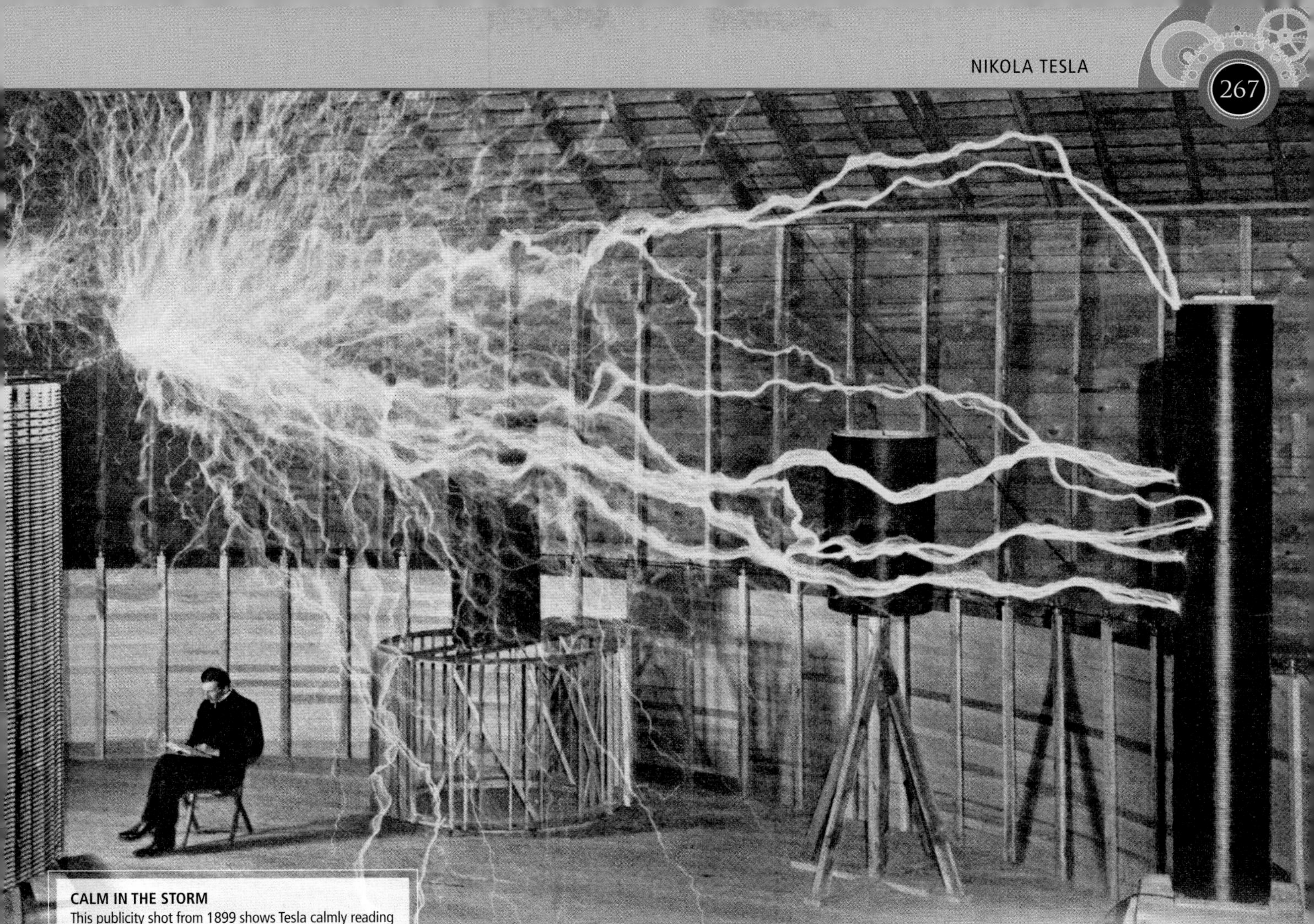

CALM IN THE STORM
This publicity shot from 1899 shows Tesla calmly reading a book while great electrical discharges arc across the room. If touched, these streamers could burn, but their low current meant that this would not normally be fatal.

at the forefront of innovation was taken by the young Marconi. In 1899, Marconi transmitted the first international radio signals across the English Channel. But the similarity between his equipment and that devised earlier by Tesla led the US patent office to turn down Marconi's patent application in 1900. But Marconi's stock was rising, literally, as his Wireless Telegraph Company began to receive significant financial backing from Andrew Carnegie and Thomas Edison. In 1901, Marconi upped the ante, transmitting radio signals across the Atlantic, but Tesla remained unfazed, commenting, "Marconi is a good fellow. Let him continue. He is using seventeen of my patents". His attitude would be considerably less relaxed in 1904, however, when the US Patent Office (probably under pressure from powerful vested interests) sensationally reversed its earlier decision and awarded Marconi the patent for the invention of the radio. Financial difficulties caused by the "War of the Currents" meant that Tesla was unable to challenge the decision. He also lost the royalties to his European patents at the outbreak of World War I.

THE "MAD SCIENTIST" YEARS

After the war, Tesla developed a severe case of obsessive-compulsive disorder, which manifested itself in an obsession with doing things in threes and an excessive devotion to pigeons, which he fed every day in Central Park. He continued to work in a range of fields, including radar and Vertical Take-off and Landing (VTOL), but increasingly became a marginalized figure. When he died in the New Yorker Hotel in 1943, he left behind significant debts, but one of the greatest inventing legacies of all time – amounting to about 300 patents worldwide – and a revered, if slightly unusual, place in popular culture.

MYTH MAN

Like Edison, Tesla was not merely a talented inventor, but also a tremendous self-publicist, more than happy to participate in the making of the myths that were as much a part of his public image when he was alive as they are now. He regularly staged spectacular public demonstrations in which he would use a Tesla coil to produce great arcs of artificial lightning bolts. He was also extremely adept at manipulating the press, forever feeding them juicy tidbits about his next project, the claims for which grew ever more outlandish the older and less successful he became (a "death ray", an "earthquake machine", and a method for splitting the world in two "like an apple" were some of the most sensational), but which were nonetheless eagerly lapped up and regurgitated by the newspapers of the day.

GUGLIELMO MARCONI

THE IMPROVER OF RADIO INVENTIONS

ITALY 1874–1937

IF NEWTON SAW FURTHER by standing "on the shoulders of giants", then Marconi used a ladder. Marconi's name is closely associated with the invention of the radio because of his success in exploiting radio's commercial potential. But it was men such as Heinrich Hertz and Nikola Tesla who were responsible for the real innovations. Marconi refined them and made them popular.

A LIFE'S WORK

- Marconi makes the **first international wireless broadcast** across the English Channel in 1899
- He **establishes the first transatlantic radiotelegraphy system in 1907**, which is initially used for ship-to-shore communication
- In 1904, **the US Patent Office rescinds Tesla's patents in favour of Marconi,** making Marconi the official "father of radio"
- He is **awarded the Nobel Prize for Physics** in 1909
- His **radiotelegraphs are credited with rescuing** many of those who survived the sinking of the *Titanic*
- His company **adopts the new technology of continuous wave transmitters in the 1920s** to produce some of the first audio-entertainment radio broadcasts in Britain
- **Joins the Italian Fascist Party in 1923** and is made a member of its grand council by Mussolini in 1930

Born in Bologna, Italy, Marconi's comfortable, home-schooled youth gave him plenty of time to pursue his own interests, which were of a practical, scientific bent. While studying at the Livorno Technical Institute, he learnt about the work of Heinrich Hertz (1857–94), who had recently established that it was possible to produce and detect electromagnetic radiation. Inspired by the discovery, Marconi thought it might be possible to create a new type of telegraph – then the most advanced communication system in the world – that did not use wires but low-frequency electromagnetic radiation: radio waves. He committed himself to the task and had soon managed to rig up a wireless device that made a bell ring on the other side of his laboratory when he pressed a button. Within a couple of years, he progressed to sending signals over distances of 2km (1.2 miles). None of these achievements were the result of his own great discoveries, however, but came through the careful assembly and improvement of existing components made by other engineers, including Hertz.

Convinced that he was on the right track, Marconi wrote to the Italian government for funding, but they did not respond. He then used his family's connections to get in touch with the Italian ambassador in London, who encouraged him to go to England to try and find investors. With the support of the chief electrical engineer of the British Post Office, Marconi continued his experiments in England, gradually increasing the distance of his transmissions, and gaining the interest of the British government. In 1897, he sent a signal 11km (7 miles) across Salisbury Plain, a feat he bettered later that year with the first radio communication across water: a 14-km (9-mile) transmission over the Bristol Channel. These achievements were followed by the first international wireless communication, sent 50km (31 miles) across the English Channel in 1899.

STATION TO STATION
This is a replica of the wireless telegraph that Marconi used to made the first international radio communication across the English Channel in 1899.

MESSAGE ACROSS THE ATLANTIC

Many experts at the time thought that Marconi's wireless system would have only a limited application because they believed that radio waves would be lost by the curvature of the Earth. To prove them wrong, Marconi embarked on his most audacious experiment. On 12 December 1901, he used a transmitter in Cornwall, England, to send a message 2,700km (1,700 miles) across the Atlantic Ocean to Newfoundland in Canada, where specially adapted antennae attached to balloons (to get them as high as possible) received the signals. The message may have comprised just a single letter, "S", but as the great physicist and radio pioneer Sir Oliver Lodge (1851–1940) stated, it represented an "epoch in history". Suddenly the idea of worldwide wireless communication was possible. Upon hearing the news, industrialist Andrew Carnegie and Thomas Edison (see pp.262–65) became interested, and began to provide Marconi with some investment capital.

By 1907, Marconi had established a regular transatlantic radiotelegraph system and his technologies were soon being used to aid communication all around the world. In 1909, Arctic explorer Robert E Peary relayed to the world that he was the first person to reach the Geographic North Pole by radio-telegraphing the words, "I have the Pole".

Marconi's success was greatly aided by his victory in a patent dispute with the other "father of radio", Nikola Tesla (see pp.266–67). Marconi's equipment included nothing that had not been originated by others. He was a refiner and an improver, not an inventor. His equipment, which consisted of a spark-gap transmitter and a coherer-receiver, used elements devised by, among others, Tesla and Lodge. As such, his initial application for a US patent for the invention of radio was turned down in 1900 in favour of Tesla's 1897 application. However, the US Patent Office reversed this decision in 1904, awarding the patent to Marconi and the profits to his rich investors (who had probably helped influence the decision).

TURNING DOWN THE SIGNAL

Awarded the Nobel Prize for his achievements in 1909, Marconi was the undisputed king of radio communication at the start of World War I. He served as a lieutenant in the Italian army and was put in charge of the country's military radio service. His post-war career, however, was considerably less pioneering. It took a long time for Marconi to be persuaded to change spark-gap transmitters (used to send radiotelegraph messages) to continuous wave transmitters, which opened up the possibility of audio communications. However, Marconi's company did adopt the new technology in 1920 – by which time it was already giving way to the vacuum-tube transmitter – to produce some of the first audio entertainment radio broadcasts in England. Marconi's empire continued to grow in the 1920s and '30s, expanding into new areas, including television and microwave transmissions.

Throughout his career, Marconi was able to use his connections to turn situations to his advantage – sometimes at the expense of more talented rivals. Indeed, it could be argued that he did not suffer a major defeat until 1943, six years after his death, when the US Supreme Court abruptly threw out most of Marconi's original patents, giving precedence once again to Tesla – a decision that may well have been a retaliatory move against the Marconi Company's decision to sue the US government for the illegal use of its patents in World War I.

SHIP-TO-SHORE STORIES

In the early 1910s, Marconi's radiotelegraph received tremendous publicity in the wake of two tragic events. The first came in 1910 with the arrest of Dr Crippen, as he disembarked from a transatlantic liner in Quebec, for the murder of his wife in London. The captain of the ship had recognized Crippen from police reports and radiotelegraphs, making the doctor the first criminal to be caught using Marconi's new technology. The second event took place two years later, following the sinking of the *Titanic*. The few people that survived did so chiefly through Marconi's telegraph service, with radio operators sending out distress signals that were picked up by RMS *Carpathia,* which rescued the survivors. Britain's postmaster general said of the event, "Those who have been saved, have been saved through one man, Mr Marconi … and his marvellous invention".

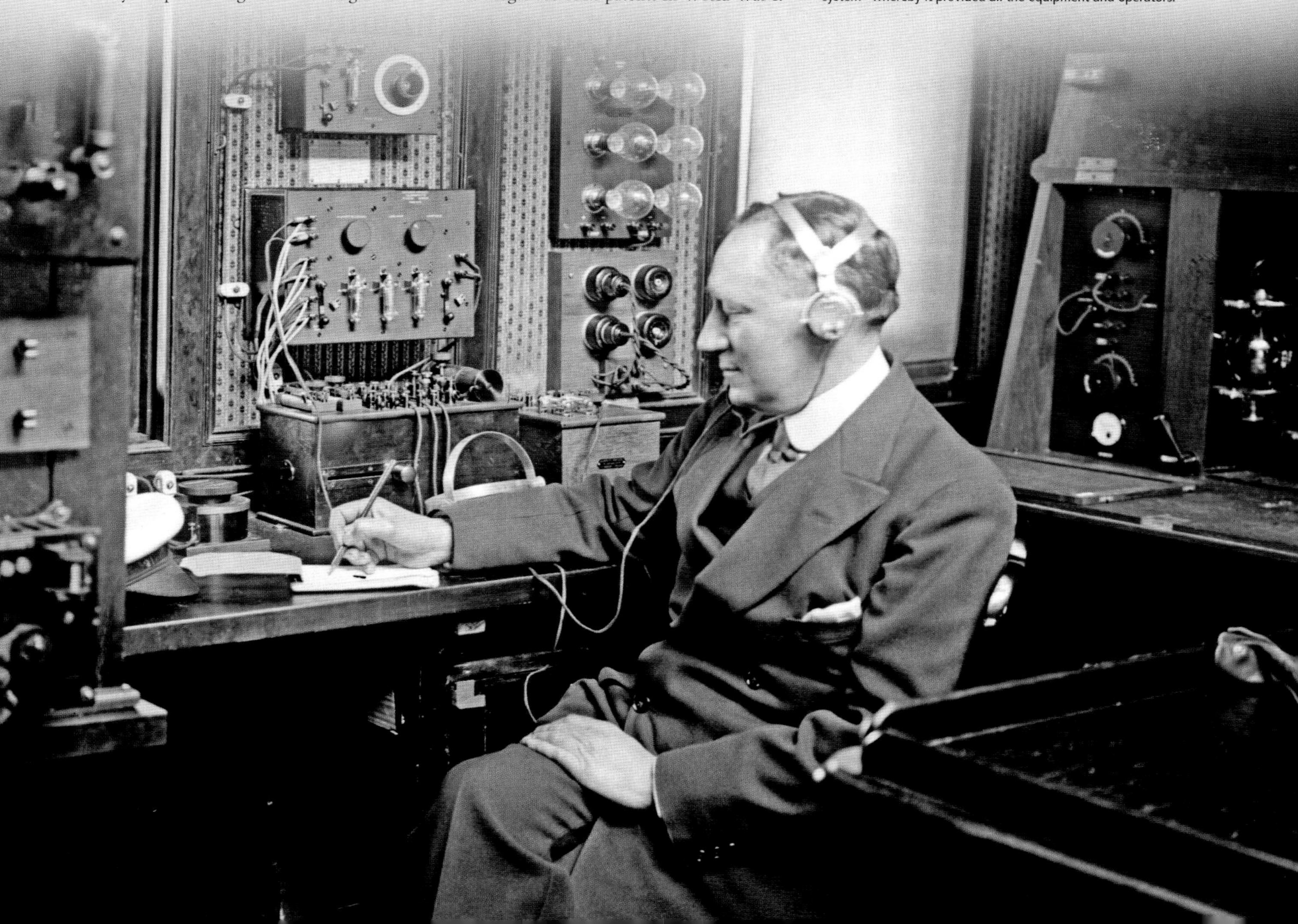

SHIP RADIO MONOPOLY
An image from 1920 shows Marconi sitting in the wireless room of a ship. His company dominated ship-to-shore radio communications in the early 20th century, operating a "closed system" whereby it provided all the equipment and operators.

JULIO CERVERA BAVIERA

PIONEER OF THE SPANISH RADIO INDUSTRY

SPAIN 1854–1929

An inventor, designer, and teacher, Cervera Baviera made major contributions to Spain's nascent radio industry, making possible one of the first radio transmissions over long distances and the development of the radiotelegraph. However, his short-lived career in the field of radio technology was overshadowed by the achievements of other inventors.

A LIFE'S WORK

- Cervera **leads a military expedition** into the uncharted reaches of the Sahara in 1886
- In 1890, he **publishes a criticism of the Spanish colonial government** in *El Imparcial* and is arrested for it
- Obtains **several patents** in England for his **refinements to the radio system**
- Obtains a patent for developing the **principle of remote control**
- Collaborates with Marconi and **establishes the world's second regular radiotelegraph service**
- In 1901, he **establishes the Spanish Wireless Telegraph and Telephone Company**, through which he initiates regular radiotelegraph services
- **Establishes a long-distance education programme** in the US in 1903, which **awards degrees** in mechanical engineering, agricultural engineering, and other subjects

Cervera's journey to the world of radio was complicated. Born just outside Valencia, he was keen on turning his early childhood interest in science into a career when he enrolled at the city's university to study physical and natural sciences. However, his life took a detour when he left his studies to join the Spanish army.

HEINRICH HERTZ

GERMANY 1857–94

The advances made by Julio Cervera Baviera and other radio pioneers would have been impossible without the contribution of German physicist Heinrich Hertz.

Hertz was the link between James Clerk Maxwell (1831–79), who formulated the classical theory of electromagnetism, and Cervera and Marconi who turned its implications into practical applications. Published in 1865, Maxwell's theory postulated that light is an electromagnetic wave – one with both electric and magnetic components – and that radio waves were a lower frequency form of electromagnetic wave. The theory was proved in 1888, when Hertz built an apparatus that could send and receive radio waves. The SI unit for frequency, the Hertz, was named in his honour.

Following his graduation from the Academy of Military Engineers in Guadalajara in the early 1880s, he was sent by the army on missions of discovery into previously unexplored parts of Morocco and the colony of Rio de Ora in the Spanish Sahara. His 900-km (560-mile) trek through the desert in 1886 is generally considered the first scientific study of that part of the Sahara. While on the trip, he also conducted government business, negotiating treaties with local chiefs on behalf of Spain.

TROUBLED TIMES

With his reputation growing, he became a military attaché to the Spanish embassy in Tangiers in 1888. However, his success made him overconfident. In 1890, he wrote an ill-judged attack on the Spanish colonial government in Morocco, which was published in the Moroccan newspaper *El Imparcial.*

The article got him into trouble. He was arrested and sent to prison on mainland Spain for two years. Nonetheless, following his release, he was able to resume his military career, serving once again in North Africa in Melilla, but also taking on assignments in the Canary Islands and Puerto Rico. In the latter country, he played a major role in the Spanish-American War of 1898, repulsing a US attack at the Battle of Guamani. In fact, his was one of the few success stories in the conflict, which in a few months saw Spain lose the remnants of its empire and suffer a huge blow to its national pride.

However, Cervera's uncompromising prose once again landed him in trouble. He wrote a pamphlet, *La Defensa de Puerto Rico*, on Spain's defeat, in which he squarely laid the blame for the defeat on Puerto Rican volunteers in the Spanish army – "I have never seen such a servile, ungrateful country … in 24 hours, the people of Puerto Rico went from being fervently Spanish to enthusiastically American … they humiliated themselves, giving in to the invader as the slave bows to the powerful lord". Cervera was challenged to a duel by the outraged soldiers of Puerto Rico, but for once managed to find the words to talk himself out of the situation.

RADIO TIMES

Just as the century was coming to a close, he embarked on his career in radio. Perhaps to keep him out of trouble, Cervera was removed from front-line duties in 1899 and sent to England to study Guglielmo Marconi's new wireless communication system (see pp.268–69) and determine whether it had any useful applications for the Spanish military. In England, Cervera's technical abilities, honed through many years of army training, were given free rein. He collaborated with Marconi on his work, helping to resolve the difficulties of making reliable radio transmissions over long distances, and obtained a number of patents for his refinements to the system. In fact, some experts now believe that the Spaniard's contributions were vital in overcoming many of the technical hurdles in the development of the

IT WAS COMMANDER CERVERA, WHO WORKED WITH MARCONI AND HIS ASSISTANT GEORGE KEMP IN 1899, WHO RESOLVED THE DIFFICULTIES OF THE WIRELESS TELEGRAPH

PROFESSOR ÁNGEL FAUS, UNIVERSITY OF NAVARRA

radiotelegraph. It is certainly true that Marconi was not a particularly able inventor, relying more on his ability to exploit and adapt the ingenuity of others.

For a while it looked as if Cervera was set to become the dominant figure in the field of radio. On returning to Spain in 1901, he founded the Spanish Wireless Telegraph and Telephone Company, through which he established regular radiotelegraph services (the earliest in the world after Marconi's) between Tarifa and Cueta, and Jávea and Ibiza. He also obtained a patent for "the production of signals for imparting motion to machines or apparatus", in other words "remote control", although there appears to be no record of him having found a practical application for this.

In fact, just as suddenly as Cervera had entered the world of radio, he left it. Perhaps because of the lack of recognition from the army or his countrymen, or because he felt that he could not compete with Marconi and his rich financiers, or even because he believed that he had contributed all he could to the field, he brought his inventive efforts to an end and began encouraging those of others instead by taking up a career in teaching.

In 1900, he was made an instructor at the Escuela Superior de Artes e Industrias de Madrid. However, unhappy with the curriculum, he left the position. In 1903, he moved to the US, where he set up a long-distance education programme, "the Institución de Enseñaza Técnica", which awarded degrees in mechanical engineering, agricultural engineering, and other subjects.

Even though Cervera's career in radio was now over, he kept his interest in engineering alive, designing tram systems for Tenerife and his hometown of Segorbe. He later returned to Spain and briefly held political office before moving to Madrid, where he died in 1929.

SPANISH RADIO
Cervera's radiotelegraph service could not make audio broadcasts, but by the time of his death, audio broadcasting was an established industry and many owned a radio.

MODERN TIMES

MODERN TIMES

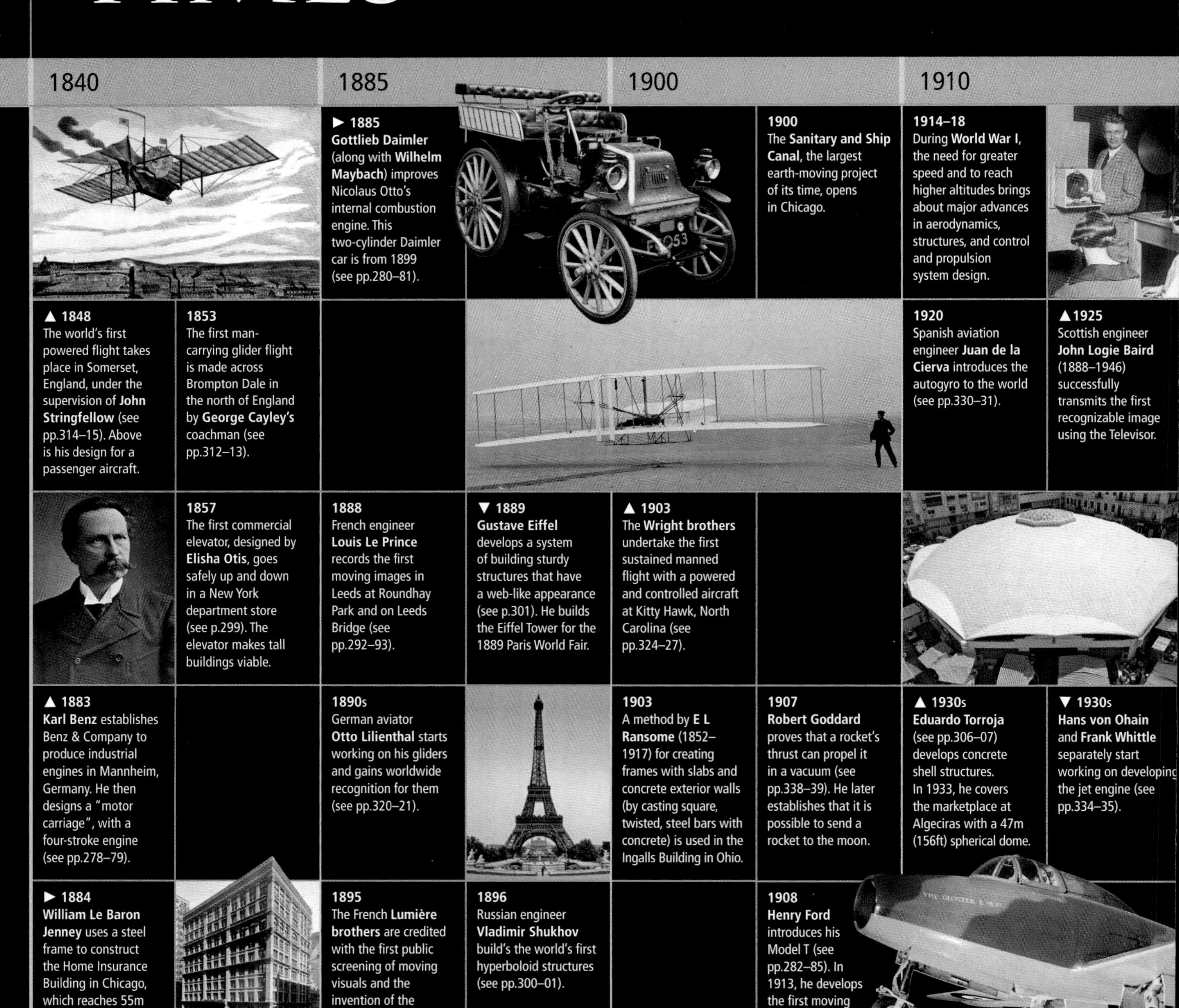

1840

1885

1900

1910

▲ **1848**
The world's first powered flight takes place in Somerset, England, under the supervision of **John Stringfellow** (see pp.314–15). Above is his design for a passenger aircraft.

1853
The first man-carrying glider flight is made across Brompton Dale in the north of England by **George Cayley's** coachman (see pp.312–13).

1857
The first commercial elevator, designed by **Elisha Otis**, goes safely up and down in a New York department store (see p.299). The elevator makes tall buildings viable.

▲ **1883**
Karl Benz establishes Benz & Company to produce industrial engines in Mannheim, Germany. He then designs a "motor carriage", with a four-stroke engine (see pp.278–79).

► **1884**
William Le Baron Jenney uses a steel frame to construct the Home Insurance Building in Chicago, which reaches 55m (180ft) (see pp.298–99). It is the world's first skyscraper.

► **1885**
Gottlieb Daimler (along with **Wilhelm Maybach**) improves Nicolaus Otto's internal combustion engine. This two-cylinder Daimler car is from 1899 (see pp.280–81).

1888
French engineer **Louis Le Prince** records the first moving images in Leeds at Roundhay Park and on Leeds Bridge (see pp.292–93).

▼ **1889**
Gustave Eiffel develops a system of building sturdy structures that have a web-like appearance (see p.301). He builds the Eiffel Tower for the 1889 Paris World Fair.

1890s
German aviator **Otto Lilienthal** starts working on his gliders and gains worldwide recognition for them (see pp.320–21).

1895
The French **Lumière brothers** are credited with the first public screening of moving visuals and the invention of the "cinematograph" (see pp.294–95).

1896
Russian engineer **Vladimir Shukhov** build's the world's first hyperboloid structures (see pp.300–01).

1900
The **Sanitary and Ship Canal**, the largest earth-moving project of its time, opens in Chicago.

▲ **1903**
The **Wright brothers** undertake the first sustained manned flight with a powered and controlled aircraft at Kitty Hawk, North Carolina (see pp.324–27).

1903
A method by **E L Ransome** (1852–1917) for creating frames with slabs and concrete exterior walls (by casting square, twisted, steel bars with concrete) is used in the Ingalls Building in Ohio.

1907
Robert Goddard proves that a rocket's thrust can propel it in a vacuum (see pp.338–39). He later establishes that it is possible to send a rocket to the moon.

1908
Henry Ford introduces his Model T (see pp.282–85). In 1913, he develops the first moving assembly line for automobiles.

1914–18
During **World War I**, the need for greater speed and to reach higher altitudes brings about major advances in aerodynamics, structures, and control and propulsion system design.

1920
Spanish aviation engineer **Juan de la Cierva** introduces the autogyro to the world (see pp.330–31).

▲ **1925**
Scottish engineer **John Logie Baird** (1888–1946) successfully transmits the first recognizable image using the Televisor.

▲ **1930s**
Eduardo Torroja (see pp.306–07) develops concrete shell structures. In 1933, he covers the marketplace at Algeciras with a 47m (156ft) spherical dome.

▼ **1930s**
Hans von Ohain and **Frank Whittle** separately start working on developing the jet engine (see pp.334–35).

◄ **PP.272–73** The Apollo service module orbits above the Moon.

ECHNOLOGICAL INNOVATION ACCELERATED during the ate 19th and early 20th centuries, greatly transforming eople's lives. Victorian inventors introduced the steam urbine and the internal combustion engine, allowing he cheap production of steel and aluminium. Electricity rovided power for new industries, and brought the world ogether through cinema and television. The rapid development of the motorcar and the aeroplane drove the economies of the industrial nations and fostered further invention. Yet it took a major war to redouble the efforts of scientists and engineers. Rockets, jet engines, electronics, nuclear power, and the computer were born out of desperate times. Space technology is still in development, but already our horizons have been expanded beyond all recognition.

1935

1960

1990

1960s
Fazlur Khan revolutionizes the design of tall buildings through both steel and concrete with his tube system, which culminates in the Sears Tower in Chicago (see pp.308–09).

1990
The **Hubble Space Telescope** is carried into orbit and begins operations. It is soon sending back images of galaxies billions of light years away.

1990
British civil engineer **Don Burland** uses new grouting technology to prevent the Leaning Tower of Pisa from falling.

1961
USSR's **Yuri Gagarin** becomes the first human to travel to Earth's orbit.

1969
Man steps on the Moon. The feat is achieved by **Neil Armstrong** and **Edwin "Buzz" Aldrin**.

◀ 1997
The first prototype of a **robotic vacuum cleaner** is unveiled.

▲ 1998
The **International Space Station** is launched into low Earth orbit. It serves as a research laboratory, with a permanent crew.

▲ 1936
he **Hoover Dam**, the irst large concrete-rch dam, is completed n the border between rizona and Nevada.

1938
Austrian engineer **Ferdinand Porsche** creates Volks-Wagen, the People's Car, on the directive of Adolf Hitler (see pp.288–89).

1970s
One of the world's largest embankment dams, the **Aswan High Dam**, across the River Nile in Egypt, is completed.

▼ 1981
The **Space Shuttle Columbia** is launched. It combines rocket launchers, an orbital spacecraft, and a re-entry spaceplane.

2004
Structural engineer **Michel Virlogeux**'s (b.1946) Millau Viaduct bridge opens in the south of France. It is the longest cable-stayed bridge and the tallest road bridge in the world.

939–45
Vorld War II witnesses he use of aircraft-etecting radar, adio-wave navigation, irborne radar, the first et fighter, the V-2 ocket, and the lying Fortress by the articipating countries.

1950s
Japanese architect **Tachu Naito** develops structural engineering and designs the lightweight Tokyo Tower, the world's tallest self-supporting steel tower (see pp.304–05).

1956
George Devol (1912–2011) invents and patents a process for recording instructions in a robotic device. In 1956, he develops Unimate, the first industrial robot.

1957
Sputnik I launches the Space Race between the Soviet Union and the US. Two years later, Luna 3 takes pictures of the dark side of the Moon.

2008
The **Large Hadron Collider**, the world's largest particle accelerator, opens 100m (300ft) beneath the Swiss-French border (see pp.348–49).

▲ 2009
The world's tallest building, the (818-m) 2,717-ft **Burj Khalifa in Dubai**, is completed.

THE MOTORCAR

IT IS IMPOSSIBLE TO OVERSTATE THE IMPORTANCE OF THE CAR, THE INVENTION THAT HAS TRANSFORMED THE LIVES OF MILLIONS. A RESULT OF NUMEROUS EXPERIMENTS, PATENTS, FLAIR, AND ENTERPRISE, THE MOTORCAR IS ONE OF THE GREATEST DEVELOPMENTS OF THE 20TH CENTURY.

A PROLIFIC PARTNERSHIP
By 1900 the development of the motorcar was well underway. The French partnership of De Bion and Bouton was producing more than 50 types of cars (with a single-cylinder petrol engine).

Mention any other invention and individual scientists spring to mind. The wireless, the locomotive, and the aeroplane for example, conjure up respectively Marconi (even though he only improved what had been invented earlier), the Stephensons, and the Wright brothers. This is not the case with the motorcar. Developed in workshops, kitchens, and garden sheds across many countries, it evolved piece by piece by various men working independently of one another. Further, when the pioneers of the motorcar began experimenting, there was no precedent. Some had experimented with the horseless vehicle propelled by steam, and for a time, steam carriages looked like the future. The invention of the automobile had to wait for the arrival of an internal-combustion engine, but by the turn of the 20th century, most aspects of the modern motorcar were in place.

TOWARDS THE AUTOMOBILE

French engineer Beau de Rochas (1815–93) had designed a four-stroke internal-combustion engine in 1862, but a dispute over the priority with Nicolaus Otto (1832–91) allowed it to be used by any manufacturer. The first vehicle to be driven by an internal-combustion engine was a close run between Frenchman Étienne Lenoir (1822–1900) and Austrian Siegfried Markus (1831–98). The latter used petrol as a fuel in a two-stroke internal-combustion engine. In 1870, he built a crude vehicle with no seats, steering, or brakes. Both Lenoir and Markus were discouraged, and gave up. In 1883, Frenchmen Edouard Delamare-Deboutteville and Léon Malandin installed

MASS PRODUCTION
By the 1920s, car manufacturers from around the world were producing vehicles in bulk, which were affordable for the average worker.

MOTOR RACES
The dawn of the 20th century saw a rapid rise in the popularity of cars as manufacturers tested faster and more powerful engines in the new sport of motor racing.

RANSOM OLDS
The man responsible for the first series-produced car, Olds invented the Curved Dash model in 1901; in its first year 425 were manufactured in America.

SETTING THE SCENE

- Thirteenth-century Franciscan Friar Roger Bacon envisions **carriages moving without animals** to pull them along. In 1478, Leonardo da Vinci sketches a **self-propelled vehicle**.
- In about 1670, Jesuit missionary Ferdinand Verbiest develops a **steam-powered trolley**.
- The earliest sufficiently light, compact, and powerful source of motive power is the **high-pressure steam engine** built by Nicholas Cugnot in 1769.
- Richard Trevithick builds a second steam carriage in 1803, which makes several runs in London. Other engineers continue to develop the steam engine. In 1831, Charles Dance sets up the **world's first passenger service** by automobiles.
- Trevithick's engines are small and powerful, and lead to the **development of steam carriages and railway locomotives**.

an internal-combustion engine powered by a tank of gas on a tricycle. The tank exploded. Undeterred, they built another contraption, but it, too, came apart during the trial.

Steam carriages and trains evolved during the first decades of the 19th century amid extreme hostility from the Church and the state. However, it was not until the end of the century that the motor car took to the road.

UP TO SPEED

It was left to Karl Benz (see pp.278–79) to persevere and claim the title of inventor of the first automobile. Benz completed his gas-powered Motorwagen in 1885 and patented it in 1886 after years of setbacks. Close on his heels was fellow German Gottlieb Daimler (see pp. 280–81). Unlike Benz, who worked alone, Daimler was helped by the genius of Wilhelm Maybach (see pp. 280–81), who, in 1890, built the first four-cylinder, four-stroke engine.

Advances in fuel technology gave a massive boost to the industry. Rudolf Diesel (see p.281) almost killed himself trying to prove that fuel could be ignited without a spark. He operated his first successful engine in 1897.

Until this point, cars looked roughly the same. The first car to break the mould emerged in 1901. Designed by Maybach and Paul Daimler for their French financier Emile Jellinek, the Mercedes had a handsome, low, pressed-steel chassis, and a long bonnet. Its four-cylinder, 35-horsepower (26-kilowatt) engine was state of the art.

The British lagged behind. While the French were whizzing about their roads and celebrating the first races, the railway lobby in the British parliament kept cars to a crawl: only 19kph (12mph). This did not hold back British firm Thorpe and Salter from engineering a speedometer in 1902 that went up to 56kph (35mph).

The Americans also had a slow start, impeded by a heavy tax on imports. A single-cylinder petrol automobile, the "Curved Dash" invented by Ransom Olds (1864–1950), proved popular. By 1901, Olds was producing ten a week. By 1904, this had increased to 100, the precursor to mass production. America scorned the hand-crafted traditions of Europe. In 1903, Cadillac sold cars with fully interchangeable parts, and, in 1908, the Model T Ford, a car that could be quickly and cheaply assembled, made its debut. It was the first car for the masses.

Designers now focused on ease, safety, and style. Driving had been largely for the hardy, or even foolhardy. Bumpers were patented by British firm Simms in 1905, and night driving became safe only in 1908 when the British completed an electrical system that included headlights, sidelights, and a tail-light. Windscreens were optional and shattered easily. In 1909, French chemist Edouard Benedictus (1879–1930) solved the problem by inventing shatter-proof glass. The pace of innovation increased just prior to and after World War II. Charles Nash (1864–1948) pioneered air conditioning in 1939, and power steering and automatic gears were introduced. Ferdinand Porsche designed super-fast cars (see pp.288–89), and, at Adolf Hitler's request, a small car for the masses: the Volkswagen Beetle.

Cars became cheaper, more comfortable, and more durable. They were no longer a luxury, but a necessity. In search of new challenges and to address growing pollution levels, the motor industry and its engineers turned to a pollution-free electric car – a quest that continues.

EARLY INNOVATIONS
This innovative vehicle was one of the first produced by American James L Packard. In his 1901 Model C he uses a steering wheel instead of a tiller. Its top speed was 40kph (25mph).

KARL BENZ

THE MAKER OF THE FIRST HORSELESS CARRIAGE

GERMAN CONFEDERATION 1844–1929

One of the most important pioneers of the automobile, Karl Benz was a mechanical engineer who was determined to replace the horse-drawn carriage with one pulled by an engine. Often impoverished and discouraged, he persevered with the help of his wife, Bertha, to play a pivotal role in the transport revolution by inventing the gasoline-powered car.

A LIFE'S WORK

- Benz **unveils his first vehicle propelled by an internal combustion engine** in 1886 to mixed reviews, and is banned from driving his "devil's carriage"; he struggles to convince backers to invest
- He **fails miserably in business and marketing**, and is saved by his wife's daring road trip to garner publicity in 1888
- Refines his designs, adding the four-wheeled Benz Victoria and the Benz Velo, and his company **becomes the world's leading automobile manufacturer** by the turn of the century
- He struggles to accept modernization and **resigns from his company in 1903, but rejoins the board as an adviser**, living to see Benz & Co merge with its competitor, Daimler

Karl Benz had an inauspicious beginning. His father, a train driver, died when he was very young and his mother struggled by on a widow's pension. She became a cook and supplemented her income by renting out rooms to university students. Aged nine, Benz was sent to the Karlsruhe Gymnasium (grammar school), where he showed remarkable aptitude for physics and chemistry – so much so, that he was made assistant to the master of physics.

Benz had inherited five timepieces from his father, and helped his mother by repairing clocks and watches. At 17, he attended the city's polytechnic to study engineering. His tutor, Professor Ferdinand Rechtenbacher, a distinguished engineer, postulated that the steam engine's days were numbered as it was too big and bulky. An enthusiastic student, Benz began to imagine whether an internal combustion engine could take its place. He dreamt that he could take his father's train off the tracks and drive it anywhere he wanted.

A ROUGH BEGINNING

The Engine Construction Company of Karlsruhe offered Benz his first job. The hours were long and the light in the factory so poor that he suffered from acute headaches and eyestrain. In his scant spare time, he sketched out designs for a horseless carriage that incorporated the internal combustion engine. In 1867, he gained a position with Johann Schweizer and Company at Mannheim, a manufacturer of cranes and wagons. After two years, in search of higher wages, he moved again, joining the bridge-building firm of Benckiser Brothers. There he met his future wife, Bertha Ringer. In 1871, Bertha invested in Karl's workshop, making it possible to carry on with a long and expensive developing process. They married in 1872. Benz used Bertha's dowry to start his own small engineering business, risking everything on his vision for the internal combustion engine. He set up a workshop making a small two-stroke gas engine suitable for driving pumps. He was penniless; on one occasion his tools were confiscated by bailiffs, and, on the night his first gas engine was finished, he and Bertha had no money for an evening meal.

Slowly, however, he managed to sell a few of his engines to local factories. The workshop became too small, and he expanded, but was thwarted again when his backers made it clear that they were only interested in using the engine for manufacturing – not a penny was to be wasted on Benz's strange idea of a horseless carriage. Benz was eventually forced to dissolve the company and return to his old workshop.

FIRST FOUR-WHEELED BENZ CAR
Karl Benz and his daughter Clara take a spin in the Benz Victoria, the first four-wheeled Benz car. With its ground-breaking steering, the automobile was to remain Benz's favourite vehicle.

BENZ & CO

In 1883, Benz found financial support in Max Rose and Friedrich Esslinger, who ran a bicycle shop in Mannheim. In October the same year, the three men founded a new company: Benz & Co. Soon static gas engines were produced there. The success of the company gave Benz the opportunity to indulge in his old passion of designing a horseless carriage. At last he could focus on his invention.

The design of his carriage evolved: it would have to be lightweight, and his gas engine would need to run at high speed to provide enough power. He used the same technology when he created an automobile. He worked on an electrical ignition, a carburettor, and a cooling system, and took out several patents. Most significantly, instead of transforming an existing carriage, he built the frame from scratch. In 1885, his crude three-wheeled contraption was ready. Karl and Bertha enjoyed listening to the sound of the

engine running for more than an hour. On 29 January 1886, he took out a patent, and the first "modern" automobile was born.

When Benz first drove the motorized wagon out of his workshop, it stalled several times. During subsequent trials, Benz drove Bertha around the neighbourhood. Gradually, the whole of Mannheim heard of the strange machine. Huge crowds followed, and Benz, worried that the police would stop them from undertaking their trips, started road-testing at night.

He was right to feel anxious. The year 1888 was full of setbacks. Benz was banned from driving the contraption because the church did not approve; it was called the "devil's carriage" and it frightened the horses. He was discouraged by lack of enthusiasm in his own country; Gottlieb Daimler (see pp.280–81) seemed to be hogging the headlines. He tried exhibiting in Paris, but failed miserably to drum up interest.

MERITED SUCCESS

Bertha, however, had faith. She was convinced that her husband's invention would be a success, but only if people got to hear about it. She and their two sons hatched a plot to drive from Mannheim to Pforzheim, a distance of 120km (75 miles), to prove the car's reliability. She set off in August 1888 and their journey, the first joyride in history, was a triumph (see box, right).

Soon, a syndicate was opened in France to sell Benz cars and business came flooding in. Between 1894 and 1901, the Benz "Velo" was built, which was a light, reasonably priced vehicle for two people. At the turn of the century, Benz & Co had grown into the world's leading automobile manufacturer.

Designers across Europe were catching up, and Benz cars began to look dated. Benz resisted modernizing his machines and resigned from the company in 1903. For some years, it was clear that his health was failing. On 2 April 1929, several hundred motor cars proceeded along Kenz Platz in Ladenburg to pay homage to the great man and acknowledge his work as the inventor of the automobile. Benz was too frail to view the procession, and died two days later.

BERTHA'S ROAD TRIP

In August 1888, Bertha Benz left her husband a note, stole the car, and set off. Her road trip from Mannheim to Pforzheim and back was not only extremely risky, it revealed how the vehicle could be improved. En route she cleared a fuel blockage with her hat pin, and used her garters to insulate an ignition wire that short-circuited. On the way to Pforzheim, the car had difficulty climbing hills, so Bertha drove up in reverse. When she got home, she told Karl the car needed another gear.

ON THE JOYRIDE TO PFORZHEIM, BERTHA BENZ RESOLVED MANY TECHNICAL GLITCHES WITH THE MOTORWAGEN, EVEN BUYING LIGROIN FROM A PHARMACY TO USE AS FUEL.

KARL BENZ ON HIS MOTORWAGEN
Karl Benz sits on his Motorwagen, the first official automobile (1885). By the time this picture was taken in the 1920s, Benz's invention was already a historical curiosity. It survived a crash, and hostility from neighbours, but by 1888 it was in production and became the first car ever to be sold to the public. Initially, though, buyers were few on the ground.

GOTTLIEB DAIMLER

PIONEER OF THE MODERN AUTOMOBILE INDUSTRY

GERMAN CONFEDERATION 1834–1900

Motor-vehicle pioneer Gottlieb Daimler has an assured place in history as one of the founding fathers of the automobile age. The inventor of the carburettor, he was an engineer, industrialist, and designer. A perfectionist and tireless worker, he developed engines that powered boats, airships, trams, and cars. He founded motor factories all around the globe.

A LIFE'S WORK

- Daimler meets **Wilhelm Maybach** and they begin a close and **productive partnership**, sharing a mission to develop an **efficient engine**
- In 1883, the pair build the first-ever **high-speed four-stroke engine** in their garden workshop. They apply it to a bicycle frame, making the world's **first motorized bike**
- They make refinements and fit the engine to a riding carriage in 1887; three years later the **Daimler Motor Company** is established, and the single-cylinder engine is used for boats, airships, trucks, and trams
- A Daimler-powered car wins the first international car race, the **1894 Paris–Rouen**
- In 1898, they manufacture the **first four-cylinder car**, and design the **Mercedes**, which is delivered after Daimler's death in 1900

Gottlieb Daimler was not of engineering stock. Born in the village of Höllgasse, in Schorndorf, his father came from a long line of master-bakers. His parents sent him to a school that was known for its strict discipline, and wanted him to become a civil servant. Daimler, however, showed little aptitude for such a vocation and was more interested in mechanics.

EARLY HARD WORK

On leaving school, he took up an apprenticeship with a gun-maker named Hermann Reythel, as engineering opportunities were thin on the ground. Reythel was impressed with his new charge. Daimler swiftly moved on, completed his apprenticeship with another gun-maker, and went with him to Stuttgart. This was a turning point in Daimler's life, for he came into contact with the most influential people in the city. In 1853, he joined a machine-tool factory in Grafenstadt, near Strasburg, which had links with a famous mechanical training college. In 1857, he managed to get a place there to study engineering. He was persuaded to travel to England, where he was employed at the Coventry works of Sir Joseph Whitworth, (see pp.152–53) but he was restless, and never seemed to stay long in one place. He went to Belgium, France, and then back to Germany, where he worked with the young engineer Wilhelm Maybach (1846–1929), but soon moved on again, this time to the Karlsruhe Engineering Works. There he met and married Emma Kurz, the daughter of a chemist, and stayed at the company for five years.

WOODEN MOTORCYCLE
This motorcycle, produced by Daimler in 1885, had an internal combustion engine, wooden frame, and steel-rimmed wooden wheels.

A REMARKABLE PARTNERSHIP

In 1872, Nikolaus Otto (1832–91), the inventor of the four-stroke internal combustion engine, established a factory near Cologne. He offered Daimler the position of Technical Director. It was a dynamic partnership, during which the factory was restructured, and soon gained international attention. Maybach became the chief designer. Then followed 10 years of hard, intense labour, during which Daimler never took a holiday.

In 1881, Daimler and Otto fell out, and Daimler and Maybach became convinced that Otto did not appreciate the potential of the internal combustion engine as a means of transport. They resigned, and set up a business of their own in a converted greenhouse in Daimler's back garden. They worked for long hours, in strictest secrecy – so much so that one night they were raided by police on suspicion of manufacturing fake coins.

The first engine Daimler and Maybach made in 1883 ran on gasoline. It had an efficient ignition system and surface carburettor. They mounted the engine on a wooden bicycle and tested it – the world's first "motored bike".

Daimler built his engine onto a small boat, but prospective passengers were afraid that the boat would blow up, and refused to sail in it. In 1885, they refined the engine and placed it under the seat of a new motorbike frame. When Daimler's son Paul took it for a ride, however, the seat caught fire. Working on the practicality of the engine, the pair focused on the construction of a four-wheeled car. In 1887, their first gasoline-powered vehicle, looking like a horse-drawn carriage with the shafts removed and a steering wheel fitted, was taken out at night to sputter down the streets of Cannstatt. Finally, Daimler and Maybach had produced something with real commercial potential and they set up the Daimler Motor Company in 1890.

IMMENSE PERSEVERANCE

Daimler and Maybach never stopped inventing. In 1892, they produced a new engine, the Phoenix. Maybach's pioneering invention of a spray-nozzle carburettor in 1893 became the model for carburettors for decades to come. In 1898, he demonstrated a fire-extinguishing pump, driven by a Daimler engine, and fitted the first engine to an airship. They also improved the design of their motorized carriage, producing a twin-cylinder vehicle with slender iron wheels in 1899.

Commercially, Daimler was thriving. Syndicates had been set up in Europe and orders flooded in. In 1896, the Daimler Company produced the first road truck. Then in Paris, in the late 1890s, he met Emile Jellinek (1853–1918), an influential industrialist who was selling Daimler vehicles on the Côte d'Azur. He asked Daimler to build a car that was lower, "less wobbly", and lighter than previous models. Maybach went back to the drawing board. Jellinek was impressed and saw a future for the new car in France, but only if the name Daimler was dropped as it sounded too Germanic. It was suggested that the new type of car be named Mercedes, after Jellinek's daughter.

Daimler did not live to see the Mercedes being delivered. He died in 1900 of heart disease. Ironically, the man who helped pioneer the modern automobile industry hated driving, and may never have driven any of his cars.

RUDOLF DIESEL

GERMANY 1858–1913

Born to an immigrant leather worker in Paris, Diesel trained in Germany as a thermal engineer, and in 1885 conceived an improvement to Otto's internal combustion engine.

His idea was to compress the air-fuel mixture in the driving cylinder to a point where it was hot enough to ignite itself. Early attempts exploded and almost killed him. He switched his attention to crude fuel oil; his engine became vastly superior and was adopted worldwide. Diesel disappeared without a trace on a ferry trip to London.

FOUR-WHEELED DAIMLER
The first four-wheeled automobile powered by a petrol engine is driven by Wilhelm Maybach and Daimler's son Paul. Daimler took a stagecoach and adapted it to hold the engine, called the "grandfather clock". The vehicle, produced in 1886, reached 16kph (10mph).

HENRY FORD

VISIONARY WHO BUILT CARS FOR THE MASSES

UNITED STATES 1863–1947

FOR ALL HIS MECHANICAL TALENTS, Henry Ford was primarily a manufacturer whose genius made him the father of 20th-century industry. He introduced mass production, and with it mass consumption, changing the face of the United States. Stubborn and dictatorial, he alienated many, yet was hailed by ordinary Americans as a hero.

Sailing from Ireland in 1847, William Ford left behind a country decimated by the Great Potato Famine to seek a better life in America. He chose Dearborn, Michigan, where land was cheap and filled with timber. The farm he cultivated with his wife Mary covered hundreds of acres, and as their children grew, they were expected to contribute. Chores came first. The eldest son, Henry, loathed farm work, and dragged his feet with milking and apple-picking. He preferred mathematics, tinkering with machines, and playing practical jokes.

Two events shaped Henry's life dramatically. He was given a watch for his twelfth birthday, and this began a lifelong passion for building and making things. A year later, shortly after the death of his mother, he took a trip with his father in their horse-drawn wagon and came across a steam engine. These were being increasingly used by farmers, but this monster of a machine was different. Instead of being pulled by horses, it chugged along by its own power. Ford was captivated.

SEEING THE WORLD

In 1876, he visited the Centennial Exposition Fair in Philadelphia. Packed with the latest in design and technology, it was an inspiring experience. He saw an exhibition by inventor Thomas Edison (see pp.262–65), and was awed by a two-storey-high Corliss Steam Engine, the biggest ever built. Suddenly, "science" and "invention" took on a whole new meaning.

Although proud of his son's mechanical prowess, Ford's father still hoped he would commit to the family farm. But at the age of 16, Ford left home and walked 15km (9 miles) to Detroit to begin life as an apprentice machinist. His first job lasted only six days. His father came to the rescue, introducing him to friends who owned James Flowers & Brothers machine shop. There, Ford lasted longer, managing a nine-month stint before joining the Detroit Dry Dock Company, where he learned more about tools and machinery.

In 1884, Ford turned 21 and his father, still anxious to establish him on the farm, gave him 16 hectares (40 acres) of land. Determined to avoid the drudgery of manual labour, Ford became an expert

FORD'S FIRST CAR
Ford sits proudly in his first car, the Quadricycle. Built in his workshop in 1896, it has a simple frame with four bicycle wheels and a twin-cylinder gasoline engine.

A LIFE'S WORK

- Ford **completes his first automobile, the "Quadricycle"**, in 1896, and drives it through the streets of Detroit to the annoyance of pedestrians
- In 1908, he **begins manufacturing** his famous **Model T**, and moves operations to a factory at Highland Park, Michigan
- Introduces the first **assembly-line production** and **conveyor belt** in 1913; **one complete car** leaves the production line **every ten seconds**
- Alarms the industrial world by raising the **minimum wage to $5** and slashing the **working day to eight hours**
- Begins construction of an **industrial plant** on the Rouge River, Dearborn, which would house everything needed to make his cars

WORKERS ON SHIFT
The afternoon shift at the largest industrial complex in the world at Highland Park, Michigan, is pictured here in the mid-1920s. Ford introduced a revolutionary eight-hour day for employees. His plant ran for 20 hours a day, and workers were constantly changing shifts.

MOST PEOPLE SPEND MORE TIME AND ENERGY GOING AROUND PROBLEMS THAN IN TRYING TO SOLVE THEM

HENRY FORD

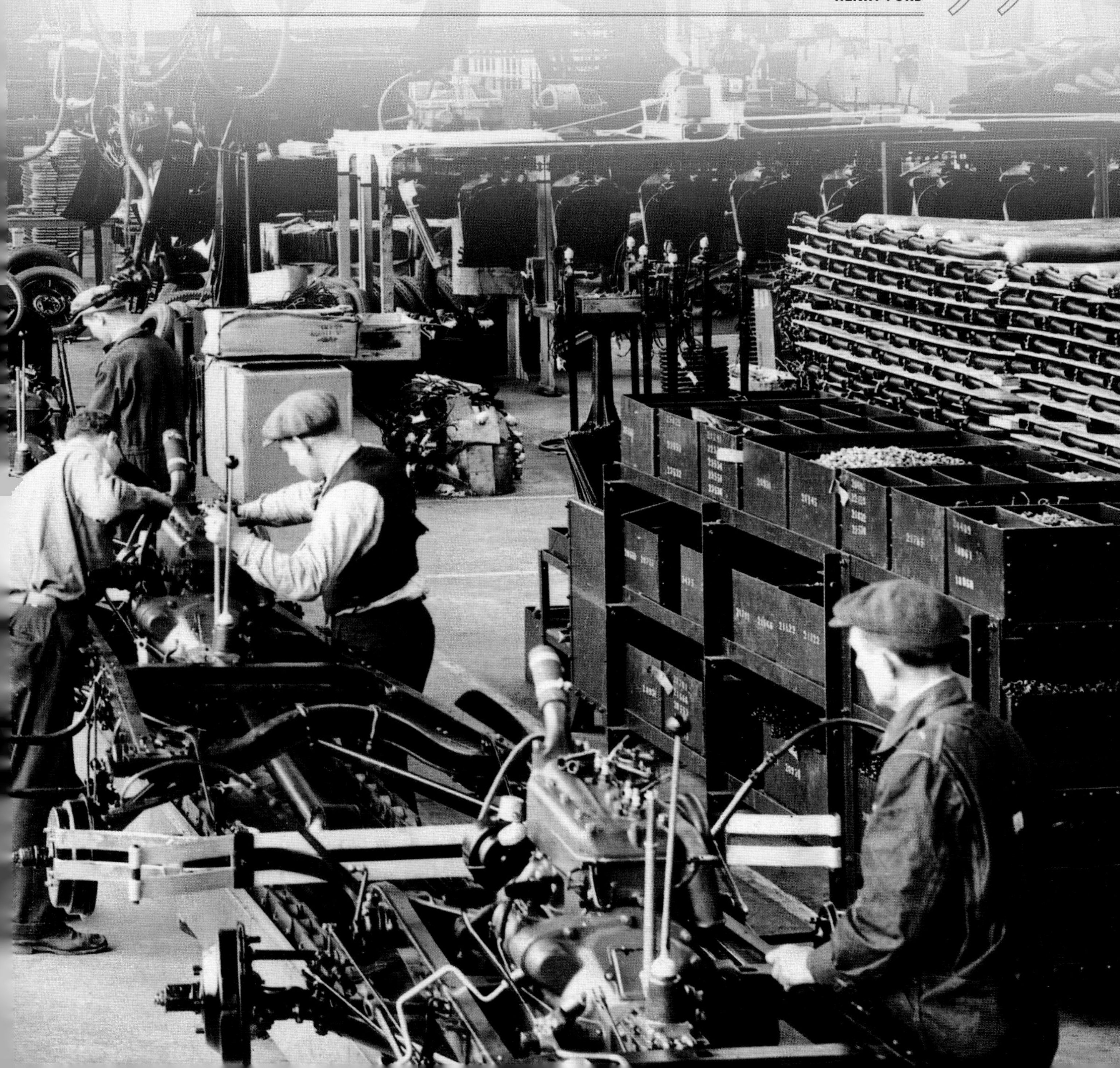

on the Westinghouse Portable Steam Engine, and demonstrated it to farmers across Michigan in his belief that machines could make life substantially easier. Ford was convinced that gas engines could replace the noisy steam engine. Soon after, he met and married Clara Bryant, and, in 1891, they moved to Detroit, where Ford joined the Edison Illuminating Company as an engineer.

DABBLING WITH DESIGNS

Working 12 hours a day and earning $45 dollars a month, Ford built an engine in his kitchen – a spark plug attached to the ceiling light socket. This led to the development of the "Quadricycle". His first trip in the vehicle was to Edison's, at an impressive 32kph (20mph). This endeavour sold for $200, which enabled him to build his second vehicle.

By this time, Ford was thinking big. With financial backing from prominent men of the city, including the mayor, William Maybury, he set up the Detroit Automobile Company in 1899, appointing himself the mechanical superintendent. However, in 1901 the company was wound up. Parts were inferior, the workmen not up to scratch, and Ford failed to make up his mind about the design of the cars.

Some of the backers remained faithful to Ford but, to their dismay, his next grand idea was a racing car. The 730-kg (1,609-lb) machine was relatively light compared to other racing cars, and Ford did make a name for himself, but ultimately he failed to impress his investors, who did not share his vision. Ford, furious with their lack of faith, vowed to be his own boss.

Ford's last venture into car manufacturing began inauspiciously. Finances were tight, but the board of Ford Motor Company, founded in 1903, discovered, to everyone's relief, that there was a brisk demand for their first car, the Model A. Ford told his investors that "the way to make automobiles is to make one automobile like another automobile". It was a vision that would make him the world's first billionaire.

A FRIENDSHIP OF TITANS

Nobody was more of an inspiration to Ford than the great inventor Thomas Edison. At the turn of the century, Edison had endorsed Ford's pursuit of an efficient, gas-powered car during a chance meeting. After the Model T's enormous success, the two men from Michigan became good friends – a relationship that lasted more than 30 years – often working on each other's projects. Together with businessman Harvey Firestone, and naturalist Luther Burbank, they would set off into the wild on camping trips in Ford's new cars. In one of his last public speeches, Edison said of Ford, "I can only say that in the fullest meaning of the term, he is my friend".

ASSEMBLY LINE
Ford was the father of the assembly line. Here, workers at his Rouge Plant in Michigan create products swiftly and cheaply. The cars are moved along conveyor belts where each job is broken down into its simplest form, from the chassis to the finished car.

CARS FOR THE MASSES

The conventional wisdom of the early 1900s was that money would be made from selling luxury cars. Ford had other ideas. He made cars that were lightweight, simple to run, and relatively cheap. His first car manufactured at the Ford Motor Company, the Model A, did good business. Other models followed, each given a letter. By 1905, the company was averaging 25 vehicles a day and employed 300 men.

With each design, the cars became more reliable, lighter, stronger, and more affordable. Scorning the hand-crafting traditions of Europe, Ford was on a mission to "build a motor car for the great magnitude". In 1908, the Model T, a car revolutionary in its durability and rapid assembly, made its debut. It sold for $850, still more than the average American could afford.

By 1913, however, Ford had installed a moving conveyor belt to perfect his assembly-line system, producing his millionth car two years later. Workers remained in one place, each adding one component to the car as it passed by. It revolutionized production and the company became the largest automobile manufacturer in the world. By 1918, half of all cars on American roads were Model Ts.

Ford understood his market – he pushed for gas stations and better roads. His car now sold for less than $400, was rugged enough to cope with the roads of the day, and was simple enough for rural folk to fix.

As demand grew, the company built the world's largest industrial complex along the banks of the Rouge River in Dearborn, Michigan. It contained everything needed to make and build cars, embodying Ford's ideal vision of mass production.

FORD MODEL T
This is Ford's supreme creation and a car for the masses – the Model T. Built in 1908, it was produced non-stop until 1927, by which time 15 million Model Ts had rolled off the production line.

ENGINEERING TIMELINE

1863–90

Born in 1863 near Dearborn, Michigan, Ford takes up his first job at the Michigan Car Company

Ford works as a machine apprentice and also as a watch repairman in a shop in his own time

1891–1902

In 1891, he moves to Detroit and starts work with the Edison Illuminating Company, developing his knowledge of tools and machinery

1903–08

In 1903, he forms the Ford Motor Company and manufactures the two-cylinder Model A, assembled at the Mack Avenue Plant in Detroit

Ford begins manufacturing his famous Model T in 1908, and moves operations to a factory at Highland Park

This organizational genius had an impact that went beyond the creation of a major industry. In 1914, Ford introduced the eight-hour working day, and raised the minimum wage to $5 a day.

THE GREAT PATERNALIST

Ford became an international figure, a wealthy man, and achieved phenomenal results in car production. His private ventures, however, were less successful. In 1915, dismayed at the appalling carnage of World War I, he sent a peace ship, *Oskar II*, to Europe to attempt to put an end to hostilities. The press scoffed at his naïve mission. Ford caught a chill and hastily returned to the US, where he got some popular approval in recognition of his efforts. In 1917, the US entered the war, and Ford's promise that he would give back any profits made on military equipment (something he failed to do) silenced the critics. He was often accused of being opinionated, ignorant, and uneducated, and got caught up in a number of libel suits, but his autobiography, produced in 1922, became a bestseller. Americans loved him; he was one of them.

Ford was stubborn in his beliefs and insisted on doing everything his way. He relied on his Model T, and lagged behind other manufacturers in updating his brands. Despite introducing the Ford V-8 in 1932, Ford's sales slipped to third position in the industry. He loathed the unions, and did not sign a union contract until 1941, yet his reputation soared once again during World War II, when his rapid production of bomber planes was hailed as "miraculous". Paternalistic and philanthropic, he founded the Henry Ford Hospital in Detroit, established schools across the country, and constructed the Ford Museum and Greenfield Village in Dearborn.

Ford was eccentric and full of contradictions. Throughout his life he gave different versions of his story, and baffled many. He died aged 83, by old-fashioned candlelight, exactly 100 years after his father had left Ireland to pursue the "American dream".

SOICHIRO HONDA

JAPAN 1906–91

From humble beginnings, Soichiro Honda built an empire. He dropped out of school at 13 and developed an interest in engines. After a serious car accident, Honda decided to set up his own motor company.

During World War II, he bought a small surplus of army engines and adapted them to make power-bikes. Within a decade, he was the world's leading manufacturer of motorbikes.

Honda was the quintessential "hands-on" boss who nearly went bankrupt five times, but never gave up. From the outset, he aimed to dominate the world market, not just the Japanese. He could be difficult, but the high standards he set were soon emulated by other manufacturers around the world.

1912–17

He introduces the first assembly line and conveyor belt in 1913, and one complete car leaves the production line every 10 seconds

In 1917, Ford begins construction of an industrial plant on the Rouge River, which would house everything needed to make his cars

1918–27

In December 1918, his son Edsel succeeds him as the President of the Ford Motor Company

A drop in the sales of the Model T prompts Ford to make a new model; the result is the introduction of the new Model A in 1927, which is a success

1928–47

After his son's death in 1943, Ford hands over the presidency of the company to his grandson in 1945; Ford dies in 1947

IN THE FAST LANE
In 1913, Model Ts were leaving Henry Ford's Highland Park Plant near Detroit, Michigan, at an unprecedented rate thanks to his new assembly lines. Here, the automobile bodies are sent down a ramp where they are lowered onto the chassis arriving from a separate assembly line below. Inside the plant, hundreds of workers assemble the cars' engines and other parts on moving belts. The system allowed a vast increase in production. By 1914, Ford was selling 250,000 cars per year.

FERDINAND PORSCHE

PROLIFIC AUTOMOTIVE DESIGNER AND ENGINEER

AUSTRIA-HUNGARY 1875–1951

He made engines for aircraft, and flew the planes himself; he designed sports cars, and raced them. The pioneer of hybrid cars and four-wheel drive, Ferdinand Porsche had a staggering capacity for innovation. In a career that spanned half a century and endured two world wars, his inventive genius never flagged.

A LIFE'S WORK

- Porsche designs the **first hybrid car**, which takes the industry by storm
- Develops **four-wheel drive** and **front-wheel brakes**, and takes to the air in airships and balloons
- Wins the **prestigious Poetting Prize** as Austria's most outstanding automotive designer
- During World War I, he **constructs aeroplane engines** for Austro-Daimler and works on supercharged high-performance vehicles, which he races himself
- In the 1930s, he branches out on his own and **designs the prototype for a small car** (which catches the eye of Adolf Hitler) and a Grand Prix racing car
- Throughout World War II, he **designs "jeeps"** and **amphibious vehicles**
- After World War II, his luxury sports cars and **mass production of the Volkswagen Beetle** ensure that the Porsche name lives on

Born in the small town of Maffersdorf, Bohemia, (now part of the Czech Republic), Ferdinand Porsche spent his youth tinkering with machinery and poring over books on mechanical theory. His father, the owner of a plumbing workshop, expected Porsche to take over the family business, but the young man had other ambitions.

EARLY SUCCESSES

Porsche was fascinated by electricity and showed prodigious talent. At the age of 16, for instance, he designed, built, and installed an electric light system in his parents' house.

His mechanical aptitude earned him a position with the electrical company Bela Egger in Vienna. He spent his evenings attending classes at the Technical University to learn everything he could about engineering. In 1898, he landed his first job in the car industry with Jacob Lohner, a manufacturer of electric cars. Porsche wasted no time. Aged only 23, he designed the "Lohner-Porsche", an electric carriage with wheel-hub motors driving the front wheels. The car was exhibited at the most prestigious car exhibition of its day, L'Exposition Universelle de Paris, in 1900. It caused a sensation, setting several Austrian land-speed records and boasting an impressive 56kph (35mph). This was just the beginning. The car sold well, enabling Porsche to design other prototypes: a four-wheel drive with an electric motor in each wheel, the petrol engine, and front-wheel brakes. By 1905, Porsche was considered one of the most famous automotive engineers in Europe, winning speed records, international acclaim, and the coveted Poetting Prize as Austria's most outstanding automotive designer.

ROAD, RAIL, AND AIR

In 1906, Porsche moved with his family to Weiner Neustadt, a city in northeast Austria, where he joined Austro-Daimler as its chief designer. There, he was able to develop his interest in faster, more powerful, streamlined cars. Porsche also saw opportunities in the sky. He worked on engines for airships, and, in 1909, had a number of adventures in his Excelsior gas balloon. Once he almost got blown away to Hungary.

As World War I loomed, Porsche turned his attention towards designing vehicles for the Austro-Hungarian army. One major innovation, the "Landwehrzug", could move on both road and rail tracks, and was used by Emperor Franz Josef to send supplies to his troops. He also designed highly efficient aircraft engines, which earned him an honorary doctorate from the Technical University of Vienna.

After World War I, Porsche renewed his passion for designing sports cars, and built the first two-seater, AD "Sascha", which, predictably enough, won the famous Targa Floria, an endurance race held in the mountains of Sicily. In 1926, Daimler merged with Benz (see pp.278–79), and Porsche was able to work on sporty Mercedes models with high-performance engines. The 1928 Mercedes SSK, with 165 kilowatts (225 horsepower), was considered the most powerful sports car in the world. It seemed as if Porsche could do no wrong.

THE GREAT DEPRESSION

Porsche's short temper, however, did not impress the company's board of directors. His push for a small, light, Daimler-Benz was the last straw. He left the company in 1929 for a brief stint at car manufacturer Steyr, but the Great Depression

PORSCHE ELEGANCE
A small sports car based on the design of the Volkswagen Beetle, the Porsche 356 appeared in 1948 and was the Porsche company's first automobile.

BEETLE PRODUCTION
Here, workers assemble the new Volkswagen Beetle, designed by Ferdinand Porsche. Produced by the British military caretaker government in West Germany in the late 1940s, the Beetle became a phenomenal success.

was making its mark and Steyr collapsed. Aged 55, Porsche was unemployed. Further, his reputation for stubbornness meant that gaining employment would not be easy. He had only one option: to begin his own business.

THE "PEOPLE'S CAR"

In 1931, Porsche launched his own consulting firm. He worked tirelessly and slept little, and often had to be reminded to eat lunch. The German economy was declining and projects became scarce. Porsche decided to design a new small car, and funded it with a loan against his life insurance. He built three prototypes, but after several manufacturers backed out, his project lay dormant. The little car looked destined to stay in the workshop until the arrival of Adolf Hitler. Germany's newly elected chancellor decided that every family needed a small, low-cost car. Porsche's streamlined vehicle with an air-cooled engine fit the bill perfectly. The "Volks-Wagen", or "People's Car", was born.

The arrival of World War II, however, shifted Hitler's priorities. Only a handful of cars were made before the new factory was given over to make military vehicles based on the chassis of the Volkswagen.

After the war, Porsche was arrested as a suspected war criminal and imprisoned in Dijon, France. He was freed 20 months later, a shattered man, and in poor health. He died of a stroke, aged 75. His "People's Car" eventually started production under the auspices of British troops. Although scorned by a British motor industry commission, and by American car manufacturer Henry Ford (see pp.282–85), the Volkswagen Beetle went on to become the best-selling car of all time. The success of the Volkswagen, and the designs that made Ferdinand Porsche synonymous with luxury sports cars, ensured the company's future.

ROBERT BOSCH

GERMANY 1861–1942

One of the leading innovators of automobile technology, Robert Bosch was only 25 when he opened his first workshop in November 1886 in Stuttgart, Germany.

A precision mechanic, Bosch designed a low-tension magneto-ignition for cars that transformed the car industry. It sold worldwide, enabling him to expand, producing spark plugs, horns, batteries, servo brakes, and windscreen wipers. One of the first employers in Germany to introduce high wages and an eight-hour working day, his life was not without tragedy. He lost his son to multiple sclerosis in 1921 and suffered the consequent breakdown of his marriage.

MOTION PICTURES

ALTHOUGH PHOTOGRAPHY HAD REVOLUTIONIZED THE VISUAL MEDIUM, IT ALSO LAID THE FOUNDATION FOR AN EVEN GREATER SHIFT – TO CINEMA. BY THE LATE 19TH CENTURY, THE INVENTION OF CELLULOID FILM AND MOVING-PICTURE CAMERAS HERALDED THE ARRIVAL OF THE MOVIE INDUSTRY.

LOUIS DAGUERRE
This French inventor created the daguerreotype, the forerunner of modern photography, which eventually led to the development of moving pictures.

Attempts to project images and display motion predated the invention of photography. In the 17th century, magic lanterns used a bright light to project an image painted on glass – slides with hand-drawn illustrations changed quickly and successively, and seemed "magical". By the 19th century, there were inventions that took advantage of the eye's persistence of vision – any image on the retina lasts for a fraction of a second. If the image is replaced by another that is closely similar, and then another, the brain will interpret the result as a moving picture, as long as the images follow one another quickly enough. Before that, however, illustrations were still being manipulated in several ways to give viewers the impression they were watching an object move.

For example, the phenakistoscope, developed by Joseph Plateau (1801–83) in 1832, spun pictures around on a wheel, producing a "moving" image that could be seen in a mirror. Plateau realized that the speed of 16 frames per second was the minimum needed to produce an illusion of continual motion. His work was crucial in the development of early animation. This device was followed by the Zoetrope, developed by English mathematician William George Horner (1786–1837). The Zoetrope was a cylinder that was like a lampshade with a dozen vertical slits, and a sequence of pictures between the slits on the inside. When it was spun round, a viewer looking in through the slits could see what appeared to be a moving picture. When Louis Daguerre (1787–1851) introduced still photography in 1839, the challenge was to incorporate these new kinds of images into moving pictures.

> **The desire to display motion is an old one. First there were magic lanterns and multiple cameras. With the invention of the celluloid film and *cinématographe*, it became possible to replay images at a high speed, and modern cinema was born.**

CAPTURED MOTION

British photographer Eadweard Muybridge (1830–1904) had become interested in the subject of motion. He used multiple cameras in 1877 to capture moving images – the most famous was the sequence he shot of a horse's gait. He screened them on a rotating glass disc called a Zoopraxiscope, and the device made it look as if the horse was galloping. Étienne-Jules Marey (1830–1904) took Muybridge's idea a step further and developed a type of camera that could take a rapid series of photographs. This "photographic gun" could shoot up to 12 pictures a second and capture the movement of its subject across the frames. Meanwhile, in New Jersey, Hannibal Goodwin (1822–1900), began to experiment with celluloid. An amateur photographer, he wanted to produce his own slides for his talks, and in 1887, he discovered that cellulose nitrate could be applied to photographic emulsions, a critical step in moving away from the use of plates (metal or glass sheets on which photographic images were recorded). Soon US inventor George Eastman (1854–1932) became involved, and, in 1888, created a hand-held box camera called the Kodak. The following year, he began to produce roll film that used celluloid.

EDISON'S KINETOSCOPE
This is Thomas Edison's motion picture machine from 1894. Viewers looked in through the eyepiece at the top and saw about 20 seconds of moving images.

PROJECTED TO THE PUBLIC

At about the same time in England, French inventor Louis Le Prince (see pp.292–93) was experimenting with the use of strips of film in his own moving picture cameras. In 1888, he shot some film sequences around the city of Leeds – Roundhay Garden Scene and Leeds Bridge Scene – and became the first person to record moving images and replay them. Meanwhile, US inventor Thomas Edison (see pp.262–65) developed a device called the kinetoscope, which was both a camera and a viewing instrument that used film. A strip of film was passed between a lens and a light, while the viewer looked through a peephole in the top of the device. An electrically driven wheel moved the film and a shutter controlled the light, giving the impression of motion. It was an instant success

MUYBRIDGE'S HORSES
The images from Eadweard Muybridge's study of a galloping horse were revolutionary. This experiment was crucial in understanding how to capture motion.

LUMIÈRE'S FIRST FILM
This is a still from "Workers Leaving the Lumière Factory" by French brothers Louis and Auguste Lumière, a film that lasted less than a minute.

SETTING THE SCENE

- **Magic lanterns** are developed in the 17th century. They work by lighting images painted on glass that are then projected onto a wall – stories can be told by changing the panes of glass.
- By the 19th century, devices are invented that make **hand-drawn images look as if they are moving**, such as the phenakistoscope.
- Louis Daguerre **develops a still photography process** in 1839. By the 1870s, multiple cameras are used on moving objects, and **cameras with faster shutter speeds** are developed.
- The application of celluloid to sensitized photograph paper allows for the **evolution of rolls of photographic film** by the 1880s.
- Inventors work on ways of turning these film rolls into **moving pictures**, and the Lumière brothers build the *cinématographe*, which can **record and project celluloid film**. Soon hundred of films are being made and shown.

when it made a public debut in New York in 1894. However, there was a serious drawback – only one viewer could watch at a time.

French brothers Auguste and Louis Lumière (see pp.294–95) went to see the device when it was on display in Paris. They were impressed, and immediately wanted to develop their own projector that could show these moving pictures to a larger audience. The brothers worked in the photography business, and with the arrival of reel film, they capitalized on this new technology. Soon they had developed their *cinématographe* – a camera and a projector that used roll film and was capable of replaying it back at 16 frames per second. They gave their first public screening in 1895, and modern cinema was born.

After their success, Edison began to develop his own projector, as did many other inventors, including French magician Georges Méliès (1861–1938) who saw the possibilities of film for story-telling, and when the Lumière brothers refused to sell him a *cinématographe*, he designed his own camera. He made 500 films between 1896 and 1913, including his famous 1902 *Le Voyage dans la lune* ("A Trip to the Moon"), and set up the Star Film Company to distribute his films to eager American audiences. When he developed techniques such as stop motion and slow motion, the era of silent films had truly begun.

LE VOYAGE DANS LA LUNE
"A Trip to the Moon", a 14-minute film released in 1902 that was written and directed by Georges Méliès, delighted audiences with its innovative special effects.

LOUIS LE PRINCE

EARLY INNOVATOR OF FILM TECHNOLOGY

FRANCE 1842–90

THE DEATH OF FILM INNOVATOR Louis Le Prince could have come from the plot of a movie. He was last seen boarding a train in Dijon, France, and never heard from again. The mystery of his disappearance remains unsolved. Le Prince, until this ill-fated train journey, had been doing ground-breaking work on motion picture cameras, and became the first person to record moving images.

A LIFE'S WORK

- While growing up Le Prince **studies art** and learns about the new practice of **photography**
- Moves to Leeds to work as a **designer in a brass foundry**
- He **opens a school for technical arts** in Leeds, but leaves in 1881 to work for a brass foundry in New York
- Begins to experiment with **designs for a moving-picture** machine, and **develops a 16-lens camera**
- Is granted a **patent for his 16-lens camera** in the US in 1888
- Develops a **single-lens device** that can capture moving images and then play them back
- Is credited with the **world's first motion picture**, *Roundhay Garden Scene*

Le Prince's father was a major of artillery in the French army and a close friend of photography inventor Louis-Jacques-Mandé Daguerre (see pp.290–91), who gave young Louis his early lessons in the craft. Le Prince also undertook art lessons in Paris and started oil painting and pastel portraits, as well as the painting and firing of pottery. He did his postgraduation in chemistry, however, at Leipzig University, which was useful for his future career.

In 1866, he moved to England at the invitation of his friend John Whitely, who owned a brass foundry and wanted Le Prince to work as a designer. There, he fell in love with Whitely's sister, Elizabeth, and they married in 1869. This was followed by his service in the Franco-Prussian War, after which he returned to Leeds, and, with Elizabeth, established the Leeds Technical School of Arts in 1871.

THE 16-LENS CAMERA
Le Prince's 16-lens camera took a series of pictures using 16 independent shutters, fired in sequence. The camera was operated with only one glass plate, limiting the sequence to 16 images.

NEW YORK INVENTIONS

Le Prince travelled to New York in 1881 to conduct some business for his brother-in-law, and Elizabeth soon joined him. They continued to teach applied arts, and she also taught in a school for the deaf. At the school she had access to moving-picture machines (see pp.290–91), and Le Prince was inspired to apply the principles of these devices to what he knew about photography. Soon he was working on a 16-lens camera, which was also a projector, calling these two "receiver" and "deliverer", respectively, and he tested the projections of his devices at the school.

His early machine used strips of film on rollers and eight shutters, behind which were eight lenses that clicked one after another. A second set of eight lenses would do the same while a new strip of film was moved along, and it was able to do this at the rate of 16 frames per second – the slowest rate at which the eye can be "tricked" into seeing an image in motion.

Le Prince applied for a patent for his invention in the US in 1886. This was not granted quickly, and he began to suspect that the information that he had filed in his application was being leaked to other inventors (photography and the nascent motion picture industry were competitive fields) and so he pursued his work with urgency. The US patent was finally granted in 1888, although it only acknowledged the 16-lens machine, and not his other ongoing designs for devices with one or two lenses. He did, however, receive patents for those later in Britain and France.

THE BREAKTHROUGH

By 1887, Le Prince had returned to Leeds and set up a workshop employing woodworker Frederic Mason and automatic-ticket machine inventor J W Longley. Soon he had turned his attention away from the 16-lens model of the camera to focus on his single-lens designs. One of the main problems he faced in the course of his efforts to make a motion-picture machine was finding the right kind of film. The film needed to have some degree of transparency to allow for projection, but it also needed to be able to withstand the strong heat from the lamp that he wanted to use. By 1885, George Eastman (see p.290) had introduced rolls of film, but he was not yet using celluloid, so they were not quite right for Le Prince's purposes because they were not sufficiently transparent. Unfazed, Le Prince kept experimenting.

By the summer of 1888, he had made two single-lens cameras. The cameras worked by using a system of spools around which the film could wind, making them similar to modern film cameras, though these were much larger and made of wood. They took pictures on gelatine strips of film. However, projecting the images was a different matter. For this purpose he used gelatine positives on glass backing. The machine had three spiral belts that moved these glass positives around a belt. The process was noisy and slow – but it worked. His first film, in 1888, was a 2.11-second recording in his father-in-law's garden, called *Roundhay Garden Scene*, and it was followed by *Leeds Bridge Scene* the same year. The second short film captured the bustle of city life at the rate of 12–20 pictures per second, although neither film was screened publicly. The following year, he refined the camera, and employed an electrical engineer to help him install brighter electric arc lights in the device. At about the same time, Le Prince realized that the new sensitized celluloid film could be moved on sprockets (tooth-like projections on a wheel's rim that help engage the links of a chain), which was the answer to his problems.

MYSTERIOUS DISAPPEARANCE

Le Prince returned to New York in 1890, although the same year he crossed the Atlantic again to visit friends and family in France, while also planning a public display of his invention. On 16 September 1890 he stepped aboard the train at Dijon departing for Paris, but was never seen again. His mysterious disappearance has prompted claims that he could have been murdered by competitors. However, he was also heavily in debt after spending all his money on developing cameras, and may have committed suicide. In 2003, a photo was found in the Paris police archives of a victim of drowning from 1890 who resembled Le Prince, but no conclusion has ever been reached about what happened.

In the years to come, his camera was eclipsed by the work of other inventors, and constant patent struggles meant that his family were forced to fight court battles for the recognition of his work. However, over time, his legacy has been re-established and his role in laying the groundwork for modern cinema has finally been recognized.

SINGLE-LENS CAMERA

Louis Le Prince's single-lens camera had a brass plate and shutter. The plate held the film frame for exposure. For projection, a belt drove glass frames, although it could not go fast enough. However, with the introduction of celluloid, the problem pertaining to projection was resolved, and Le Prince ordered long strips of it to run back through the machine.

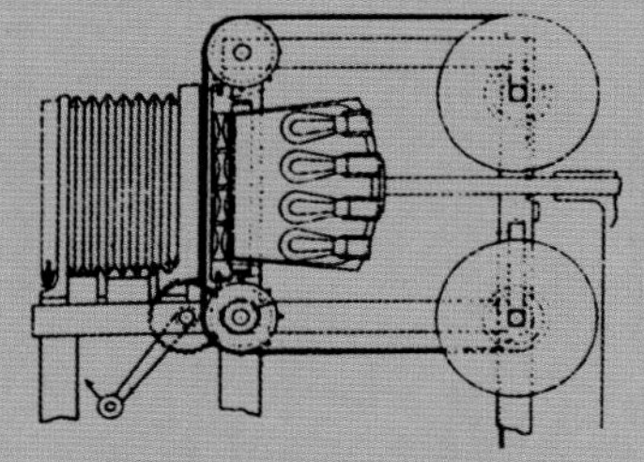

THIS ILLUSTRATION SHOWS THE DESIGN OF LOUIS LE PRINCE'S 1888 SINGLE LENS CAMERA.

MOVIE SET
By the early 1900s, footage of workers leaving a factory or of street scenes no longer had the novelty value for paying audiences. Instead, "scenes" became stories, giving rise to movie stars and film sets.

LUMIÈRE BROTHERS

INVENTORS OF CINEMA

FRANCE 1862–1954; 1864–1948

AUGUSTE LUMIÈRE LOUIS LUMIÈRE

Among the earliest film-makers, the Lumière brothers were active inventors, working on improving photographic processes, as well as medical instruments. Their *cinématographe* of 1895 showed the world's first motion picture, inventing cinema as a mass medium.

Like audiences all over the world, the brothers Auguste and Louis Lumière had been impressed by the kinetoscope, which they had seen while in Paris. Auguste said, "I imagined how marvellous it would be if we were able to project these pictures to a screen ... I decided to try to tackle this problem straight away".

For Auguste and Louis, the problem with Thomas Edison's innovative device (see pp.262–65) was that only one person at a time could watch the moving pictures through the peephole. The brothers instinctively understood how these films could have a wider popular appeal if they were projected so that an audience could watch them. In fact, the brothers not only had the business sense to realize the implications of this development – they also had the technological skill to innovate it.

FAMILY INNOVATIONS

The fortunes of the Lumière family – whose name, fittingly, means "light" in French – were tied to photography. Their father, Antoine, was a painter who shifted into the burgeoning field of photography. When that profession became too crowded, he moved to developing photographic processes in Lyons rather than taking pictures. In the meantime, the young Auguste and Louis studied optics, chemistry, and physics, and soon helped their father run his business. In 1881, Louis worked out a method of photograph development known as "dry plate", which involved using a gelatine emulsion on light-sensitive paper. It was immediately successful. The brothers also knew about George Eastman's work in the US (see pp.290–91), such as the development of the hand-held camera, and rolls of film. After seeing Eastman's celluloid film, they were eager to produce a cheaper version of his innovative product.

It was around this time that the brothers encountered the kinetoscope – their father had seen it first and recommended that the two have a look. They also knew of other photographers and inventors who were trying to create something similar. One of them was the talented photographer Eadweard Muybridge (1830–1904), who had provided photographic proof that solved a long-debated popular puzzle: whether horses lift all four feet off the ground at the same time while galloping (Muybridge's photos proved that they did). In 1877, he repeated the experiment using 24 cameras in a row, each tripped by a string across the track as the horse galloped past. His *Horse in Motion* (1878) and the later *Animal Locomotion* (1887) became some of the most iconic images of the early years of cinematography.

With the arrival of celluloid, building a cheaper version of the kinetoscope seemed possible. The brothers realized that they would need long strips of film, a camera that could take pictures and play them back at at least 16 frames per second, a drive mechanism to move the film, and a projector mechanism with a light. To add to the challenge, they wanted to make a camera that was much smaller than the bulky kinetoscope.

A LIFE'S WORK

- Louis Lumière **develops a photographic process known as "dry plate"**, which becomes very successful
- Louis and his brother, Auguste, **see the kinetoscope made by US inventor Thomas Edison**; they are inspired to design their own machine, but one that can project moving images for large audiences
- The **brothers apply for a patent** in February 1895 **for their *cinématographe*, which can take pictures and project them** – and which is much smaller than any previous device
- On 19 March 1895, **they shoot their first film** of workers leaving their family's factory
- On 28 December 1895, 33 people pay 1 franc each to attend **the world's first cinema screening**

THE *CINÉMATOGRAPHE*
Invented by Auguste and Louis Lumière, this device was able to record moving images as well as play them back for an audience through its built-in projector.

However, the brothers had trouble designing the device's drive mechanism. After months of building parts and testing, Louis came up with an answer to the problem while lying in bed with influenza. His solution partly drew from the sewing machine, which uses a wheel to move the needle and allows the cloth to pause long enough for a stitch.

The brothers adapted this idea and designed a hand-cranked driving mechanism for their camera, which had a wheel that moved a small hood that fitted into holes running along the edge of the film, and pulled it through to the shutter. It remained there for the brief exposure, and was then pulled further through. By February 1895, the brothers had a design that was ready to be patented. On 19 March 1895, they took their *cinématographe* to their family's factory and recorded the workers leaving the building. The film lasted for 50 seconds and used up 800 frames. Two days later they were able to play it back, and screened it in Paris to the Société d'Encouragement pour l'Industrie Nationale, where it was an instant hit. The projector worked on a similar basis. The shutter blocked out light, and the rotating wheel (also known as a cam) and the hooks pulled the frames down, the shutter allowed light in, and the image was projected onto a screen.

BIRTH OF MODERN CINEMA

Not long after, Antoine wanted his sons to arrange a show for the general public. He organized a screening at the Salon Indien on the Boulevard des Capucines, and guaranteed a payment of 30 francs to the owner of the place. On 28 December 1895, 33 people willing to pay 1 franc each turned up to watch *Leaving the Lumière Factory*. Soon the brothers began to make more films, including *The Sea* and *Arrival of a Train at La Ciotat*, and people flocked to see these moving pictures.

Screenings were also arranged in London (1896), and then in the US, Russia, and even China. Potential directors were soon clamouring for the rights to build a *cinématographe*, while sanctioned operators began to make films for the Lumière brothers. In 1900, Auguste and Louis organized a film projection on a 30 × 24m (100 × 80ft) screen for that year's Paris Exposition.

THE AGE OF MOVING PICTURES
Promotional material for the *cinématographe* displayed the excitement generated by moving pictures. Their film *Arrival of a train at La Ciotat* caused people to recoil in terror as the approaching train filled the screen.

DIFFERENT PURSUITS

Although the Lumière brothers oversaw the beginning of the film industry and produced many of the earliest films, they were not interested in making movies and turned their focus towards other inventions. In 1907, Louis developed the autochrome process using potato starch, which allowed for colour plates to be made on glass. This colour photography was later refined by Kodak and other film producers. Auguste, meanwhile, turned his attention towards medical and surgical instruments, a field marked by a proliferation of new products and techniques.

TOWERING STRUCTURES

FROM THE HI-RISES OF CHICAGO TO THE GLASS MONOLITHS OF THE GULF STATES, SUPER-TALL BUILDINGS ARE THE WONDERS OF THE MODERN WORLD – TOWERING SYMBOLS OF MODERNITY MADE POSSIBLE BY ADVANCES IN BUILDING MATERIALS AND ENGINEERING.

GILBERT'S GIANT
Architect Cass Gilbert designed the 232m- (791ft-) tall Woolworth Building, which was finished in 1913. It held the record for the world's tallest building at the time.

Before the second half of the 19th century, buildings could only reach a limited height. The taller the building, the thicker its lower walls needed to be. However, as cities swelled and space became tight, pressure increased to find a new material that could liberate designers and allow them to build upwards.

RISING HIGH

The problem bothered American inventor James Bogardus (1800–74), who dreamt of building high-rise structures. During the industrial revolution, scientists had begun to experiment with new materials. Iron became stronger and more durable, and could be held together by rivets. In 1848, Bogardus went back to the drawing board and conceived of a building whose weight could be supported by a cast-iron frame. Inspired, he built a five-storey factory in New York City, which was a significant step towards the development of skyscrapers.

However, it took a fire to truly propel buildings upwards. When the city of Chicago burnt down in the Great Fire of 1871, it presented a unique opportunity for engineers to experiment and take risks. Advancements in the production of steel – lighter and stronger than iron – enabled engineer William Le Baron Jenney to construct the 10-storey Home Insurance Building in Chicago – the first to use an all-steel skeleton to support the floors and walls (see pp.298–99). The world's first skyscraper, it was completed in 1885 and fitted with the latest technology – two Otis elevators, fireproofing, and electric lights.

REACHING FOR THE SKY
The Burj Khalifa Tower in Dubai, United Arab Emirates, is currently the world's tallest building at a staggering 818m (2,717ft).

> **When the Twin Towers collapsed in 2001, many predicted the end of the skyscraper. Yet the appeal of these structures has not diminished. Countries with dense commercial hubs continue to commission these super-tall "cities".**

TALLER AND STRONGER

Jenney's work inspired a new school of architects and engineers, drawn to Chicago's explosive growth and optimistic energy. One such protégé, Daniel Burnham (1846–1912), achieved international fame by designing New York City's 87m- (285ft-) tall Flatiron Building in 1902.

New building methods started emerging and concrete began to dominate. Although it had low tensile strength, it could absorb more tension once it was embedded with steel rods. The technique was pioneered by German and French engineers, notably François Hennebique (see p.243), but it was the Americans who built the world's first office block using reinforced concrete – the 15-storey Ingalls Building in Cincinnati (1903). This early phase culminated in New York's 55-storey Woolworth Building, which was followed by a spate of skyscraper construction in the late 1920s as industrialists and financiers competed to build the tallest structure in the city.

In 1928, designer H Craig Severance (1879–1941) started work on the new headquarters of the Bank of Manhattan. At the same time, his former partner, William Van Alen (1883–1954), was working on a building commissioned by car manufacturer Chrysler. Motivated to surpass Van Alen, workers completed the Bank of Manhattan, a mammoth 282m (927ft), in under a year. Van Alen, however, hoisted a 37m- (123ft-) tall stainless-steel spire to make the Chrysler Building a world-beating 318m (1,046ft). In 1929, construction stopped as the Wall Street stock market crashed. Only one gargantuan project continued, a symbol of hope in the Great Depression years – William Lamb's (1883–1952) 102-storey Empire State Building. For 40 years, it remained the world's tallest building. Skyscrapers had become expensive to build and World War II created new priorities.

After the war, huge investments were made in construction to boost employment and a major problem of high-rise living was addressed: the feeling of claustrophobia. Architects turned to a lighter cladding material – glass – which ushered in a new generation of skyscrapers that were spacious and light. New York's Lever House,

WORLD TRADE CENTER
The Twin Towers of downtown Manhattan's World Trade Center, completed in 1973, were built with a tube-frame structural system that allowed for greater internal floor space.

SKYSCRAPING IN SHANGHAI
Shanghai is home to some of the world's tallest buildings and thousands more are planned. A building boom has completely transformed the skyline of China's largest city.

SETTING THE SCENE

- In 2575BCE the **Great Pyramid of Khufu** is built with 2,300,000 stone blocks to peak at an impressive **137m (450ft)**, and is the world's **tallest structure for 4,000 years**.
- In the Middle Ages (5th–15th century), **massive cathedrals** are built in Europe. At 160m (524ft), **Lincoln Cathedral** (England) holds the title of the highest building at the time. Its height is not surpassed until the completion of the Eiffel Tower in Paris. The cathedral's spire collapsed in 1549 in a hurricane.
- In the late Middle Ages, the **wool-making centre of Bruges** becomes one of the first places where a **secular structure**, a 108-m (354-ft) belfry, **dominates nearby churches**. It takes another 400 years before other secular structures tower over religious ones.
- Until the 19th century, heights are restricted by the cost of building and a limit on the desire to climb stairs. The word **"skyscraper"**, a nautical term referring to a tall mast, or sail, on a ship, becomes popular to describe **early 10-storey buildings that seem to scrape the sky**.

completed in 1952, was the world's first glass-curtained wall building. It stood across the street from the site where Mies van der Rohe (1886–1969) would refine the fusion of steel and glass with his 160m- (525ft-) tall Seagram Building, a masterpiece of skyscraper design. Steel framing took a giant step forwards in the 1960s with the "bundled-tube" concept pioneered by Fazlur Khan in Chicago (see pp.308–09). Safety features became more sophisticated and engineers grew adept at tackling the effects of wind. A new era of skyscrapers began: the Twin Towers (New York City, 1970–72), the Sears Tower (Chicago, 1974), and the Petronas Towers (Kuala Lumpur, 1998).

SKYSCRAPER COMPETITION
By 1931, Manhattan was a forest of towers competing in height and design, ranging from the classical to the neo-Gothic, and the streamlined modernity of art deco.

WILLIAM LE BARON JENNEY

FATHER OF THE SKYSCRAPER

UNITED STATES 1832–1907

THE CITY OF CHICAGO prides itself on possessing one of the world's greatest skylines, its glittering commercial skyscrapers regarded as a symbol of human achievement and a tribute to the initiative of one man – William Le Baron Jenney. An architect and engineer, Jenney used steel skeletons to support tall buildings, and designed and built the world's first skyscraper.

The product of an old New England family who had arrived in the New World in 1623, William Le Baron Jenney was born in Fairhaven, Massachusetts. His father, the affluent owner of a fleet of whaling ships, was able to ensure that his son had a first-rate education. While still in his teens, William was encouraged to travel on board the family's vessels. He spent three months in San Francisco, where he witnessed the rapid re-development of the city from wood to brick buildings after the fire of 1850. From there he sailed to Honolulu and the Philippines, and was impressed by ingenious methods of making buildings out of lightweight bamboo frames that could withstand the impact of fierce typhoons. Having struggled to decide on a career, he set his heart on engineering.

A LIFE'S WORK

- **Jenney hones his skills during the American Civil War** and learns about metal construction
- Opens an office in Chicago in 1867 and **designs his first tall building, the First Leiter Store**, using iron columns for the interior
- Becomes the **"father of the skyscraper"** by constructing his masterpiece, the **ten-storey Home Insurance Building** (1885), in which the floors and walls are supported by a metal skeleton
- His **Horticultural Building** of 1893 is the **largest botanical conservatory ever built**
- Writes and teaches, drawing together a pool of talented architects, who become known as the **"First Chicago School"**
- His other skyscrapers **redefine Chicago's skyline** and are copied across the world

AN ENGINEER PREPARES

Jenney entered Harvard University in 1850 to study at its Lawrence Scientific School, but he was disappointed with the fledgling programme and left. He instead enrolled in the École Centrale des Arts et Manufactures in Paris. There he enjoyed observing continental innovations in metal construction, fireproofing, and drainage. He graduated with honours in 1856, a year after Gustave Eiffel (see p.301).

Any plans that he had for starting his own business, however, were thwarted by the outbreak of the American Civil War. In 1861, he enlisted with the Army Corps of Engineers, with duties under General Ulysses S Grant. He honed his engineering skills designing earthwork fortifications and demolishing bridges, and, in the process, mastered the nuances of metal structures. By the time he resigned from the army in 1866, he had risen to the rank of major.

HIS OWN MAN

Jenney headed for Chicago, where he launched his own firm, Jenney, Schermerhorn and Bogart, with his first major assignment as chief engineer of Chicago's West Parks. This was a prestigious commission. The construction of parks was seen as a way to address some of the

FIRST SKYSCRAPER
The ancestor of all high-rise structures, the Home Insurance Building is regarded as the world's first skyscraper. Completed in 1885, it had an interior metal frame to support its weight. It was demolished in 1931.

THE SAFETY ELEVATOR

Tall buildings would have been impractical without safe elevators. It was American industrialist Elisha Graves Otis (1811–61) who invented an elevator safety device – a brake – in 1852, making skyscrapers viable. Otis demonstrated his safety elevator at the 1854 Crystal Palace Exposition in New York. When the rope that held up the platform was cut, the audience gasped in horror, but the brake held, and the elevator stayed where it was. The public trusted the machine, and business orders followed.

ELISHA OTIS DEMONSTRATES HIS ELEVATOR IN THE CRYSTAL PALACE, NEW YORK, IN 1854.

problems of an increasingly urban environment and improve public health. Jenney combined elegant thoroughfares with bridle paths, fountains, and groves.

By 1871, Chicago was the fourth-largest city of the US. It was a thriving metropolis with a population of 334,000. However, its closely packed wooden structures made it a huge tinderbox. On the night of 8 October 1871, a massive fire broke out, destroying 9 sq km (3.5 sq miles) of the city, and, with it, about 17,500 buildings. The fire provided a momentous opportunity for architects and engineers to rebuild Chicago, not in its old image, but as a modern city made up of iron and steel.

GROUND-BREAKING "FIRSTS"

In 1878, Jenney designed the First Leiter Building, a department store at Washington and Wells streets in Chicago. It was five storeys high, and looked aesthetically similar to buildings built before the fire. However, it marked a milestone in architectural engineering. Its interior columns, large windows, and cast-iron beams were new concepts. Further, it featured vertical transportation via elevators (see box, above).

Two years later, Jenney began work on the Home Insurance Building at Adams and LaSalle streets. He imagined that the building's weight would be carried on an internal steel frame, on which the fabric of the building would be stretched, like skin over a human skeleton. Based on this premise, he determined that this new structure could rise as high as ten storeys. This was unprecedented. Previously, the walls of the ground floor had to bear the load of the storeys above, limiting the height to five storeys. By 1885, Chicago had its first skyscraper.

The Leiter and Home Insurance buildings were the precursor to many tall commercial structures designed by Jenney, including the Ludington Building (1891), one of the first all-steel structures; and the Manhattan Building (1891), which was a massive 16 storeys. His Horticultural Building of 1893 became the world's largest botanical conservatory. Jenney had become the undisputed leader in his field.

PASSING ON THE MANTLE

In addition to his pioneering experimentation with metal-frame skyscrapers, Jenney was a writer and lecturer, training many young architects who became known as the "First Chicago School". Inspired by Jenney's principles of economy, simplicity, and function, they developed much of Chicago's landscape, and, in doing so, helped define the image of America.

Jenney was modest, insisting that his designs were gathered from the ideas of those who came before him. He retired at the age of 73. Fittingly, the last assignment of this Civil War veteran was the State of Illinois monument on the Vicksburg Battlefield.

HORTICULTURAL BUILDING
Jenney designed this building for the 1893 World's Columbian Exposition in Chicago. The ornate structure, with a 55-m (180-ft) dome, was the largest botanical conservatory ever built.

VLADIMIR SHUKHOV

PIONEER OF LIGHTWEIGHT METALLIC STRUCTURES

RUSSIA **1853–1939**

AN INVENTOR AND DESIGNER who built structures of breathtaking beauty, Vladimir Shukhov pioneered lightweight construction. He is credited with a number of firsts: hyperboloid towers, oil reservoirs, thermal cracking, lattice-shell structures, and Russia's first oil pipeline. Yet despite being prolific, versatile, and ground-breaking, his work is little known.

A LIFE'S WORK

- Shukhov joins entrepreneur Alexander Bari in his oil business and **builds Russia's first oil pipeline**, oil tankers, oil reservoirs, and barges
- **Invents the technique of thermal cracking**, which revolutionizes the oil industry
- **Designs eight thin-shell structures** for the All-Russia Exhibition in Nizhny Novgorod in 1896, and **the world's first hyperboloid water tower**
- Stays in Russia during World War I and the **October Revolution** to build hundreds of **towers, bridges, railway stations, and lighthouses**
- Becomes famous for his **hyperboloid radio tower**, commissioned by Lenin and finished in 1922
- His **photographic work** uses an **unprecedented documentary style**
- He is elected to the USSR Academy of Sciences in 1927, and **awarded the Lenin Prize** in 1929

The Shukhovs were a respectable and prosperous family. Gregory Petrovich, father of Vladimir, came from a long line of Russian army officers and was well educated. By the time Vladimir was born, Gregory was director of the local branch of the St Petersburg state bank.

Young Shukhov showed a remarkable talent for mathematics. After graduating from the gymnasium high school of St Petersburg in 1871, he enrolled at the prestigious Imperial College of Technology in Moscow. There he produced his first invention: an ingenious steam injector for the combustion of liquid fuel.

FORTUITOUS MEETINGS

In 1876, Shukhov accepted a position as part of a Russian delegation of scientists travelling to the World Exhibition in Philadelphia, US. The fair was packed with the latest inventions and technological achievements. Shukhov met fellow engineers and architects, among them Alexander Bari, a young Russian-American entrepreneur. He also visited factories in Pittsburgh to study railroad construction. Bursting with new ideas, he returned home and worked on the Warsaw–Vienna Railway, but the contrast with his colourful experiences in the US was sharp. His employment was restrictive and rigid, with little space for creativity. Shukhov left and, taking the advice of a family friend, joined the Academy of Military Health.

Shukhov's engineering endeavours appeared to be over. However, Alexander Bari, who had returned to Russia to start an oil business and remembered his meeting with the talented young man, soon persuaded Shukhov to join his team. In 1878, Shukhov set off for Baku in Azerbaijan. Here, oil technology was primitive, but Shukhov revolutionized the industry.

BREAKTHROUGH INVENTIONS

He invented a new type of tubular boiler and constructed pipelines to transport oil, which had previously been carried in horse-drawn wagons. In 1891, he invented the thermal cracking technique (which involves heating oil to high temperatures to break up large hydrocarbon molecules, making more suitable fuels, such as petroleum), and built the first Russian oil reservoirs and tankers. The versatility of Shukhov's work took him into the worlds of shipbuilding, civil engineering, and mechanics. He also began to take photographs in a reportage style, which was ahead of its time.

His career was to take yet another turn when, in 1894, he applied for a patent for his lattice-roof design. These roof structures consisted of a grid of steel strips and angled sections. What was particularly striking about them was their double-curvature surfaces. These forms, based on non-Euclidean hyperbolic geometry, became known as "hyperboloids of revolution". They were simple and elegant, requiring minimal steel. He used the concept to construct eight halls at the Pan-Russian Exhibition in 1896 in Nizhny Novgorod. He also presented there the world's first hyperboloid tower structure. His work caused a sensation.

THE END IS THE BEGINNING

In 1913, just before the outbreak of World War I, Bari died, and his enterprise passed to his son, who decided to emigrate to the US. These

HYPERBOLOID TOWER
Shukhov's tower, a unique hyperboloid structure, was commissioned by Vladimir Lenin in 1919 to broadcast propaganda to the masses. Resembling an upturned waste paper basket, it is a remarkable feat of engineering.

were turbulent times. Shukhov, who spoke publicly against Russia's involvement in the war, was forced to turn his attention towards guns and mines. The Russian October Revolution of 1917 saw many of his contemporaries make a hasty exit from their homeland, but Shukhov refused to leave, believing that it was essential for technicians to remain in their country. Many of his works were damaged. After the revolution, everything would be radically different.

At this time, aged 64, Shukhov found himself propelled into a new era. However, he saw in the revolution an opportunity to build some of the designs that had been fermenting in his head. In 1919, Vladimir Lenin commissioned him to construct an antenna for a radio station on the outskirts of Moscow. Originally intended to be 350m (1,150ft) high, a shortage of steel stunted its growth, capping it at 150m (492ft). This brilliant structure, which broadcast propaganda across the Soviet Union, was cost-effective and graceful. Built without scaffolding or cranes, it suggested that the proletariat was thrusting its way skywards. Poets and writers were inspired – in particular Alexei Tolstoy, who wrote the book *The Hyperboloid of Engineer Garin*.

Shukhov was also at the forefront of a radical quest for modernity known as Russian Constructivism. It was followed by a building boom, which included breathtaking pylons, bridges, aqueducts, railway stations, lighthouses, and a revolving stage at the Moscow Theatre. In 1927, Shukhov was elected to the USSR Academy of Sciences, and in 1929, he was awarded the Lenin Prize.

Joseph Stalin's Great Purge of the 1930s changed the country's climate, and Shukhov, weary of the new regime, retired to spend time with his wife, Anna. One cold January night in 1939, he knocked over a candle while at his desk. His house caught fire and Shukhov suffered burns on a third of his body. He died six days later. Shukhov was honoured in Russia, with a stamp issued in his image and a university named after him. His works are scattered across the Russian landscape, yet, outside Russia, he remains largely unknown.

GUSTAVE EIFFEL

FRANCE 1832–1923

A brilliant engineer, Gustave Eiffel built daring bridges in France, Hungary, and the Indo-China region.

In 1885, he designed the wrought-iron skeleton for the interior of the Statue of Liberty in New York Harbour in the US. But it was his tower in Paris that earned him fame. The winning entry to design a "wonder" to commemorate the 100th anniversary of the French Revolution, it was erected in 1889 using Eiffel's own money. At first Parisians hated it, dismissing it as ugly and monstrous, but it attracted millions of visitors. Later in life, Eiffel became fascinated with aerodynamics, and built a laboratory for experiments at the top of the tower.

GRACEFUL ROTUNDA
Shukhov built eight pavilions for the All-Russia Exhibition in Nizhny Novgorod. They occupy an area of 27,000 sq m (291,000 sq ft). The rotunda was the world's first membrane roof and tensile structure and brought him international fame.

BREAKTHROUGH INVENTION

STEEL-FRAME BUILDING

THE LATTER HALF OF THE 19TH CENTURY was a period of tremendous change in the use of new building techniques and materials. None was more significant than steel. It could be worked easily, forged and rolled, and was strong under duress. Produced in large quantities, it became the backbone of the construction industry. Steel offered innovative solutions, allowing architects to stretch their imagination and build tall structures that transformed city skylines.

In the 19th century, engineers could only dream of building some of the challenging structures they designed. It took the refinement of the Bessemer process to transform the industry. Sir Henry Bessemer, an English metallurgist and inventor of the Bessemer converter, enabled steel to be produced cheaply and in bulk (see pp.228–31). He initiated the "steel revolution" of the 19th century, reducing the cost of steel by 80 per cent.

Mechanical engineer Alexander Holley (1832–82) persevered in mastering what was known about the Bessemer technique, and, acting as a consultant engineer, designed steel plants in the US. First used in the US in the 1860s, Holley's innovations allowed for major advances in construction. In Chicago, political pressure for a fireproof city meant that iron and steel were used instead of timber. American engineer William Le Baron Jenney devised a way to use a steel frame to support a building's weight instead of load-bearing walls (see pp.298–99). His Home Insurance Building in 1885 was the first to use structural steel.

Steel frames became popular and inspired a new generation of engineers. With the steel frame acting as a skeleton to support the building, structures could grow taller. The 242-m (792-ft) Woolworth Building in New York (1913) revealed the capabilities of steel frames, but it was the awe-inspiring Empire State Building that became the quintessential American skyscraper. It soared over the skyline at 417m (1,252ft), demonstrating that steel offered not just strength, but beauty and drama as well.

ONLY THE SKYSCRAPER OFFERS ... THE WIDE-OPEN SPACES OF A MAN-MADE WILD WEST, A FRONTIER IN THE SKY

REM KOOLHAAS, DUTCH ARCHITECTURAL THEORIST

EMPIRE STATE BUILDING
Here, riveters are at work on New York's Empire State Building in early 1930. The steel beams were bolted together manually. The building was the culmination of intense competition among the city's developers for the "world's tallest building". More than 3,000 workers completed the work in just 16 months.

A LIFE'S WORK

- Naito is **tutored by a leading authority on earthquake-resistant structures**, Toshikata Sano, at Tokyo University
- Develops a **new type of study – earthquake engineering**
- On his return to Japan from the US, he has an "epiphanic" moment and **develops the "rigid design theory"**
- **Builds several structures that survive the Great Kanto Earthquake** of 1923 and **gains the attention of engineers** worldwide
- **Engineers the Tokyo Tower**, a **symbol of national rebirth**, in 1958
- Wins **numerous plaudits and awards** for his work

TOKYO TOWER
Tachu Naito's masterpiece, the Tokyo Tower in Japan, is the world's tallest self-supporting steel tower. The iconic radio and television transmitter, once a symbol of Japanese post-war recovery, is now a popular tourist site. The tower is painted every five years with white and orange paint, as dictated by Japan's aviation code.

TACHU NAITO

FOUNDING FATHER OF QUAKE-RESISTANT DESIGN

JAPAN 1886–1970

HAILED AS "DR STEEL TOWER", architect and engineer Tachu Naito helped construct tall buildings in quake-prone Japan, and proved that they could withstand even the most disastrous tremors. He also developed new construction codes and shared his work and ideas with others to make buildings safer around the world.

THE HARDY TRUNK

The world of structural engineering might have been very different had Naito's luggage not fallen apart on his rail trip across the US in 1917. Naito collected dozens of papers during his tour of the country. To make room for them in his trunk, he removed all the interior dividers. However, the constant jostling of the trains destroyed the case. Forced to buy a new one for his return to Japan, Naito kept all the dividers inside, and the trunk survived with his luggage intact. Naito was struck by the way in which the interior "walls" of the case held the trunk together, which led to his theories on "quake-resistant walls". The trunk became his prized possession and took pride of place in his old home, now the Tachu Naito Memorial Museum. Fittingly, the design of the house is unique; it was the first to use reinforced-concrete walls, with no pillars, and, of course, was built to withstand all possible seismic activity.

Born in the village of Sakaki, in the district of Nakakoma, Tachu Naito went to the University of Tokyo to study naval architecture, but switched to conventional architecture after the Russo-Japanese War (1904–05) devastated the Japanese shipbuilding industry. Naito began his new studies in 1907 under Professor Toshikata Sano, who was a leading authority on earthquake-resistant structures. Naito graduated in 1910, and, three years later, became a professor of structural engineering at the prestigious Waseda University near Tokyo.

In 1917, he was sent to the US to enhance his knowledge of earthquake engineering. He met Professor George Swain, head of civil engineering at the Massachusetts Institute of Technology, and visited New York City and Washington DC, only to discover that no one had been able to devise quake-proof structures. On his return to Japan, he started his own structural engineering firm, and set about designing his first buildings using his "rigid design theory" of quake-resistant, reinforced-concrete shear walls. He also found that shear walls alone would not guarantee safety. Columns, floor slabs, and beams also had to be firmly connected to withstand the worst quakes.

THEORY INTO PRACTICE

His early projects deploying his "rigid design theory" were the Kabuki-Za Theatre in Tokyo and the 30m- (98ft-) tall head office of the Industrial Bank of Japan. Three months after these were completed in 1923, Japan was hit by the 7.9-magnitude Great Kanto Earthquake, which devastated Tokyo and its surroundings. Nearly 700,000 buildings were partially or completely destroyed, but Naito's buildings stood tall amidst the rubble. No wonder, then, that the engineering world sat up and took notice.

Naito realized that his ideas could help a wider community and shared them with others, particularly those working in the earthquake-prone state of California. He personally took visitors on tours through the construction sites of Japan, and was a key speaker at the 1929 World Congress on Engineering held in Tokyo. He also engineered the Old Library, a five-storey building at Waseda University, completed in 1925. Two years later, he completed the Okuma Auditorium, also at the university. In 1941, he was named Chair of the Architectural Academy, and, in 1954, he became a member of the Science Council of Japan. As his work progressed, he grew more confident of engineering taller structures. He designed 33m- (108ft-) tall radio towers using steel, including the Sapporo and Beppu towers in 1957.

This work culminated in 1958 with Naito's streamlined masterpiece, the Tokyo Tower, which is taller than the Eiffel Tower in Paris but only half the weight. Steel was in short supply; but the problem was overcome by melting down American tanks used in the 1950–53 Korean War. Construction workers came from all over Japan to build the new landmark, completed in only 15 months and using bamboo scaffolding.

> THIS WAS A HISTORIC FIRST IN EARTHQUAKE ENGINEERING … THE ENGINEERING WORLD WAS IMPRESSED BY NAITO
>
> **ROBERT REITHERMAN,** SEISMOLOGIST, ON EARTHQUAKE ENGINEERING

HANDS-ON ENGINEER

Although bold and adventurous in spirit, Naito always used modest engineering tools. He kept with him a pocket-sized slide rule that his tutor Sano had given him. It was soon made obsolete by the calculator but Naito distrusted the apparent precision of the digital machine, and preferred his slide rule, especially for earthquake calculations, where he was always grappling with the imprecise and unknown.

A zealous worker, Naito continued to remain active in his later years, working on the engineering challenges of nuclear power plants. He retired in 1957, after being appointed trustee of Waseda University. He became a member of the Japan Academy in 1960, was awarded the Distinction for Cultural Merit in 1962, and the second-class Order of the Rising Sun in 1964.

EDUARDO TORROJA

EXEMPLARY DESIGNER OF CONCRETE SHELL STRUCTURES

SPAIN 1899–1961

One of the most outstanding structural engineers of the 20th century, Eduardo Torroja was a master at pioneering new construction techniques. An artist in his use of concrete, reinforced and pre-stressed, he created new forms with shells and spatial structures, and found new uses for old forms, building vast domes. He developed new ways of looking at structures as well as ways to increase the strength of the structures without diminishing their aesthetics. He was posthumously ennobled in recognition of his outstanding work.

A LIFE'S WORK

- Torroja **develops concrete shell construction** and covers the marketplace at Algeciras, Madrid, with a vast spherical dome, pioneering a golden age of thin shell structures
- Evolves the use of reinforced concrete and **designs a cover for the Zarzuela Racecourse**, Madrid, using breathtaking curves
- **Founds the Technical Institute for Construction Science** in 1934 and designs bridges, churches, water towers, aqueducts, and the monumental roof for the Frontón Recoletos
- Is appointed professor in 1939 and **director of the Central Laboratory for Testing of Construction Materials** in 1941
- Writes two seminal books about his work and **begins to teach and extend his work** internationally, exerting a fundamental influence on the world of construction in post-war Europe
- **Receives the prestigious National Architecture prize** alongside architect Manuel Sánchez Arcas

As a child, Eduardo Torroja lived and breathed mathematics. His father, Eduardo Torroja Caballé, had revolutionized the teaching of mathematics analysis in Spain. He was the vice president of both the Spanish Mathematical Society and the Spanish Association for the Advancement of Science. His son inherited the same interests, and graduated in 1923 from Madrid University after studying civil engineering.

TURBULENT TIMES

Europe was left in ruins after World War I. Unemployment was high, economic recovery slow, and political unrest rife. A military dictatorship took power in Spain in the same year Torroja graduated. It was during this turbulent time that Torroja joined Spain's Civil and Hydraulic Construction Company led by José Eugenio Ribera, who had been grappling with a woeful shortage of materials, particularly steel. It was under Ribera's careful tutelage that Torroja began exploring the use of concrete, which was strong and comparatively cheap.

In 1927, Torroja set up his own engineering company and as a consultant joined the Construction Board for the City University campus in Madrid. King Alfonso XIII of Spain told the board, "it is necessary, with proud aspirations, to try to carry out this great work so that this university is one of the foremost of the world ... My golden dream is to see in Madrid, created during my reign and for the good of our country's culture, a university renowned for being a model educational centre". In response, the government sent a committee of experts to the US to look at the designs of 15 institutions of higher learning as paradigms for the new university of Spain. Torroja was then commissioned to engineer the central thermal power station that would supply all the heating and hot water.

FLAIR FOR EXCELLENCE

It was during this productive period that Torroja developed his own aesthetic concepts. He began accepting projects that other engineers turned down. He had already undertaken a challenge with the Tempul Aqueduct, completed in 1926 and considered an outstanding example of modern design. Replacing an earlier structure

IN ORDER TO SUCCESSFULLY CONCEIVE AND PLAN A STRUCTURE OF ANY KIND, IT IS NECESSARY TO INVESTIGATE ITS REASONS FOR EXISTENCE, ITS CAPACITIES TO RESIST, AND TO BEAR ”

EDUARDO TORROJA, *PHILOSOPHY OF STRUCTURES*

that had been destroyed by flooding in 1917, it was one of the first examples of a cable-stayed bridge. Yet Torroja dreamed of even greater concrete structures. In 1933, collaborating with Manuel Sánchez Arcas, he covered the Algeciras marketplace with a 47.5-m (156-ft) vaulted roof, which was to be Torroja's masterpiece. The roof, supported by eight pillars, was a wafer-thin 9cm (3½in) thick.

In 1934, he designed the cover for the main grandstand at the Zarzuela Racecourse near Madrid. A soaring concrete roof made of a cantilevered thin shell stretching over 13m (42ft), it was only 5cm (2in) thick at the edges. Torroja's desire for more ambitious structures did not stop and new shapes were designed. His roof for the El Frontón Recoletos sports centre, built in Madrid in 1935, was a huge technical challenge. The building, commissioned to host Basque pelota, a ball sport invented in the 19th century in the Basque region of Spain, incorporated unprecedented features due to its vast size and height. At each stage of the design the stresses of the reinforced concrete had to be measured to prevent buckling, cracking, and collapse. The effects of wind and snow were also factored in. All these calculations had to be done by hand, in an era when computer technology was not yet available. Torroja worked tirelessly. He designed many aqueducts, including his most recognized one at Alloz, spanning the Salado River, and small churches and shrines in the Pyrenees Mountains including the Pont de Suert, and the mountain refuge of Sancti Spíritus, making modern reinforced concrete look strikingly beautiful. Torroja also used steel in his trussed-arch bridges, and brick, most notably in hyperboloid cooling towers.

Uniting other leading architects and engineers, Torroja founded the Technical Institute for Construction Science in 1934. The rising consciousness of a new community of nations in post-war Europe led to the creation of international institutions to coordinate research, vital for the reconstruction of Europe's destroyed infrastructure. In 1950, the eminent architect Frank Lloyd Wright (1867–1959) called him “the greatest living engineer”. Torroja died 11 years later while working at his desk.

FRONTÓN RECOLETOS

Torroja worked through extraordinarily turbulent times. He began his career after World War I, and helped pick up the pieces of a destroyed Europe post World War II. But it was the Spanish Civil War (1936–39) that posed the greatest challenge. Many of his ideas had to be abandoned, and some of his completed works were bombed. The roof of the Frontón Recoletos was finished in 1936, but three years of war left the shell seriously damaged. Torroja conceived of a way to repair the structure, but the roof collapsed just as the work began.

THE SLENDER SHELL ROOF OF THE FRONTÓN RECOLETOS HAD A SPAN OF 54M (180FT) AND WAS ONLY 8.12CM (3¼IN) THICK.

HIPÓDROMO DE LA ZARZUELA
The elegant racetrack grandstand at Zarzuela near Madrid is one of Torroja's most admired structures. Designed in 1936, it was not opened until 1941, having been damaged by artillery fire during the Spanish Civil War.

FAZLUR KHAN

ARCHITECT OF SUPER-TALL SKYSCRAPERS

INDIA 1929–82

His audacious and innovative approach to tall building design, married to a sharp eye for aesthetic detail, earned Fazlur Khan a reputation as the "Einstein of Structural Engineering". Khan understood the flaws that were holding modern architecture back, and with his revolutionary "bundled tube" system helped shape urban skylines around the world, making very tall buildings economically and viable for the first time. Khan's respect for culture and environment shines through in many of his designs.

A LIFE'S WORK

- Khan pioneers the use of **lightweight concrete** for the One Shell Plaza Building in Houston and develops a **diagonal-framed tube system** for Chicago's John Hancock Center
- His **"bundled tube" system** leads to the building of the tallest skyscraper in the world at the time, the **Sears Tower**, and proves that tall buildings can be economically viable
- Takes a lead role in **humanitarian relief efforts** and founds the Chicago-based **Bangladesh Emergency Welfare Appeal** to help people in his homeland
- He designs **innovative lightweight roof structures** – such as the tensile structure for the **Hajj Terminal in Saudi Arabia**
- In 1972, he is awarded the title **"Construction's Man of the Year"**

Growing up in Dhaka in Bengal (in modern-day Bangladesh), Fazlur Khan was constantly reminded of the need for a good education. His father, Addur Khan, a renowned mathematician, had been granted the title "Khan Bahadur" by the government for services to public education. Khan was a star pupil with an aptitude for both physics and engineering. His father suggested engineering as a career choice and Khan agreed.

The promising student graduated top of his class at Shibpur Engineering College in Calcutta, India, in 1950, before teaching at the Ahsanullah Engineering College for two years. In 1952, he was awarded both the Fulbright and Ford Foundation scholarships for graduate studies, and set off for the University of Illinois in the US. In just three years, Khan earned two Master's degrees, one in structural engineering and one in theoretical and applied mechanics, and also gained a doctorate in structural engineering.

HAJJ PASSENGER TERMINAL
The stunningly modern passenger terminal of Saudi Arabia's King Khalid International Airport, engineered by Khan in six years from 1974 to 1980, was designed to accommodate 80,000 people at a time and resembles a Bedouin tent.

Khan decided to settle in Chicago and was employed with Skidmore, Owings and Merrill (SOM), a leading architectural firm. He returned briefly to his native country, then East Pakistan, and won a prominent position as executive engineer of the Karachi Development Authority, but felt his creative genius and innovative design ideas were being thwarted, so he returned to Chicago.

GIVING HEIGHT TO CHICAGO

It was a challenging time for architects. Cities were running out of space and the population was expanding, but constructing high buildings was not considered commercially feasible. Floor space was more expensive to rent in tall structures than in smaller buildings, for example, and tall buildings also required huge amounts of steel. The higher they went, the stronger and stiffer the tower's structural framework needed to be.

In response to these problems, Khan started work on a completely new design concept – the "framed-tube" structure, in which columns were closely spaced around the perimeter of the building, allowing for fewer interior columns and more floor space, while stiff spandrel beams connected the columns at every floor level. It revolutionized the construction of tall buildings. Khan first implemented this design in 1964 in the 43-storey DeWitt-Chestnut Apartments.

Developing the concept further, along with SOM colleague Bruce Graham, Khan introduced the highly efficient diagonal-framed tube system, which led to the construction of the 100-storey John Hancock Center in Chicago. Widely spaced exterior columns, with diagonals on all four sides, decorated the tallest building in the world. He had also succeeded in making a tall building economically viable. It was not without risk, though. Khan had calculated the tower's sway in high winds, but no one knew what it would

> I PUT MYSELF IN THE PLACE OF THE WHOLE BUILDING, AND VISUALIZE THE STRESSES ... A BUILDING UNDERGOES
>
> **FAZLUR KHAN**

actually feel like for the tenants. Lacking finances for a major study, Khan performed an experiment at Chicago's Museum of Science and Industry, placing eight volunteers on a rotating exhibit. The test confirmed that the sway would be within acceptable limits.

SKY-HIGH MASTERPIECE

A few years later, Khan unveiled another ground-breaking structural system – the "bundled tube". He decided that a building's external covering could, given enough trussing and bracing, be the structure itself. He used the concept to build what many believe to be his masterpiece – Chicago's 110-storey Sears Tower (see box, below). At 442m (1,451ft), it was spacious, and rose higher than the Empire State Building, yet it cost less to rent out. Only the Pentagon accommodated more space at the time. Khan transformed the Chicago landscape and accolades poured in. In 1973, he was elected to the National Academy of Engineering.

Khan never designed a building without considering the societal habits of the people who would use it. As the building boom in the West stalled, Khan responded to demands in the Middle East and Asia for large-scale building projects. For the roof of the Hajj Terminal in Jeddah, Saudi Arabia, he designed a fabric roof structure that stretched over 170,000 sq m (1.8 million sq ft) yet resembled the simplicity of a Bedouin tent.

Khan died of a heart attack in 1982. After his death, Chicago named the intersection located at the foot of the Sears Tower the "Fazlur R Khan Way" in honour of his outstanding achievements.

SEARS TOWER

The Chicago skyline boasts an array of impressive designs, but nothing attracts the eye as purposefully as the Sears Tower (now Willis Tower). Completed in 1975, it was the tallest building in the world, earning Khan a place in engineering history. His "bundled tube" concept saved Sears Roebuck and Co about US $10 million in steel costs. In recognition of his genius, renowned Spanish artist Carlos Marinas designed a large sculpture featuring a bust of Khan set alongside a representation of Chicago's skyline. The sculpture found a home in the Willis Tower lobby.

THE SEARS TOWER, NOW WILLIS TOWER, REMAINED THE TALLEST BUILDING IN THE WORLD UNTIL 1998.

TRUSSED-TUBE MASTERPIECE
The John Hancock Center, completed in 1968, was a milestone for Khan. The 100-storey, 343.5m- (1,127ft-) high building was the tallest in the world outside New York City. It demonstrated Khan's "trussed tube" structural system and his innovative concept of cross-bracing, seen clearly here on the building's exterior.

FLYING MACHINES

THE DESIRE TO FLY IS AN ANCIENT ONE, AS THE MYTH OF ICARUS TESTIFIES. THIS FASCINATION WAS REVIVED DURING THE RENAISSANCE WITH DESIGNS FOR FLYING MACHINES, BUT IT WOULD BE ANOTHER 400 YEARS BEFORE ENGINEERS MADE HEAVIER-THAN-AIR FLIGHT A REALITY.

POWER OF STEAM
American aviation pioneer Samuel Langley created an unmanned machine powered by steam that was launched in 1896 and managed to fly for about a minute and a half.

WAR IN THE AIR
These German Albatross fighter planes were deployed in World War I. The development of aircraft changed the nature of warfare.

LILIENTHAL IN FLIGHT
Otto Lilienthal, a German flight pioneer, advanced the study of aerodynamics with his graceful glider designs, making more than 2,000 test flights, the last of which killed him.

ACROSS THE CHANNEL
This photograph shows French pilot Louis Blériot leaving Calais to fly across the English Channel on 25 July 1909. His flight showed that aircraft could have a wider use.

SETTING THE SCENE

- During the Renaissance, **Leonardo da Vinci** draws hundreds of **designs for flying devices**.
- By the 19th century, **George Cayley** develops **ideas about concepts such as lift**, which would become crucial for airline design.
- **Alberto Santos-Dumont** flies his *14-bis* aircraft before a crowd of Parisian onlookers in 1906.
- The **Wright brothers** manage to make their **flying machine airborne in 1903**.
- **Louis Blériot** flies **across the English Channel** in 1909.
- **World War I** leads to the **proliferation of aircraft** for bombing and reconnaissance.
- In 1919, **John Alcock** and **Arthur Whitten Brown** become the first pilots to **fly across the Atlantic**.
- **Charles Lindbergh crosses the Atlantic** – solo – in 1927, flying from New York to Paris.

As continental Europe was gripped by a mania for taking up hot-air "ballooning", an English landowner, George Cayley, had turned to the question of wings (see pp.312–13). He began to experiment with gliders and puzzled over design aspects such as wing shape and rudders – parts that would be indispensable to the development of an aircraft. In 1849, he built a model glider, and, a few years later, made a full-scale aircraft with wings. At about the same time, French inventor Félix du Temple de la Croix (1823–90) designed parts for a machine that would later influence aircraft design. In 1874, he and his brother, Louis, built a monoplane with 12m (40ft) wings, a propeller, and a compact steam-powered engine. It managed to stay airborne briefly after being launched from a ramp, but was quickly brought back down to the ground. Meanwhile, in Russia, Alexander Mozhaysky (1825–90) designed a machine that suffered a similar fate. He had the approval of the government to design a powered flying machine and developed a model in 1884. But when it was launched from a ramp it only made a short hop and quickly fell back to earth.

Across Europe, inventors were trying to make their flying machines stay in the air for more than a second. But when it finally happened, it was on the other side of the Atlantic, through the efforts of US aviators Samuel Langley (1834–1906) and Octave Chanute (see p.316–17). However, it took Orville and Wilbur Wright to provide the crucial breakthrough (see pp.324–29). They kept their machine aloft in Kill Devil Hills in North Carolina on 17 December 1903, reaching an altitude of 3m (10ft) and travelling 37m (120ft), landing 12 seconds later – making the dream of flight a reality.

Although human beings have not been able simply to strap on wings and take to the skies as Renaissance polymath Leonardo da Vinci had hoped, the development of airplanes has allowed them to come quite close.

TAKING TO THE SKY

Soon, other engineers and designers were building their own planes. In France, Louis Blériot (1872–1936) made numerous designs for machines, and also put some of them to the test. Spurred on by a competition in the British *Daily Mail* newspaper that offered £1,000 to anyone who could fly across the English Channel, Blériot created a machine that accomplished this feat in 37 minutes.

In doing so, another use of the aircraft became clear – aerial attack. Air combat quickly became part of World War I in Europe. Aircraft production boomed, and went alongside changes to the design and engineering of planes.

Once peace had returned, aviation challenges focused on crossing the Atlantic. John Alcock (1892–1919) and Arthur Whitten Brown (1886–1948) flew a twin-engined Vimy in 1919 from Newfoundland to Ireland, making the first-ever crossing. They set out on 14 June in poor weather – eventually Brown had to climb out on the wings as the plane was thousands of feet in the air to clear off ice around the engine. After flying for 16 hours and 27 minutes, they landed in Ireland, having flown 3,400km (1,890 miles) at an altitude of 1,200m (4,000ft) for much of the journey. Soon, the next big challenge was to cross the Atlantic alone, which Charles Lindbergh (1902–74) did when he flew the *Spirit of St Louis* monoplane, taking 31 hours to cover the 5,810-km (3,610-mile) distance from New York to Paris. Not long after, US pilot Amelia Earhart (1897–1937) began to set records for speed, 291kph (181mph) in 1930; and altitude, 5,623m (18,451ft) in 1931; as well as crossing the Atlantic solo the following year.

Throughout the rest of the 20th century, the fledgling airline industry was transformed into a transport powerhouse and aircraft quickly evolved from biplanes to the huge jet carriers of modern times.

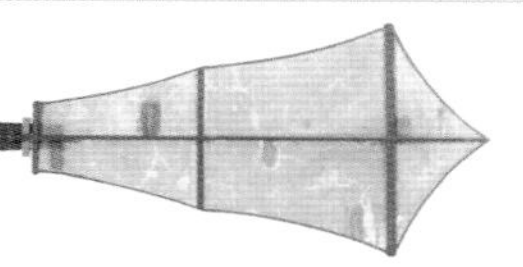

STEAM FLYING MACHINE
English inventor John Stringfellow (1799–1883) developed many designs for aircraft, and in 1848 built one that flew about 10m (33ft), powered by a steam engine.

GEORGE CAYLEY

THE FATHER OF AERONAUTICS

ENGLAND 1773–1857

The first significant advances in heavier-than-air flight are credited to George Cayley. He identified the four forces of flight – lift, weight, thrust, and drag – and built the first successful human-carrying gliders. Cayley's ground-breaking work was critical in preparing the way for the development of the first aeroplanes half a century later.

A LIFE'S WORK

- Cayley **designs an airship**, a type of **caterpillar truck**, and an **engine fuelled by gunpowder**
- **Investigates streamlined shapes** by studying trout
- Identifies the **forces of flight**
- **Studies birds** to find out how they fly and realizes that they **arch their wings into a curved shape**, now known as aerofoil, for gliding and that these **curved wings create more lift** than flat wings
- **Builds the first model aeroplane** with the modern layout of body, wings, and tail
- **Builds the first human-carrying glider** that flies a distance of 275m (900ft)
- Is **elected as a Member of Parliament** for Scarborough in 1832

Cayley was born on 27 December 1773 in a house named "Paradise" in Scarborough, England. His mother, Isabella Seton Cayley, was from a well-to-do Scottish family, while his father, Sir Thomas Cayley, was a member of the British landed gentry. Young Cayley spent most of his early years away from home, particularly at the family house at Helmsley, owing to his father's chronic ill health.

It was at Helmsley that he first showed signs of being interested in mechanical things, and was often seen with the village watchmaker. He was also an inquisitive boy. As a teenager, he had discovered that it took 100 days for his thumbnail to grow 1.5cm (½in). However, as was usual in the 18th century, Cayley did not receive any formal education and was primarily tutored at home by George Walker, a mathematician and fellow of the Royal Society, and George Morgan, a scientist and lecturer on electricity. Both teachers had a considerable impact on developing Cayley's inquisitive and inventive spirit.

EARLY INVENTIONS

When Cayley was 19 his father died, passing on to him the title of baronet, along with the family estates and a grand house called Brompton Hall in Yorkshire. He found much of the family's land in very poor condition, being prone to flooding. He therefore installed a drainage system that was so successful that he was able to grow crops on the land years before his neighbours. While he was carrying out this work, he invented and patented the forerunner of the caterpillar tractor.

Chief among Cayley's interests was heavier-than-air flight. At the end of the 18th century, the only successful manned flights had been made by lighter-than-air balloons, and the nine-year-old Cayley was inspired by the Montgolfier brothers (see pp.172–73). One of Cayley's first experiments involved making a small model of a helicopter. This was a copy

FLYING A CAYLEY GLIDER
In 1973, a modern replica of Cayley's 1853 glider was flown by pilot Derek Piggot in Brompton Dale, the site of the original flight. The replica was towed into the air first by a group of people and then, at faster speeds, by a car. It reached a maximum altitude of 12m (40ft).

of a 1784 design by a French naturalist called Christian de Launoy and a mechanic called Bienvenu. A small rotor made of feathers was fitted to each end of a shaft. A bow, similar to a small archery bow, provided the power to make the shaft spin. The bowstring was wound around the shaft, providing tension to the bow. When it was released, the bowstring unwound, making the rotors spin.

As early as 1799, Cayley designed an aircraft with the modern aeroplane layout – a fuselage, wings, and a cruciform (cross-shaped) tail. It also had a cockpit for the pilot to sit in and moving vanes for propulsion. He made a series of model gliders to test this layout, and launched them from the top of the staircase at Brompton Hall.

OBSERVING NATURE

Cayley studied birds to determine how they managed to stay airborne even when they were not continuously flapping their wings. He noticed that for gliding they arched their wings into a curved shape, now known as an aerofoil. He deduced that curved wings produced more lift than flat wings, writing, "I am apt to think that the more concave the wing, to a certain extent, the more it gives support". He used a whirling arm device to measure the lift and drag produced by surfaces moving through the air at different angles. In 1804, using data from the whirling arm experiments, Cayley built what was probably the first fixed-wing aircraft in history, a glider about 1.5m (5ft) long, with wings and a cruciform tail. However, the care he took with his experiments was not necessarily appreciated until his notebook was discovered in 1933.

LIGHTER THAN AIR

By 1809, Cayley had studied the lift produced by aerofoils, aircraft stability, streamlining, and the effect of a rudder. This was amazingly advanced research for its time. He described his work and findings in a paper titled "On Aerial Navigation", which was published in three parts in William Nicholson's *Journal of Natural Philosophy, Chemistry and the Arts* in 1809–10.

For the next few years, Cayley turned his attention to lighter-than-air craft. In 1816, he designed an airship 131m (432ft) long, powered by a steam engine. This was 36 years before French engineer Henri Giffard (1825–82) would be credited with inventing the airship. Cayley calculated that his airship would have a range of 1,500km (932 miles) in calm air, travelling at 32kph (20mph). It incorporated an ingenious system for condensing steam from the steam engine and reusing it, thereby reducing the amount of water that had to be carried. In 1837, he wrote an astonishing paper, published in *Mechanics Magazine*, in which he discussed propelling an airship by a stream of air produced by a turbine driven by an engine – which sounds remarkably like a form of jet engine.

MANNED GLIDERS

In the 1840s, Cayley redirected his attention to heavier-than-air flight. In 1843, he designed an extraordinary aircraft that today would be called a "convertiplane". It had four revolving rotors for vertical takeoff and propellers for propulsion. When it was airborne, the rotors flattened out and became circular wings. Unfortunately, he never built it. In 1849, he built a triplane glider – a glider with three wings, one above the other. It looked like a box kite, with a boat-shaped gondola hanging underneath. The 10-year-old son of one of his servants sat in it while it was towed along. It took off and glided for a short distance, making the unidentified boy the first person in history to fly in a heavier-than-air craft.

Meanwhile, in 1843, English inventor William Samuel Henson (1812–88) designed a steam-powered aeroplane based on Cayley's research. Henson and his business partner, John Stringfellow (see pp.314–15), imagined a fleet of steam-powered aerial carriages flying passengers all over the world. However, he failed to attract investors and the project foundered due to the lack of funds. Stringfellow went on to design and build, in 1848, the first engine-driven aeroplane model that actually flew. It was a monoplane made from wood and silk with a 3-m (10-ft) wingspan, powered by a small steam engine.

Five years later, Cayley built a glider that was larger than his earlier model. This time, he persuaded his coachman, John Appleby, to sit in it. Other staff members pulled the glider along at the end of the ropes. As it accelerated down a slope, it took off and flew a distance of about 275m (900ft). When it landed, Appleby is said to have got out and resigned, complaining that he was hired to drive, not fly! Had a sufficiently lightweight and powerful engine been available in Cayley's time, he would very likely have made the first successful powered flight half a century before the Wright brothers (see pp.324–27). As it was, his work inspired the next generation of aviation pioneers. Wilbur Wright acknowledged Cayley's remarkable achievements by stating, "Cayley carried the science of flight to a point which it had never reached before".

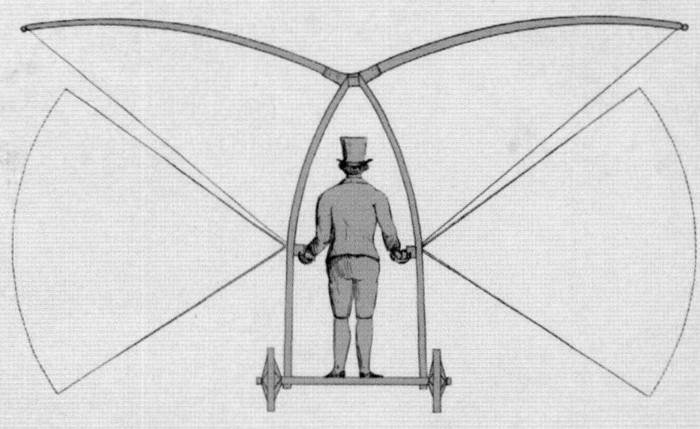

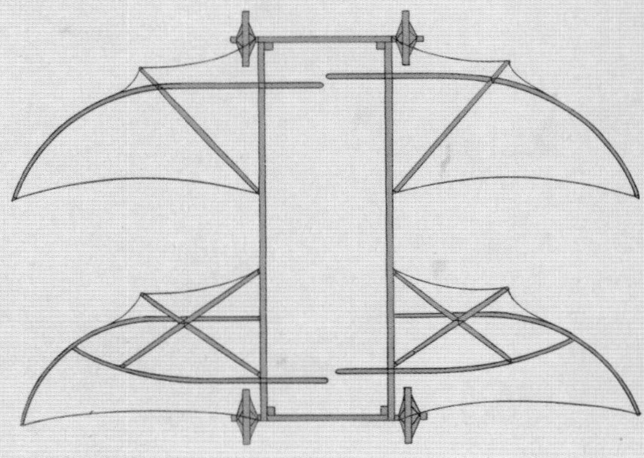

MUSCLE-POWERED FLIGHT
In 1853, Cayley sketched this design for a human-powered aircraft. It was impractical, because materials with the required combination of strength and lightness for human-powered flight were not available in Cayley's time.

AERIAL NAVIGATION WILL FORM A MOST PROMINENT FEATURE IN THE PROGRESS OF CIVILIZATION

GEORGE CAYLEY

JOHN STRINGFELLOW

BUILDER OF THE WORLD'S FIRST POWERED AIRCRAFT

ENGLAND 1799–1883

A LIFE'S WORK

- Stringfellow **works closely with William Henson** to study and develop airplanes
- He **builds the world's first powered aircraft** with Henson
- Creates a small steam-powered aircraft **to make the world's first powered-flight** in 1848
- He builds a **steam-powered triplane** with his son Fred, **which looks remarkably like the first triplanes** of 50 years later, missing only the vertical tailplane; it is the **world's first triplane**

FIFTY-FIVE YEARS BEFORE the Wright brothers made their historic first, sustained, controlled powered flight in 1903, an English lacemaker succeeded in getting an aircraft to fly under its own power in Somerset, England. His name was John Stringfellow, and the machine that provided the power to propel the aircraft was, amazingly, a miniature steam engine.

A native of Sheffield, John Stringfellow was trained as a machine engineer in the textile industry. However, Luddite campaigns against mechanization were making mechanics unpopular in his home town, and he moved to Chard in Somerset to make bobbins in a lace factory.

After about 15 years in Chard, he became friends with a talented young engineer named William Samuel Henson (1812–88), who, aged 23, had already obtained a patent for an ingenious lace-making machine. Stringfellow and Henson were delighted to discover a mutual interest in the possibilities of winged flight, perhaps inspired by the work of George Cayley (see pp.312–13).

In the late 1830s, Stringfellow and Henson would go out into the hills around Chard to observe birds flying, and try to understand how their wings worked. They also conducted experiments with stuffed birds. They were particularly interested in rooks, and calculated that to fly at 30kph (20mph) a rook needed 0.3m (1ft) of wing span to lift 0.2kg (½lb) of weight.

They realized, however, that while flapping wings worked for a bird, a flying machine must have fixed wings. Cayley had shown how a wing that slopes backwards slightly is forced to rise by the natural resistance of the air. Henson and Stringfellow's crucial insight was to see that by adding power to drive a sloping wing through the air, this natural lift can be sustained.

To test this, they boarded a train to London in 1841, armed with small wings of different shapes and sizes. To the bafflement of fellow passengers, they flew their various designs on twine above an open truck, using the slipstream of the moving train rather like modern aircraft designers use a wind tunnel.

SCREWS AND STEAM

In a moment of genius, they also realized that the new screw propeller devised for ships by Swedish inventor John Ericsson (1803–89)

FLYING STEAM CARRIAGE
To raise money for their steam carriage, Henson and Stringfellow published a brochure in which they showed their vision for the machine – with their steam carriage (right) bearing passengers through the skies just as airliners do today.

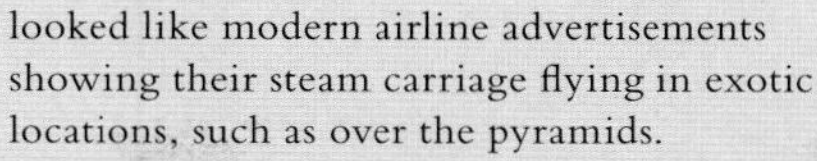

might provide the propulsion through the air that they needed. However, to turn the propeller, they would need an incredibly light, powerful steam engine. The skilful Stringfellow built a tiny engine with a paper-thin copper boiler, heated not by a coal furnace but a spirit lamp.

By now the duo had created the classic aeroplane design, with an engine, propellers, and fixed wings. They drew up plans for a prototype and submitted a patent for their "Aerial Steam Carriage", designed to "convey letters, goods and passengers from place to place through the air". They were so excited by the prospect that they formed the Aerial Steam Transit Company in 1843 to attract funding. The company prospectus insisted that everyone in the world would want their carriage, and they drew up posters that looked like modern airline advertisements showing their steam carriage flying in exotic locations, such as over the pyramids.

Investors were not forthcoming, however, and, in 1847, the two committed their own money to build a trial model. It was the world's first powered aircraft. Built from silk stretched over bamboo and flat, wooden spars, it had a 6-m (20-ft) wingspan and a propeller on the back of each wing. Worried about their ideas being poached, they tested it on the downs near Chard before sunrise. But the machine would do little more than hop. Stringfellow later realized that the engine was too heavy and the wings had become saturated with dew and lost the necessary rigidity. A disappointed Henson emigrated to America.

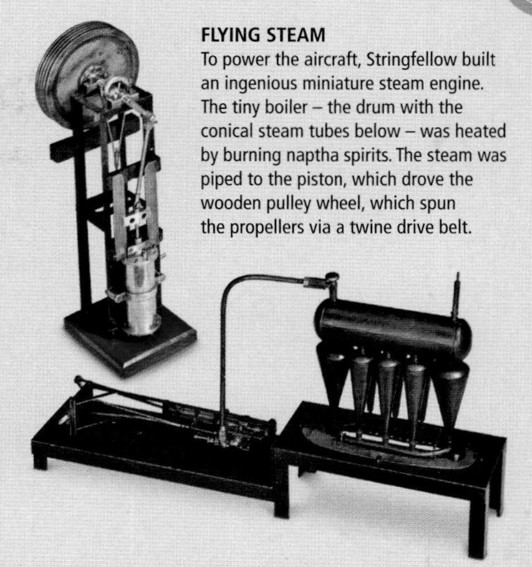

FLYING STEAM
To power the aircraft, Stringfellow built an ingenious miniature steam engine. The tiny boiler – the drum with the conical steam tubes below – was heated by burning naptha spirits. The steam was piped to the piston, which drove the wooden pulley wheel, which spun the propellers via a twine drive belt.

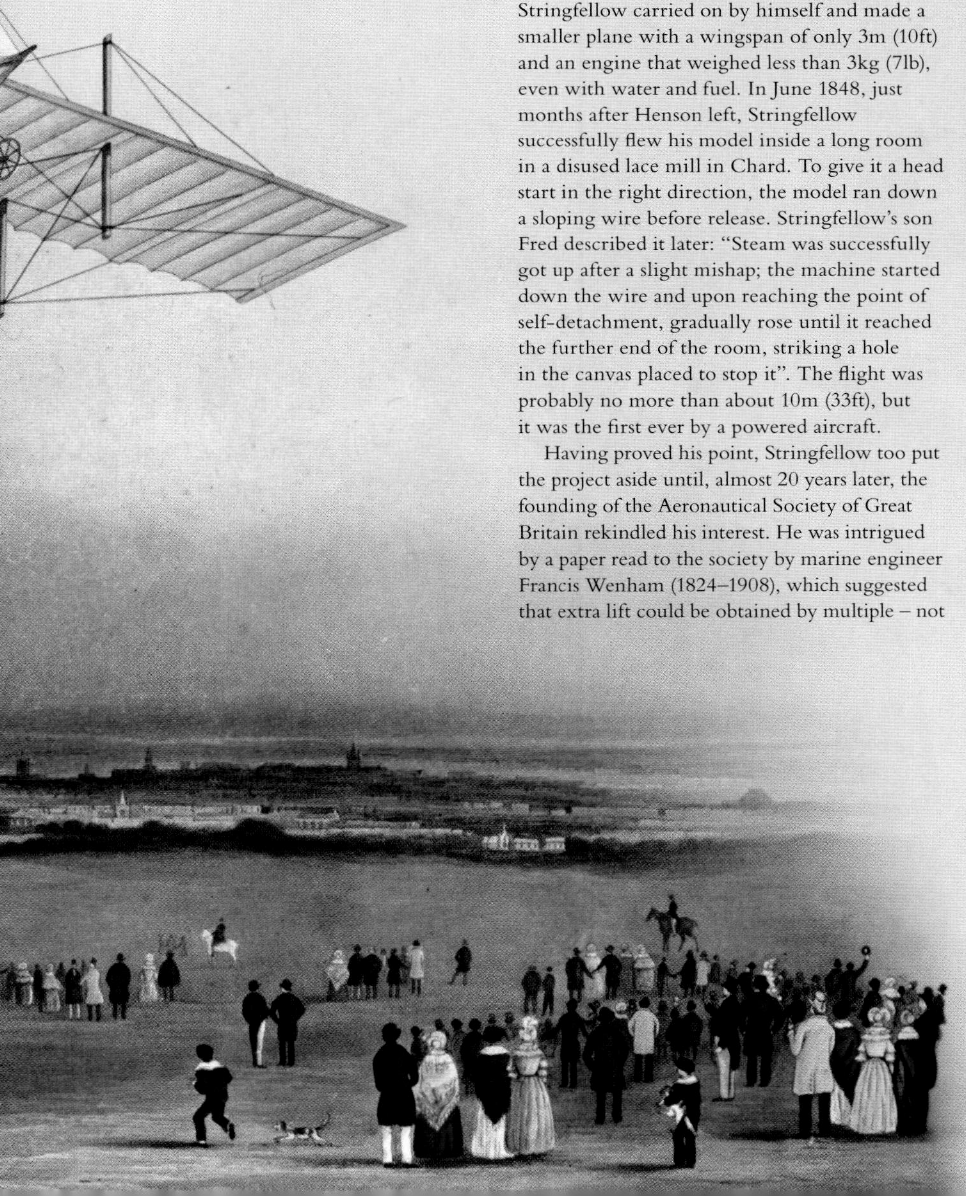

THE FIRST POWERED FLIGHT

Stringfellow carried on by himself and made a smaller plane with a wingspan of only 3m (10ft) and an engine that weighed less than 3kg (7lb), even with water and fuel. In June 1848, just months after Henson left, Stringfellow successfully flew his model inside a long room in a disused lace mill in Chard. To give it a head start in the right direction, the model ran down a sloping wire before release. Stringfellow's son Fred described it later: "Steam was successfully got up after a slight mishap; the machine started down the wire and upon reaching the point of self-detachment, gradually rose until it reached the further end of the room, striking a hole in the canvas placed to stop it". The flight was probably no more than about 10m (33ft), but it was the first ever by a powered aircraft.

Having proved his point, Stringfellow too put the project aside until, almost 20 years later, the founding of the Aeronautical Society of Great Britain rekindled his interest. He was intrigued by a paper read to the society by marine engineer Francis Wenham (1824–1908), which suggested that extra lift could be obtained by multiple – not single – wings. Therefore, with his son Fred, Stringfellow built a steam-powered triplane, which looks remarkably like the first triplanes of half-a-century later, missing only the vertical tailplane. The triplane was run in the Crystal Palace Exhibition in 1868.

Stringfellow now wanted to make a machine that could carry a human pilot, and equipped a building to construct it, but his health had suffered. He died in 1883, his vision unrealized.

WILLIAM SAMUEL HENSON

ENGLAND 1812–88

Like his partner Stringfellow, Henson was born in the north of England in Nottingham and moved south to join the lace-making industry in Chard, Somerset.

Henson was an ingenious young man, and, by the age of 23, he had invented a machine for automatically making ornaments on lace. For 10 years, he worked on the aerial steam carriage with Stringfellow. He had helped Stringfellow bring the project to the very point of fruition before getting married and heading for the US in 1849, having already spent a small fortune on the venture. Before Henson left, though, he made one popular invention in 1847 – the T-handled safety razor. Once he was in New Jersey, Henson carried on his inventive strain, working first as a mechanic and then as a civil engineer. He developed knitting looms, steam engine governors, and even an ice-making machine.

OCTAVE CHANUTE

FATHER OF THE MULTI-WINGED FLYING MACHINE

UNITED STATES 1832–1911

French-born engineer Octave Chanute was one of America's leading railroad and bridge engineers until he retired in 1890. He then turned his attention to aviation and had such an impact – developing biplane and triplane flying machines – that the Wright brothers believed they owed their success to him. Their *Flyer* was based on Chanute's designs.

A LIFE'S WORK

- In 1865, Chanute **designs the gigantic Union Stockyard** in Chicago, which soon began processing 2 million animals annually
- In 1871, he builds another **gigantic stockyard at Kansas City,** and also builds the **first bridge** in Kansas City **across the Missouri River**
- He becomes the world's **leading authority on flying machines** after beginning his career as a **railroad and bridge engineer**
- Creates a team that includes Albert Herring and William Avery **to build and test gliders**
- **Develops the biplane design** and becomes a mentor to the **Wright brothers**
- At his death he is hailed as the **father of aviation** and the **heavier-than-air flying machine**

Until Octave Alexandre Chanute became involved in aviation in the 1890s, the development of flying machines had seemed like something for dreamers. But after a distinguished career as an engineer, Chanute brought credibility to the field.

His entire approach introduced solidity to aeronautics. First, he collated all that was known on the subject. Then he brought together a highly qualified team to test existing designs, and only then did he move on to new designs – and even these he would set up to deliver maximum information. This step-by-step approach at last put aviation on a firm foundation.

EARLY YEARS

Octave was born on 18 February 1832 in Paris, France. But when he was six, his father Joseph Chanut was offered a post as history professor in New Orleans, and so the little boy was brought up in the American South. When Octave was 14, the family moved to New York, and the month-long journey on the steamboat up the Mississippi and Ohio rivers, then on the new railroads to New York, made such an impression that he was determined to be an engineer.

At that time, there was no college training available for engineers. Therefore Octave applied for a job in the Hudson River Railroad. When they turned him down, he offered to work for free, and after a few weeks of volunteer work, the railroad employed him on a full-time basis. To make his progress easier, he anglicized his name from Chanut to Chanute and dropped his French middle name.

RAILROAD ENGINEERING

Chanute's progress was rapid and he quickly became a leading railroad engineer, surveying the constructions and upgrading lines all over the country. In 1865, aged only 33, he designed the gigantic Union Stockyard in Chicago – a meat packing district – that soon started processing two million animals annually. In 1871, he built the Kansas City Stockyards.

Six years later he built another gigantic stockyard at Kansas City. In Kansas City, too, he built the first bridge across the Missouri River, the Hannibal Bridge, and went on to design many other bridges, including the Illinois River Bridge at Peoria.

FLYING HERRING
One of Chanute's test pilots, Augustus Herring (1867–1926), prepares to launch a Chanute biplane.

Concerned about the effect of rail construction on forests, he introduced measures to ensure that the wood used in railway sleepers was not wasted – preserving them with creosote and using "date nails" – nails stamped with the date to ensure future maintenance is done only at the right time.

UNDERSTANDING AVIATION

Chanute was fascinated by developments in flying in Europe, and when he retired in 1891, he could at last indulge his interest. His first step was to gain a better understanding of the subject. "My general idea is to pass in review what has hitherto been experimented," he wrote, "with a view to accounting for the failures, clearing away the rubbish, and pointing out some of the elements of success". He published his review in the book *Progress in Flying Machines* in 1894, which became the definitive guide for every would-be aviator.

Then he assembled a team to help him with tests, including Albert Herring and William Avery. They began with an Otto Lilienthal-style glider (see pp.320–21), but after a few disappointing tests, decided to try new designs.

MULTI-WINGED FLIGHT

Following the ideas of British marine engineer Francis Wenham (1824–1908) about multiple wings, Chanute designed the *Katydid* glider with five sets of wings. These could be moved up and down the fuselage for experiments. The design was continually revised over 200 test flights along Miller Beach, on the shores of Lake Michigan.

After carefully analysing the results from the *Katydid*, the team constructed more gliders, including a biplane built to Chanute and Herring's design. Significantly, the Chanute-Herring biplane's wings were tied together with strong, light wire-bracing called Pratt trussing that Chanute was familiar with from his bridge-building days. This system set the pattern for all multi-winged planes in the future.

To get away from the attention of reporters, the group moved tests from Miller Beach to the remote Dune Park. The biplane was originally a triplane, but tests at Dune Park showed that two wings worked better. The team was getting closer to a satisfactory design for flight, although, despite experiments with movable wings and tail-control surfaces, they still needed to find a way to achieve control in the air.

In 1897, Herring began to work with another sponsor, Matthias Arnot, and the following year built another biplane on which he installed a petrol engine and propellers.

Chanute himself built one more biplane glider, flown by Avery at the St Louis World's Fair in 1904. Ingeniously, the glider was launched from a level field by using an electric winch to tug the glider forward like a kite until the pilot was high enough aloft to release the cable.

AN INSPIRATION

In 1899, Chanute was contacted by the Wright brothers (see pp.324–29) for advice, and became a mentor to them, providing guidance every step of the way as they developed their *Flyer* – with Chanute travelling to Kitty Hawk in North Carolina each summer to see how the brothers were progressing.

But in time they fell out. Chanute, the Wrights felt, did not appreciate the originality of their wing-warping technique for achieving control in the air, and wanted them to share it with the world, which they were unwilling to do. They tried to resolve their differences over the years and, when Chanute died on 23 November 1910, it was Wilbur Wright who read the eulogy at his funeral.

THE RUBBER-BAND MAN

Very few engineering breakthroughs are ever the work of one inspirational genius. Most are achieved through the slow build-up of small contributions by lesser known heroes. In the development of aircraft, one of these forgotten heroes is Frenchman Alphonse Pénaud (1850–80). Instead of getting bogged down with engines, Pénaud had the brilliantly simple idea of experimenting with model flying machines powered by twisted rubber bands. Before he committed suicide at the tragically young age of 30, Pénaud made significant contributions to achieving stability in flight using his little models.

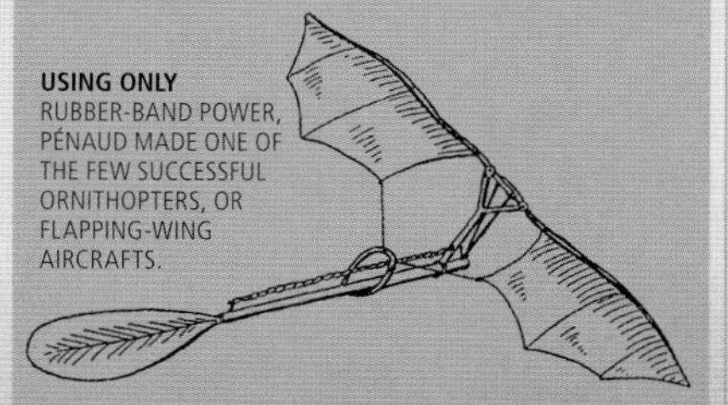

USING ONLY RUBBER-BAND POWER, PÉNAUD MADE ONE OF THE FEW SUCCESSFUL ORNITHOPTERS, OR FLAPPING-WING AIRCRAFTS.

CHANUTE *KATYDID*
This photograph of 1896 shows a multiple-winged Chanute glider launching from a cliff, probably on Miller Beach on Lake Michigan. Designs with fewer wings were soon found to work better.

FALSE START

Orville Wright took this photograph of his brother Wilbur on *Flyer I* at Kitty Hawk, North Carolina, on 14 December 1903. The aircraft's gasoline engine, which the brothers had made in their bicycle shop, stalled after takeoff, allowing a flight of only three seconds, and its lightweight construction of spruce wood and muslin covering was damaged as it hit the ground. Three days later, on 17 December, Orville made the historic first flight, taking off into a freezing headwind for 12 seconds and flying for 37m (120ft).

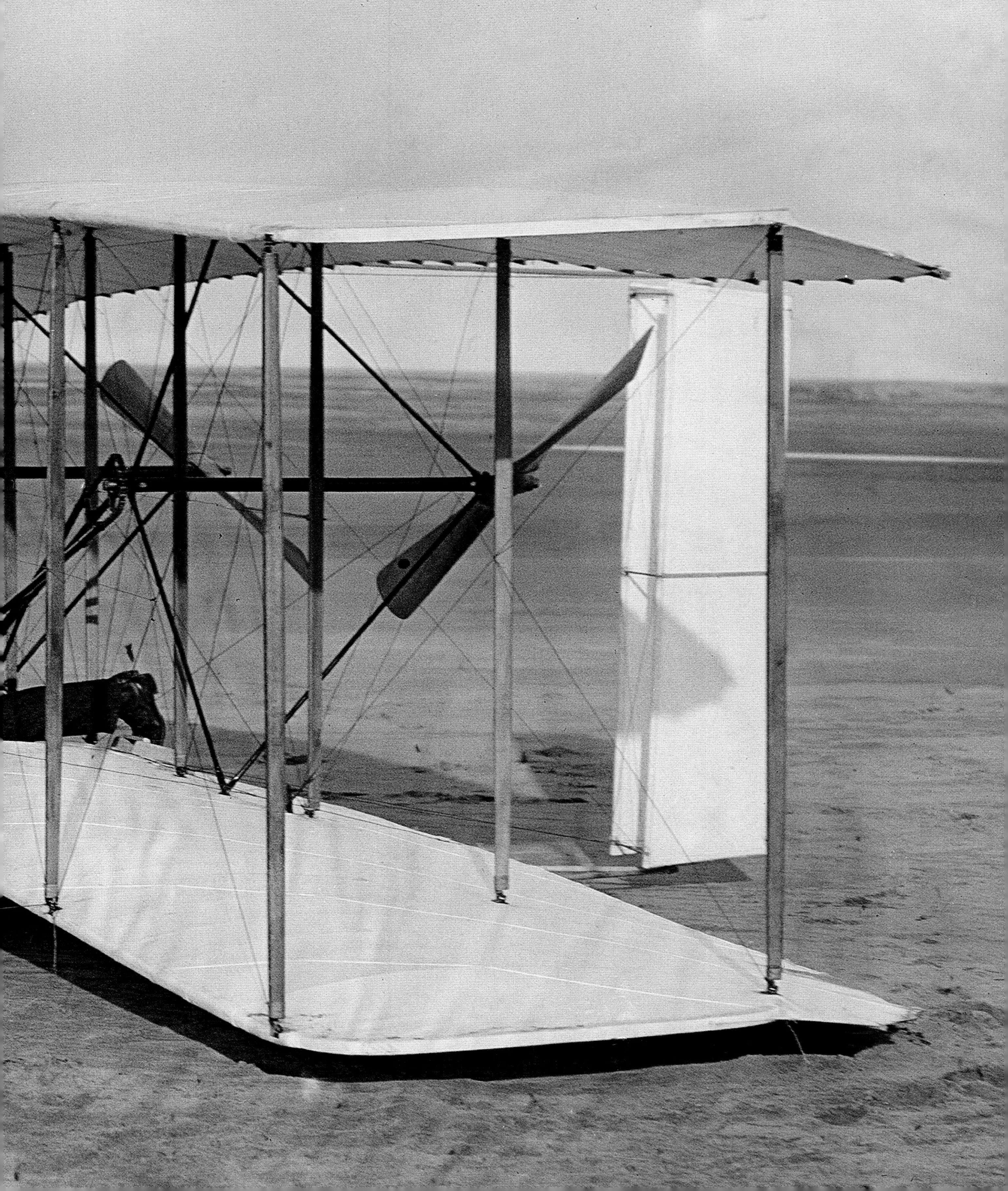

OTTO LILIENTHAL

DARING HERO OF EARLY AVIATION

PRUSSIA 1848–96

AVIATION PIONEER Otto Lilienthal was the first to show the world that human flight was not a novelty or a fluke but could be repeated on a routine basis. Unfortunately, he was killed in a flying accident, but not before he had made thousands of safe and hugely informative flights in his beautifully designed gliders, the forerunners of modern hang-gliders.

A LIFE'S WORK

- Otto Lilienthal **works closely with his brother Gustav** throughout his life to develop his flying machines
- As a teenager attempts to **build an "ornithopter"**
- **Studies mechanical engineering** at Potsdam and Berlin universities
- Becomes a **construction engineer** and sets up his **own engineering company**
- Extensively **studies how birds fly and writes a book** on the subject
- **Manufactures 18 gliders** within five years; is credited with the **first production-run of an aircraft**
- He makes more than **2,000 successful glider flights**

Combining scientific rigour and personal bravery, Lilienthal was the first man to fly on wings more than once – and in some ways it was he, not the Wrights, who introduced the world to the age of aviation, showing that humans could actually use wings to fly safely in a controlled way.

BIRD MAN

Otto Lilienthal was born in Anklam in 1848 in the province of Pomerania in what was then Prussia and is now Germany. The tragic flying accident that killed him has led to him being portrayed as one of those heroic but foolish birdmen who contributed little to the real progress of human flight, but this could not be further from the truth. Through years of theoretical study of how wings work, followed by crucial practical experiments, Lilienthal sowed the seeds of the aviation age, aided by his brother, Gustav (1849–1933).

Lilienthal became interested in flight from an early age, and as a teenager he and his brother experimented with the idea of an "ornithopter" – a flying machine that worked by flapping its wings like a bird. The brothers built contraptions with wings of thin birch veneer, which they would strap onto themselves and run downhill, trying, unsuccessfully, to take off.

Unfazed by the ridicule they must have faced, the Lilienthal brothers remained serious about flying, and Otto studied mechanical engineering at Potsdam and Berlin universities to arm himself with the necessary scientific and practical grounding to build on his work.

After a brief interlude of national service in the Franco-Prussian War of 1870–71, he began to work as a construction engineer with Berlin-based machine manufacturer C Hoppe. Ten years later, he set up his own factory in Berlin that made boilers and steam engines. But even as he built up his business and his engineering expertise, he continued to develop his ideas on flying. Over the next 20 years, Lilienthal tried experiments with ornithopters, but he mostly concentrated on trying to understand the theoretical basis of flight.

He tested various wing shapes and studied in detail how birds fly, publishing a seminal book, *Bird Flight as the Basis of Aviation,* in 1880. He began to appreciate the balance between wing area and lift, and the importance of the curve, or camber, of the wing in generating the air resistance on which lift depends.

"ARTIFICIAL ALBATROSS"
Like Lilienthal, Jean-Marie le Bris, too, drew inspiration from the flight of birds and designed his gliders based on the wings of an albatross. His "Artificial Albatross" gliders were the first gliders that soared into the air rather than simply descending slowly.

He particularly admired the stork and the seagull for their ability to glide and, while not abandoning the idea of "flapping flight", he realized that the way forward lay in perfecting the art of gliding. Aeronautical historian Charles Gibbs-Smith divided flight pioneers into two kinds: "Chauffeurs of the Air", who concentrated on achieving sheer power to drive through the air as if they were driving a car, regardless of lift and control; and "Airmen", who wanted to float on the air and achieve lift and control with the least amount of effort. Lilienthal was an Airman and, when he finally began his experiments in 1891, they were with gliders, even though other pioneers of the time, such as Frenchmen Clément Ader (see p.323), were already taking off into the air with the aid of engines.

THE LURE OF THE ALBATROSS

The Lilienthal brothers were not alone in their fascination for birds. After working as a sailor for 20 years and watching albatrosses sweep across the sky, Frenchman Jean-Marie le Bris (1817–72) decided to build gliders based on the wings of an albatross that he had killed. His "Artificial Albatross" gliders had a 16-m (52-ft) wingspan and were made from wood and silk. Interestingly, they also had hand-levers and pedals to alter the angle of the wings and a tail to provide him with control in the air. In 1856, le Bris mounted one of the gliders on a horse cart and made the cart-driver accelerate across a beach. His large artificial "bird" took off and

HANG-GLIDER SIMPLICITY
It was Lilienthal's thousands of flights in his beautifully simple and well-designed hang-gliders that showed how human beings really could fly. The news of his tragic death inspired the Wright brothers to try and continue his work.

climbed 100m (328ft), the first time a glider had ever climbed, before landing 200m (650ft) away. He made another partially successful flight, which ended in a crash in which he broke his leg. Twelve years later, he built another glider and managed a few more flights.

"THE GLIDER KING"

Between 1891 and 1896, Otto Lilienthal built 18 gliders, 15 monoplanes, and 3 biplanes, and made over 2,000 successful flights. What is fascinating about the man is his patient and systematic approach. For his initial flights, he and Gustav constructed a glider consisting of just a simple bamboo framework with thin cotton stretched above for the wings. But he started by jumping off a springboard no higher than a table and slowly built it up to a launchpad as high as a wardrobe only once he had learnt how to control the glider in the air by shifting his weight. Today hang-glider fliers know all about these techniques, but Lilienthal had to discover them entirely for himself – and the available materials meant that his gliders were heavy and required immense strength to manoeuvre.

After a few years of experimenting, Lilienthal constructed a small hill, topped by an earth-covered shed for storing his machines. He called it the *Fliegeberg* ("flight mountain") and it enabled him to take off into the wind, no matter what its direction. By 1893, he was regularly achieving flights of up to 45m (148ft), and was learning all he could about controlling his glider in the air. He also tried building a machine with flapping wings, driven by a small motor. The following year, he constructed what became known as his "standard" glider, and moved his tests to the Rhinower Hills 10km (6 miles) northwest of Berlin. There, he was able to fly up to about 350m (1,000ft).

A DOOMED FLIGHT

By now, the exploits of "The Glider King" had made him famous. Aviation enthusiasts and experts from around the world were commissioning him to build similar gliders. Altogether, he built eight copies of the standard glider – the first-ever production-run of an aircraft. On 9 August 1896, a gust of wind made his glider stall in mid-air, so that it plummeted by 20m (66ft). The impact broke his spine. He died in hospital the next day saying, "Small sacrifices must be made!".

> TO INVENT AN AIRPLANE IS NOTHING; TO BUILD ONE IS SOMETHING; BUT TO FLY IS EVERYTHING
>
> **OTTO LILIENTHAL**

HIRAM MAXIM

PIONEER OF POWERED HEAVIER-THAN-AIR FLIGHT

UNITED STATES 1840–1916

HIRAM STEVENS MAXIM is best known for inventing the Maxim machine gun, but he was also an aviation pioneer. The Maxim gun made him a fortune, which enabled him to pursue other interests, including heavier-than-air flying machines. He is also remembered for his Captive Flying Machine, a fairground ride, through which he hoped to generate public interest in flying.

Maxim was born near Sangerville, Maine, but spent most of his life in Britain. After only five years of schooling, he started working on his father's farm, and then as an apprentice carriage-maker. After several jobs in various trades, he met the founder of the US Electric Lighting Co, Spencer D Schuyler, who was so impressed with Maxim that he hired him as his chief engineer. Maxim represented the company at the Paris Exposition of 1881.

Later that year, Maxim moved to England, where, in 1885, he invented his famous Maxim gun, the first fully portable machine gun (see box, below). As a child, Maxim had been knocked over by a rifle's recoil, and he realized that the recoil force could be used to operate a gun automatically.

BUILDING A GIANT

Maxim's interest in flying machines was stirred at the age of 16, when his father sketched an idea for an aircraft. It was a small platform, lifted by two rotors. When Maxim was wealthy enough to start thinking about building his own flying machine, he considered a machine similar to his father's. However, Maxim's calculations suggested that such an aircraft, using the materials and engines of the day would be too heavy to take off. He turned his attention to building an aircraft with wings.

Maxim noted that birds use their wings for both lift and propulsion, but he thought it best to separate these two functions in a man-made flying machine. Wings would produce lift, while propellers would provide propulsion. He tested model wings by attaching them to the end of a rotating arm to see how much lift they produced. He then built a giant test rig for a full-size aircraft.

The work began in 1891. The whole machine was 36.5m (120ft) long and 32m (104ft) wide, and weighed about 3.6 tonnes (4 tons). It consisted of a wheeled platform carrying two 134-kilowatt (180-horsepower) steam engines, which could turn two huge propellers – nearly 5.5m (18ft) in diameter. The aircraft ran along a pair of steel rails 550m (1,800ft) long and 3m (9ft) apart.

Maxim knew that getting an aircraft off the ground and controlling it in the air were two different challenges. He concentrated on getting the aircraft airborne, designing the test rig in a way that would let his aircraft take off, but stop it from flying away. The aircraft had lightweight wheels at the front and heavy cast-iron wheels at the back. If it started to take off, the lighter front would rise first, while the heavier back would stay on the rails. However, the great weight of the cast iron made it difficult to start and stop the machine quickly, and on one test run, a gust of wind lifted the whole machine off the track and turned it over.

Maxim then adapted his track to hold the machine down more securely. He added a pair of wooden safety rails, which were 9m (30ft) apart and higher than the steel rails. He also added outrigger wheels to the aircraft. If the aircraft tried to take off, the outrigger wheels would come up underneath the safety rails, which would hold the machine down.

TAKING FLIGHT

On 31 July 1894, Maxim made the first test run. His giant flying machine showed no signs of taking off. The steam pressure was increased for the next run. This time the machine lifted off the steel rails momentarily on several occasions. A third run was made at full power. After about 180m (600ft), at a speed of about 68kph (42mph), the aircraft took off. The safety rails held the aircraft down, preventing it from flying away, as they were designed to do. After another 90m (300ft), the back end of the aircraft

A LIFE'S WORK

- Maxim has **hundreds of inventions and patents in his name** in Britain and the US
- He **receives his first patent** at the age of 26, **for a hair-curling iron**
- His **other inventions include** a mousetrap, an electric lamp, shoe heel protectors, a locomotive headlamp, **his famous Maxim machine gun, smokeless gunpowder,** and a **variety of engines**
- In 1894, he **makes a brief flight in a powered, heavier-than-air flying machine** of his own design
- In 1900, he **becomes a British citizen,** and **is knighted in 1901** for his contributions
- He designs a popular **fairground ride, the Captive Flying Machine,** which is installed in coastal resorts across Britain and the US

MAXIM'S MACHINE GUN

Maxim developed his design for an automatic machine gun using an action that would close the breech and compress a spring – by storing the recoil energy released by the previous shot – to prepare the gun for its next shot. He thoughtfully ran announcements in the local press, warning that his neighbours should keep their windows open to avoid the danger of glass breaking as he would be experimenting with the gun in his garden. Versions of the Maxim gun were used extensively by both the Allied and Central Powers during World War I. In his later years, Maxim became deaf – his hearing damaged by years of exposure to the noise of his guns.

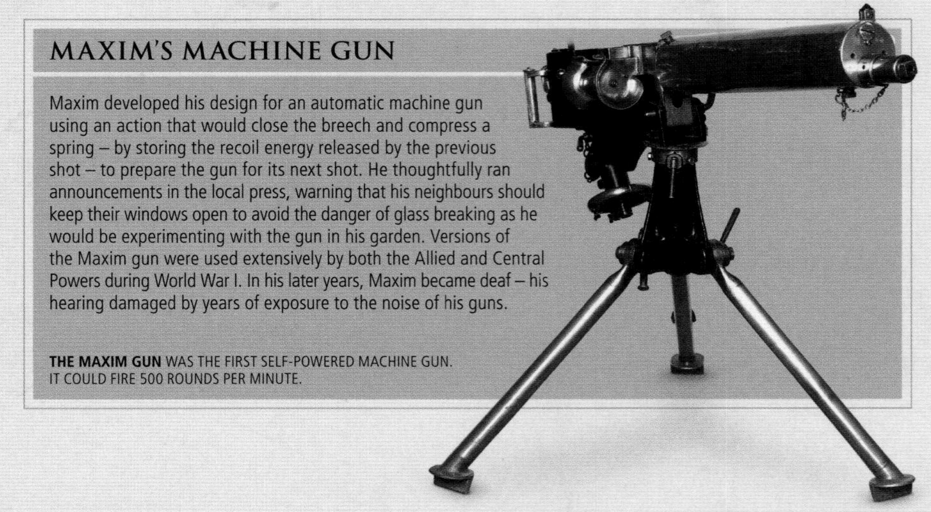

THE MAXIM GUN WAS THE FIRST SELF-POWERED MACHINE GUN. IT COULD FIRE 500 ROUNDS PER MINUTE.

broke free from the safety rails and rose higher. About 30m (100ft) further on, the front broke free too. The giant machine was now flying with Maxim and his three assistants onboard. They closed the steam valve, cutting off power to the propellers, and the machine crash-landed. It had flown a distance of about 30m (100ft) at a height of 60cm (2ft). Maxim had achieved a powered take-off – nine years before the Wright brothers' historic flight (see pp.324–27).

CAPTIVE AIRCRAFT

Maxim repaired his machine, and used it to raise money for charity by charging people for rides. However, the aircraft was aerodynamically unstable and it was dismantled in 1895. In 1900, Maxim became a British citizen and received a knighthood the following year.

Maxim built another flying machine to raise funds for further flight experiments, but this time it was a fairground ride. He called it his Captive Flying Machine, and demonstrated it to the press in 1904. The passengers sat in ten carriages hanging from cables. When the ride started revolving, the carriages picked up speed and swung out further and further from the centre. This ride was built in several sizes, and was installed in London, and in coastal resorts elsewhere in Britain and the US, and soon became very popular. By then, other aviation pioneers had overtaken Maxim, who lost interest in flight research.

Maxim's experiments are important because they showed that flight in a heavier-than-air machine was possible – something that was not universally accepted at the time. Indeed, the year after Maxim's machine broke free from its track and took off, the distinguished physicist and engineer William Thomson (see pp.232–33) declared that, "Heavier-than-air flying machines are impossible". He did not know it then, but Maxim had already proved him wrong.

CLÉMENT ADER

FRANCE 1841–1926

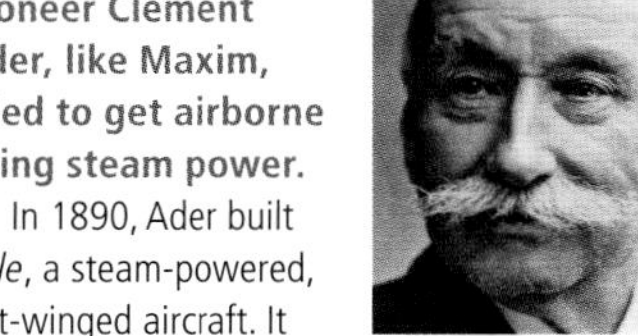

Self-taught engineer, inventor, and aviation pioneer Clément Ader, like Maxim, tried to get airborne using steam power.

In 1890, Ader built *Éole*, a steam-powered, bat-winged aircraft. It made a 50-m (160-ft) uncontrolled hop into the air. With financial help from the French War Office, Ader built a bigger twin-propeller aircraft called Avion III. However, the War Office was not impressed, and stopped funding it. In 1909, Ader published his book, *L'Aviation Militaire* ("Military Aviation"), which predicted air warfare and aircraft carriers.

MAXIM'S FLYING MACHINE
Visitors in elegant Victorian dresses gather at Baldwin's Park, Bexleyheath, Kent, in 1894 to see Maxim's impressive flying machine, and perhaps, to take a ride on it. The giant machine, standing on its rails, towers over them.

ISN'T IT ASTONISHING THAT ALL THESE SECRETS HAVE BEEN PRESERVED FOR SO MANY YEARS JUST SO WE COULD DISCOVER THEM!

ORVILLE WRIGHT

A LIFE'S WORK

- **As a child, Orville makes kites** and sells them to other children to raise extra pocket money
- **Orville publishes a weekly newspaper** in Dayton, Ohio, called the ***West Side News***
- While **working at Kitty Hawk**, the Wright brothers **begin gliding experiments**
- As late as 1901, **Wilbur is so discouraged** by setbacks that he **predicts man will not fly "within a thousand years"**
- **Flyer III, the first practical aircraft** invented by the brothers in 1905, is **disassembled** after the US, Britain, and France refuse to buy it
- In 1909, the **Wright Company** is founded, which **establishes a factory** at Dayton **and a flying school** at Huffman Prairie

WRIGHT BROTHERS

LEGENDARY PIONEERS OF POWERED FLIGHT

UNITED STATES 1871–1948; 1876–1912

ORVILLE WRIGHT

WILBUR WRIGHT

Wilbur and Orville Wright were just two bicycle mechanics from Dayton, Ohio, until 17 December 1903. On that day, they did something that changed history and subsequently affected the travel of millions of people. They built an aircraft that made the first controlled, powered flight.

The two brothers, interested in all things mechanical from an early age, became fascinated by flight in 1878, when their father gave them a toy helicopter. At the age of seven, when questioned by his teacher while playing with pieces of wood, Orville had said that he was making a flying machine and if it worked he would build a bigger one that he could fly with his brother – prophetic words.

Wilbur and Orville's father, Bishop Milton Wright, encouraged his children to think for themselves. In 1892, the two brothers opened a bicycle shop and later went on to manufacture their own bikes. Reports of the exploits of early aviation pioneers rekindled their interest in flying machines and motivated them to build one.

THE CONTROL PROBLEM

The brothers determined that the main problem lay not in getting a machine airborne, which several people had already done, but in controlling it once it took off. Their bicycle experience proved useful. A bicycle rider steers by leaning to one side, so they considered making an aircraft lean, or roll, in the same way. While Wilbur was holding a long thin box containing a bicycle inner tube, he started twisting it in his hands. It reminded him of the way a bird's wings twist in flight. He and Orville worked out a way for a pilot to pull wires to twist an aircraft's wings and make the plane roll. They called it "wing-warping". In 1899, they tested it with a kite – it worked. The next step was to build a bigger wing-warping glider that could carry a pilot.

They searched for a suitable site for their test-flights and asked the US Weather Bureau for places where the wind blew strongly. The closest site to their home in Dayton, Ohio, was the town of Kitty Hawk on the North Carolina coast.

In 1900, they built a glider with the help of data produced by the German aviation pioneer, Otto Lilienthal (see pp.320–21). It was a biplane with a wingspan of 5.4m (17ft 6in). At first, they flew it as a kite. They added a horizontal surface called an elevator at the front. Tilting the elevator made the glider fly higher or lower. The wing-warping system worked too, making the glider roll to one side or the other. Then Wilbur climbed

WRIGHT CYCLE COMPANY
The Wright brothers' bicycle shop was situated in Dayton, Ohio. In 1936, car manufacturer Henry Ford bought the building and physically moved it to an outdoor museum in Dearborn, Michigan.

KITE-FLYING DUO
The Wright brothers tested their gliders by flying them first as kites. If the kite flights were successful, they tried manned flights next, with the pilot lying on the lower wing. Here, they are testing their second glider in 1901 at Kill Devil Hills, near Kitty Hawk in North Carolina.

onboard and made several successful test-flights from the sand dunes at Kill Devil Hills, near Kitty Hawk. When it was time to return home, they were already thinking about an improved design for their next glider.

EARLY SETBACKS

Their 1901 glider was larger than the earlier machine, but it did not fly as well. They suspected that Lilienthal's data, which they had used to design the glider, was not correct. So they decided to source their own data. Back home in Dayton, they cut small wing shapes from thin metal sheets and mounted each of them on the rim of a bicycle wheel attached to a bicycle's handlebars. The wheel was free to rotate. By cycling so that air flowed over each wing, they could see which shapes worked best. Next, they built a wind tunnel to test the best wings.

In 1902, the brothers built a new glider with a wingspan of 9.7m (32ft). They added a pair of tail fins to stop it slipping sideways in turns, and Wilbur taught Orville to fly it. After several crashes, they changed the fins to a moveable rudder linked to the wing-warping wires. The modified glider flew better. Now that they could control a glider in the air, they were ready to install an engine.

During 1903, the Wright brothers designed and built their first powered aircraft, called Flyer. They first thought they would be able to buy an engine for it, but they could not find a suitable model at an affordable price. So they built their own engine, a simple water-cooled four-cylinder powerplant. The bulk of this work was done by their mechanic, Charlie Taylor.

Now they needed propellers. They assumed that marine propeller designers would be able to provide the necessary design equations, but no such equations existed. Undeterred, they set about analysing how propellers work. They treated each propeller blade as a wing, creating thrust instead of lift. From this, they designed a pair of twin-bladed propellers 2.6m (8.5ft) in diameter.

The engine sat on the plane's lower wing to the right of the pilot, who lay in a cradle. Sliding the cradle to one side or the other pulled the wing-warping wires and made the plane roll, an arrangement they had used successfully in their 1902 glider. A lever operated the front elevators. The plane was 6.4m (21ft) long and had a wingspan of 12.3m (40.5ft).

FIRST FLIGHT

Wilbur attempted the first flight on 14 December 1903, but the plane stalled and crashed into the sand. They tried again on 17 December with Orville at the controls. This flight was successful. It lasted only 12 seconds, but it was the first controlled flight in a fixed-wing aircraft.

The brothers wanted to be able to sell their aircraft, so they knew they would have to develop more controllable planes. Their first was Flyer II, built in 1904. As they were now less dependent on strong winds for flying, they moved their test-flights to Huffman Prairie near their home in Dayton. They made more than 100 flights lasting up to about five minutes in Flyer II.

After that, they dismantled it and used some of its parts to build Flyer III. After a crash on 14 July 1905, they made the elevator and rudder bigger and moved them further away from the wings. They also disconnected the rudder from the wing-warping wires and gave it a separate control. When flights resumed, they found there was an immediate improvement, and they could easily fly for more than 20 minutes at a time.

SHOWING THE WORLD

The Wrights received a patent for their invention on 23 May 1906, and, in 1908, they signed a contract to supply aircraft to the US Army. Orville stayed in the US to carry out test-flights for the army, while Wilbur travelled to France to demonstrate a Flyer there. As they had not flown for more than two years, they returned to Kitty Hawk to make some practice flights. The plane, now called a Type A Flyer, had two seats. On 14 May 1908, Wilbur took mechanic Charles Furnas for a flight. It was the first ever passenger flight.

Wilbur demonstrated one of their planes over a race-course at Hunaudières, France, on 8 August 1908. The spectators were stunned by how advanced and controllable it was compared to experimental aircraft in Europe. Orville's demonstration flights for the US Army went equally well, but ended in tragedy. On 17 September 1908, Orville took off from Fort Myer, near Washington DC, with Lieutenant Thomas Selfridge onboard. During the flight, one of the propellers broke free from its supports. Orville tried to land, but the plane

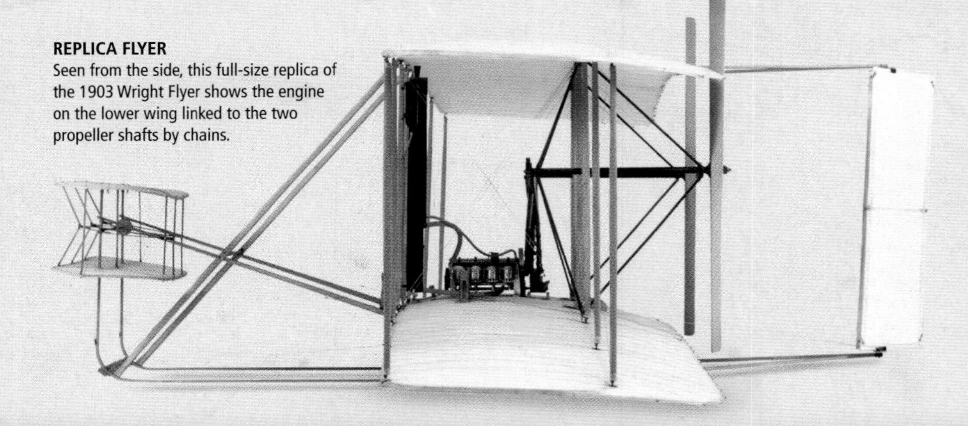

REPLICA FLYER
Seen from the side, this full-size replica of the 1903 Wright Flyer shows the engine on the lower wing linked to the two propeller shafts by chains.

ENGINEERING TIMELINE

1878–91	1892–94	1895–99	1900
	Orville and Wilbur open their own bicycle repair shop and there begin experiments with propellers and wings		The duo begin their first field experiments with a glider in North Carolina

crashed, killing Lieutenant Selfridge. It was the first fatal crash in powered aviation. Orville was badly injured, but survived.

THE BUSINESS OF FLIGHT

The Wright brothers were now international celebrities. The Wright Company was set up in 1909 to build Wright aircraft in the US, with Wilbur as its president. When Wilbur died on 30 May 1912 from typhoid fever, Orville succeeded him. Wright aeroplane factories and flying schools were established on both sides of the Atlantic. One of the flying schools was located at their old Huffman Prairie testing ground. The aviation pioneers had become businessmen.

Orville lived to see the development of military aircraft during World War I and the first commercial passenger planes in the 1920s. By the time he died, in 1948, the first jet aircraft had flown and the Bell X-1 rocket-plane had made the first supersonic flight. The two bicycle mechanics from Dayton, Ohio, had made it all possible.

DEMONSTRATION FLIGHTS
The Wright brothers made demonstration flights in the US and Europe in their Type A aircraft. The Type A had seats for a pilot and a passenger, and the sliding hip cradle of the earlier Flyers had been replaced by a hand-operated roll control.

1901

1902
The brothers construct a new glider with tail fins for stabilization that is tested successfully

1903–04
The Wright brothers fly into history with the first powered heavier-than-air flight

1905

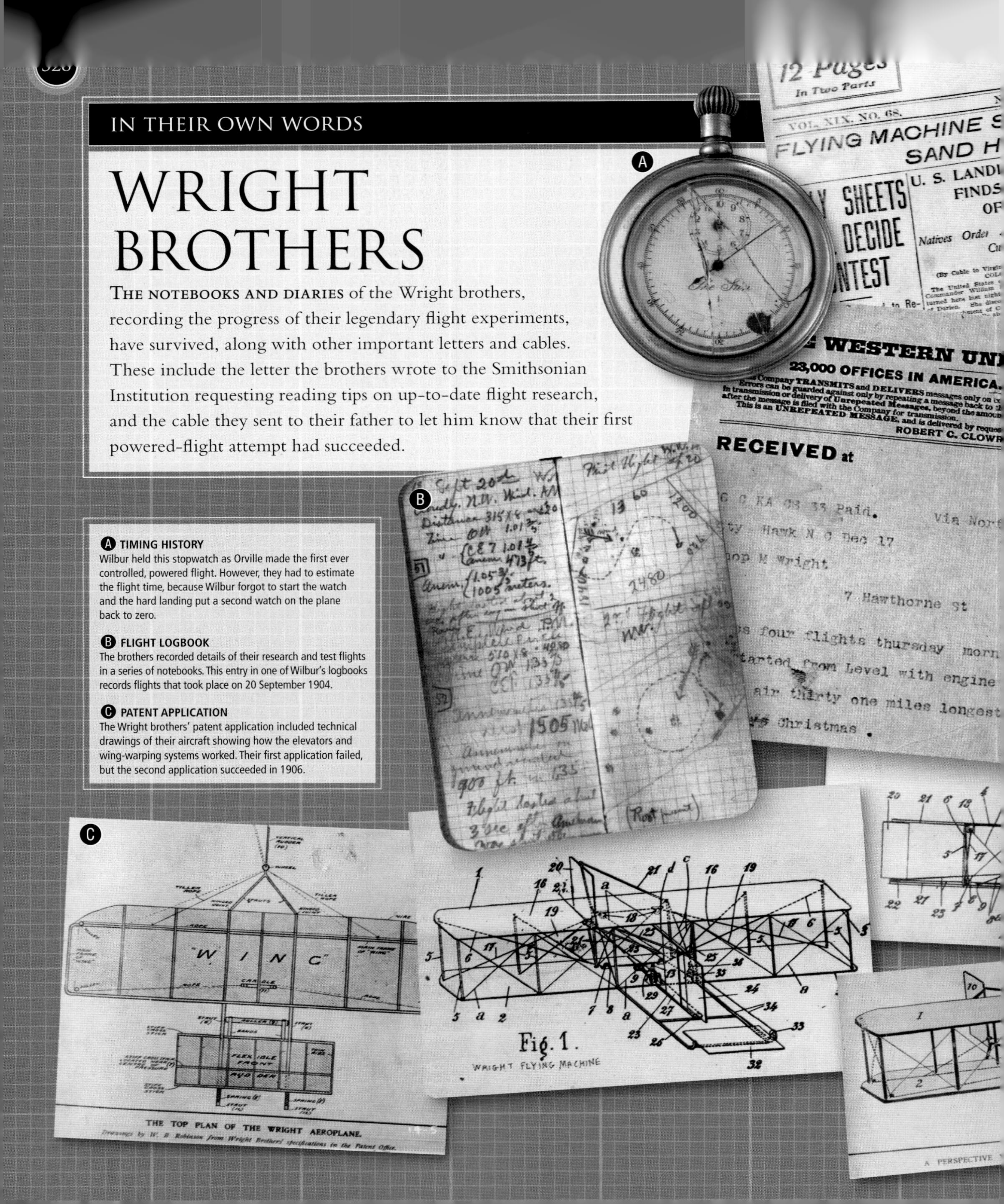

WRIGHT BROTHERS

The notebooks and diaries of the Wright brothers, recording the progress of their legendary flight experiments, have survived, along with other important letters and cables. These include the letter the brothers wrote to the Smithsonian Institution requesting reading tips on up-to-date flight research, and the cable they sent to their father to let him know that their first powered-flight attempt had succeeded.

A TIMING HISTORY
Wilbur held this stopwatch as Orville made the first ever controlled, powered flight. However, they had to estimate the flight time, because Wilbur forgot to start the watch and the hard landing put a second watch on the plane back to zero.

B FLIGHT LOGBOOK
The brothers recorded details of their research and test flights in a series of notebooks. This entry in one of Wilbur's logbooks records flights that took place on 20 September 1904.

C PATENT APPLICATION
The Wright brothers' patent application included technical drawings of their aircraft showing how the elevators and wing-warping systems worked. Their first application failed, but the second application succeeded in 1906.

D NEWS SENSATION
A story about the brothers' first powered flight appeared in the *Virginian-Pilot* newspaper on the day after the event, although it was highly inaccurate. Other newspapers multiplied the inaccuracies, prompting the brothers to release their own press statement a few weeks later.

E ASKING FOR HELP
On 30 May 1899, Wilbur wrote a letter to the Smithsonian Institution, asking for advice on what to read before attempting flight experiments.

F SENDING NEWS HOME
Orville cabled his father, Bishop Wright, from Kitty Hawk on 17 December 1903, with news that he and Wilbur had succeeded in making four powered flights that morning.

G TRAGEDY AT FORT MYER
This photograph shows Lieutenant Thomas E Selfridge sitting in a Wright plane that is being prepared for take-off. Sadly, the flight ended in a crash and Selfridge died later from his injuries.

H GUGGENHEIM MEDAL
In 1929, Orville became the first person to receive the Guggenheim Medal. It is awarded for the most notable achievements in the advancement of aeronautics.

JUAN DE LA CIERVA

INVENTOR OF THE ORIGINAL HELICOPTER

SPAIN 1895–1936

Juan de la Cierva was a Spanish civil engineer who invented a completely new type of aircraft, the autogyro – a single-rotor aircraft. His invention grew from a desire to build a safer aircraft that would be less likely to stall and crash. Now almost forgotten, the autogyro contributed important features that made the helicopter possible.

A LIFE'S WORK

- Cierva designs his first aircraft, **a bomber**, in 1918
- He becomes a member of the **Spanish parliament** in 1919
- He **invents the autogyro** and **receives a patent** for it in 1920
- **Moves to England** in 1925
- Builds the **C.6 model of the autogyro** that has a four-blade rotor with flapping hinges but relies on conventional aircraft controls for pitch, roll, and yaw; **British financiers** are so impressed that they **help him form the Cierva Autogiro Company** in 1926
- **Cierva autogyros are manufactured commercially** in England and under licence in France, Germany, Japan, Russia, and the US
- He supports **General Franco** at the outbreak of the **Spanish Civil War**, in which his brother is killed
- He **dies in a plane crash** in London in 1936

Born in Murcia, Spain, Cierva was interested in aviation as a teenager, and began experimenting with model gliders. He attended an engineering college in Madrid from 1911 and earned a degree in civil engineering. He would probably have preferred to study aeronautical engineering, but there were no such courses in Spain at the time. While he was still at college, he and some friends bought the wreckage of a crashed biplane and rebuilt it.

In 1918, the Spanish government held a design competition for a new military aircraft. The prize was almost $10,000, a huge sum at the time. Cierva entered with a design for a bomber, which had a wingspan of 25m (82ft), weighed 5 tonnes (5.5 tons) and was powered by three 168 kilowatt (225hp) engines. Unfortunately, during a test flight, it stalled (suddenly lost lift) and crashed, ending his chances of winning the competition.

THE GYRODYNE

The autogyro led to another new type of aircraft called a gyrodyne. Invented by the Cierva Autogiro Company's chief engineer, Dr James Allan Jamieson Bennett, its rotor was powered for takeoff and landing, enabling it to hover, which an autogyro could not do. This led to the invention of the Cierva C.41 Gyrodyne, the Fairey FB-1 Gyrodyne, and the Fairey Rotodyne. The Fairey Rotodyne's rotor blades were powered by jets at the tips of the blades. Propulsion was provided by two turboprop engines on stub wings. Despite interest in developing the Fairey Rotodyne as a commercial airliner and military aircraft, the British Government withdrew financial support in 1962 and the project was cancelled.

CIERVA'S BIG IDEA

Following the crash of his bomber, Cierva wondered if it might be possible to design an aircraft that would not stall even when flying at slow speeds. He realized that if the wings rotated to generate lift, the aircraft would not be able to stall, however slowly it flew. He designed an aircraft with an autorotating (freewheeling) rotor on top. While the spinning rotor created lift, a propeller drove the aircraft forwards. The stream of air flowing through the rotor kept it spinning. Cierva coined a name for the new aircraft – *autogiro* (autogyro in English) – and received a patent for it in 1920. He built his first prototype, the C.1, in 1920, followed by the C.2 in 1921, and the C.3 in 1922. Small models of the aircraft worked, but the full-size prototypes always rolled over to one side as soon as they took off.

Cierva realized that the rotor blades on one side of the aircraft moved forwards into the oncoming air and produced more lift than the retreating blades on the other side. He solved the problem by attaching the blades to the central hub by hinges. These "flapping hinges" let the blades rise and fall, equalizing the lift on both sides of the aircraft. He built a new aircraft with these hinges, his fourth full-size prototype – the C.4. It was a success and made the first flight at Cuatro Vientos airfield in Madrid on 9 January 1923, with Lieutenant Alejandro Gomez Spencer at the controls. A few days later, its engine failed after takeoff, but the aircraft descended slowly and made a safe landing, just as Cierva had designed it to do.

All this pioneering work was carried out in Cierva's native Spain. Then, in 1925, Cierva brought a later model, the C.6, to England, and demonstrated it to the Air Ministry at Farnborough, Hampshire. In 1927, Cierva added drag hinges to the rotor to let the blades move forwards and backwards a little, easing the bending forces acting on the blade roots. He made the first crossing of the English Channel by a rotorcraft in 1928 when he flew

a C.8 – a more powerful autogyro – from London to Paris, and then went on to Berlin, Brussels, and Amsterdam.

AMERICAN AUTOGYROS

In the late 1920s, an American businessman and pilot called Harold Pitcairn (1897–1960) had several meetings with Cierva and bought the rights to build autogyros in the US. On 18 December 1928, a Cierva C.8 became the first autogyro to fly in the US.

Pitcairn then developed a new feature called a "pre-rotator". Until then, autogyros had to taxi on the ground to make the rotor spin. Since the pre-rotator used engine power to get the rotor up to speed before takeoff, an autogyro that was fixed with a pre-rotator would not need to taxi first to be able to take off. It was such a good idea that Cierva incorporated pre-rotators into his own autogyros.

Autogyros were equipped with wings until Cierva designed a new control system that removed the need for wings. In doing so, he developed the cyclic and collective controls that were later used in helicopters. The first production autogyro, the C.30, was built in 1933. In the same year, Cierva was awarded the Daniel Guggenheim Medal for the most notable achievement in aviation. The introduction of jump takeoff was another major improvement. The rotor was accelerated in no-lift pitch until the rotor speed required for flight was achieved, and then declutched. The loss of torque (the force that causes rotation) caused the blades to swing forwards on angled drag hinges, making the aircraft leap into the air. With all the engine power now applied to the forward thrusting propeller, it was possible to continue in forward flight with the rotor in autorotation. The C.40 was the first production jump-takeoff autogyro.

END OF THE AUTOGYRO

Given that Cierva had striven to build safer aircraft, it is ironic that he himself died in an aircraft accident, travelling as a passenger on a KLM Douglas DC-2 flying from London to Amsterdam. On 9 December 1936, the plane took off in fog from Croydon Airfield; visibility was only about 50m (160ft). The plane veered off the correct takeoff line, clipped the chimney of a house, and then crashed into another house on the other side of the road. Cierva, along with 12 passengers and two of the four crew members, died. He was posthumously awarded the Royal Aeronautical Society's Gold Medal.

Autogyros were widely used until World War II, when they were replaced by helicopters. Although commercial autogyros and their manufacturers disappeared in the 1940s, there are enthusiasts who still build and fly autogyros today. Cierva's legacy lives on, as helicopters incorporate many of the developments in rotor design and controls that were made first in autogyros by Cierva and Pitcairn.

ELIZABETH MACGILL

CANADA 1905–80

Elizabeth "Elsie" MacGill was the first Canadian woman to earn a degree in electrical engineering (1927), the first woman in North America to obtain a degree in aeronautical engineering (1929), the first woman to be employed as a chief aeronautical engineer (1938), and the first woman to design an aircraft, the Maple Leaf II trainer (1939).

She supervised the production of Hawker Hurricane fighters in the 1940s, earning her the nickname, "Queen of the Hurricanes". She designed modifications for the Hurricane to equip it for cold-weather flying. After World War II, she set up in business as an aeronautical consultant. She campaigned on women's issues and was a member of the Ontario Status of Women Committee, for which she was given the Order of Canada in 1971. Elsie MacGill was one of the most highly respected women in aeronautical engineering.

TEST FLIGHT
Test pilot Frank Courtney (on the right, wearing goggles) prepares to make a flight in a Cierva autogyro at the Farnborough Aerodrome in Hampshire, England, in 1925. The men on the left hold the rope they will pull to start the rotor turning.

HUGO JUNKERS

DRIVING FORCE OF GERMAN AVIATION

GERMAN CONFEDERATION 1859–1935

BEST REMEMBERED as one of the most innovative engineers of early aviation, German industrialist and inventor Hugo Junkers revolutionized the design and construction of aircraft. He pioneered the development of all-metal monoplanes at a time when most aircraft were flimsy biplanes made up of wood, fabric, and wire.

A LIFE'S WORK

- Junkers is **granted a patent** in 1892 for his **invention of a "calorimeter"** designed to measure the calorific value of combustible gases
- After working as an engineer, he begins **experiments in aerodynamics and aircraft construction** at the age of 49
- Establishes successful aircraft and aero-engine factories and **one of the first airlines**
- He develops several **successful all-metal aircraft**
- The **Ju 52 trimotor** served as a passenger aircraft, a cargo carrier, a troop transporter, and a bomber
- His aircraft division **builds the engines** that power the only operational **jet aircraft of World War II**

Born on 3 February 1859 in Rheydt, Rhine Province, Hugo Junkers was the third of seven children of the proprietor of a weaving mill and brickworks. After studying mechanical engineering in Berlin and Aachen, Junkers became technical manager of his father's textile company in 1883. He returned to university for further studies in electrical engineering and, in 1888, joined a company developing gas engines in Dessau. Two years later, he co-founded a gas-engine laboratory and, in 1892, he established an engineering business. In 1895, he started Junkers & Co at Dessau, which produced steam boilers and heating equipment.

METAL AIRCRAFT

It was not until Junkers was almost 50 years old that he discovered a passion for aviation. In 1897, he was appointed professor of mechanical engineering at a technical institute in Aachen. A fellow professor, Hans Reissner (1874–1967), was designing aircraft at that time. In about 1906, he asked Junkers to help him in the venture. In 1910, Junkers patented a design for a flying wing, an aircraft that consisted of a single large wing with no separate body. The following year, the two professors built an aeroplane. Junkers made the wings from corrugated iron, which prompted him to state that future planes would be all-metal monoplanes.

At the outbreak of World War I in 1914, Junkers founded a research institute to continue his experiments with aircraft. The German government even ordered an aircraft from him. The result was the first practical, all-metal aircraft, J1, in 1915. It was nicknamed the "Blechesel" (Tin Donkey). J1's flight tests encouraged the government to place another order for an all-metal aircraft, the J2 fighter. J2 was quick in level flight, but could not climb fast enough. Junkers concluded that the iron he had used to build J2 was too heavy and that he would have to switch to a lighter metal, such as duralumin, an aluminium alloy used for building airships.

In 1917, the German government forced Junkers to work on military aircraft with the Dutch aircraft manufacturer Anthony Fokker (1890–1939). The combination of Junkers' research and Fokker's experience in aircraft manufacturing proved fruitful. They produced the first all-metal fighter to enter military service, the Junkers DI, a two-seat fighter called CLI, and a highly successful all-metal ground-attack aircraft called JI. The JI aircraft was the first all-metal aircraft to be produced in bulk, with 227 being manufactured.

After the war, Junkers and Fokker went their separate ways. Junkers set up a series of new businesses, including an airline, a pilot-training school, and an aerial-photography unit. He was also active in aero-engine development through another of his companies, Junkers Motorenwerke. He produced the F13, the

NEW AIRLINES

The first airlines to use planes instead of airships were established in the 1920s. In 1923, Hugo Junkers predicted that as many as those who crossed the Atlantic Ocean by ship would soon do so by plane. This came to pass in the late-1950s. Junkers promoted the formation of new airlines in Germany and other countries in the hope that they would buy his planes. He set up his own airline, Junkers Luftverkehr, in 1921. In 1926, it merged with another airline, Deutscher Aero Lloyd, to form Deutsche Luft Hansa, the forerunner of the modern Lufthansa airline. The Junkers G24 was a popular German passenger aircraft of the 1920s. In 1926, the newly formed Luft Hansa airline flew two G24s 20,000km (12,400 miles) from Berlin to Beijing with 10 stops on the way.

first all-metal passenger aircraft. In 1922, with the support of the German government, Junkers signed an agreement with the Soviets to develop military aircraft in Russia.

CRISIS AND CRASH

Three years later, in 1925, the German government withdrew its backing and demanded the repayment of loans. Junkers had to sell two-thirds of his aircraft construction company and 80 per cent of his airline to the government to clear his debts. He was also forced out of his management positions. His dispute with the government was settled in court in 1926, with Junkers giving up the rest of his airline but regaining control of his aircraft and aero-engine companies. In 1929, he built the Junkers G38, the world's largest airliner at the time; passengers were seated both inside the wings and the body.

The German government requested two new Junkers aircraft, which became the Ju 52 and Ju 60 passenger planes. Ju 60, the last plane to feature the Junkers-trademark corrugated-metal wings, did not progress beyond the prototype stage, because the faster Heinkel He 70 was preferred by the Luft Hansa airline. In contrast, Ju 52 was very successful. Originally underpowered, it was redesigned with three engines, making it one of the most popular German transport planes of the 1930s and '40s.

The worldwide economic crash of 1930–31 pushed several companies that owed money to Junkers into bankruptcy. As a result, Junkers had to file lawsuits to reach a settlement with his creditors. He sold off nearly all his assets to save his aircraft and aero-engine plants.

Just as he had resolved his latest crisis, the Nazi party came to power in Germany. The new government forced him to transfer more than 170 privately owned patents to his companies. He was also forced to give up more than half of his shares to the government and leave his management position. In 1934, he was placed under house arrest, banned from contacting his companies, and prohibited from being visited by anyone without the presence of police officers. His health deteriorated and he died the following year. His wife received a fraction of the value of his remaining assets from the government. The Nazi government used Junkers' factories to build military aircraft during World War II, including Stuka dive bombers and Ju 88 fighters. Junkers' engine division built the jet engines that powered the only operational jet aircraft of the war, the Messerschmitt Me-262 fighter.

IGOR SIKORSKY

RUSSIA 1889–1972

Igor Sikorsky was an aviation pioneer who designed the first practical and successful helicopters.

His aviation career began in Russia, where, after experimenting with helicopters, he made his first solo flight in 1910 in a fixed-wing plane that he had designed and built himself. In 1913, he built the first four-engine plane, which was followed by the first four-engine bomber. He moved to the US in 1919 and built a series of successful planes such as the flying boats that Pan American Airways used to establish air routes across the Atlantic and Pacific oceans. He also designed the VS-300 helicopter and piloted it himself on its first flight in 1939, wearing his trademark fedora hat. Lastly, he developed the world's first mass-produced helicopter, the R-4.

JUNKERS G38 AIRLINER
Passengers and spectators crowd around a Junkers G38, which has just arrived at an airfield in England on 11 June 1931. At that time, the G38 was the world's largest aeroplane. Two of these giants were operated by the German airline, *Luft Hansa*, in the 1930s. They carried up to 34 passengers in comfortable cabins designed to rival the commercial Zeppelins of the day.

WHITTLE AND VON OHAIN

PIONEERS OF THE JET AGE

ENGLAND 1907–96
GERMANY 1911–98

FRANK WHITTLE

HANS VON OHAIN

Frank Whittle invented the jet engine despite the many obstacles placed in his way by politicians and other engineers. While Whittle's research progressed at a snail's pace without backing or support, German engineer Hans von Ohain developed his own jet engine.

A LIFE'S WORK

- At the age of 15, Whittle **decides to join the Royal Air Force** and become a pilot
- He joins the RAF on his **third attempt**, having been **rejected twice because of his short stature** – he was only 1.5m (5ft) tall
- He invents and develops the jet engine **despite indifference, poor funding, and interference** from the authorities
- Von Ohain **develops a jet engine independently** of Whittle
- After World War II, **von Ohain is brought to the US** under ***Operation Paperclip***, which recruited the best scientists from Nazi Germany
- He works on the design of **nuclear-powered rockets, producing technology** that was also used in pumps and centrifuges, and **uses the same principles to design a jet engine** with moving parts
- He **invents the jet wing**, which uses air from a jet engine's compressor **to generate extra lift from a wing**

In January 1923, having passed the RAF entrance examination, Frank Whittle reported as an aircraft apprentice. He lasted for only two days: just 1.5m (5ft) tall and with a small chest measurement, he failed the medical. He did, however, finally get in. Then in 1928, when he was still a Royal Air Force cadet, he wrote a paper entitled "Future Developments in Aircraft Design". He argued that planes would have to go higher to fly faster. However, the piston engines that powered aircraft at that time did not work well in the thin air at high altitudes, so a new propulsion system would be needed. He considered using rockets, or gas turbine engines driving propellers. By 1929, he had started to think about propulsion by jet thrust. He envisaged an engine that would suck in air, compress it, and heat it with burning fuel, making it expand so rapidly that it would blast out of the back of the engine as a jet.

He offered the idea to Britain's Air Ministry, but they turned it down as impractical. Whittle then patented the idea himself. The RAF sent him on an engineering course and he was such an exceptional student that he was allowed to take a more advanced engineering course at Cambridge University. He did the three-year course in two years, while at the same time developing his jet engine design – an extraordinary achievement. Still at Cambridge, Whittle could ill afford the £5 renewal fee for his jet engine patent, and when it lapsed, the Air Ministry refused to pay the fee required to renew it.

BATTLING BUREAUCRACY

In 1935, he formed a company called Power Jets to develop jet engines. The first experimental engine was tested in 1937, but it had an alarming habit of running out of control. It was modified and rebuilt twice, but scarce funds slowed the work. When war broke out, the Air Ministry requested a more powerful engine and asked Gloster Aircraft to build an experimental plane for it. Just as all the bureaucratic and funding obstacles seemed to have been resolved, the Air Ministry bypassed Power Jets and offered engine contracts to two other companies, British Thomson-Houston and Rover. Power Jets was to be used only for research. This placed a great strain on Whittle, whose health suffered. Fortunately, the Minister of Aircraft Production, Lord Beaverbrook, stepped in and personally assured Whittle that his work would continue. The new engine was ready by April 1941.

The first British jet-powered flight was made on 15 May by a Gloster E.28/39 aircraft. When a colleague said, "Frank, it flies", Whittle answered, "Well, that's what it was bloody well designed to do, wasn't it?" Over the following few days, the plane reached a speed of 595kph (370mph). Pilots who saw the flight were astonished. They had never seen or heard a jet engine before. When the US government learned of Whittle's work, a Power Jets team

WHITTLE'S JET ENGINE
A Whittle jet engine has ten tubular combustion chambers arranged around the outside of the engine, which produce impressive thrust. In this rear view, the chambers surround the large exhaust nozzle.

THE IMPRESSION WHITTLE MADE WAS OVERWHELMING

LANCELOT LAW WHYTE, INVESTMENT BANKER AT O T FALK & PARTNERS

was flown to the US to brief engineers from General Electric. The result was the first US jet aircraft, the P-59 Airacomet fighter. It was flying by October 1942, but never saw action because of poor performance. The first British jet fighter, the Gloster Meteor, entered service with the RAF in July 1944. Whittle felt that if he had not had to spend years battling his own government for support and funds, Britain could have had operational jet fighters at the beginning of the war in 1939.

Meanwhile in Germany, unaware of Whittle's work, Hans von Ohain had developed his own theory of jet propulsion in 1933, while studying for a doctorate in physics and aerodynamics at the University of Göttingen. He was granted a patent in 1936. The aircraft manufacturer Ernst Heinkel met von Ohain and talked about his ideas. Von Ohain and automotive engineer Max Hahn had already built their first jet engine, but it did not work very well. Von Ohain and Hahn went to work for Heinkel. The next engine they built, the HeS-1 (Heinkel Jet Engine 1), was more successful. It was built and running in 1937, just after Whittle's first engine. A later version of this engine, the HeS 3b, powered a Heinkel He 178 for the first ever jet-powered flight on 27 August 1939 – nearly two years before the first British jet flight. Heinkel continued to develop further models of jet engines, but it was the Jumo 004 engine developed by Dr Anselm Franz at Junkers that was chosen to power the first operational jet fighter, the Messerschmitt Me-262.

AFTER THE WAR

Whittle retired from the RAF in 1948 after a bout of ill health, leaving with the rank of Air Commodore. In the same year, he was granted a sum of £100,000 by the Royal Commission on Awards to Inventors in recognition of his work on the jet engine, and two months later he was made a Knight of the Order of the British Empire (KBE). He joined the British Overseas Airways Corporation (BOAC) airline as a technical advisor on aircraft gas turbines and travelled extensively over the next few years, viewing jet engine developments in the US, Canada, Africa, Asia, and the Middle East. He then worked for Shell Oil, where he invented a drill powered by a turbine driven by mud pumped down the drill-hole.

Von Ohain moved to the US after the war and worked at Wright-Patterson Air Force Base. In 1963, he was appointed chief scientist at the base's Aero Propulsion Laboratory. He retired in 1979 and became a consultant to the University of Dayton Research Institute. Whittle, too, moved to the US in 1976 and became a research professor at the US Naval Academy at Annapolis. Whittle and von Ohain met in the 1970s and became friends. They toured the US together giving talks. Although none of von Ohain's designs entered production, his contributions to the development of the jet engine in Germany are invaluable.

JET ENGINES

Whittle's paper of 1928 entitled "Future Developments in Aircraft Design" was amazingly prescient. He was absolutely right about the need for aircraft to fly higher in order to fly faster, which would require a new type of engine. Jet engines evolved, and became ever more powerful, and today, more than 80 years later, the majority of the world's aircraft have jets. The inertia and incompetence of the British government might have lost them the war – in 1939, Whittle's company, Power Jets, had a staff of only 10 and was almost bankrupt – but fortunately for the British, German research was just too far behind, and the Me-262, although fast, was not sufficiently manoeuvrable or reliable to turn the tide of the war.

HEINKEL HE 178
In 1939, the Heinkel He 178 became the first aircraft to fly solely under turbojet power. It was a jet engine developed by German engineer Hans von Ohain.

SPACE ENGINEERING

THROUGHOUT HISTORY, PEOPLE HAVE DREAMED OF LEAVING THE EARTH AND FLYING TO THE MOON, BUT IT WAS ALWAYS CONSIDERED AN UNATTAINABLE FANTASY. THEN IN THE EARLY 1900s, A HANDFUL OF ENGINEERS DARED TO WONDER IF SPACEFLIGHT MIGHT BE POSSIBLE.

FROM MISSILE TO SPACECRAFT
The V-2 missile, developed during World War II, inspired the design of the first space rockets. This V-2, brought to the US after the war, is being test-launched at Cape Canaveral.

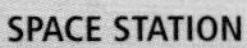

SPACE STATION
The International Space Station is an orbiting laboratory as large as an American football field and weighs about 390 tonnes (430 tons). It has been manned continuously since 2000.

The story of modern rocketry and space travel began in 1903 with the publication of an article "The Exploration of Cosmic Space by means of Reaction Devices" in a Russian science journal. It was the result of years of theoretical work by a schoolteacher, Konstantin Tsiolkovsky, who had developed the scientific principles of spaceflight.

Tsiolkovsky's theories and the works of other pioneers, such as Hermann Oberth in Germany, were taken up by the next generation of rocket scientists. Oberth was a member of a rocket society where he met a young enthusiast called Wernher von Braun (see pp.344–45), who was destined to become the leading rocket designer and engineer of the US manned space programme. Meanwhile, Robert Goddard (see pp.338–39) had built the first liquid-fuel rocket.

NEW PROFESSIONALS

Engineers such as these were ridiculed as dreamers divorced from reality. Most of them worked in isolation with meagre funds. But World War II harnessed their passion and transformed it from a hobby into the industrial production of military weapons with terrifying destructive power. The time of the rocket engineer had come.

The development of large liquid-fuel rockets in Germany during World War II created the technology necessary to reach space. After the war, most of Germany's rocket engineers went to the US, where they set about developing rockets for spaceflight. However, it was Soviet engineers, principally Sergei Korolev (see pp.342–43), who made the first breakthrough. The Soviet Union shocked the world by launching the first satellite, Sputnik 1, in 1957. It was followed by missions

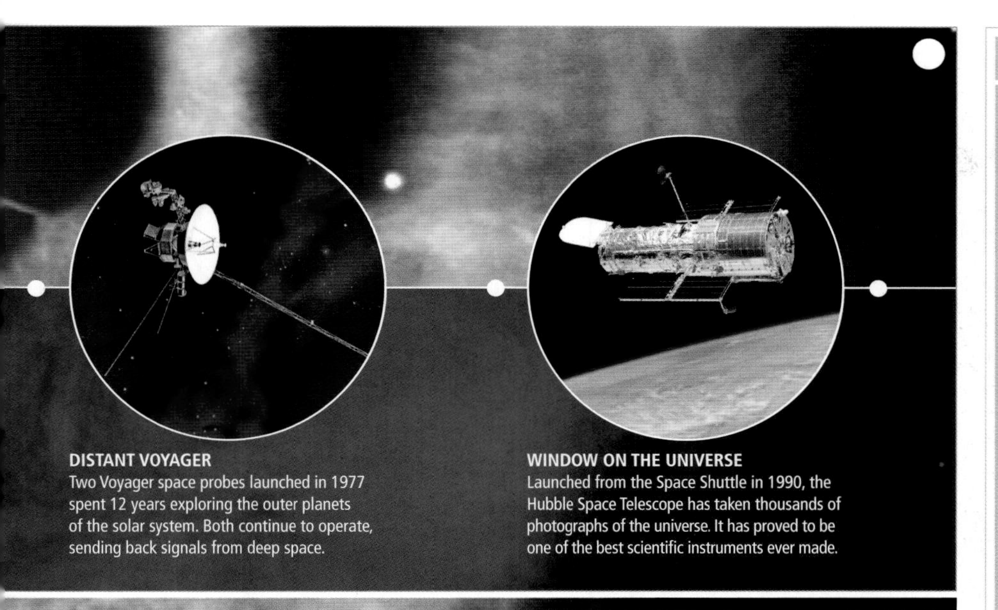

DISTANT VOYAGER
Two Voyager space probes launched in 1977 spent 12 years exploring the outer planets of the solar system. Both continue to operate, sending back signals from deep space.

WINDOW ON THE UNIVERSE
Launched from the Space Shuttle in 1990, the Hubble Space Telescope has taken thousands of photographs of the universe. It has proved to be one of the best scientific instruments ever made.

SETTING THE SCENE

- Modern rocketry is born in the rocket clubs of the 1920s and 1930s, where **enthusiasts meet in their spare time to build and launch rockets.**
- Military funding for rockets and missiles during World War II **enables engineers to develop bigger rockets**. These are adapted to make the first space launchers in the 1950s.
- Press and television coverage of manned spaceflights **transforms the first astronauts** from unknown military pilots **into global superstars**.
- **Spacecraft sent to the Moon and planets send back stunning photographs** that revolutionize knowledge of the solar system.
- Scientific instruments in space, such as the Hubble Space Telescope, **gather huge amounts of information about the Earth**, the rest of the solar system, and the furthest reaches of the universe.
- **International communications, television broadcasting, news coverage, navigation, weather forecasting, and environmental monitoring** are all made possible or improved by satellites.

sending the first living creature and then the first human being into space. The US then set out to match them in a Space Race that ended with the Apollo 11 landing on the Moon in 1969. The US and Soviet Union cooperated on the Apollo-Soyuz Test Project in 1975. Later, 16 nations joined forces to develop the International Space Station, the largest structure ever built in space.

Many of the earliest rocket engineers were inspired by reading the works of science-fiction authors. One name that appears time and again is Jules Verne (1828–1905). His book *From the Earth to the Moon* (1865) was particularly influential.

ELECTRONICS AND COMPUTERS

Space travellers have to carry out complex navigation and propulsion calculations, for which computers are essential. The first programmable digital computer was the Z3. It was built by Konrad Zuse (1910–95), in 1941. Such computers were built from large glass vacuum tubes (valves). Not only were these very fragile, but the computers were huge and heavy. The miniaturization of electronic components made possible by the invention of the transistor in 1947 and, just over ten years later, the integrated circuit, or chip, enabled engineers to build computers small and light enough to be carried by spacecraft. The US Gemini craft was the first manned spacecraft to have an onboard computer in 1965.

After Project Mercury – the US's first human spaceflight programme, which spurred Project Gemini and the Apollo Moon-landing programme – the Gemini spacecraft enabled astronauts to manoeuvre in space, change orbit, and dock with other spacecraft – all necessary features for a mission to the Moon. Soviet space engineers, directed by Korolev, were also moving towards a Moon-landing. When the US's Apollo 11 reached the Moon first, the Soviet Union turned its attention to building space stations. Meanwhile, space probes had been sent to fly past, orbit, or land on the planets. After Apollo 11, the politicians who controlled spaceflight funding saw no reason to continue with spectacular space projects. By the third Moon-landing mission in 1970, US television networks chose not to broadcast transmissions from the spacecraft on its way to the Moon. Then something happened that gripped a global audience and brought engineers to the fore.

When Apollo 13 was 322,000km (200,000 miles) from Earth, an explosion ripped through it, crippling its power and oxygen systems. Astronauts and engineers on the ground worked round the clock to make the spacecraft's remaining resources last long enough to bring the crew home alive. It was a catastrophic accident, but it was also a great triumph for the engineers who saved the day.

The next four Apollo missions proceeded without serious incidents. The permanent Moon-bases and manned flights to Mars that were expected to follow Apollo did not happen. The Space Shuttle and International Space Station were developed, but astronauts have not ventured beyond Earth's orbit since the last Apollo Moon-landing in 1975. Instead, NASA has sent the Voyager probes to explore the outer solar system, a series of robotic landers to Mars, and launched the Hubble Space Telescope, which continues to reveal the most distant reaches of the universe. More recently, a new generation of commercial spacecraft has emerged since Burt Rutan (see pp.346–47) produced the first privately funded spacecraft, SpaceShipOne, in 2004.

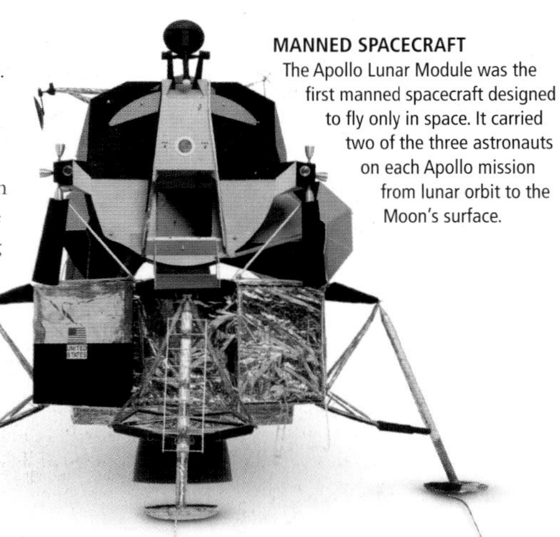

MANNED SPACECRAFT
The Apollo Lunar Module was the first manned spacecraft designed to fly only in space. It carried two of the three astronauts on each Apollo mission from lunar orbit to the Moon's surface.

ROBERT GODDARD

FATHER OF MODERN ROCKETRY

UNITED STATES 1882–1945

A BRILLIANT SCIENTIST, engineer, and inventor, Robert Hutchings Goddard is credited with creating, building, and launching the world's first liquid-fuel rocket. He also proved that rockets could work in the vacuum of space, contrary to common belief, and therefore that space exploration by rockets was possible. His work in the field was revolutionary.

A LIFE'S WORK

- Goddard develops the **mathematical theory of rocket propulsion** in 1912, and also accepts a position as physics research fellow at Princeton University
- **Receives patents for solid- and liquid-fuel rockets**, and multi-stage rockets, in 1914
- Writes **"A Method for Reaching Extreme Altitudes"** in 1919
- Marries **Esther Christine Kisk** in 1924
- Launches the **first liquid-fuel rocket** in 1926
- Using a grant from **Daniel Guggenheim**, he sets up an **experimental rocket station** in New Mexico
- He is **granted 214 patents**, out of which 131 are awarded after his death

Robert Goddard was born on 5 October 1882, in Worcester, Massachusetts. He was fascinated with science from an early age. His father encouraged his interest by buying him a telescope, a microscope, and a subscription to *Scientific American* magazine. He was also an avid reader of science fiction stories. At the age of 16 he read *The War of the Worlds* by H G Wells, which inspired in him a lifelong interest in spaceflight.

After high school, he attended Worcester Polytechnic Institute and came up with a theory about how transportation in the future would be done through hovering, by the forces of electromagnets. While still an undergraduate, Goddard wrote a paper proposing a method for "balancing aeroplanes". He submitted the idea to *Scientific American*, which published the paper in 1907. Goddard later wrote in his diaries that he believed his paper was the first proposal of a way to automatically stabilize aircraft in flight. His proposal came at about the same time as other scientists were making breakthroughs in developing functional gyroscopes. His first writing on the possibility of a liquid-fuel rocket came on 2 February 1909. Goddard had begun to study several ways of increasing a rocket's efficiency by using methods other than conventional, powder rockets. He wrote in his journal about using liquid hydrogen as a fuel with liquid oxygen as the oxidizer. He believed greater efficiency could be achieved with liquid fuel. Goddard worked as a physics instructor at Worcester Technical University, where he earned a PhD in 1911. By 1912, he had developed the mathematical theory of rocket propulsion.

During World War I, Goddard investigated the use of rockets as weapons. He developed small, solid-fuel rockets that could be launched from tubes. These led to the bazooka and other rocket-powered weapons used in World War II. In 1915, he proved by experiment that a rocket could produce thrust in a vacuum. He would later provide the groundwork for the design of modern rockets.

ROCKETS IN SPACE

When Goddard suggested sending rockets to the Moon, he was ridiculed. People thought that rockets would not work in space, because there was no ground or atmosphere for them to push against. In fact, a rocket works according to Newton's Third Law of Motion – every action has an equal and opposite reaction. A rocket pushes gas out in one direction and the gas pushes back, forcing the rocket in the opposite direction. Therefore, a rocket generates thrust in a vacuum. Goddard proved this by conducting experiments in vacuum chambers.

ROBERT GODDARD (STANDING ON THE LEFT) CHECKS THE FUEL PUMP OF ONE OF HIS ROCKETS AT ROSWELL, NEW MEXICO, IN 1940.

REACHING FOR THE MOON

In the early 1920s, Goddard began to investigate liquid-fuel rockets. These rockets are essential for controlled spaceflight, because, unlike solid-fuel rockets, they can be throttled, turned off, and restarted. Goddard knew that liquid hydrogen was the best fuel, but he used gasoline because it was easier to handle. He applied to the Smithsonian Institution for funding and was awarded a $5,000 grant. The results of his research were published in a report entitled "A Method of Reaching Extreme Altitudes" in 1919.

Goddard was ridiculed in the press for suggesting that a manned rocket travelling at 11.2km per second (6.95 miles per second) could escape from the Earth's gravity and head out towards the Moon and planets. He was labelled a dreamer and compared with the science fiction author Jules Verne. It was commonly believed at that time that rockets could not possibly work in space, because there was nothing for them to push against. The shy Goddard was stung by the criticism, especially as his work was firmly based on science. The Smithsonian was criticised, too, for funding him. Nevertheless, Goddard continued with his research.

On 16 March 1926, Goddard launched the world's first liquid-fuel rocket from his aunt Effie's farm in Auburn, Massachusetts. The gasoline-and-oxygen rocket, nicknamed Nell, stood about 3m (10ft) high. When he launched it, it rose to a height of 12.5m (41ft) and landed 56m (184ft) away. The whole flight lasted for only about 2.5 seconds. This modest flight signalled the beginning of the modern age of rocketry. Because of the mauling Goddard had received in the press, he made no announcement of his success. When news eventually leaked out and reporters contacted him, he simply told them, "Work in progress, there is nothing to report". As a result, most people outside rocketry circles had no idea who Goddard was or what he had achieved.

On 17 July 1929, Goddard launched the first rocket to carry instruments – barometer, thermometer, and camera. It crashed, starting a fire. People living nearby complained to the police and Goddard was banned from launching rockets in Massachusetts. Even worse, when newspapers found out what had happened, they printed the story under the headline "Moon rocket misses target by 238,799½ miles". It could have been the end of Goddard's research, but for the intervention of a famous supporter.

THE DREAM OF YESTERDAY IS THE HOPE OF TODAY AND THE REALITY OF TOMORROW

ROBERT GODDARD

Charles Lindbergh (1902–74) had become world-famous in 1926 by making the first solo nonstop flight across the Atlantic Ocean. When he learned of Goddard's work, he wondered if it might hold the key to aviation in the future. In 1930, he persuaded the wealthy Guggenheim family to fund Goddard's research.

ROCKETS AT WAR

Goddard used the money to set up an experimental base near Roswell, New Mexico. There, he built bigger rockets and refined their design. He developed ways of steering rockets and using gyroscopes to keep them on course. He published his work in 1936 in a paper called "Liquid-Propellant Rocket Development". Rocketry enthusiasts from other countries often contacted Goddard with technical questions, which he answered. He noticed that most of the questions came from Germany, until 1939. War in Europe was imminent and Goddard worried that Germany might be developing rocket-powered weapons. He offered his rocket technology to the US Army, which politely declined. Five years later, German V-2 ballistic missiles were landing on London.

In 1941, he succeeded in launching a rocket to an altitude of 2,743m (9,000ft). Soon after this, he went to work for the Navy, where he developed jet-assisted takeoff (JATO) rockets to help heavily laden aircraft get off the ground. At the end of the war, he finally got to see a captured V-2 missile. He recognized its design instantly. It was clear that German rocket engineers had read his patents, which were readily available, although the V-2 was more advanced than Goddard's rockets.

In 1960, the US Department of Defense and the US space agency, NASA, paid his estate $1 million for the use of his rocket patents. Goddard did not live to see the Space Age. He died of throat cancer in 1945, but he contributed a great deal to making space exploration a reality.

During his lifetime, Goddard became protective of his privacy and his work. Years after his death, at the dawn of the Space Age, he came to be recognized as one of the founding fathers of modern rocketry. He was the first, not only to recognize the scientific potential of missiles and space travel, but also to bring about the design and construction of the rockets needed to implement those ideas.

LIQUID-FUEL ROCKET
Robert Goddard stands in a Massachusetts field beside the world's first liquid-fuel rocket in 1926. The rocket rests on its launch frame. Its motor is at the top, fed with fuel and oxygen from the tanks at the bottom.

ENGINEERING INNOVATIONS

THE ROCKET

The history of rocketry can be traced back to ancient China. Early rockets resembled large fireworks – with a tube of gunpowder at the end of a long stick. For more than 1,000 years there were few advances in rocket design. Then, in the 20th century, engineers built the first rockets capable of reaching space. The Space Race between the United States and the Soviet Union spurred a rapid technological development of even more powerful rockets, culminating in the giant Saturn V rocket that sent astronauts to the Moon.

CELEBRATIONS AND WAR

The first rockets were fireworks sparked off at religious festivals in China more than 1,000 years ago. There are reports, however, that by the 13th century, rockets were being used as weapons in battles in China, Arabia, and eastern Europe. Their use spread slowly across Europe, reaching Italy by 1500. But they eventually caught the imagination of the continent. A book about gunnery published in London in 1647 contained 43 pages on rocketry.

In the 1790s, British forces came under rocket attack at the two battles of Seringapatam (1792 and 1799) in India. This prompted English artillery officer and inventor William Congreve to develop rockets for the British army. The British fired 25,000 rockets during an attack on Copenhagen in 1807. Five years later, they used rockets against American forces at the battle of Bladensburg during the War of 1812. By 1846, the Americans had their own rocket brigade

1241 – EUROPEAN WARS
Rockets are used in Europe for the first time, when Mongols use them against Magyar forces.

1232 – ROCKETS FOR WAR
The Chinese use rockets against the Mongols, who are besieging the city of Kai-fung-fu.

1696 – FIRST ROCKETRY MANUAL
Englishman Robert Anderson publishes a treatise on how to make rockets and their propellants.

1792 – SERINGAPATAM BATTLES
Tipu Sultan, son of the king of Mysore Hyder Ali, uses rockets against the British in the battles of Seringapatam.

1803 – BRITISH ROCKETS
William Congreve (1772–1825) develops rockets for the British army.

1807 – BURNING COPENHAGEN
William Congreve attacks Copenhagen with 25,000 rockets.

1846 – STICKLESS ROCKET
British engineer William Hale (1797–1870) invents a stickless rocket by making the rocket spin for stability.

1862 – AMERICAN WARFARE
During the American Civil War, Confederate forces fire rockets at Union troops at Harrison's Landing, Virginia.

1903 – DOCUMENTING ROCKETS
Russian schoolteacher Konstantin Tsiolkovsky (1857–1935) writes the first serious scientific papers on rockets, liquid propellants, staged rockets, space travel, and spacesuits.

1926 – LIQUID-FUEL ROCKET
Robert H Goddard (see pp.338–39) launches the first liquid-fuel rocket.

1927 – ROCKET SOCIETY
The Verein für Raumschiffart (VfR) rocket society is formed in Germany.

CE 1200 1350 1500 1650 1800 1825 1850 1875 1900 1920

The first recorded experiment in manned rocketry is carried out in China in about 1500, when a Chinese government official called Wan Hu fastens 47 rockets to a chair. Then he sits in the chair and orders the rockets to be lit. When the smoke and flames clear, the chair and Wan Hu have disappeared. They are never seen again.

c.1500 WU HAN'S EXPERIMENT

Military rockets are introduced to Europe by the British army. They are developed by British artillery officer and inventor William Congreve. A Congreve rocket is an iron case containing an explosive charge fixed to the end of a 4.6-m (15-ft) pole. In 1806, hundreds of Congreve rockets are fired from boats at French forces in Boulogne.

1806 CONGREVE'S ROCKETS

and used rockets in the Mexican War (1846–48) and in the War Between the States, or the Civil War (1861–65).

AIR ATTACK

During World War I (1914–18), aircraft used rockets to attack enemy observation balloons and airships, but they were rarely successful. Improved air-to-air rockets were developed during World War II (1939–45) for use by fighters against bombers. Between the wars, the first experimental liquid-fuel rockets had been built in the US. This led to new, more powerful military rockets in World War II, principally the German V-2. The first rocket-powered aircraft were also built during the war.

Military and civilian rocket development continued apace after 1945. For example, the rocket-powered Bell X-1 plane made the first supersonic flight in 1947. Rocket-powered guided missiles were developed at this time as well. They steered themselves towards a target by homing in on heat and radar signals, or reflected laser light. Increasingly large missiles were developed to deliver high-explosive and nuclear warheads. Meanwhile, civilian rockets of various sizes and types were developed for scientific research and space exploration. The rocket has come a long way since its ancient beginnings. Today, rockets power ejector seats in fighter planes, help launch spacecraft into the Earth's orbit, and send space probes to the outer reaches of the solar system.

X-15 ROCKET PLANE

In the 1950s, North American Aviation built for NASA an experimental rocket-plane called the X-15, which made 199 flights between 1959 and 1968. It set an altitude record of 108km (67 miles) on 22 August 1963, and a speed record of 7,273kph (4,519mph) on 3 October 1967. The rocket-plane provided invaluable flight data, which contributed to the design of all US manned spacecraft.

THE X-15 ROCKET-POWERED PLANE PIONEERED HIGH-ALTITUDE HYPERSONIC FLIGHT RESEARCH.

1944 – V-2 ATTACK V-2 rockets are launched by Germany against London and Paris.

1947 – BELL X-1 The rocket-powered Bell X-1 plane makes the first supersonic flight, with American Charles "Chuck" Yeager at the controls.

1953 – REDSTONE MISSILE German Wernher von Braun (see pp.344–45) launches the first Redstone missile.

1957 – FIRST ARTIFICIAL SATELLITE The Soviet Union launches the first artificial satellite called Sputnik-1.

1958 – EXPLORER 1 The US launches the first American satellite, Explorer 1, on a Jupiter-C rocket.

1962 – FIRST US MAN IN SPACE The US launches John Glenn into orbit in a Mercury capsule on top of an Atlas-D booster.

1968 – BEYOND EARTH'S ORBIT A Saturn V rocket carries astronauts outside the Earth's orbit and around the Moon for the first time during the Apollo 8 mission.

1969 – APOLLO 11 Apollo 11 lands the first astronauts on the Moon – Neil Armstrong and Edwin "Buzz" Aldrin.

1979 – ARIANE 1 European Ariane 1 rocket makes its maiden flight.

1981 – SPACE SHUTTLE The Space Shuttle reusable manned space transport system is launched from Cape Canaveral on its first space mission using solid- and liquid-fuel rockets.

2004 – SPACEPLANE The SpaceShipOne spaceplane is launched.

2008 – PRIVATE ROCKET The privately funded SpaceX Falcon 1 rocket is launched and successfully reaches the Earth's orbit.

2011 – RETIREMENT The Space Shuttle is retired after its last mission.

1940 1960 1980 2000

The first manned spacecraft is launched by the Soviet Union in 1961. Called Vostok 1, it carries the first human to orbit the Earth, Soviet air force pilot Yuri Gagarin. A Vostok spacecraft is a metal sphere 2.3m (7.5ft) across, connected to a cylindrical equipment module. The capsule separates from the rest of the craft for re-entry. During its descent, the cosmonaut ejects and lands by parachute.

1961 VOSTOK SPACE CAPSULE

A Proton rocket launches the first part of the International Space Station in 1998. The rocket makes its maiden flight in 1965. The latest version, the Proton-M, can launch payloads weighing up to 22 tonnes (24 tons) into a low Earth orbit. Proton rockets have been used for more than 360 launches, with a success rate of 95 per cent.

1998 PROTON ROCKET

SERGEI KOROLEV

CREATOR OF ASTRONAUTICS

RUSSIA 1907–66

FROM ITS ADVENT after World War II, Soviet rocketry was dominated by Sergei Korolev until his death in the 1960s. As a result of Soviet secrecy, his identity was not known in the West until after his death. He belatedly received the credit he deserves for his great achievements in the field of astronautics, made despite continual political interference.

A LIFE'S WORK

- Korolev studies at Moscow's Bauman High Technical School, the **best engineering college in Russia**
- **Works on the design of a bomber** while in prison
- Becomes the **leading Soviet rocket designer**
- Designs the rockets that **launch the first artificial satellite**, the first living creature in space, and **the first cosmonaut**
- Is credited with **designing the Vostok and Soyuz spacecraft**
- Begins work on a **Moon rocket, N-1**
- A Moscow street, **a city, a crater on the Moon**, a crater on Mars, and **an asteroid are named after him**

Sergei Korolev was born on 12 January 1907 in the city of Zhitomir in present-day Ukraine. A brilliant student, he progressed through several schools and colleges, arriving at Moscow's elite Bauman High Technical School in 1926. In 1932, he became chief of the Central Aero and Hydrodynamics Institute's Jet Propulsion Group, one of the earliest Soviet rocket-development centres. When the military significance of rockets was recognized, this group was transformed into the Jet Propulsion Research Institute, and Korolev was appointed its deputy chief. The institute developed rocket-propelled missiles and gliders. One of the gliders, SK-9, was developed into the first Soviet rocket-powered aircraft, RP-318, in 1936.

Korolev's steady upward progress came to a shuddering halt in 1938, when he was arrested during Joseph Stalin's Great Purge (1936–38). Korolev was accused of sabotage, a common allegation, sent to a concentration camp, and put to work in a gold mine in the Kolyma region of eastern Siberia. Thousands of workers died every month in the freezing Kolyma camps. Korolev was lucky to survive.

SPUTNIK 1
This is the first artificial satellite, Sputnik 1, which was launched into low Earth orbit by a Sputnik rocket developed from Korolev's successful R-7 Intercontinental Ballistic Missile (ICBM) in 1957.

IMPRISONMENT AND BOMBERS

In 1940, he was transferred to a *sharashka*, a design bureau run by scientists and engineers in prison. *Sharashkas* were set up in 1939, when Stalin recognized the importance of aeronautical engineers to the war effort. Korolev's *sharashka* in Omsk, headed by aircraft designer Andrei Tupolev (1888–1972), worked on the design of a plane that became the Tupolev Tu-2 bomber. Then Korolev was moved to another *sharashka* in the city of Kazan, developing rocket-assisted boosters for aircraft. His chief there was Valentin Glushko (1908–89), who later became the leading Soviet rocket-engine designer.

In 1945, Korolev was sent to Germany to inspect A-4 ballistic missiles, better known as V-2s. A year later, he was finally freed and appointed chief of a new design team based near Moscow to develop Soviet missiles, drawing on German rocket technology. He began by building a copy of V-2, called R-1, using Soviet parts. This led to his development of the R-7 missile, a two-stage rocket that could carry nuclear warheads. It was the world's first intercontinental ballistic missile. R-7 itself was developed into a series of space launchers.

The drive to build a space rocket began in 1952, when the International Council of Scientific Unions designated 1 July 1957 to 31 December 1958 as the International Geophysical Year (IGY). Then, in 1954, the council called for the first artificial satellites to be launched during the IGY. The following year, US President Dwight D Eisenhower announced the US intention to launch a satellite. Meanwhile, Korolev campaigned for the Soviet Union to launch its own satellite. In 1956, the government gave him the go-ahead.

ROCKETS, MAN, AND MOON

On 4 October 1957, Sputnik 1, a metal sphere 58cm (23in) in diameter, containing a radio transmitter, was launched into low Earth orbit on top of a modified R-7 missile. News of Sputnik's launch surprised the world. The Soviet Union had beaten the US. To make matters worse, the first US attempt to launch a satellite, on 6 December 1957, ended in a humiliating failure in front of a worldwide television audience, when the Vanguard rocket exploded on the launchpad. It would take the US until January the following year to put a satellite, Explorer 1, into orbit. Meanwhile, Korolev's team had already launched the first living creature into space – a dog called Laika.

On 12 April 1961, the first human space traveller, Russian Yuri Gagarin, was sent into orbit in a small Vostok capsule launched on top of another rocket developed from R-7. Korolev then developed the Vostok capsule into the bigger, three-person Soyuz spacecraft. His rockets evolved gradually, too, with an approach that maximized safety.

In 1962, Korolev started work on a giant rocket called N-1 for manned landings on the Moon. The project ran into trouble almost immediately, when Korolev and engine designer

THE FUTURE OF SPACE EXPLORATION IS LIMITLESS

SERGEI KOROLEV

Valentin Glushko had a major disagreement. Korolev wanted to use cryogenic (extremely low temperature) propellants, namely liquid hydrogen and oxygen, whereas Glushko favoured highly toxic hypergolic (self-igniting) propellants that could be stored at room temperature. Korolev would not agree and Glushko refused to produce the engines. Korolev turned to an aircraft engine designer, Nikolay Kuznetsov (1911–95), who had no experience in rocket engines. N-1 suffered from serious technical problems. In its first stage, it had 30 rocket engines, which proved unreliable and difficult to control.

BITTER RIVALRIES

The US space programme was placed under the direction of one organization, the National Aeronautics and Space Administration (NASA), with a single aim – to land an astronaut on the Moon by the end of the 1960s. In contrast, the Soviet space programme was split between design teams led by Sergei Korolev, Valentin Glushko, Vladimir Chelomei (1914–84), and Mikhail Yangel (1911–71). The teams continually competed with each other, criticized each other's work, and campaigned for their own projects. In addition, the Soviet leadership pressed designers for spectacular "space firsts" to make headlines.

Korolev died on 14 January 1966 following routine colon surgery. His identity and leading role in the Soviet space programme became widely known only after his death. Until then, he was known simply as "Chief Designer". The development of the N-1 rocket continued, but it never achieved a successful launch. Five years after the Apollo 11 landed on the Moon in 1969, the Soviet lunar-landing programme was quietly cancelled. In spite of these setbacks, however, Korolev is remembered as a brilliant engineer who overcame extraordinary hardship, opposition, and rivalry to achieve greatness.

SOYUZ BLASTS OFF
A Soyuz rocket takes off from the Baikonur Cosmodrome. Updated versions of the Soyuz rocket and spacecraft developed by Korolev and his team in the 1960s still ferry astronauts to the International Space Station today.

LIFT OFF
Flames pour from Launch Pad 39A at the Kennedy Space Center as a Saturn V rocket blasts off, carrying Apollo 11 on its way to the first manned Moon-landing in 1969. The giant Saturn V was rocket engineer Wernher von Braun's greatest achievement.

WERNHER VON BRAUN

LEADING ROCKET ENGINEER OF THE 20TH CENTURY

GERMANY 1912–77

The towering figure of modern rocketry, Wernher von Braun doggedly pursued his passion for space exploration from Germany in the 1930s to the Moon landings of the late 1960s and 1970s. He was the first rocket engineer to become a public figure known to millions. His lifelong passion for space exploration made him the pre-eminent rocket engineer of the 20th century.

A LIFE'S WORK

- Von Braun joins **the Verein für Raumschiffarht (VfR)** rocket society as a teenager
- **Joins the German Army** and develops military missiles, including the V-2
- Is **brought to the US after World War II,** where he works for the US Army; in 1950, he is moved to the **Redstone Arsenal** near Huntsville, Alabama
- Marries **Maria Luise von Quistorp**, his cousin, in 1947, and they have three children
- Becomes a **US citizen** in 1955
- **Transfers to NASA in 1960**, where he is appointed **director of NASA's Marshall Space Flight Center** near Huntsville, Alabama
- Works for **Fairchild Industries** until his retirement due to ill health in 1976
- Is awarded the **1975 National Medal of Science**, one of many awards and honours he was given during his career, but is too ill to attend the presentation ceremony in 1977

Von Braun was born in Wirsitz, which was then in Germany but is now in Poland. As a young man he read scientific works on rocketry and astronautics by the engineer Hermann Oberth (1894–1989), which encouraged him to master calculus and trigonometry. When Germany's Nazi government came to power in 1933, civilian rocket tests were outlawed. Only military rocket research was permitted, so von Braun joined the army to continue his work, and soon earned a PhD in aerospace engineering from the University of Berlin.

In 1937, von Braun joined the Nazi party and became an officer in the SS. He later said that as technical director of the Army Rocket Centre, it was impossible for him to refuse to join the party or he would have had to abandon his rocketry work. Others have disputed his account.

His work with the army culminated in the A-4 rocket, subsequently renamed V-2, the world's first long-range ballistic missile. The rocket incorporated discoveries that had been made by US rocket pioneer Robert Goddard (see pp.338–39), whose papers and patents were freely available. It stood 14m (46ft) high, weighed 12.7 tonnes (14 tons), and delivered a 980-kg (2,200-lb) warhead at a speed of up to 5,600kph (3,500mph). Its engine was 17 times more powerful than any rocket engine built until then.

Development work took place at a secret base at Peenemünde on the Baltic coast. Operational V-2s were built by slave labour in a factory called Mittelwerk, housed in tunnels under the Kohnstein mountain near Nordhausen. The workers came from the Mittelbau-Dora concentration camp. Of the 60,000 prisoners who worked on V-2 production, more than 20,000 died. Between September 1944 and the end of the war, London and Antwerp were attacked by more than 2,800 V-2s.

AMERICA'S ROCKET MAN

As World War II neared its end, von Braun surrendered to the Allies. He was brought to the US, where he developed ballistic missiles for the army. He worked at Fort Bliss, Texas, and launched rockets from the White Sands Proving Ground in New Mexico. In the 1950s, von Braun realized that if his vision of manned space exploration were ever to become a reality, he would have to win the support of the public. He wrote articles for *Collier's* magazine, which had a circulation of 4 million. He advised Walt Disney on the production of three films about spaceflight, the first of which, *Man in Space*, was watched by 42 million people. He also made plans for manned missions to Mars. He convinced Americans that spaceflight need not be a fantasy, and that the US could be a space-faring nation.

Von Braun's team developed the Redstone and Jupiter-C missiles. The Jupiter-C launched the first US satellite, Explorer 1, on 31 January 1958, and a modified Redstone missile launched the first two US astronauts in 1961. By 1960, von Braun and his team had been transferred to the newly formed National Aeronautics and Space Administration (NASA). There, he was given the job of designing the giant Saturn V rocket to send astronauts to the Moon. The pinnacle of his career was the Apollo 11 Moon landing, made possible by his Saturn V, one of the most complex and powerful vehicles ever built. It weighed almost 3,000 tonnes (3,300 tons) and contained 3 million parts.

Von Braun moved to Washington DC in 1970 to work on NASA's strategic planning, but retired less than two years later, disappointed with the cancellation of the last three Apollo missions and the government's reluctance to fund any further manned missions beyond the Earth's orbit. He died in 1977 from pancreatic cancer.

AFTER THE WAR

During World War II, German scientists and engineers developed a host of new technologies, including jet fighters, unpiloted aircraft, and guided missiles. The nations that defeated Germany wanted these technologies, and searched for the people behind them. The US campaign was called Operation Paperclip because files on the most desirable "assets" were marked by a paperclip. About 700 individuals, including every leading German aircraft engineer such as Ohain (see pp.334–35) and von Braun's rocket team, were brought to the US under Operation Paperclip.

A REPLICA OF A V-2 ROCKET STANDS AT THE PEENEMÜNDE MUSEUM, GERMANY.

BURT RUTAN

PIONEER OF SPACE TRAVEL

UNITED STATES B.1943

Aerospace engineer and designer Burt Rutan has spent a lifetime designing and building some of the most unusual and innovative aircraft. He is best known for winning the $10 million Ansari X-Prize for developing the spaceplane that made the first privately funded manned spaceflights, and is noted for his originality in designing light, strong, energy-efficient aircraft.

Born in 1943 in Estacada, Oregon, southeast of Portland, and raised in Dinuba, California, Elbert Leander "Burt" Rutan became a flight-fanatic; in fact he spent his entire life designing and building aircraft. Rutan started at a young age, and by the time he was eight years old, he was building model planes. He was a keen flier too, and made his first solo flight at the age of 16, piloting an aircraft in an Aeronca Championship in 1959. He also studied aeronautics academically, and, graduating third in his class, earned a degree in aeronautical engineering from the California Polytechnic State University in 1965.

Rutan went on to work as a flight test project engineer at the famous Edwards Air Force Base in California, where he worked on US Air Force research projects. In 1972 he moved to Newton, Kansas, and worked as director of development for Bede Aircraft. He spearheaded the construction of the Bede BD-5, a single-seat homebuilt aircraft.

Returning to California in June 1974, Rutan started his own company, the Rutan Aircraft Factory. In this business he designed and developed prototypes for several kit planes, mostly intended for amateur builders. His first design, executed while he was still at Bede, was the VariViggen, a two-seat pusher single-engine aircraft. Other designs included the VariEze and Long-EZ planes. They shared a canard design, which features quite often in Rutan's aircraft. Most planes have a large main wing at the front and small horizontal stabilizers, or tail-planes, at the back. A canard aircraft reverses this. It has small wing-like surfaces called canards at the front and the main wing at the back.

COLD RE-ENTRY

Manned spacecraft normally need a substantial heat shield to protect them from the searing, white-hot heat of re-entry. The Space Shuttle was covered with carbon blocks, ceramic tiles, and heat-resistant blankets to protect it from re-entry temperatures as high as 1,650°C (3,000°F). In contrast, SpaceShipOne needed very little thermal protection. It returned from its sub-orbital spaceflights at less than one-eighth the speed of an orbital spacecraft, greatly reducing the heating effect it experienced. Rutan designed a novel feature that reduced the re-entry temperature even more. As SpaceShipOne fell back to Earth, its tail sections swung upwards, changing its streamlined shape and creating considerable drag. It fell back into the atmosphere like a shuttlecock. This rapidly slowed the craft down and reduced its maximum skin temperature to just 260°C (500°F).

ROUND-THE-WORLD VOYAGER

Rutan was approached by his brother, Dick, about designing an aircraft that could make a non-stop, round-the-world flight without refuelling, something that had never been done before. This inspired Rutan to come up with his most ambitious design – the Voyager aircraft. The plane was made from carbon fibre and weighed only 426kg (939lb). It had to carry so much fuel that it was nicknamed the flying fuel tank. Two piston engines raised the weight to 1,021kg (2,250lb), but the fuel weighed more than three times as much as the plane. The aircraft's pusher engine ran continuously, and the tractor engine was used for takeoff and the initial climb to altitude, and then stopped.

On 14 December 1986, Jeana Yeager and Burt's brother took off in Voyager from Edwards Air Force Base in California. They did not land for another nine days, three minutes, and 44 seconds. When they touched down on the dry lakebed at Edwards Air Force Base, they had flown 40,211km (24,986 miles) around the world. Burt Rutan was awarded the FAI Gold Medal, the Collier Trophy, and the Society of Experimental Test Pilots' Doolittle Trophy.

In April 1982, Rutan formed a company called Scaled Composites to develop more aircraft made from composite materials. One of the first was the Beech Starship, an innovative business aircraft with canards on its nose and no regular tail. In 1985, while the Starship was being developed, Rutan sold Scaled

A LIFE'S WORK

- After finishing college, Rutan works on **nine Air Force research projects** in seven years
- In 1974 he starts his own company, **the Rutan Aircraft Factory**, which produces a variety of ground-breaking aircraft and spacecraft
- A jet-powered version of the tiny **BD-5 aircraft developed by him** at Bede Aircraft appears in the James Bond film, *Octopussy*
- A few days before SpaceShipOne's first spaceflight, the **Mojave Air and Space Port is licensed** as the first commercial spaceport of the US
- SpaceShipOne's spaceflights lead to the formation of a **space tourism company** using his spacecraft
- Is included in *Time* magazine's list of the **"100 most influential people in the world"**

Composites to Beech, and stayed on as chief executive. Scaled Composites progressed to become one of the world's pre-eminent aircraft design and prototyping facilities.

WINNING THE X-PRIZE

In 1996, entrepreneur Peter Diamandis announced a prize (later called the Ansari X-Prize for Suborbital Spaceflight) of $10 million for the first team that could build a three-person craft and launch it to an altitude of 100km (60 miles) twice within two weeks. No government funding was allowed. Rutan's answer was a spaceplane called SpaceShipOne.

SpaceShipOne was powered by a rocket engine, but it was not a solid-fuel rocket or liquid-fuel rocket. It was a hybrid rocket. The fuel was solid rubber and the oxidizer was nitrous oxide. Unlike a solid-fuel rocket, which is unstoppable once it is lit, a hybrid rocket can be turned off, restarted, and throttled. It is also simpler, cheaper, and safer than a liquid-fuel rocket.

Instead of blasting off from the ground on top of a rocket, SpaceShipOne was designed to be carried aloft under the belly of a strange-looking aircraft called White Knight. At an altitude of 15,000m (50,000ft), White Knight released SpaceShipOne, which fired its rocket and soared away into space at up to Mach 3 – three times the speed of sound. Then it fell back into the atmosphere and glided down to land on a runway.

During its test flights, SpaceShipOne went supersonic for the first time on 17 December 2003, exactly 100 years after the first powered flight by the Wright brothers (see pp.324–27). It went on to score a series of firsts – it was the first privately funded aircraft to exceed Mach 2 and Mach 3, the first privately funded craft to exceed an altitude of 100km (60 miles), and the first privately funded reusable manned spacecraft. It flew with the equivalent weight of three persons, and did so while reusing at least 80 per cent of the aircraft's hardware.

EXOTIC AIRCRAFT
Burt Rutan is noted for his unusual aircraft designs. The three flying together here are (from top to bottom) the Vari-EZE, Defiant, and Boomerang.

On 21 June 2004, SpaceShipOne made the first privately funded manned spaceflight with Mike Melvill at the controls, making Melvill the first US commercial astronaut. Melvill piloted the craft again for the first of the Ansari X-Prize spaceflights on 29 September 2004. Just five days later, on 4 October, Brian Binnie made the second spaceflight that secured the prize. British entrepreneur Richard Branson saw the potential of a spaceplane like SpaceShipOne for taking passengers on suborbital spaceflights. He set up a company called Virgin Galactic to offer space tourism flights in a bigger version of SpaceShipOne, named SpaceShipTwo.

In 2005, another of Burt Rutan's revolutionary aircraft hit the headlines. Aviation record-breaker Steve Fossett flew the Rutan-designed Virgin Atlantic GlobalFlyer on the first solo non-stop round-the-world flight without refuelling. It was a jet-powered aircraft made of carbon fibre with a cockpit in a central pod between two booms full of fuel. The following year, Fossett flew the GlobalFlyer again to make the longest ever aircraft flight – 41,467km (25,766 miles).

Rutan retired from Scaled Composites in 2011 with a remarkable body of work behind him, having designed some of the most innovative and unconventional aircraft and spacecraft ever built. He has left an aircraft design legacy from 45 years of work, designing aircraft that are often quite dissimilar from their predecessors.

HITCHING A RIDE
The White Knight aircraft takes off with SpaceShipOne tucked underneath its fuselage on 29 September 2004. At a height of 15,000m (50,000ft), White Knight dropped SpaceShipOne, which then rocketed into space on the first of the two flights that won the Ansari X-Prize.

WORLD'S LARGEST PARTICLE ACCELERATOR

LARGE HADRON COLLIDER

THE LARGE HADRON COLLIDER (LHC) is a giant scientific instrument used by scientists to probe the tiniest particles of matter. It was built inside a circular tunnel that runs for 27km (17 miles) at a depth of 100m (330ft) under the French-Swiss border by the European nuclear research organization, CERN. It accelerates beams of particles almost to the speed of light and then smashes them together. A battery of detectors records the tracks of particles created in the collisions. Scientists study these for evidence of how the universe was formed.

▲ **TUNNEL VISION**
A technician cycles past the blue tube containing the LHC's beam pipes. Inside it, two beams of particles fly in opposite directions so fast that they circle the 27-km (17-mile) instrument 11,000 times per second.

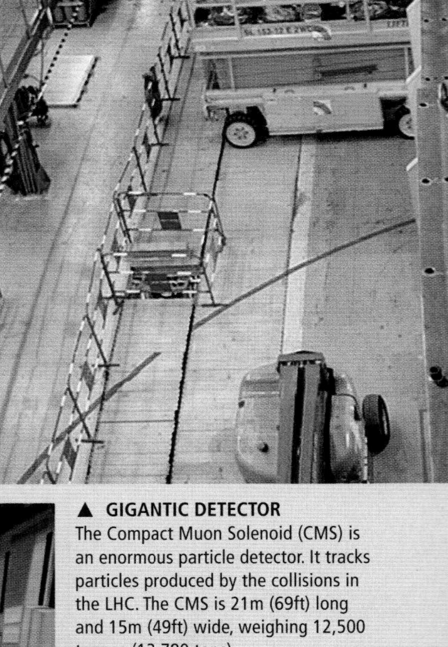

▲ **GIGANTIC DETECTOR**
The Compact Muon Solenoid (CMS) is an enormous particle detector. It tracks particles produced by the collisions in the LHC. The CMS is 21m (69ft) long and 15m (49ft) wide, weighing 12,500 tonnes (13,780 tons).

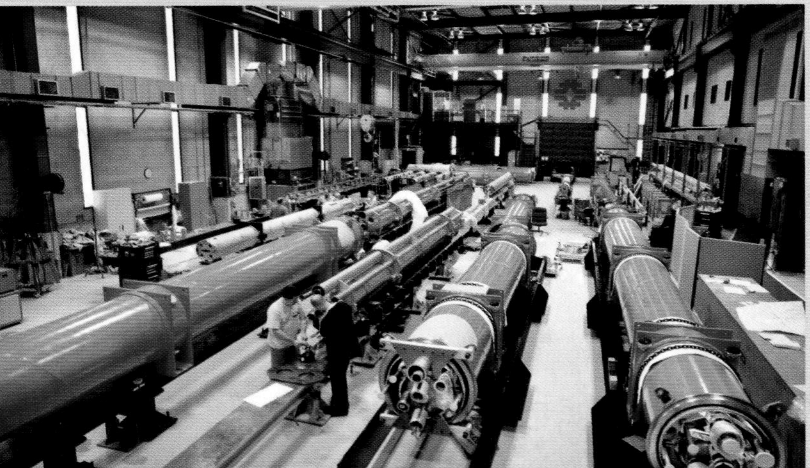

◄ **MULTIPLE MAGNETS**
These red cylinders are some of the 1,624 magnets that steer particles around the LHC. Some of the magnets bend the beams of particles into a circular path. Others keep the particles in each beam clustered tightly together and guide the beams into head-on collisions.

▲ ATLAS – THE PARTICLE DETECTOR
A second particle detector, called ATLAS (**A** **T**oroidal **L**HC **A**pparatu**S**), confirms the results obtained by the CMS. ATLAS is about half the size of Notre Dame Cathedral in Paris, and as heavy as the Eiffel Tower.

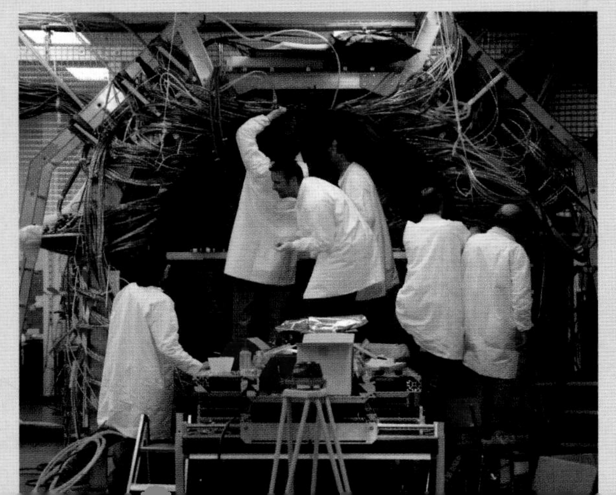

◄ BUILDING A GIANT
Technicians swarm over part of the Compact Muon Solenoid (CMS) during its construction. More than 2,000 physicists in 35 countries worked on its construction.

► CATCHING PARTICLES
The domed ends of ATLAS are designed to spot particles called muons, which are similar to electrons. The muons are created in the explosion of new particles generated by collisions between particle beams in the middle of ATLAS.

INDEX

Main entries for engineers are in **bold**.

Q

R

S

T

U

V

W

X Y

Z

ACKNOWLEDGMENTS

Dorling Kindersley would like to thank the following people for their help in the preparation of this book: Satu Fox for editorial assistance, Lucy Wright for design assistance, Phil Gamble for the icon artworks, and Romaine Werblow and Martin Copeland for image research.

Tall Tree Ltd would like to thank Chris Bernstein for the index.

The publisher would like to thank the following for their kind permission to reproduce their photographs:

Key:
a-above; b-below/bottom; c-centre; f-far; l-left; r-right; t-top.
CC License - Creative Commons Attribution-Share Alike 2.0 Generic license - http://creativecommons.org/licenses/by-sa/3.0/legalcode; DK - Dorling Kindersley; DT - Dreamstime.com; GI - Getty Images; LoC: Library of Congress, Washington, D.C.; SM - Courtesy of the Science Museum, London

Jacket images: *Front:* **Corbis: LuckyPix** (ca); *Back:* **Dorling Kindersley: SM** (c)

1 GI: Hulton Archive / Imagno. **2-3 Corbis:** Bettmann. **4 DT:** Tupungato (tc). **GI:** SSPL (tr). **5 Corbis:** Bettmann (tl, tr). **GI:** Archive Photos (tc). **6-7 Corbis:** Heritage Images (t/3); Michael Nicholson (t/10). **DT:** Empehun (b); Georgios Kollidas (t/4); Richie Lomba (t/5). **GI:** Hulton Archive (t/6, t/7); Roger Viollet Collection (t/1); Stock Montage (t/2). **Franz Krüger:** (t/8). **LoC:** (t/11, t/12); National Photo Company Collection (t/9). **8-9 Corbis:** Mouton Laurence. **10 Corbis:** Yann Arthus-Bertrand (crb); Jason Hawkes (c); Design Pics / Patrick Swan (br). **DT:** 3drenderings (bc). **GI:** De Agostini / G. Dagli Orti (cla). **Wikipedia:** Ricardo Liberato/The file, 'All_Gizah_Pyramids', is licensed under the CC License (bl). **William Henry Goodyear**, *A History of Art: For Classes, Art-Students, and Tourists in Europe*, **1889:** (cra). **11 Conrad Cichorius**, *The Reliefs of Trajan's Column*, **1896:** (bl). **Corbis:** Bettmann (c); EPA / Tolga Bozoglu (crb). **DK:** Donks Models - modelmaker / Clive Streeter (clb). **Wikipedia:** Dantla/The file, 'Bridge_Alcantara', is licensed under the CC License (cra). **12-13 Wikipedia:** Danichou (t/Background). **12 Alamy Images:** The Art Archive (bc). **Corbis:** Arcaid / Will Pryce (tr). **13 Corbis:** Yann Arthus-Bertrand (b). **DT:** Pietro Basilico (tc); Iuliia Kryzhevska (tl). **14 Corbis:** The Art Archive / Alfredo Dagli Orti (bc). **GI:** SSPL (cla). **15 Corbis:** Ecoscene / Sally A. Morgan (b). **Wikipedia:** Einsamer Schütze/The file, 'Statue-of-Hemiun', is licensed under the CC License (tr). **16 Alamy Images:** Imagebroker (bl). **17 akg-images:** Hervé Champollion (t). **Corbis:** Araldo de Luca (br). **18-19 DT:** George Bailey (b). **19 Corbis:** The Art Archive / Alfredo Dagli Orti (crb). **20 akg-images:** (cla). **20-21 GI:** SSPL (tc). **21 Wikipedia:** Akajune/The file, 'Caldarium', is licensed under the CC License (crb). **22 Alamy Images:** David Lyons (bl). **Corbis:** Mimmo Jodice (br). **23 Corbis:** Paul Panayiotou (bl). **DT:** Tupungato (tr). **GI:** Stone / Silvia Otte (br). **24 Wikipedia:** Gun Powder Ma/The file, 'Apollodorus_of_Damascus,_Greek_Architect_and_Engineer._Pic_01', is licensed under the CC License (cla). **25 Corbis:** Peter M. Wilson (b). **26 Corbis:** Vittoriano Rastelli (bl). **DT:** Oleksandr Masliuchenko (r). **GI:** De Agostini / A. Dagli Orti (c, bc). **27 Corbis:** Vittoriano Rastelli (tc, bl). **GI:** De Agostini / A. Dagli Orti (cr). **28-29 akg-images:** Nimatallah (t/Background). **28 Corbis:** Christel Gerstenberg (tr). **GI:** Stone / Michael Townsend (bl). **29 Alamy Images:** Mary Evans Picture Library (tc). **GI:** De Agostini / A. Dagli Orti (br). **Wikipedia:** The file, 'NAMA_Machine_d_Anticythère_1', is licensed under the CC License (tl). **30 akg-images:** (r). **GI:** Hulton Archive (cla). **31 GI:** SSPL (br). **32 GI:** Hulton Archive (cla). **Hartmann Schedel**, *Schedelsche Weltchronik*, **1493:** (tr). **33 The Art Archive:** Musée Municipal Sémur en Auxois / Gianni Dagli Orti (b). **DK:** SM / Clive Streeter (tc). **34 GI:** SSPL (bc). **State Post Bureau of the People's Republic of China:** (cla). **35 GI:** SSPL (b). **36-37 123RF.com. 38 GI:** Roger Viollet Collection (cla). **39 GI:** Hulton Archive. **40 Corbis:** Bettmann (tr). **40-41 Corbis:** Mouton Laurence (b/Background). **GI:** SSPL (c). **41 GI:** Hulton Archive (cr). **42-43 Corbis:** Yann Arthus-Bertrand. **44 The Bridgeman Art Library:** The Makins Collection (bl). **45 GI:** SSPL (b). **Wikipedia:** Caiguanhao/The file, 'Cultural_Zone_in_Foshan_University_(Zu_Chongzhi_and_Confucius)', is licensed under the CC License (tr). **46-47 GI:** SSPL. **48 Alamy Images:** A & J Visage (cr). **Corbis:** Atlantide Phototravel (clb); PoodlesRock (cla); The Gallery Collection (cb). **Wikipedia:** Shizhao/The file, 'Guo_Shoujing-beijing', is licensed under the CC License (cl). **49 Corbis:** Bettmann (br); Heritage Images (bc). **DK:** Rough Guides / Mark Thomas (c). **GI:** Hulton Archive (cr); SSPL (cl, ca); Universal Images Group (bl). **50-51 Corbis:** Historical Picture Archive (t/Background). **50 DK:** Courtesy of the National Maritime Museum, London / James Stevenson (bc). **GI:** Hulton Archive (tr). **51 Alamy Images:** The Art Archive (br); EmmePi Travel (tl). **GI:** Universal Images Group (tc). **52 GI:** SSPL (bl). **53 al-Jazari**, *The Book of Knowledge of Ingenious Mechanical Devices*, **1315:** (r). **54 Corbis:** Burstein Collection. **55 Alamy Images:** The Art Archive. **56-57 Corbis:** The Gallery Collection (c). **58 GI:** The Bridgeman Art Library. **60-61 DT:** Cosmin - Constantin Sava (t/Background). **60 Corbis:** National Geographic Society (tr). **61 GI:** Hulton Archive (tc); Stone / Garry Gay (tl); SSPL (br). **62-63 Imaginechina:** Wang Zirui (b). **62 Wikipedia:** Shizhao/The file, 'Guo_Shoujing-beijing', is licensed under the CC License (cla). **63 Marilyn Shea/University of Maine at Farmington:** (br). **64 Mariano di Jacopo detto il Taccola**, *De ingeneis*, **1433:** (bl). **65 The Bridgeman Art Library:** Biblioteca Marciana, Venice / Giraudon (b). **67 GI:** Gallo Images / Danita Delimont (b). **68 Corbis:** Bettmann (cla). **GI:** The Bridgeman Art Library (bl). **69 Corbis:** Bettmann (b). **70-71 Alamy Images:** Dennis Hallinan. **72-73 DT:** Mikhail Kokhanchikov (t/Background). **72 DK:** SM / Dave King (bl). **GI:** Hulton Archive (tr). **73 GI:** SSPL (tl). Wikipedia: (b); Racklever/The file, 'Harrison_s_Chronometer_H5', is licensed under the CC License (tc). **74 Christoffel van Sichem** (cla). **75 Alamy Images:** Pat Behnke (tr); John Cairns (b). **76 Corbis:** Bettmann (br). **GI:** De Agostini / G. Dagli Orti (bl). **77 Corbis:** The Gallery Collection (tr). **DK:** SM / Nick Lepine (bl). **GI:** Hulton Archive (br). **78 DK:** SM / Dave King (bl); (cla). **79 Alamy Images:** INTERFOTO (tr). **GI:** Hulton Archive (br). **80 Corbis:** Bettmann (cla). **GI:** SSPL (br). **81 GI:** SSPL (t). **82 National Maritime Museum, Greenwich, London:** (bl). **83 The Bridgeman Art Library:** The Worshipful Company of Clockmakers' Collection (t). **Corbis:** Hulton-Deutsch Collection (cla). **84-85 Science Museum / Science & Society Picture Library:** Science Museum Pictorial. **86-87 Corbis:** Hemis / Camille Moirenc (t/Background). **86 Alamy Images:** The Art Archive (tr). **Corbis:** Hulton-Deutsch Collection (bl). **87 Alamy Images:** Hemis (br). **GI:** SSPL (tl, tc). **88 GI:** Hulton Archive (cla). **89 Alamy Images:** Didier Zylberyng (t). **GI:** The Bridgeman Art Library (br). **90-91 Corbis:** Historical Picture Archive. **92 Alamy Images:** Travelib History (bl). **Corbis:** Hulton-Deutsch Collection (cla). **93 GI:** Hulton Archive (b). **94 GI:** SSPL (cla, bc). **95 GI:** De Agostini (t). **96 Corbis:** Michael Nicholson (cla). **GI:** AFP / Martyn Hayhow (tr). **97 GI:** SSPL (b). **98-99 GI:** Archive Photos / Kean Collection (tc). **100 GI:** SSPL (cr); Universal Images Group (cla/Smeaton). **Adam Hart-Davis:** (cla/Winstanley). **101 GI:** SSPL (r). **102 DT:** Stéphane Bidouze (br). **103 Corbis:** Jean Guichard (tr). **DK:** SM / Dave King (bl). **LoC:** George Grantham Bain Collection (br). **104-105 DT:** Rambolin (t/Background). **104 Corbis:** Heritage Images (b). **GI:** Hulton Archive (tr). **105 Alamy Images:** Jeffrey Blackler (tc). **GI:** Hulton Archive (br); SSPL (tr). **106 The Bridgeman Art Library:** Conservatoire Nationale des Arts et Metiers, Paris / Archives Charmet (bl). **107 GI:** Hulton Archive (tr, b). **108 GI:** SSPL (cla, bl). **109 GI:** SSPL (clb, r). **111 GI:** Hulton Archive (bl); SSPL (r). **112-113 The Bridgeman Art Library:** Neue Nationalgalerie, Berlin. **114 Corbis:** Bettmann (bc). **DT:** Lance Bellers (bl). **GI:** The Bridgeman Art Library (c); SSPL (cr, clb). **LoC:** (crb). **Wikipedia:** Audrius Meskauskas/The file, 'Shuttle_with_bobin', is licensed under the CC License (cla). **115 DK:** SM / Clive Streeter (c). **GI:** SSPL (cl, cb, bl); Universal Images Group (br). **Science Photo Library:** Sheila Terry (cr). **116-117 GI:** SSPL (t/Background). **116 GI:** Hulton Archive (tr); SSPL (b). **117 GI:** SSPL (tl, tc, br). **118 GI:** Stock Montage (cla). **Francis Eginton:** (bl). **119 Corbis:** Stefano Bianchetti. **120 DK:** SM / Dave King (cl). **120-121 DT:** Sean Pavone (b/Background). **121 GI:** Hulton Archive (tr); SSPL (tl). **122 GI:** SSPL (c, bc). **123 GI:** SSPL (l, tc, cr, br). **124 LoC:** (c). **W.G Jackman:** (cla). **125 Corbis:** Bettmann (b). **126 LoC:** (cla). **Wikipedia:** Jim Kuhn/The file, 'Perkins_D_cylinder_printing_press_in_the_British_Library', is licensed under the CC License (bl). **127 GI:** SSPL (b). **128 Corbis:** Heritage Images (cla). **DK:** Courtesy the Science Museum, London / Clive Streeter (bl). **129 DK:** Courtesy of the National Motor Museum, Beaulieu / Simon Clay (tr); SM (b). **130-131 GI:** SSPL. **132-133 Alamy Images:** The Print Collector (b). **134-135 DT:** Soldeandalucia (t/Background). **134 GI:** Hulton Archive (tr); SSPL (b). **135 Corbis:** Stefano Bianchetti (tl). **GI:** Hulton Archive (tc); SSPL (br). **136 DT:** Richie Lomba (cla). **136-137 GI:** Hulton Archive (c). **137 The Franklin Institute, Inc., Philadelphia:** Peter Harholdt (br). **138 Corbis:** Stefano Bianchetti (br). **GI:** SSPL (cla). **139 DK:** SM / Clive Streeter (bc). **140 DK:** SM / Clive Streeter (bl). **DT:** Georgios Kollidas (cla). **141 Corbis:** Bettmann (b). **142 DK:** SM / Clive Streeter (br). **GI:** SSPL (tr). **Smithsonian Institution, Washington, DC, U.S.A.:** (cla). **143 LoC:** Currier & Ives (r). **144-145 GI:** Hulton Archive (t/Background). **144 GI:** SSPL (tr, b). **145 GI:** SSPL (tl, tc, br). **146 DT:** Georgios Kollidas (cla). **147 Corbis:** Chris Hellier (tr); Heritage Images (tl). **DK:** SM / Dave King (br). **148 Corbis:** Bettmann (cla). **GI:** Archive Photos / Kean Collection (bl). **149 GI:** Archive Photos / MPI (b); SSPL (tr). **150 The Art Archive:** Eileen Tweedy (cla/Bramah). **GI:** SSPL (cla/Maudslay, bc). **151 GI:** SSPL (b). **152 Corbis:** Chris Hellier (cla/Nasmyth); Heritage Images (cla/Whitworth). **153 GI:** SSPL (tl, br). **154-155 Corbis:** Heritage Images. **156-157 Corbis:** TH-Foto (t/Background). **156 Charles-Émile-Auguste Carolus-Duran,** *Portrait of Emily Warren Roeblin*, **1896:** bc. **Barabás Miklós,** *Laying the Foundation Stone for Chain Bridge*, **1864:** tr. **157 Corbis:** Hemis / Franck Guiziou (b); Underwood & Underwood (tc). **GI:** Hulton Archive (tl). **158 John Nichols,** *Literary Anecdotes of the Eighteenth Century*, **1815:** (cla). **158-159 GI:** SSPL (b). **159 Alamy Images:** Mary Evans Picture Library (cla). **160 GI:** Hulton Archive (cla). **161 Wikipedia:** Bencherlite/The file, 'Menai_Suspension_Bridge_Dec_09', is licensed under the CC License. **162 GI:** SSPL (tr). **Barabás Miklós,** *William Tierney Clark*, **1842:** (clb). **162-163 DT:** Sean Pavone (b/Background). **163 Wikipedia:** Akke Monasso/The file, 'WalesC0047', is licensed under the CC License (c). **164 GI:** Hulton Archive (cla). **165 Corbis:** Charles O'Rear (t). **Wikipedia:** Patrick Giraud/The file, 'Locomotive_Seguin_01', is licensed under the CC License (br). **166 GI:** Archive Photos / Museum of the City of New York. **167 Amédée Burat,** *Géologie appliquée ou ƒraité de la recherche et de l'expoitation des Mines*, **1846:** bc. **Franz Krüger:** cla. **168-169 LoC.** **170-171 Digital Vision:** (t/Background). **170 DK:** Judith Miller / Huxtins (bl). **LoC:** Tissandier Collection (tr). **171 GI:** SSPL (tc). **LoC:** (tl). **US Naval History & Heritage Command:** (br). **172 Corbis:** Bettmann (cla/Joseph-Michel); Stefano Bianchetti (cla/Jacques-Étienne). **LoC:** Tissandier Collection (tr). **173 GI:** Hulton Archive (r). **175 GI:** Hulton Archive (t). **Jean-Baptiste Marie Meusnier de la Place:** (br). **176 Corbis:** Michael Nicholson (cla). **176-177 GI:** Hulton Archive (c). **177 Alamy Images:** Pictorial Press Ltd (br). **178 LoC:** (bc); George Grantham Bain Collection (cla). **179 GI:** Archive Photos (b). **German Federal Archive (Bundesarchiv):** Georg Pahl (Bild 102-06800) (tr). **180-181 Corbis:** Bettmann. **182-183 GI:** Hulton Archive / Douglas Miller. **184 Corbis:** (cra). **GI:** Archive Photos / Robert Howlett (crb); SSPL (cla, cr, bl); Hulton Archive (cl); Universal Images Group (clb); Oxford Scientific / Nick Wild (br). **185 Alamy Images:** Mary Evans Picture Library (cra). **Corbis:** PoodlesRock (cb); Underwood & Underwood (crb). **GI:** SSPL (cl, clb). **Wikipedia:** Sven Teschke/The file, 'Johann_Philipp_Reis-Gelnhausen', is licensed under the CC License (cla). **186-187 DT:** Alexander Lvov (t/Background). **186 GI:** SSPL (bl). **187 Corbis:** Underwood & Underwood (b). **Jack Spurling,** *Ariel and Taeping*, **1926:** (tc). **U.S. Department of Transportation:** Carl Rakeman (tl). **188 GI:** SSPL (cla). **188-189 LoC:** (bc). **189 LoC:** Detroit Publishing Company (tr). **190 Corbis:** Heritage Images (bl); Hulton-Deutsch Collection (cla/George, cla/Robert). **191 Corbis:** Hulton-Deutsch Collection (b). **U.S. Army:** Army Art Collection, Washington, D.C. (tr). **192 DT:** Konstantin32 (ca). **GI:** SSPL (tr, cb, bl). **192-193 GI:** Photographer's Choice / Kevin Summers (Background); SSPL (bc). **193 GI:** SSPL (t/Background Map, tl, tc, c, cra). **194 GI:** Hulton Archive (cla). **195 GI:** The Bridgeman Art Library. **196 Alamy Images:** 19th era 2 (clb). **GI:** Archive Photos / Robert Howlett (ca). **196-197 DT:** Bevanward (b/Background). **197 Corbis:** Hulton-Deutsch Collection (c). **198-199 GI:** Archive Photos. **200 GI:** Hulton Archive (br); SSPL (bl). **201 Corbis:** Bettmann (bl); Imaginechina (br). **DT:** Nick Stubbs (tr). **202 LoC:** (cla). **202-203 LoC:** J.A. Stewart (b). **203 Corbis:** Transtock / Steve Crise (crb). **204-205 Corbis:** Bettmann. **206 Bovington Tank Museum:** (cla). **Wikipedia:** Andrew Skudder/The file, 'Little_Willie', is licensed under the CC License (bl). **207 Alamy Images:** Classic Image (r). **208-209 DK:** Imperial War Museum, London. **210 DK:** Courtesy of the National Cycle Collection / Gary Ombler (bl). **211 GI:** Archive Photos (r). **212-213 GI:** SSPL (t/Background). **212 Corbis:** Bettmann (b). **GI:** Hulton Archive (tr). **213 GI:** Archive Photos (tl). **LoC:** U.S Office of War Information / Alfred T. Palmer (tc). **214-215 GI:** Archive Photos (bc). **216 GI:** Hulton Archive (bc); SSPL (cla). **217 GI:** SSPL (b). **218 DT:** Georgios Kollidas (cla). **GI:** SSPL (b). **219 The Bridgeman Art Library:** The Stapleton Collection (tr). **220 GI:** Chicago History Museum (cla); Hulton Archive / Henry Guttmann (bc). **221 GI:** Hulton Archive (t). **222 Corbis:** Reuters / Darren Zammit Lupi (bl). **GI:** Hulton Archive (cla). **223 LoC:** George Grantham Bain Collection (b). **224 GI:** Archive Photos / Kean Collection (cla). **LoC:** Liljenquist Family Collection of Civil War Photographs (bc). **225 Corbis:** (br). **DK:** Courtesy of the Board of Trustees of the Royal Armouries / Gary Ombler (t). **226-227 GI:** The Bridgeman Art Library. **228 Corbis:** Hulton-Deutsch Collection (cla). **DK:** Judith Miller / Sloan's (bc). **229 GI:** Hulton Archive (b). **230-231 Corbis:** Bettmann. **232 GI:** SSPL (bl). **LoC:** T. & R. Annan & Sons (cla). **233 GI:** SSPL (t). **234 Courtesy Siemens:** (cla). **234-235 Corbis:** Bettmann (c). **235 GI:** Hulton Archive (crb). **237 GI:** Chicago History Museum (b); SSPL (cra). **238 GI:** SSPL (bc). **Science Photo Library:** Peter Menzel (cla). **239 Rex Features:** Times Newspapers Ltd (r). **240-241 GI:** Bloomberg. **242-243 DT:** Dellyne (t/Background). **242 Corbis:** National Geographic Society / Portland Cement Co. (b). **GI:** Hulton Archive (tr). **243 DT:** Stephen Finn (tl). **244 Wayss & Freytag Ingenieurbau AG:** (cla/Freytag, bl). **245 GI:** SSPL (b). **Joseph Monier,** *Flowerpot with reinforced concrete*, **1850:** (tr). **246 GI:** Hulton Archive (cla). **247 GI:** Hulton Archive (crb); SSPL (b). **248-249 Digital Vision:** (t/Background). **248 Corbis:** The Art Archive / Alfredo Dagli Orti (tr). **GI:** SSPL (bc). **249 Corbis:** (tc); Hulton-Deutsch Collection (b). **GI:** SSPL (tl). **250 GI:** Hulton Archive (cla). **251 Corbis:** Stefano Bianchetti (crb); Leonard de Selva (t). **252 GI:** Hulton Archive (cla). **253 Corbis:** Michael Nicholson (cr). **GI:** Oxford Scientific / Nick Wild (tl). **254 Australian War Memorial:** (bc). **Corbis:** Bettmann (cla). **255 GI:** SSPL (b). **256-257 DT:** Clearviewstock (t/Background). **256 GI:** SSPL (bl); Universal Images Group (tr). **257 Corbis:** Bettmann (b); Stefano Bianchetti (tc). **LoC:** Brady-Handy Collection (tl). **258 GI:** SSPL (bl). **LoC:** (cla). **259 GI:** Archive Photos (t). **LoC:** National Photo Company Collection (cr). **260 Corbis:** Profiles in History (tr). **DK:** SM / Dave King (c). **GI:** Universal Images Group (bl). **261 Corbis:** Bettmann (bc); Science Faction / National Archives (l/Background, r); Heritage Images (c). **GI:** PhotoQuest (tc). **262 Corbis:** Bettmann (cla). **DK:** SM / Dave King (bl). **263 LoC:** Brown Bros.. **264 Corbis:** (c). **DK:** SM / Dave King (tr). **264-265 DT:** Bevanward (b/Background). **265 LoC:** Byron, N.Y. (r). **266 Corbis:** Science Photo Library / Victor Habbick Visions (bc). **GI:** SSPL (cla). **267 Corbis:** Bettmann (t). **268 DK:** SM / Clive Streeter (bl). **LoC:** (cla). **269 Corbis:** Hulton-Deutsch Collection (b). **270 Science Photo Library:** LoC (cla). **271 GI:** Archive Photos (r). **272-273 DK:** NASA. **274 Corbis:** Bettmann (cra, bl); Hulton-Deutsch Collection (cl). **GI:** SSPL (ca, br); Universal Images Group (cla). **LoC:** (cb); John T. Daniels (c). **Wikipedia:** Falconaumanni/The file, 'Mercado_torroja', is licensed under the CC License (cr). **275 DT:** Nicunickie (crb). **NASA:** (cra); Kennedy Space Center (bc). **Wikipedia:** Larry D. Moore/The file, '651px-Roomba_original', is licensed under the CC License (c). **United States Bureau of Reclamation:** (cla). **276-277 Wikipedia:** LSDSL/The file, 'Benz_Patent_Motorwagen_Engine', is licensed under the CC License (t/Background). **276 Corbis:** (tr);

Hulton-Deutsch Collection (b). **277 GI:** Car Culture Collection (br); Gamma-Keystone (tl); Hulton Archive (tc). **278 Corbis:** Bettmann (bl); Hulton-Deutsch Collection (cla). **279 Corbis:** Bettmann (b). **280 GI:** Hulton Archive (cla). **Wikipedia:** Joachim Köhler/The file, 'ZweiRadMuseumNSU_Reitwagen', is licensed under the CC License (bl). **281 GI:** Hulton Archive (b); SSPL (tr). **282 Corbis:** Bettmann (bl). **LoC:** National Photo Company Collection (cla). **283 Corbis:** Rykoff Collection. **284 Corbis:** Bettmann (tc). **284-285 Corbis:** Gordon Osmundson (b/Background). **285 GI:** AFP / Pierre Clement (cr). **286-287 Corbis:** Bettmann. **288 Corbis:** Bettmann (cla). **DK:** Matthew Ward (bl). **289 GI:** Hulton Archive (t). **290-291 DT:** (t/Background). **290 GI:** SSPL (bc). **Jean-Baptiste Sabatier-Blot:** (tr). **291 Alamy Images:** Moviestore Collection Ltd (b). **GI:** Hulton Archive (tc). **LoC:** (tl). **292 GI:** Hulton Archive (cla); SSPL (bl). **293 The Kobal Collection:** (b). **294 GI:** SSPL (cla/Auguste, cla/Louis, br). **295 Marcellin Auzolle:** (t). **296-297 GI:** Bloomberg (t/Background). **296 DT:** Pavel Losevksy (bc). **LoC:** Harris & Ewing Collection (tr). **297 Corbis:** Bettmann (tl); Reuters (tc); Underwood & Underwood (b). **298 GI:** Archive Photos / Chicago History Museum (cla). **298-299 GI:** Archive Photos / Chicago History Museum (c). **299 Corbis:** Bettmann (tr). **GI:** Archive Photos / Field Museum Library / W.B. Conkey Company (br). **300 Wikipedia:** The file, 'Shukhov_tower_shabolovka_moscow_02', is licensed under the CC License (bc). **301 GI:** The Bridgeman Art Library (tr). **Andrei Osipovich Karelin:** (b). **302-303 GI:** Archive Photos / Lewis W. Hine. **304 DT:** Neale Cousland. **305 GI:** Sankei Archive (cla). **306-307 Alberto Minio Paluello:** (b). **308 Alamy Images:** Mike Goldwater (bl). **309 Corbis:** Richard Cummins (bl). **DT:** Ian Whitworth (r). **310-311 GI:** SSPL (t/Background). **310 Corbis:** Hulton-Deutsch Collection (b). **LoC:** Harris & Ewing Collection (tr). **311 Corbis:** Heritage Images (tl). **DK:** SM / Dave King (cr). **LoC:** George Granthan Bain Collection (tc). **312 GI:** Hulton Archive (cla); SSPL (b). **313 LoC:** Tissandier Collection (tr). **314 GI:** SSPL (cla). **314-315 Corbis:** Lebrecht Music & Arts / Derek Bayes Aspect (c). **315 GI:** SSPL (tr). **316 GI:** Hulton Archive (bl). **LoC:** (cla). **317 GI:** Hulton Archive (b). **318-319 LoC. 320 Corbis:** Bettmann (cla). **GI:** Hulton Archive / Henry Guttmann (bl). **321 GI:** SSPL (r). **322 DK:** Imperial War Museum, London / Andy Crawford (bc). **LoC:** (cla). **323 Corbis:** Bettmann (tr). **GI:** SSPL (b). **324 LoC. 325 Corbis:** Smithsonian Institution (bc). **LoC:** (ca/Orville, ca/Wilbur). **326 DK:** Courtesy of the Shuttleworth Collection, Bedfordshire / Martin Cameron (clb). **326-327 Corbis:** Gordon Osmundson (b/Background). **327 Corbis:** Underwood & Underwood (t). **328 LoC:** (c); George Grantham Bain Collection (b). **National Air and Space Museum, Smithsonian Institution:** Eric Long (tc). **329 Corbis:** (tl). **LoC:** (cl). **National Air and Space Museum, Smithsonian Institution:** Gift of Hugh L. Dryden (cb). **Smithsonian Institution, Washington, DC, U.S.A.:** (ca). **330 Corbis:** Bettmann (cla). **331 GI:** Hulton Archive (b). **Library and Archives Canada:** Elsie Gregory MacGill Fonds / R4349-0-6-E (tr). **332 LoC:** George Grantham Bain Collection (cla). **333 Corbis:** Bettmann (tr). **GI:** Gamma-Keystone (b). **334 DK:** SM / Dave King (bc). **GI:** Hulton Archive / Topical Press Agency (cla/Whittle). **U.S. Air Force:** Air Force Research Laboratory Propulsion Directorate (cla/Ohain). **335 U.S. Air Force:** (b). **336-337 NASA:** ESA and H.E. Bond (STScl) (t/Background). **336 NASA:** (bl); Kennedy Space Center (tr). **337 DK:** Getty / Stocktrek Images (tl). **NASA:** (tc). **338 NASA:** (cla, bc). **339 NASA:** (b). **340 NASA:** (bl, br). **341 DK:** Courtesy of Bob Gathany / Andy Crawford (bl). **NASA:** (br); Dryden Flight Research Center / U.S. Air Force (tr). **342 Alamy Images:** RIA Novosti (cla). **343 Corbis:** Roger Ressmeyer (r). **344 NASA:** Kennedy Space Center. **345 NASA:** (cla). **Wikipedia:** AElfwine/The file, 'Fusée_V2', is licensed under the CC License (bc). **346 GI:** Win McNamee (cla). **347 Corbis:** Gene Blevins (b); Jim Sugar (tr). **348-349 Corbis:** EPA / Martial Trezzini (tc). **348 Corbis:** Science Faction / Peter Ginter (bc). **Adam Hart-Davis:** (cl). **349 Adam Hart-Davis:** (tr, bl, br)

All other images © Dorling Kindersley
For further information see:
www.dkimages.com

ABOUT THE EDITOR-IN-CHIEF

Adam Hart-Davis is a writer, broadcaster, and photographer, and one of the world's most popular and respected "explainers" of science. His TV work includes *What the Romans, Victorians, Tudors and Stuarts,* and *Ancients Did For Us, Tomorrow's World,* and *Science Shack,* and he is the author of more than 25 books on science, invention, and history. He is an Honorary Fellow of the British Science Association and Member of the Newcomen Society for the History of Engineering and Technology. He has also received medals from the Royal Academy of Engineering, the Sir Henry Royce Memorial Foundation, and from the Institution of Incorporated Engineers.